# HYDROGENATION

Edited by **Iyad Karamé**

**Hydrogenation**
http://dx.doi.org/10.5772/3208
Edited by Iyad Karamé

**Contributors**

Tsuneo Imamoto, Bogdan Štefane, Franc Požgan, Werner Bonrath, Jonathan Medlock, Jan Schütz, Bettina Wüstenberg, Thomas Netscher, Christos S. Karaiskos, Dimitris Matiadis, John Markopoulos, Olga Igglessi-Markopoulou, Makoto Hashimoto, Yuta Murai, Geoffery D. Holman, Yasumaru Hatanaka, Domingo Liprandi, Edgardo Cagnola, Cecilia Lederhos, Juan Badano, Mónica Quiroga, M. Juliana Maccarrone, Gerardo C. Torres, Cecilia Lederhos, Carolina Betti, Juan M. Badano, Mónica Quiroga, Juan Yori, Rogelio Sotelo-Boyás, Fernando Trejo-Zárraga, Felipe de Jesús Hernández-Loyo, A. Infantes-Molina, A. Romero-Pérez, D. Eliche-Quesada, J. Mérida-Robles, A. Jiménez-López, E. Rodríguez- Castellón, Wan-Hui Wang, Yuichiro Himeda, T.F. Sheshko, Yu. M. Serov, Shigeru Kohtani, Eito Yoshioka, Hideto Miyabe, Ken Tokunaga

**Published by InTech**

Janeza Trdine 9, 51000 Rijeka, Croatia



**Publishing Process Manager** Vedran Greblo
**Technical Editor** InTech DTP team
**Cover** InTech Design team

Additional hard copies can be obtained from orders@intechopen.com

Hydrogenation, Edited by Iyad Karamé
p. cm.
ISBN 978-953-51-0785-9

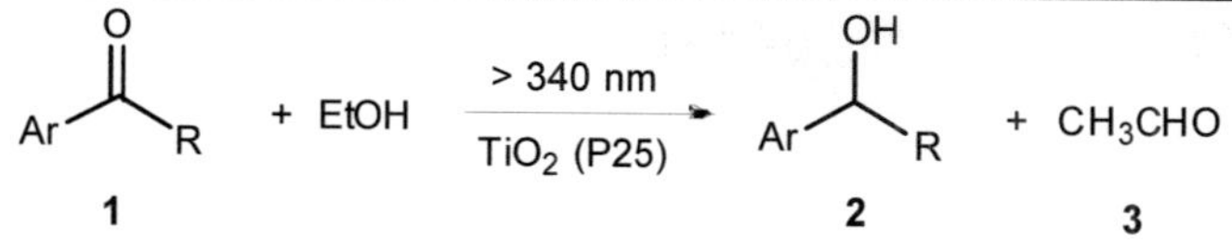


| Substrates (1) Ar | R | Time (h) | Conversion (%) | Yield of 2 (%) | Yield of 3 (%) | $-E_{red}$ (V)[b] |
|---|---|---|---|---|---|---|
| $C_6H_5$ | H | 1 | 100 | 76±2 | 57 | 1.99 |
| $C_6H_5$ | Me | 4 | 100 | 97±2 | >99 | 2.13 |
| $C_6H_5$ | Et | 8 | 91 | 87 | >99 | 2.18 |
| $C_6H_5$ | *i*-Pr | 8 | 24 | 25 | 32 | 2.21 |
| $C_6H_5$ | *t*-Bu | 8 | 17 | 7 | 42 | 2.24 |
| $C_6H_5$ | $CF_3$ | 8 | 70 | 63 | 10 | 1.59 |
| $C_6H_5$ | $C_6H_5$ | 4 | 82 | 78 | 97 | 1.83 |
| 2-Me$C_6H_4$ | Me | 8 | 66 | 63 | 84 | 2.16 |
| 4-Me$C_6H_4$ | Me | 8 | 100 | >99 | >99 | 2.15 |
| 2-F$C_6H_4$ | Me | 3 | 100 | 99 | >99 | 1.86 |
| 3-F$C_6H_4$ | Me | 3 | 100 | 97 | >99 | 2.04 |
| 4-F$C_6H_4$ | Me | 4 | 100 | >99 | >99 | 2.15 |
| 2,6-$F_2C_6H_4$ | Me | 5 | 100 | 96 | 94 | - |
| $C_6F_5$ | Me | 2 | 100 | >99 | >99 | 1.83 |
| 1-$C_{10}H_7$ | H | 1 | 95 | 76±3 | 77 | 1.66 |
| 1-$C_{10}H_7$ | Me | 10 | 100 | >99 | 93 | 1.86 |
| 2-$C_{10}H_7$ | H | 1 | 94 | 32±7 | 93 | 1.70 |
| 2-$C_{10}H_7$ | Me | 5 | 100 | 94±4 | >99 | 1.73 |
| 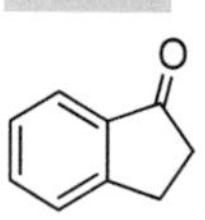 | | 8 | 100 | >99 | >99 | 2.06 |
| 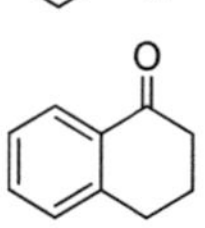 | | 8 | 100 | 91 | >99 | - |
| 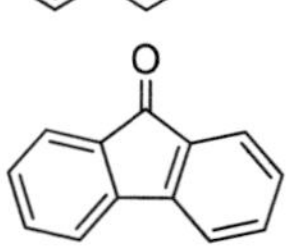 | | 2 | 82 | 81 | 98 | 1.42 |

[a] Carried out for a mixture of the substrate **1** (3 mmol) and the P25 $TiO_2$ (0.1 g) in deaerated EtOH (30 $cm^3$) under UV irradiation (>340 nm) at 32 ℃. [b] Reduction potential vs. SCE in $CH_3CN$ containing $Bu_4NClO_4$ electrolyte.

**Table 3.** Photocatalytic hydrogenation of aromatic carbonyl compounds on P25 $TiO_2$ [37].[a]

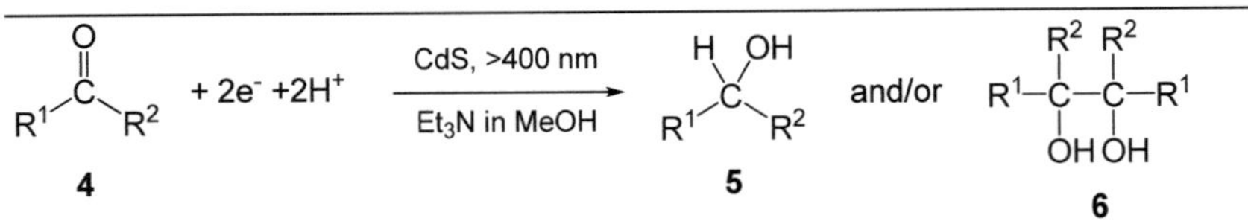


| Substrates (4) | | Time | Conversion of | Yield of | Yield of | $-E_{red}$ (V)[a] |
|---|---|---|---|---|---|---|
| R[1] | R[2] | (h) | 4 (%) | 5 (%) | 6 (%) | |
| $p$-$CNC_6H_4$ | $C_6H_5$ | 2 | 100 | 95 | 0 | 1.17 |
| $p$-$ClC_6H_4$ | $C_6H_5$ | 3 | 98 | 70 | 12 | 1.32 |
| $p$-$ClC_6H_4$ | $p$-$ClC_6H_4$ | 3 | 98 | 90 | 4 | 1.35 |
| $C_6H_5$ | $C_6H_5$ | 3 | 100 | 95 | 5 | 1.55 |
| $p$-$CH_3OC_6H_4$ | $C_6H_5$ | 3 | 97 | 33 | 56 | 1.56 |
| $p$-$CH_3OC_6H_4$ | $p$-$CH_3OC_6H_4$ | 3 | 52 | 15 | 40 | 1.56 |
| $C_6H_5$ | $CH_3$ | 3 | 10 | trace | 80 | 2.00 |

[a] Polarographic half-wave reduction potential vs. SCE in MeOH.

**Table 4.** Photocatalytic hydrogenation of aromatic ketones on nanocrystallized CdS [25].

1,2-Diketones such as camphorquinone, 1-phenyl-1,2-propanedione, and benzil were hydrogenated to the corresponding $\alpha$-hydroxyketones in moderate to good yields on the UV-irradiated P25 $TiO_2$ as shown in Scheme 4 [41, 42]. The yields and stereoselectivities were increased in the presence of water or TEA as a sacrificial electron donor in methanol solvent. The endo-hydroxycamphors were formed much more favorably than the exo-products, though there was little selectivity between 2 and 3 positions.

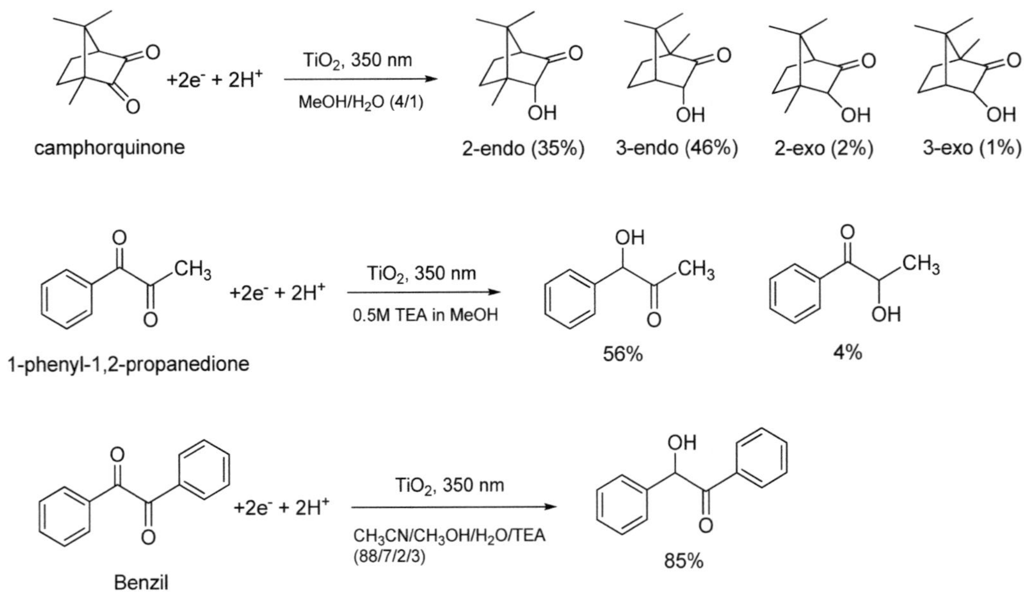


**Scheme 4.** Photocatalytic hydrogenations of diketone compounds upon the P25 $TiO_2$ powder.

### 3.4. Hydrogenation of carbon dioxide

With increasing concerns about rising atmospheric $CO_2$ concentration, the need for research to utilize $CO_2$ in chemical synthesis has been increased greatly [43]. Therefore, the photocatalytic hydrogenation of $CO_2$ upon semiconductor powders has received much attention [9-11, 44-56]. Among the earliest studies on the photocatalytic hydrogenation of $CO_2$, Inoue et al. examined a wide range of semiconductors ($WO_3$, $TiO_2$, ZnO, CdS, GaP, and SiC in 200 – 400 mesh) in aqueous solution [44]. The results indicated that $CO_2$ was reduced to HCOOH, HCHO, and $CH_3OH$, and the product yields correlated to the position of the CB level of the semiconductors. In the case of $WO_3$, since the CB level was insufficiently negative to reduce $CO_2$, no reduction products were observed. In contrast, SiC having the most negative CB level gave the highest yields. Later, Henglein and Gutiérrez reported that high quantum yield of 0.20 was achieved in the hydrogenation of $CO_2$ to formate using ZnS colloidal suspension under UV light irradiation [45]. Here, the solvent was a mixture of water and ethanol, or 2-propanol as the hole scavenger. After the earliest works, a great deal of effort has been devoted to studying the photocatalytic reduction of $CO_2$. Most of works are summarized in recent reviews [9-11], in which the use of $TiO_2$, $SrTiO_3$, ZrO, ZnS, or CdS is reported. Recently, visible-light-driven new materials were developed and applied to the photocatalytic reduction of $CO_2$ under visible light irradiation: for examples, $InTaO_4$ [46], $LaCoO_3$ [47], $BiVO_4$ [48], $ZnGaO_4$ [49], $Zn_2GeO_4$ [50, 51], $Bi_2WO_6$ [52] etc.

The photocatalytic reduction of $CO_2$ on semiconductor powders gives several reduction products such as CO, HCOOH, HCHO, $CH_3OH$, $CH_4$, and various hydrocarbons. Among those, methanol is the most valuable product because it can be directly used as a fuel or a building block. High efficiency and selectivity in the preparation of methanol were obtained by the use of Ti-oxide/Y-zeolite catalysts containing highly dispersed tetrahedral titanium oxide species under UV irradiation [53, 54]. The charge-transfer excited state of these species played an important role in the selectivity for producing $CH_3OH$, in contrast to the different selectivity giving $CH_4$ on bulk $TiO_2$ photocatalyst. Ti-incorporated mesoporous silica also exhibited high activity in the photoreduction of $CO_2$ with water to generate $CH_3OH$ and $CH_4$ under UV irradiation [55]. Methanol was also selectively produced by employing $CaFe_2O_4$ [56] and NiO-loaded $InTaO_4$ [47] under visible light irradiation.

### 3.5. Hydrogenation of Nitrate and nitrite ions, and nitroaromatics

The use of artificial fertilizers in agriculture has caused a great deal of concern regarding water pollution induced by production of nitrate ($NO_3^-$) and nitrite ($NO_2^-$) ions from the fertilizers [57]. Therefore, chemical processes for elimination or hydrogenation of $NO_3^-$ ions have been extensively studied photocatalytically. $TiO_2$ photocatalyst has been used for the hydrogenation of $NO_3^-$ and $NO_2^-$ ions to ammonia [58-60]. Other photocatalysts such as $SrTiO_3$ [61], $K_4NbO_{17}$ [61], tantalate oxides ($K_3Ta_3Si_2O_{13}$, $BaTa_2O_6$, $KTaO_3$, $NaTaO_3$) [62], and ZnS [63], have also been reported as active photocatalysts for the hydrogenation of nitrate ions under UV light irradiation. It is worth noting that Kudo et al. developed a 0.1% nickel doped ZnS ($Zn_{0.999}Ni_{0.001}S$) photocatalyst to extend the photocatalytic response of ZnS toward visible region [64], and applied it into the reduction of $NO_3^-$ and $NO_2^-$ ions under visible light irradiation [65]. The amounts of products and electrons consumed in the reduction of $NO_3^-$ and $NO_2^-$ are listed in Table 5. The ratios of the amounts of $NH_3/NO_2^-$ and $NH_3/N_2$ increased in the presence of the platinum cocatalyst, indicating that selectivity for the $NH_3$ production was improved by the Pt loading on $Zn_{0.999}Ni_{0.001}S$. More recently, Yamauchi and co-workers reported highly selective ammonia synthesis from nitrate with photocatalytically generated hydrogen on CuPd nanoalloys loaded on $TiO_2$, on which ammonia was selectively produced up to 78% yield with hydrogen evolution under UV light irradiation [66].

| Catalysts Type | Reactant (initial conc./ $molL^{-1}$) | Amounts of products (electron consumed)/μmol | | | |
|---|---|---|---|---|---|
| | | $H_2$ [b] | $NO_2^-$ | $NH_3$ | $N_2$ |
| non-loaded | no $NO_3^-$ nor $NO_2^-$ (0) | 214 (428) | (-) | (-) | (-) |
| non-loaded | $NO_3^-$ (1.0) | 60 (120) | 250 (500)[c] | 21 (168)[d] | 2.7 (27)[e] |
| 1wt% Pt-loaded | $NO_3^-$ (1.0) | 212 (424) | 14 (28)[c] | 11 (88)[d] | 0.79 (7.9)[e] |
| non- loaded | $NO_2^-$ (0.01) | 37 (74) | (-) | 41 (246)[f] | 2.4 (14)[g] |
| 1wt% Pt-loaded | $NO_2^-$ (0.01) | 96 (192) | (-) | 65 (390)[f] | 9.5 (57)[g] |

[a] Carried out for a mixture of reactant ($NO_3^-$ or $NO_2^-$) and photocatalyst powder (0.5 g) in aqueous $CH_3OH$ solution (6.25 vol%) under visible light irradiation (>420 nm) for 20 h.
[b] A by-product from the reduction of water: $2H^+ + 2e^- \rightarrow H_2$.
[c] $NO_3^- + 2H^+ + 2e^- \rightarrow NO_2^- + H_2O$. [d] $NO_3^- + 9H^+ + 8e^- \rightarrow NH_3 + 3H_2O$. [e] $2NO_3^- + 12H^+ + 10e^- \rightarrow N_2 + 6H_2O$. [f] $NO_2^- + 7H^+ + 6e^- \rightarrow NH_3 + 2H_2O$. [g] $2NO_2^- + 8H^+ + 6e^- \rightarrow N_2 + 4H_2O$.

**Table 5.** Reduction of $NO_3^-$ and $NO_2^-$ on $Zn_{0.999}Ni_{0.001}S$ photocatalysts [65].[a]

Several organic nitroaromatics can be easily hydrogenated to afford corresponding amino compounds in the presence of sacrificial hole scavenger upon the UV irradiated $TiO_2$ as firstly reported by Li and co-workers in 1993 [67]. Since then, the photocatalytic hydrogenation of nitro compounds using some kinds of semiconductor photocatalysts was extensively studied by a number of researchers. The hydrogenation reactions proceeded almost quantitatively. The details are summarized in recent reviews [9-11]. Recent progress in this research has been directed toward the development of visible light response photocatalysts such as nitrogen-doped $TiO_2$ [68] and dye-sensitized $TiO_2$ loaded with transition metal nanoparticles [69].

## 4. Advantages and disadvantages using this method

Compared to the conventional hydrogenation methods as mentioned in other Chapters, the photocatalytic hydrogenation on semiconductor particles has some great advantages: (1) The most important merit is that particular reducing agents (e.g. $H_2$ gas etc.) are not necessary in this method. In most case, the reductants are conventional solvents such as water, alcohols, and amines which concurrently act as hole scavengers. Therefore, this method allows us to avoid both the use of harmful and dangerous chemical reagents and the emission of harmful waste. (2) The reactions mostly proceed under mild conditions, e.g.

under ordinary temperature and pressure, and therefore are safety. (3) In the case of $TiO_2$ or other stable metal oxide photocatalysts, the materials are chemically stable, easily removal, and reusable. These three significant advantages imply that this method holds great promise to become an alternative "green" synthetic method.

On the other hand, a disadvantage of this method is to be unsuitable for a large-scale synthesis, because the rate of surface reaction on photocatalysts under irradiation are limited by electron-hole recombination, smaller surface area, lesser adsorptive and diffusive properties of substrates compared to those of the conventional catalysis. Therefore, up to now, scaling-up of the semiconductor photocatalysis has been successfully applied only to wastewater treatment, in which solar photocatalytic degradation of water contaminants, persistent toxic compounds and cyanide etc., is carried out on low concentration of the contaminants [70, 71]. In contrast, highly chemoselective and stereoselective hydrogenations have been made on a pilot plant scale and even fewer are available commercially on a multi kg scale by the conventional heterogeneous or homogeneous hydrogenation methods [72].

## 5. Concluding remarks and future directions

The photocatalytic hydrogenation on semiconductor particles has developed as the highly efficient and selective reaction during the past three decades. For examples, the selective hydrogenations were reported for the reactions of $CH_3C{\equiv}CH$ to $CH_3CH{=}CH_2$ on Pt/$TiO_2$ (rutile) [17], the aromatic ketones to the corresponding secondary alcohols on the P25 $TiO_2$ [37], the aliphatic ketones to the corresponding alcohols on the ZnS nano-crystallite [39], and nitroaromatics to the corresponding amino-compounds on $TiO_2$ [67-69]. In addition, the selective formation of methanol from $CO_2$ [53-56] and $NH_3$ from $NO_3^-$ [65, 66] has been received much attention to solve the environmental issues. Recent progress in the reductive hydrogenation has been directed toward the development of visible light response photocatalysts to utilize solar energy effectively [46-52, 65, 68, 69].

One of the most significant features of semiconductor photocatalysis is that we can utilize both oxidation and reduction in one-pot processes as mentioned in the deaminocondensation reactions [27-30]. The combination of redox reactions can afford several unique reactions, which are not achieved by conventional reaction techniques. Thus, the semiconductor photocatalysis is one of promising methods in fine chemical synthesis for high value pharmaceuticals etc. In order to achieve this method, development of highly stereoselective photocatalysis will be indispensable, although less known about such semiconductor photocatalysts so far. Therefore, particular attention should be directed toward the development of new enantioselective semiconductor catalyzed reactions in the future.

## Author details

Shigeru Kohtani*, Eito Yoshioka and Hideto Miyabe
*Department of Pharmacy, School of Pharmacy, Hyogo University of Health Sciences, Kobe, Japan*

* Corresponding Author

## Acknowledgement

The work in ref. [37] (summarized in Table 3) was supported by a Grant-in-Aid for Scientific Research (no. 21590052 and no. 19590005) from the Japan Society for the Promotion of Science, a grant from Prof. Nishihara (HUHS), and Astellas Foundation for Research on Metabolic Disorders.


## 6. References


[1] Fujishima A, Honda K (1972) Electrochemical Photolysis of Water at a Semiconductor Electrode. Nature 238: 37-38.

[2] Kudo A, Miseki Y (2009) Heterogeneous Photocatalyst Materials for Water Splitting. Chem. Soc. Rev. 38: 253-278.

[3] Fujishima A, Zhang X, Tryk D.A (2007) Heterogeneous Photocatalysis: From Water Photolysis to Applications in Environmental Cleanup. Int. J. Hydrogen Energy 32: 2664-2672.

[4] Hoffmann M.R, Martin S.T, Choi W, Bahnemann D.W (1995) Environmental Applications of Semiconductor Photocatalysis. Chem. Rev. 95: 69-96.

[5] Fujishima A, Rao T.N, Tryk D.T (2000) Titanium Dioxide Photocatalysis. J. Photochem. Photobiol. C. Photochem. Rev. 1: 1-21.

[6] Sakata T (1985) Photocatalysis of Irradiated Semiconductor Surfaces: Its Application to Water Splitting and Some Organic Reactions. J. Photochem. 29: 205-215.

[7] Fox M.A (1987) Selective Formation of Organic Compounds by Photoelectrosynthesis at Semiconductor Particles. Top. Curr. Chem. 142: 72-99.

[8] Kisch H (2001) Semiconductor Photocatalysis for Organic Synthesis. Adv. Photochem. 26: 93-143, and references therein.

[9] Palmisano G, Augugliaro V, Pagliaro M, Palmisano L (2007) Photocatalysis: a Promising Route for 21st Centyry Organic Chemistry. Chem. Commun. 3425-3437, and references therein.

[10] Shiraishi Y, Hirai T (2008) Selective Organic Transformation on Titanium Oxide-Based Photocatalysts. J. Photochem. Photobiol. C: Photochem. Rev. 9: 157-170, and references therein.

[11] Palmisano G, García-López E, Marcí G, Toddo V, Yurdakal S, Augugliaro V, Palmisano L (2010) Advances in Selective Conversion by Heterogeneous Photocatalysis. Chem. Commun. 46: 7074-7089, and references therein.

[12] Boonstra A.H, Mutsaers C.A.H.A (1975) Photohydrogenation of Ethyne and Ethene on the Surface of Titanium Dioxide. J. Phys. Chem. 79: 2025-2027.

[13] Redmond G, Fitsmaurice D (1993) Spectroscopic Determination of Flatband Potentials for Polycrystalline $TiO_2$ Electrodes in Nonaqueous Solvents. 97: 1426-1430.

[14] Yun C, Anpo M, Kodama S, Kubokawa Y (1980) U.V. Irradiation-Induced Fission of a C=C or C≡C Bond Adsorbed on $TiO_2$. J. Chem. Soc., Chem. Commun. 609.

[15] Anpo M, Aikawa N, Kodama S, Kubokawa Y (1984) Photocatalytic Hydrogenation of Alkynes and Alkenes with Water over $TiO_2$: Hydrogenation Accompanied by Bond Fission. J. Phys. Chem. 88: 2569-2572.

[16] Anpo M, Aikawa N, Kubokawa Y (1984) Photocatalytic Hydrogenation of Alkynes and Alkenes with Water over $TiO_2$: Pt-Loading Effect on the Primary Processes. J. Phys. Chem. 88: 3998-4000.

[17] Anpo M, Shima T, Kodama S, Kubokawa Y (1987) Photocatalytic Hydrogenation of $CH_3CCH$ with $H_2O$ on Small-Particle $TiO_2$: Size Quantization Effects and Reaction Intermediates. J. Phys. Chem. 91: 4305-4310.

[18] Yamataka H, Seto N, Ichihara J, Hanafusa T, Teratani S (1985) Reduction of C-C Multiple Bonds Using an Illuminated Semiconductor Catalyst. J. Chem. Soc., Chem. Commun. 788-789.

[19] Baba R, Nakabayashi S, Fujishima A, Honda K (1987) Photocatalytic Hydrogenation of Ethylene on the Bimetal-Deposited Semiconductor Powders. J. Am. Chem. Soc. 109: 2273-2277.

[20] Kodama S, Nakaya H, Anpo M, Kubokawa Y (1985) A Common Factor Determining the Features of the Photocatalytic Hydrogenation and Isomerization of Alkenes over Ti-Si Oxides. Bull. Chem. Soc. Jpn 58: 3645-3646.

[21] Anpo M, Nakaya H, Kodama S, Kubokawa Y, Domen K, Onishi T (1986) Photocatalysis over Binary Metal Oxides: Enhancement of the Photocatalytic Activity of $TiO_2$ in Titanium-Silicon Oxides. J. Phys. Chem. 90: 1633-1636.

[22] Yamashita H, Kawasaki S, Ichihashi Y, Takeuchi M, Harada M, Anpo M, Louis C, Che M (1998) Characterization of Ti/Si Binary Oxides Prepared by the Sol-Gel Method and Their Photocatalytic Properties: The Hydrogenation and Hydrogenolysis of $CH_3CCH$ with $H_2O$. Korean J. Chem. Eng. 15: 491-495.

[23] Lin L, Kuntz R.R (1992) Photocatalytic Hydrogenation of Acetylene by Molybdenium-Sulfur Complex Supported on $TiO_2$. Langmuir 8: 870-875.

[24] Kuntz R.R (1997) Comparative Study of $Mo_2O_xS_y(cys)_2^{2-}$ Complexes as Catalysts for Electron Transfer from Irradiated Colloidal $TiO_2$ to Acetylene. Langmuir 13: 1571-1576.

[25] Shiragami T, Ankyu H, Fukami S, Pac C, Yanagida S, Mori H, Fujita H (1992) Semiconductor Photocatalysis: Visible Light Induced Photoreduction of Aromatic Ketones and Electron-deficient Alkenes Catalysed by Quantised Cadmium Sulfide. J. Chem. Soc., Faraday Trans. 88: 1055-1061.

[26] Shiragami T, Fukami S, Wada Y, Yanagida S (1993) Semiconductor Photocatalysis: Effect of Light Intensity on Nanoscale CdS-Catalyzed Photolysis of Organic Substrates. J. Phys. Chem. 97: 12882-12887.

[27] Nishimoto S, Ohtani B, Yoshikawa T, Kagiya T (1983) Photocatalytic Conversion of Primary Amines to Secondary Amines and Cyclization of Polymethylene-$\alpha,\omega$-diamines by an Aqueous Suspension of $TiO_2$/Pt. J. Am. Chem. Soc. 105: 7180-7182.

[28] Ohtani B, Goto Y, Nishimoto S, Inui T (1996) Photocatalytic Transfer Hydrogenation of Schiff Bases with Propan-2-ol by Suspended Semiconductor Particles Loaded with Platinum Deposits. J. Chem. Soc. Faraday Trans. 92: 4291-4295.

[29] Pal B, Ikeda S, Kominami H, Kera Y, Ohtani B (2003) Photocatalytic Redox-Combined Synthesis of L-Pipecolinic Acid from L-Lysine by Suspended Titania Particle: Effect of Noble Metal Loading on the Selectivity and Optical Purity of the Product. J. Catal. 217: 152-159.

[30] Ohtani B, Pal B, Ikeda S (2003) Photocatalytic Organic Synthesis: Selective Cyclization of Amino Acids in Aqueous Suspensions. Catal. Surv. Asia 7: 165-176, and references therein.

[31] Künneth R, Feldmer C, Knoch F, Kisch H (1995) Semiconductor-Catalyzed Photoaddition of Olefins and Enol Ethers to 1,2-Diazenes: A New Route to Allylhydrazines. Chem. Eur. J. 1: 441-448.

[32] Warrier M, Lo M.K.F, Monbouquette H, Garcia-Garibay M.A (2004) Photocatalytic Reduction of Aromatic Azides to Amines Using CdS and CdSe Nanoparticles. Photochem. Photobiol. Sci. 3: 859-863.

[33] Cuendet P, Grätzel M (1987) Direct Photoconversion of Pyruvate to Lactate in Aqueous $TiO_2$ Dispersions. J. Phys. Chem. 91: 654-657.

[34] Joyce-Pruden C, Pross J.K, Li Y (1992) Photoinduced Reduction of Aldehydes on Titanium Dioxide. J. Org. Chem. 57: 5087-5091.

[35] Matsushita Y, Kumada S, Wakabayashi K, Sakeda K, Ichimura T (2006) Photocatalytic Reduction in Microreactors. Chem. Lett. 35: 410-411.

[36] Matsushita Y, Ohba N, Kumada S, Sakeda K, Suzuki T, Ichimura T (2008) Photocatalytic Reactions in Microreactors. 135S: S303-S308.

[37] Kohtani S, Yoshioka E, Saito K, Kudo A, Miyabe H (2010) Photocatalytic Hydrogenation of Acetophenone Derivatives and Diaryl Ketnoes on Polycrystalline Titanium Dioxide. Catal. Commun. 11: 1049-1053.

[38] Yanagida S, Ishimaru Y, Miyake Y, Shiragami T, Pac C, Hashimoto K, Sataka T (1989) Semiconductor Photocatalysis. ZnS-Catalyzed Photoreduction of Aldehydes and Related Derivatives: Two-Electron-Transfer Reduction and Relationship with Spectroscopic Properties. J. Phys. Chem. 93: 2576-2582.

[39] Yanagida S, Yoshiya M, Shiragami T, Pac C, Mori H, Fujita H (1990) Semiconductor Photocatalysis. Quantitative Photoreduction of Aliphatic Ketones to Alcohols Using Defect-Free ZnS Quantum Crystallites. J. Phys. Chem. 94: 3104-3111.

[40] Yanagida S, Ogata T, Shindo A, Hosokawa H, Mori H, Sakata T, Wada Y (1995) Semiconductor Photocatalysis: Size Control of Surface-Capped CdS Nanocrystallites and the Quantum Size Effect in Their Photocatalysis. Bull. Chem. Soc. Jpn. 68: 752-758.

[41] Park J.W, Hong M.J, Park K.K (2001) Photochemical Reduction of 1,2-Diketones in the Presence of $TiO_2$. Bull. Korean Chem. Soc. 22: 1213-1216.

[42] Park J.W, Kim E.K, Park K.K (2002) Photochemical Reduction of Benzil and Benzoin in the Presence of Triethylamine and $TiO_2$ Photocatalyst. Bull. Korean Chem. Soc. 23: 1229-1234.

[43] Aresta M (2010) Carbon Dioxide - Utilization Options to Reduce Its Accumulation in the Atmosphere. In: Aresta M, editor. Carbon Dioxide as Chemical Feedstock. Weinheim: WILEY-VCH pp. 1-13.

[44] Inoue T, Fujishima A, Konishi S, Honda K (1979) Photoelectrocatalytic Reduction of Carbon Dioxide in Aqueous Suspension of Semiconductor Powders. Nature 277: 637-638.

[45] Henglein A, Gutiérrez M (1983) Photochemistry of Colloidal Metal Sulfides. 5. Fluorescence and Chemical Reactions of ZnS and ZnS/CdS Co-Colloids. Ber. Bunsenges. Phys. Chem. 87: 852-858.

[46] Pan P-W, Chen Y-W (2007) Photocatalytic Reduction of Carbon Dioxide on $NiO/InTaO_4$ under Visible Light Irradiation. Catal. Commun. 8: 1546-1549.

[47] Jia L, Li J, Fang W (2009) Enhanced Visible-light Active C and Fe Co-Doped LaCoO3 for Reduction of Carbon Dioxide. Catal. Commun. 11: 87-90.

[48] Liu Y, Huang B, Dai Y, Zhang X, Qin X, Jiang M, Whangbo M-H (2009) Selective Ethanol Formation from Photocatalytic Reduction of Carbon Dioxide in Water with BiVO4 Photocatalyst. Catal. Commun. 11: 210-213.

[49] Yan S.C, Ouyang S.X, Gao J, Yang M, Feng J.Y, Fan X.X, Wan L.J, Li Z.S, Ye J.H, Zhou Y, Zou Z.G (2010) A Room-Temperature Reactive-Temperature Route to Mesoporous $ZnGa_2O_4$ with Improved Photocatalytic Activity in Reduction of $CO_2$. Angew. Chem. Int. Ed. 49: 6400-6404.

[50] Liu Q, Zhou Y, Kou J, Chen X, Tian Z, Gao J, Yan S, Zou Z (2010) High-Yield Synthesis of Ultralong and Ultrathin $Zn_2GeO_4$ Nanoribbons toward Improved Photocatalytic Reduction of $CO_2$ into Renewable Hydrocarbon Fuel. J. Am. Chem. Soc. 132: 14385-14387.

[51] Zhang N, Ouyang S, Li P, Zhang Y, Xi G, Kako T, Ye J (2011) Ion-Exchange Synthesis of a Micro/Mesoporous $Zn_2GeO_4$ Photocatalyst at Room Temperature for Photoreduction of $CO_2$. Chem. Commun. 47: 2041-2043.

[52] Zhou Y, Tian Z, Zhao Z, Liu Q, Kou J, Chen X, Gao J, Yan S, Zou Z (2011) High-Yield Synthesis of Ultrathin and Uniform $Bi_2WO_6$ Square Nanoplates Benefitting from Photocatalytic Reduction of $CO_2$ into Renewable Hydrocarbon Fuel under Visible Light. Appl. Mater. Interfaces 3: 3594-3601.

[53] Anpo M, Yamashita H, Ichihashi Y, Fujii Y, Honda M (1997) Photocatalytic Reduction of $CO_2$ with $H_2O$ on Titanium Oxides Anchored within Micropores of Zeolites: Effects of the Structure of the Active Sites and the Addition of Pt. J. Phys. Chem. B 101: 2632-2636.

[54] Anpo M, Kim T-H, Matsuoka M (2009) The Design of Ti-, V-, Cr- Oxide Single-Site Catalysts within Zeolite Frameworks and their Photocatalytic Reactivity for the Decomposition of Undesirable Molecules – The Role of their Excited States and Reacition Mechanisms. Catal. Today 142: 114-124.

[55] Shioya Y, Ikeue K, Ogawa M, Anpo M (2003) Synthesis of Transparent Ti-Containing Mesoporous Silica Thin Film Materials and their Unique Photocatalytic Activity for the Reduction of $CO_2$ with $H_2O$. Appl. Catal. A: General 254: 251-259.

[56] Matsumoto Y, Obata M, Hombo J (1994) Photocatalytic Reduction of Carbon Dioxideon p-Type $CaFe_2O_4$ Powder. J. Phys. Chem. 98: 2950-2951.

[57] Burt T.P, Howden N.J.K, Worrall F, Whelan M.J (2010) Long-Term Monitoring of River Water Nitrate: How Much Data Do We Need? J. Environ. Monit. 12:71-79.

[58] Kudo A, Domen K, Maruya K, Onishi T (1987) Photocatalytic Reduction of $NO_3^-$ to Form $NH_3$ over Pt-$TiO_2$. Chem. Lett. 1019-1022.

[59] Ranjit K.T, Viswanathan B (1997) Photocatalytic Reduction of Nitrite and Nitrate Ions over Doped $TiO_2$ Catalysts. J. Photochem. Photobiol. A Chem. 107: 215-220.

[60] Li Y, Wasgestian F (1998) Photocatalytic Reduction of Nitrate Ions on $TiO_2$ by Oxalic Acid. J. Photochem. Photobiol. A Chem. 112: 255-259.

[61] Kudo A, Domen K, Maruya K, Onishi T (1992) Reduction of Nitrate Ions into Nitrite and Ammonia over Some Photocatalysts. J. Catal. 135: 300-303.

[62] Kato H, Kudo A (2002) Photocatalytic Reduction of Nitrate Ions over Tantalate Photocatalysts. 4: 2833-2838.

[63] Ranjit K.T, Krishnamoorthy R, Viswanathan B (1994) Photocatalytic Reduction of Nitrite and Nitrate on ZnS. J. Photochem. Photobiol. A Chem. 81: 55-58.

[64] Kudo A, Sekizawa M (2000) Photocatalytic $H_2$ Evolution under Visible Light Irradiation on Ni-Doped ZnS Photocatalyst. Chem. Commun. 1371-1372.

[65] Kudo A, Hamanoi O (2002) Reduction of Nitrate and Nitrite Ions over Ni-ZnS Photocatalyst under Visible Light Irradiation in the Presence of a Sacrificial Reagent. Chem. Lett. 838-839.

[66] Yamauchi M, Abe R, Tsukuda T, Kato K, Tanaka M (2011) Highly Selective Ammonia Synthesis from Nitrate with Photocatalytically Generated Hydrogen on CuPd/$TiO_2$. J. Am. Chem. Soc. 133: 1150-1152.

[67] Mahdavi F, Bruton T.C, Li Y (1993) Photoinduced Reduction of Nitro Compounds on Semiconductor Particles. J. Org. Chem. 58: 744-746.

[68] Wang H, Yan J, Chang W, Zhang Z (2009) Practical Synthesis of Aromatic Amines by Photocatalytic Reduction of Aromatic Nitro Compounds on Nanoparticles N-Doped $TiO_2$. Catal. Commun. 10: 989-994.

[69] Füldner S, Mild R, Siegmund H.I, Schroeder J.A, Gruber M, König B (2010) Green-Light Photocatalytic Reduction Using Dye-Sensitized $TiO_2$ and Transition Metal Nanoparticles. Green Chem. 12: 400-406.

[70] Malato S, Blanco J, Vidal A, Richter C (2002) Photocatalysis with Solar Energy at a Pilot-Plant Scale: an Overview. Appl. Catal. B: Environ. 37: 1-15.

[71] Marugán J, Grieken R, Cassano A.E, Alfano O.M (2009) Scaling-up of Slurry Reactors for the Photocatalytic Oxidation of Cyanide with $TiO_2$ and Silica-Supported $TiO_2$ Suspensions. Catal. Today 144: 87-93.

[72] Blaser H-U (2010) The Development and Application of Industrially Viable Catalysts for the Selective Hydrogenation of Complex Molecules. Top. Catal. 53: 997-1001.

Chapter 13

# Hydrogenation of Fullerene $C_{60}$: Material Design of Organic Semiconductors by Computation

Ken Tokunaga

Additional information is available at the end of the chapter

http://dx.doi.org/10.5772/48534

## 1. Introduction

### 1.1. Motivation

Carrier mobility [1] in organic semiconductors is one of the most important properties in the performance of organic light-emitting diodes (OLEDs), organic field-effect transistors (OFETs), and organic solar cells, which are expected to be used in next-generation technologies [2]. Organic semiconductors have the advantages of lightness, flexibility, and low cost. Therefore, research and development of new materials with chemical and thermal stability has recently been very active [3–6]. However, the wide variety of organic materials, which is generally a major advantage of these materials, has hindered the systematic research and development of novel materials. Thus, the establishment of design guidelines for new organic materials is a matter of great urgency.

Up till now, much effort has been made to understand theoretically the relationship between the structure and the carrier-transport properties of these materials [7–10]. Theoretical investigations can give reliable guidelines for the development of such new organic semiconductors. We have also been studying the quantum-chemical design of organic semiconductors based on fullerenes from both scientific and technological viewpoints [11–19]. $C_{60}$ derivatives (Figure 1) are very interesting from the viewpoint of practical use. Some types of $C_{60}$ derivatives are shown in the figure, and $C_{60}$ derivatives of types (a) and (b) have already been studied in our research group [11–19].

It is well known that $C_{60}$ is chemically and thermally stable, and its method of synthesis is also established. Carrier mobility of amorphous silicon is about 1 $cm^2V^{-1}s^{-1}$, so that a mobility above this value is desirable for organic semiconductors. However, hole mobility in $C_{60}$ film is in the order of $10^{-5}$ $cm^2V^{-1}s^{-1}$ [20], so that $C_{60}$ has not been used as a hole-transport material (p-channel semiconductor). Thus, the main purpose of our previous studies [11–14, 17–19] was to improve the hole mobility of $C_{60}$ by chemical methods. There is the possibility

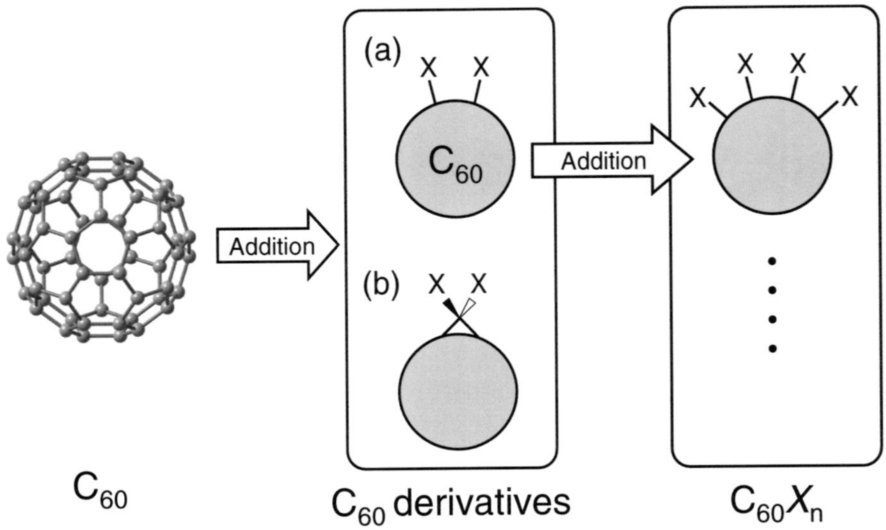


**Figure 1.** Fullerene $C_{60}$ and some examples of $C_{60}$ derivatives.

that the addition of hydrogen to $C_{60}$ would result in a considerable modification of the $C_{60}$ materials. This possibility originates from the fact that $C_{60}$ has an electronic degeneracy in its cationic state because of its high symmetry $I_h$ [21, 22], but $C_{60}$ hydrides usually do not because of the reduction in symmetry by the addition of hydrogen. Electron mobility in $C_{60}$ film is about 1 $cm^2V^{-1}s^{-1}$ [20, 23, 24] so that $C_{60}$ is one of the most useful electron-transport materials. However, to achieve low-cost production and large-area devices, it is necessary for $C_{60}$ materials to have a solution-processable form, such as [6,6]-phenyl $C_{61}$-butyric acid methyl ester (PCBM) [25]. Therefore, in organic electronics, the electronic properties of $C_{60}$ derivatives rather than those of the original $C_{60}$ are of much interest and importance.

Although ambipolar transport in [6,6]-phenyl-$C_{71}$-butyric acid methyl ester ([70]PCBM) was reported recently, its hole mobility ($2\times10^{-5}\,cm^2V^{-1}s^{-1}$) is much smaller than its electron mobility ($2\times10^{-3}\,cm^2V^{-1}s^{-1}$) [26]. Therefore, the enhancement of hole mobility is necessary for the practical use of fullerene derivatives as ambipolar transistor materials. Very recently, conduction-type control of fullerene $C_{60}$ films from *n*- to *p*-type by doping with molybdenum(VI) oxide ($MoO_3$) was demonstrated [27]. Thus, analysis of the hole-transport properties of $C_{60}$ derivatives is very important for the practical use of $C_{60}$ materials.

## 1.2. Our previous publication: $C_{60}X_n$ ($n$ = 2, 4, and 6)

In our previous publication [11], the effect on carrier-transport properties of chemical addition of $X$ (X=H, R, R-COOH, and R-SH, where R is an alkyl chain) to $C_{60}$ was systematically discussed from the viewpoint of reorganization energy ($\lambda$) using Marcus theory [28]. We focused on the $C_{60}$ derivatives of type (a) in Figure 1, $C_{60}X_n$, where $X$ is the added group and $n$ is the number of added $X$. There are many isomers for $C_{60}X_n$, so that the position of $X$ was also a subject of investigation. The dependence of carrier-transport properties on the type or chemical nature of $X$ was discussed from the results of $C_{60}X_2$. The dependence of carrier-transport properties on the number of added groups was discussed from the results

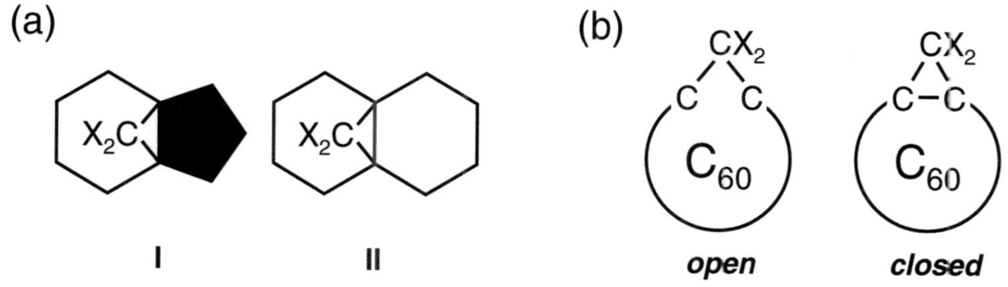


**Figure 2.** (a) Two addition patterns of $CX_2$. (b) Open and closed structures of $C_{60}CX_2$.

of $C_{60}H_n$ ($n$ = 2, 4, and 6). From these discussions, guidelines for effective design of carrier-transport materials were proposed:

- Carrier-transport properties of $C_{60}X_2$ are quite different from those of $C_{60}$ [13–15, 18].
- Chemical addition can improve hole mobility of $C_{60}$ for some isomers [13, 14, 18]. Conversely, electron mobility of $C_{60}$ is not influenced or is decreased by the chemical addition [15].
- The values of reorganization energies of both hole transport and electron transport are almost independent of the type (chemical nature) of addition group $X$, but are strongly dependent on the position and the number of addition groups [15]. Therefore, reorganization energies of other types of $C_{60}X_2$ will be approximately estimated from the results of $C_{60}H_2$ [15].
- Hole and electron mobilities are closely related to the distribution patterns of the HOMO (highest occupied molecular orbital) and the LUMO (lowest unoccupied molecular orbital), respectively [13–15]. Delocalized orbitals give high carrier mobility (small reorganization energy) and localized molecular orbitals produce a low carrier mobility (large reorganization energy).
- There is a clear linear relationship between the reorganization energy and the geometrical relaxation upon carrier injection.

These results will also be applied to other types of $X$, and give us a guideline for efficient design of novel materials based on $C_{60}$ from both experimental and theoretical approaches. For example, in the experimental viewpoint, we can freely select an $X$ that is appropriate for thin-film formation and is easily synthesized without considering the influence of $X$ on the electronic properties. From the theoretical viewpoint, this result enables us to save computation time and resources in the material design because various types of $C_{60}X_2$ can be simplified to $C_{60}H_2$ in the prediction and discussion of these properties.

### 1.3. Our previous publication: $C_{60}CX_2$

In another of our publications [12], the effect of methylene bridging of $C_{60}$ by $-CX_2$ ($X$ = H, halogen, R, R-COOH, and R-SH, where R is an alkyl chain) on carrier-transport properties was systematically discussed. There are two isomers for $C_{60}CX_2$, **I** and **II** (Figure 2). Systematic analyses of reorganization energies of methylene-bridged fullerenes $C_{60}CX_2$ give us very important knowledge for the efficient design of useful $C_{60}$ materials:

- Hole mobility of $C_{60}$ is strongly influenced by methylene bridging. $C_{60}CBr_2$ (**II**) has the smallest reorganization energy (93 meV). On the other hand, $C_{60}CX_2$ (**I**) isomers with R, R-COOH, and R-SH chains have very large reorganization energies of about 500 meV.
- Electron mobility of $C_{60}$ is not influenced or is decreased by methylene bridging [16].
- Values of reorganization energies of both hole transport and electron transport are dependent on the type (chemical nature) of *X*. Therefore, especially for the case of hole transport, reorganization energies of other types of $C_{60}CX_2$ will not be easily predicted from the results of $C_{60}CH_2$. This result is quite different from the case of $C_{60}H_2$ and $C_{60}X_2$.
- There is a clear linear relationship between the reorganization energy and the change in the distance between bridged $C \cdots C$ atoms.
- Hole and electron mobilities are closely related to the distribution patterns of the HOMO and the LUMO between bridged carbon atoms, respectively [16]. Small carrier mobility and large reorganization energy result from strong antibonding character between bridged carbon atoms.

One of the important findings in this work is that the properties of $C_{60}CX_2$ are dependent on the type (chemical nature) of *X*. This is because $CX_2$ addition directly changes the carbon network around the addition position. The possibility of a transformation between a *closed* structure and an *open* structure upon carrier injection is one of the reasons why the reorganization energy of $C_{60}CX_2$ is strongly dependent on the type of *X*. Therefore, the molecular design of type (b) molecules is a little more difficult compared with the molecular design of type (a) molecules, $C_{60}X_2$.

We also found that electron-transport properties are little influenced by the methylene bridging, so that we can freely select an *X* that is appropriate for thin-film formation and is easily synthesized without considering the effect of *X* on the electronic properties. On the other hand, when constructing high-mobility hole-transport materials, the use of isomer **I** of $C_{60}CX_2$, which includes an alkyl chain with $-CH_3$ and -COOH terminals, should be avoided.

## 1.4. This chapter

In this chapter, the effect of further hydrogenation of $C_{60}$ on hole-transport properties is systematically discussed from the viewpoint of reorganization energy. We again focus on the $C_{60}$ derivatives of type (a) in Figure 1. The dependence of hole-transport properties on the number and the position of hydrogen atoms is discussed from the results of $C_{60}H_n$ ($n$ = 2, 4, 6, 8, 52, 54, 56, 58, and 60). From these discussions, guidelines for the effective design of high-performance carrier-transport materials of type (a) are proposed.

This chapter is organized as follows: In Section 2, the definition of $\lambda$ and computational details of $\lambda$ are presented. Synthesis methods of hydrogenated $C_{60}$ are reviewed in Section 3. Structures of hydrogenated $C_{60}$ molecules studied in this chapter are shown in Section 4. In Section 5, calculated results of $\lambda$ for hole transport are shown. $\lambda$ for electron transport is not

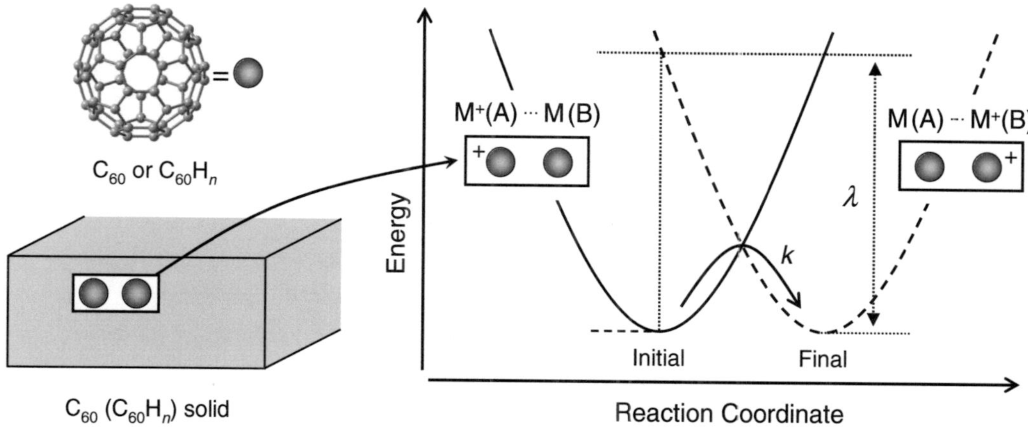


**Figure 3.** Schematic reaction diagram of a hole-transfer reaction $M^+(A) \cdots M(B) \rightarrow M(A) \cdots M^+(B)$.

discussed because it is expected that $\lambda$ for electron transport is independent of the position and the number of hydrogen atoms. From the systematic discussion of these results, the dependence of $\lambda$ on the position and the number of hydrogen atoms is shown in Section 6. Summarizing these discussions, simple guidelines for the efficient design of useful $C_{60}H_n$ semiconductors are proposed in Section 7.

## 2. Hole-transport mechanism

### 2.1. Hopping mechanism

The hole mobility of single-crystal $C_{60}$ is around $10^{-5}\,\mathrm{cm^2V^{-1}s^{-1}}$ at a maximum. Materials having a mobility of $0.1-1\,\mathrm{cm^2V^{-1}s^{-1}}$ are categorized in the boundary region between the hopping and the band-transport mechanisms [29–31]. Therefore, in this chapter, only the hopping mechanism is considered because the hole mobility of $C_{60}$ is much smaller than $0.1\,\mathrm{cm^2V^{-1}s^{-1}}$.

A schematic picture of hole hopping in an organic solid is shown in Figure 3. $C_{60}$ and $C_{60}H_n$ molecules are represented by spheres in Figure 3. In the treatment of the hopping mechanism, two neighboring molecules are chosen from the solid. Then, a hole hopping between these two molecules, that is, the hole-transfer reaction, is considered. Repeating such a hopping between neighboring molecules, a carrier travels from one edge to the other edge of the organic solid. The initial and final states of a hole-transfer reaction are represented as $M^+(A) \cdots M(B)$ and $M(A) \cdots M^+(B)$. The system is fluctuating around the bottom of the potential curve as a result of molecular vibrations in its initial state. When the system occasionally reaches the transition state at which energies of the initial state and the final state are the same, the system jumps from the initial state to the final state with a rate constant $k$. This reaction is written as

$$M^+(A) \cdots M(B) \xrightarrow{k} M(A) \cdots M^+(B). \tag{1}$$

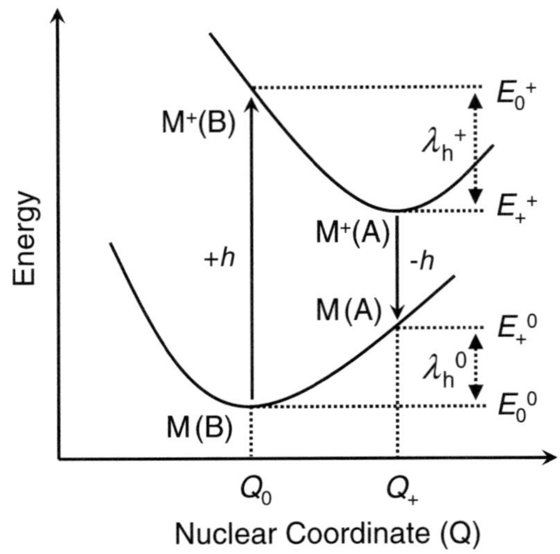


**Figure 4.** Schematic potential energy surfaces of molecules related to the hole-transfer reaction. $Q_0$ and $Q_+$ mean the nuclear coordinates of stable structures in neutral and cationic states, respectively. Subscripts on the right-hand side of $E$ mean the geometrical structure of the molecule and superscripts on the right-hand side of $E$ mean the charge on the molecule.

A localized hole on one molecule (A) jumps to the neighboring molecule (B). From the Marcus theory [28], the hole-transfer rate constant $k$ of a homogeneous carrier-transfer reaction can be estimated from two parameters, the reorganization energy ($\lambda$), and the electronic coupling element ($H$) between adjacent molecules:

$$k = \frac{4\pi^2}{h} \frac{H^2}{\sqrt{4\pi\lambda k_B T}} \exp\left(-\frac{\lambda}{4k_B T}\right), \tag{2}$$

where $k_B$ is the Boltzmann constant, $h$ is the Planck constant, and $T$ is the temperature of the system. We can see that a small $\lambda$, large $H$, and high temperature $T$ result in a fast hole hopping.

## 2.2. Reorganization energy

From Figure 3, we can see that the reorganization energy is the difference between "the energy in the final state of the system with a stable nuclear configuration in the *final* state" and "the energy in the final state of the system with a stable nuclear configuration in the *initial* state". For the calculation of $\lambda$, potential energy diagrams of the hole-transfer reaction are shown in Figure 4. The values of $\lambda$ are obtained by the following procedure: First, the geometries of neutral $C_{60}$ and $C_{60}H_n$ were fully optimized, giving a nuclear coordinate $Q_0$ and energy $E_0^0$. At $Q_0$, single-point energy calculations of cations give $E_0^+$. Next, the structures were fully optimized in their cationic states, giving $Q_+$ and $E_+^+$. Single-point energy calculations of the neutral states with geometry $Q_+$ give $E_+^0$. The reorganization energy of the hole-transfer reaction ($\lambda_h$) is defined as the sum of $\lambda_h^+$ and $\lambda_h^0$:

$$\lambda_h = \lambda_h^+ + \lambda_h^0, \tag{3}$$

where $\lambda_{\mathrm{h}}^{+}$ and $\lambda_{\mathrm{h}}^{0}$ are relaxation energies in the cationic and neutral states, respectively, estimated from

$$\lambda_{\mathrm{h}}^{+} = E_{0}^{+} - E_{+}^{+}, \tag{4}$$

$$\lambda_{\mathrm{h}}^{0} = E_{+}^{0} - E_{0}^{0}. \tag{5}$$

## 2.3. Rate constant

In this chapter, the calculation and analysis of $\lambda_{\mathrm{h}}$ are focused on. The electronic-coupling element for hole transport ($H_{\mathrm{h}}$) can be approximated by one half of the molecular orbital energy splitting between the HOMO and the next HOMO of the neutral dimer [3, 32]. However, the values of $H_{\mathrm{h}}$ are dependent on both the distance and the relative orientation between the two molecules. Therefore, $H_{\mathrm{h}}$ is assumed to be the same for $C_{60}$ and all $C_{60}H_{n}$ for ease of discussion in this work [33]. Therefore, we cannot know the numerical values of the rate constants $k_{\mathrm{h}}$. Instead, we define $k_{\mathrm{h}}'$ as a ratio of the rate constant of $C_{60}H_{n}$ to that of $C_{60}$ at $T = 300\,\mathrm{K}$. The values of $k_{\mathrm{h}}'$ for hole transport are calculated as

$$\begin{aligned} k_{\mathrm{h}}' &= \frac{k_{\mathrm{h}}^{\mathrm{C_{60}H}_n}}{k_{\mathrm{h}}^{\mathrm{C_{60}}}} \\ &= \left(\frac{H_{\mathrm{h}}^{\mathrm{C_{60}H}_n}}{H_{\mathrm{h}}^{\mathrm{C_{60}}}}\right)^2 \cdot \sqrt{\frac{\lambda_{\mathrm{h}}^{\mathrm{C_{60}}}}{\lambda_{\mathrm{h}}^{\mathrm{C_{60}H}_n}}} \cdot \exp\left(-\frac{\lambda_{\mathrm{h}}^{\mathrm{C_{60}H}_n} - \lambda_{\mathrm{h}}^{\mathrm{C_{60}}}}{4k_{\mathrm{B}}T}\right) \\ &\approx \sqrt{\frac{\lambda_{\mathrm{h}}^{\mathrm{C_{60}}}}{\lambda_{\mathrm{h}}^{\mathrm{C_{60}H}_n}}} \cdot \exp\left(-\frac{\lambda_{\mathrm{h}}^{\mathrm{C_{60}H}_n} - \lambda_{\mathrm{h}}^{\mathrm{C_{60}}}}{4k_{\mathrm{B}}T}\right) \end{aligned} \tag{6}$$

on the supposition that $H_{\mathrm{h}}^{\mathrm{C_{60}H}_n} \approx H_{\mathrm{h}}^{\mathrm{C_{60}}}$. Hereafter, we regard $k_{\mathrm{h}}'$ as the hole mobility. The values of $k_{\mathrm{h}}'$ are usually calculated using Equation 6 in this chapter.

All calculations (geometrical optimizations and self-consistent field (SCF) energy calculations) necessary to obtain the values of the energies in Figure 4 were performed by a quantum-chemical method, namely, density functional theory (DFT) using the B3LYP functional. For the calculation of $\lambda_{\mathrm{h}}$, the 6-311G($d$, $p$) basis set was adopted. All neutral (ionic) systems were calculated in singlet (doublet) states. Calculations were performed using the GAUSSIAN 03 [34] program package.

# 3. Hydrogenation of fullerene

Although the main topic of this chapter is a theoretical discussion of hydrogenated fullerenes, synthesis methods of hydrogenated fullerenes are also important for practical use. Up to now, many types of hydrogenated fullerenes have been synthesized. Hydrogenated fullerenes, $C_{60}H_2$, $C_{60}H_4$, and $C_{60}H_6$, have been prepared by hydroboration [35, 36], hydrozirconation [37], rhodium-catalyzed hydrogenation [38], diimide [39] and hydrazine [40] reduction, dissolving metal reduction [41, 42], photoinduced-electron-transfer reduction with 10-methyl-9,10-dihydroacridine [43, 44], and ultrasonic irradiation in decahydronaphthalene [45]. In many cases, a mixture of $C_{60}$, $C_{60}H_2$, and $C_{60}H_n$ ($n > 2$) is obtained [46]. Birch reduction [47] and transfer hydrogenation [48] of $C_{60}$ produce $C_{60}H_{18}$ and $C_{60}H_{36}$. In the ruthenium-catalyzed hydrogenation, other types of $C_{60}H_n$ ($n$ = 10, 12, 34, 36, 38, and 40) were observed using a field-desorption (FD) mass spectrometer [49]. Direct hydrogenation of $C_{60}$ was achieved without the use of a catalyst by exposing solid-phase fullerenes to high-pressure hydrogen gas, and many types of $C_{60}H_n$ ($n = 2-18$) were identified by laser-desorption Fourier-transform mass spectrometry [50]. Unfortunately, highly hydrogenated fullerenes discussed in this chapter, $C_{60}H_{52} - C_{60}H_{60}$, have not been synthesized.

## 4. Isomers of hydrogenated fullerenes

### 4.1. Low hydrogenation: $C_{60}H_2 - C_{60}H_8$

There are 23 isomers for $C_{60}H_2$ [51]. 11 isomers of Figure 5(a) with a small formation energy were selected to consider the possibility of synthesis. The ground state of these isomers is the singlet state. Other isomers that are not considered in this chapter have triplet ground states in semiempirical calculations [51]. The initial hydrogen atom of $C_{60}H_2$ is already shown in the figure. The second hydrogen atom is added to one of the carbon atoms labeled 1–11. These isomers are named **1–11** in boldface. As predicted by Matsuzawa *et al.* [51] from quantum-chemical calculations, two isomers, **1** from 1,2-addition and **5** from 1,4-addition, have been synthesized [35, 39]. Furthermore, isomer **1** has been synthesized by many different methods [52].

Although there are a total of 4190 isomers for $C_{60}H_4$ [53], we consider eight isomers originating from two $H_2$ additions to [6,6]-ring fusions (see Figure 5 (b)) [54]. In other words, these isomers result from 1,2-addition to $C_{60}H_2$-**1**. In Figure 5 (b), the second $H_2$ pair is added to one of the carbon atom pairs labeled 1–8 and these isomers are named **1–8**. Experimentally, some of the eight isomers have been synthesized, and four isomers (**1**, **4**, **6**, and **8**) among them were identified [36, 39–41, 55–57].

Among a total of 418470 isomers of $C_{60}H_6$ [53], we consider only 16 isomers in Figure 5 (c) originating from one $H_2$ addition to $C_{60}H_4$-**1** at a [6,6]-ring fusion. One hydrogen pair is added to one of the carbon atom pairs labeled 1–16 and these isomers are named **1–16**.

We consider only six isomers of $C_{60}H_8$ in Figure 5 (d) originating from $H_2$ addition to $C_{60}H_6$-**1** at a [6,6]-ring fusion. One hydrogen pair is added to one of the carbon atom pairs labeled 1–6 and these isomers are named **1–6**.

### 4.2. High hydrogenation: $C_{60}H_{52} - C_{60}H_{60}$

As highly hydrogenated fullerenes, we consider $C_{60}H_{52}$, $C_{60}H_{54}$, $C_{60}H_{56}$, $C_{60}H_{58}$, and $C_{60}H_{60}$. $C_{60}H_{60}$ has $I_h$ symmetry. We consider 11, 8, 16, and 6 isomers for $C_{60}H_{58}$, $C_{60}H_{56}$, $C_{60}H_{54}$, and $C_{60}H_{52}$, respectively. For these highly hydrogenated isomers, hydrogenated carbon atoms in Figure 5 are to be considered the nonhydrogenated carbon atoms.

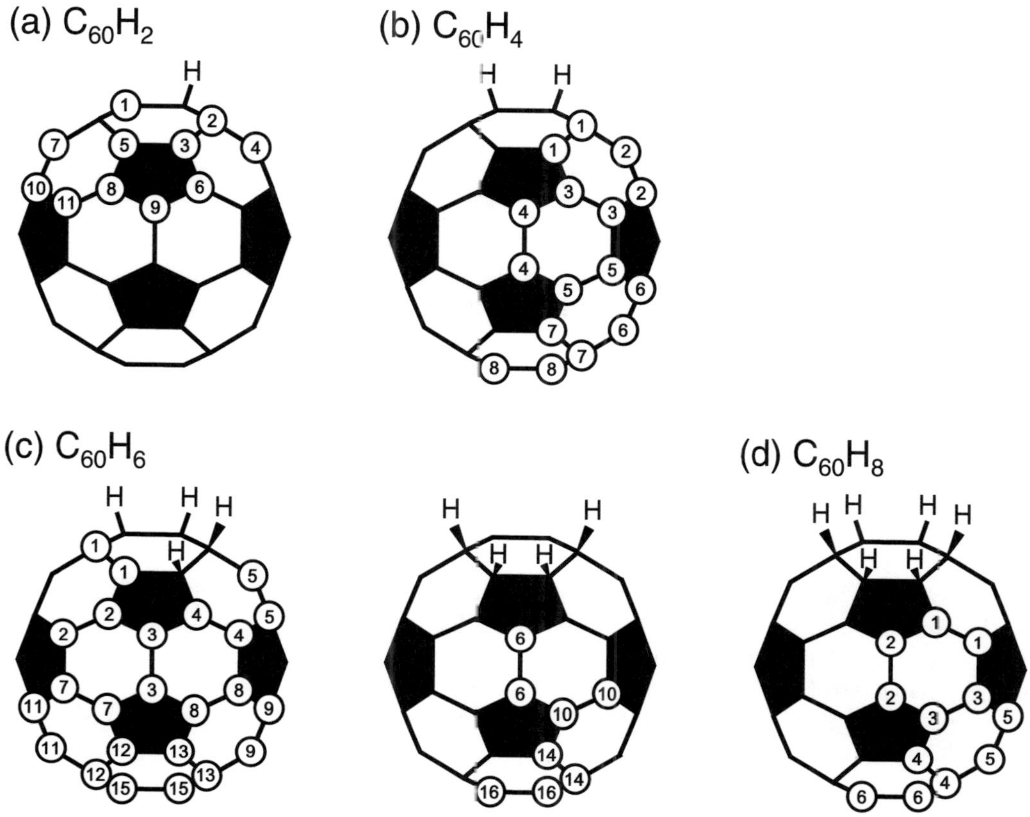


**Figure 5.** Isomers of (a) $C_{60}H_2$, (b) $C_{60}H_4$, (c) $C_{60}H_6$ (2 figures), and (d) $C_{60}H_8$. Hydrogen atoms are bonded to the labeled carbon atoms. Figures (a), (b), (c), and (d) also correspond to isomers of $C_{60}H_{58}$, $C_{60}H_{56}$, $C_{60}H_{54}$, and $C_{60}H_{52}$, respectively.

## 5. Reorganization energies and rate constants

### 5.1. $C_{60}$

The electronic structure of $C_{60}^{+}$ has been investigated both experimentally and theoretically because of its electronic degeneracy [58, 59]. $C_{60}^{+}$ has an $H_u$ degenerate electronic state so that symmetry lowering because of the Jahn–Teller effect [21] stabilizes $C_{60}^{+}$. The most stable structure of $C_{60}^{+}$ was calculated as having $D_{5d}$ symmetry, and its $\lambda_{\mathrm{h}}^{+}$ is 95 meV in our calculation. The result of $\lambda_{\mathrm{h}}^{+}$ is qualitatively consistent with previous works based on the static Jahn–Teller effect [60–62] in which the values of $\lambda_{\mathrm{h}}^{+}$ were calculated as 351 [60], 35 [61], and 71 meV [62]. The value of $\lambda_{\mathrm{h}}^{0}$ of $C_{60}^{+}(D_{5d})$ was calculated as 74 meV. Thus, the reorganization energy of the hole transport, $\lambda_{\mathrm{h}}$, is 169 meV. It should be noted that the value of $\lambda_{\mathrm{h}}^{+}$ is much larger than that of $\lambda_{\mathrm{h}}^{0}$. This result comes from the fact that geometrical relaxation in the ionic state is very large because of Jahn–Teller distortion [21]. The potential curves of the ionic states are expected to have larger curvature around the minima than that of the neutral state.

## 5.2. $C_{60}H_2 - C_{60}H_8$

On average, $\lambda_h^+$ of $C_{60}H_2$ is 84 meV, which is smaller than that of $C_{60}$ by 11 meV and $\lambda_h^0$ of $C_{60}H_2$ is 90 meV, which is larger than that of $C_{60}$ by 16 meV [12, 13]. The average value of $\lambda_h$ of $C_{60}H_2$ (174 meV) is almost as large as that of $C_{60}$ (169 meV). However, only the addition of two H atoms leads to the large difference in $\lambda_h$, 101–257 meV. It is interesting that six isomers of $C_{60}H_2$ (**1**, **4**, **5**, **6**, **8**, and **9**) have a smaller $\lambda_h$ than $C_{60}$. In particular, isomer **6** has the smallest $\lambda_h$ (101 meV), which is over 40% less than that of $C_{60}$. In addition, the $k_h'$ of **6** is about 2.5 times as large as that of $C_{60}$. From the viewpoint of practical use, it should be noted that the values of $\lambda_h$ for the two synthesized isomers, **1** and **5**, are 133 and 142 meV, respectively, which are about 20% smaller than that of $C_{60}$. The values of $k_h'$ for these isomers are about 1.5 times as large as that of $C_{60}$. These results indicate that hydrogenation can be an effective method for modifying the hole-transport properties of $C_{60}$. Two synthesized isomers of $C_{60}H_2$, **1** and **5**, have potential utility as hole-transport materials. For almost all isomers of $C_{60}H_2$, $\lambda_h^+$ is almost equal to $\lambda_h^0$. This means that the potential curves of the ionic and neutral states have almost the same curvature around the minima because the hydrogenation removes the electronic degeneracy in ionic states of $C_{60}$.

On average, $\lambda_h^+$ of $C_{60}H_4$ is 65 meV, which is smaller than that of $C_{60}$ by 30 meV and $\lambda_h^0$ of $C_{60}H_4$ is 68 meV, which is smaller than that of $C_{60}$ by 6 meV [12, 14]. The average value of $\lambda_h$ of $C_{60}H_4$ (134 meV) is much smaller than that of $C_{60}$ (169 meV) and is almost as large as that of $C_{60}H_2$-**1** (133 meV). The addition of two H atoms to $C_{60}H_2$-**1** results in the large difference in $\lambda_h$, 83–183 meV. Seven isomers of $C_{60}H_4$ (**1**, **3**, **4**, **5**, **6**, **7**, and **8**) have smaller $\lambda_h$ than $C_{60}$. Remarkably, the major product **1** has the smallest $\lambda_h$ (83 meV), which is over 50% less than that of $C_{60}$. In addition, $k_h'$ of **1** is 3.28 times as large as that of $C_{60}$, and more than twice as large as that of the synthesized $C_{60}H_2$-**1** [13]. Isomer **7** with the second smallest $\lambda_h$ has a value for $k_h'$ that is 2.83 times as large as that of $C_{60}$. Other identified isomers **4**, **6**, and **8** also have small $\lambda_h$ (138, 150, and 126 meV, respectively), and $k_h'$ of these isomers are respectively 1.49, 1.26, and 1.74 times larger than that of $C_{60}$. Synthesized isomers of $C_{60}H_4$, especially the major product **1**, have potential utility as useful hole-transport materials.

On average, $\lambda_h^+$ of $C_{60}H_6$ is 61 meV, which is much smaller than that of $C_{60}$ by 34 meV and $\lambda_h^0$ of $C_{60}H_6$ is 66 meV, which is smaller than that of $C_{60}$ by 8 meV [12]. The average value of $\lambda_h$ for $C_{60}H_6$ (127 meV) is much smaller than that for $C_{60}$ (169 meV); however, it is much larger than that for $C_{60}H_4$-**1** (83 meV). Further addition of two H atoms to $C_{60}H_4$-**1** leads to a large difference in $\lambda_h$, 71–182 meV. 14 of the 16 isomers have smaller $\lambda_h$ than $C_{60}$. Isomer **1** has the smallest $\lambda_h$ (71 meV), which is about 60% less than that of $C_{60}$. In addition, $k_h'$ of **1** is 3.94 times as large as that of $C_{60}$. Isomer **4** with the second smallest $\lambda_h$ has a value of $k_h'$ that is 3.22 times as large as that of $C_{60}$.

$\lambda_h$ and $k_h'$ of $C_{60}H_8$ are shown in Table 1. On average, $\lambda_h^+$ of $C_{60}H_8$ is 51 meV, which is much smaller than that of $C_{60}$ by 44 meV, and $\lambda_h^0$ of $C_{60}H_8$ is 62 meV, which is smaller than that of $C_{60}$ by 12 meV. The average value of $\lambda_h$ for $C_{60}H_8$ (113 meV) is much smaller than that for $C_{60}$ (169 meV); however, it is much larger than that for $C_{60}H_6$-**1** (71 meV). Further addition of two H atoms to $C_{60}H_6$-**1** leads to a large difference in $\lambda_h$, 81–175 meV. Five of the six isomers have smaller $\lambda_h$ than $C_{60}$. Isomer **5** has the smallest $\lambda_h$ (81 meV), and $k_h'$ of **5** is 3.39 times as large

| $C_{60}H_8$ | $\lambda_h^+$ | $\lambda_h^0$ | $\lambda_h$ | $k_h'$ |
|---|---|---|---|---|
| ($C_{60}$) | 95 | 74 | 169 | 1.00 |
| **1** | 57 | 72 | 128 | 1.69 |
| **2** | 37 | 50 | 86 | 3.09 |
| **3** | 81 | 94 | 175 | 0.92 |
| **4** | 55 | 68 | 123 | 1.82 |
| **5** | 36 | 45 | 81 | 3.39 |
| **6** | 41 | 42 | 84 | 3.23 |
| Average | 51 | 62 | 113 | 2.09 |

**Table 1.** Reorganization energies ($\lambda_h$ in meV) and rate constants ($k_h'$) for hole transport in $C_{60}H_8$. Values for the original $C_{60}$ and an averaged value over the six $C_{60}H_8$ isomers are also shown.

as that of $C_{60}$. Isomer **6** with the second smallest $\lambda_h$ has a value of $k_h'$ that is 3.23 times as large as that of $C_{60}$.

### 5.3. $C_{60}H_{52}$ – $C_{60}H_{58}$

$\lambda_h$ and $k_h'$ of $C_{60}H_{52}$ are shown in Table 2. On average, $\lambda_h^+$ of $C_{60}H_{52}$ is 122 meV, which is much larger than that of $C_{60}$ by 27 meV, and $\lambda_h^0$ of $C_{60}H_{52}$ is 132 meV, which is much larger than that of $C_{60}$ by 58 meV. The average value of $\lambda_h$ of $C_{60}H_{52}$ (254 meV) is much larger than that of $C_{60}$ (169 meV). Hydrogenation leads to a large difference in $\lambda_h$, 175–441 meV. Isomer **6** has $\lambda_h$ as large as $C_{60}$.

| $C_{60}H_{52}$ | $\lambda_h^+$ | $\lambda_h^0$ | $\lambda_h$ | $k_h'$ |
|---|---|---|---|---|
| **1** | 163 | 174 | 337 | 0.14 |
| **2** | 204 | 237 | 441 | 0.04 |
| **3** | 100 | 104 | 204 | 0.65 |
| **4** | 86 | 94 | 180 | 0.87 |
| **5** | 92 | 96 | 188 | 0.78 |
| **6** | 86 | 89 | 175 | 0.92 |
| Average | 122 | 132 | 254 | 0.36 |

**Table 2.** Reorganization energies ($\lambda_h$ in meV) and rate constants ($k_h'$) for hole transport in $C_{60}H_{52}$. The average value over six $C_{60}H_{52}$ isomers is also shown.

Values of $\lambda_h$ and $k_h'$ for $C_{60}H_{54}$ isomers are shown in Table 3. On average, $\lambda_h^+$ of $C_{60}H_{54}$ is 176 meV, which is much larger than that of $C_{60}$ by 81 meV, and $\lambda_h^0$ of $C_{60}H_{54}$ is 181 meV, which is much larger than that of $C_{60}$ by 107 meV. The average value of $\lambda_h$ of $C_{60}H_{54}$ (357 meV) is much larger than that of $C_{60}$. It is interesting that the values of $\lambda_h$ are in the range 254–397 meV and there is not so large a difference.

Values of $\lambda_h$ and $k_h'$ for $C_{60}H_{56}$ isomers are shown in Table 4. On average, $\lambda_h^+$ of $C_{60}H_{56}$ is 140 meV, which is much larger than that of $C_{60}$ by 45 meV, and $\lambda_h^0$ of $C_{60}H_{56}$ is 154 meV, which is much larger than that of $C_{60}$ by 80 meV. The average value of $\lambda_h$ for $C_{60}H_{56}$ (294 meV) is

| $C_{60}H_{54}$ | $\lambda_h^+$ | $\lambda_h^0$ | $\lambda_h$ | $k_h'$ |
|---|---|---|---|---|
| **1** | 122 | 132 | 254 | 0.36 |
| **2** | 182 | 190 | 372 | 0.09 |
| **3** | 192 | 196 | 388 | 0.08 |
| **4** | 187 | 177 | 363 | 0.10 |
| **5** | 182 | 187 | 368 | 0.10 |
| **6** | 214 | 184 | 397 | 0.07 |
| **7** | 174 | 186 | 360 | 0.11 |
| **8** | 176 | 180 | 356 | 0.11 |
| **9** | 158 | 170 | 328 | 0.15 |
| **10** | 188 | 194 | 382 | 0.08 |
| **11** | 183 | 187 | 369 | 0.10 |
| **12** | 178 | 187 | 364 | 0.10 |
| **13** | 164 | 174 | 338 | 0.14 |
| **14** | 160 | 185 | 345 | 0.13 |
| **15** | 184 | 179 | 363 | 0.10 |
| **16** | 179 | 188 | 367 | 0.10 |
| Average | 176 | 181 | 357 | 0.12 |

**Table 3.** Reorganization energies ($\lambda_h$ in meV) and rate constants ($k_h'$) for hole transport in $C_{60}H_{54}$. The average value over 16 $C_{60}H_{54}$ isomers is also shown.

much larger than that for $C_{60}$. The values of $\lambda_h$ are in the range 230–501 meV. Isomer **2** has the largest $\lambda_h$ of more than 500 meV.

| $C_{60}H_{56}$ | $\lambda_h^+$ | $\lambda_h^0$ | $\lambda_h$ | $k_h'$ |
|---|---|---|---|---|
| **1** | 180 | 186 | 366 | 0.10 |
| **2** | 217 | 284 | 501 | 0.02 |
| **3** | 126 | 141 | 267 | 0.31 |
| **4** | 128 | 121 | 249 | 0.38 |
| **5** | 119 | 123 | 242 | 0.41 |
| **6** | 123 | 129 | 252 | 0.37 |
| **7** | 107 | 123 | 230 | 0.47 |
| **8** | 118 | 124 | 242 | 0.41 |
| Average | 140 | 154 | 294 | 0.23 |

**Table 4.** Reorganization energies ($\lambda_h$) and rate constants ($k_h'$) for hole transport in $C_{60}H_{56}$. The average value over eight $C_{60}H_{56}$ isomers is also shown.

$\lambda_h$ and $k_h'$ of $C_{60}H_{58}$ are shown in Table 5. On average, $\lambda_h^+$ of $C_{60}H_{58}$ is 606 meV, which is much larger than that of $C_{60}$ by 511 meV and $\lambda_h^0$ of $C_{60}H_{58}$ is 468 meV, which is much larger than that of $C_{60}$ by 394 meV. The average value of $\lambda_h$ of $C_{60}H_{58}$ (1074 meV) is much larger than that of $C_{60}$. It is interesting that the values of $\lambda_h$ are in the range 392–2411 meV. Isomers **1**, **2**, and **4** have smaller $\lambda_h$.

| $C_{60}H_{58}$ | $\lambda_h^+$ | $\lambda_h^0$ | $\lambda_h$ | $k'_h$ |
|---|---|---|---|---|
| **1** | 205 | 263 | 467 | 0.03 |
| **2** | 196 | 214 | 409 | 0.06 |
| **3** | 1323 | 1089 | 2411 | 0.00 |
| **4** | 184 | 208 | 392 | 0.08 |
| **5** | 1235 | 607 | 1842 | 0.00 |
| **6** | 704 | 478 | 1182 | 0.00 |
| **7** | 693 | 513 | 1206 | 0.00 |
| **8** | 585 | 463 | 1049 | 0.00 |
| **9** | 543 | 458 | 1001 | 0.00 |
| **10** | 444 | 366 | 810 | 0.00 |
| **11** | 556 | 488 | 1044 | 0.00 |
| Average | 606 | 468 | 1074 | 0.00 |

**Table 5.** Reorganization energies ($\lambda_h$) and rate constants ($k'_h$) for hole transport in $C_{60}H_{58}$. The average value over 16 $C_{60}H_{58}$ isomers is also shown.

These results mean that highly hydrogenated $C_{60}H_{52}$ – $C_{60}H_{58}$ are generally not suited for hole-transport materials.

## 5.4. $C_{60}H_{60}$

$C_{60}H_{60}^+$ has nine electrons in fivefold degenerate $h_u$ orbitals. Thus, $C_{60}H_{60}^+$ has an $H_u$ degenerate electronic state, and symmetry lowering because of the Jahn–Teller effect [21] stabilizes $C_{60}H_{60}^+$. The most stable structure of $C_{60}^+$ was calculated as having $D_{3d}$ symmetry, and its $\lambda_h^+$ is 101 meV in our calculation. The value of $\lambda_h^0$ for $C_{60}^+(D_{3d})$ was calculated as 40 meV. Thus, the reorganization energy for hole transport, $\lambda_h$, is 140 meV. Similar to $C_{60}$, $\lambda_h^+$ is much larger than $\lambda_h^0$ because of the Jahn–Teller effect. $k'_h$ of $C_{60}H_{60}^+$ is 1.44 times as large as that of $C_{60}$.

## 5.5. Summary

Figure 6 shows minimum and average values of $\lambda_h$ for $C_{60}H_n$. $C_{60}$ and $C_{60}H_{60}$ have only one isomer so that the minimum values are equal to the average values. From the systematic discussion through $C_{60}H_n$ ($n$ = 2, 4, 6, and 8), it was found that hydrogenation has a large effect on the improvement of hole-transport properties ($\lambda_h$). The minimum values of $\lambda_h$ decrease as the number of hydrogen atoms increases: $C_{60}$ (169 meV) → $C_{60}H_2$-**6** (101 meV) → $C_{60}H_4$-**1** (83 meV) → $C_{60}H_6$-**1** (71 meV). However, $\lambda_h$ increases for $C_{60}H_8$-**5** (81 meV). The average values of $\lambda_h$ change as $C_{60}$ (169 meV) → $C_{60}H_2$ (174 meV) → $C_{60}H_4$ (134 meV) → $C_{60}H_6$ (127 meV) → $C_{60}H_8$ (113 meV). The average $\lambda_h$ of $C_{60}H_4$ (134 meV) is almost as large as that of $C_{60}H_2$-**1** (133 meV), but the average $\lambda_h$ of $C_{60}H_6$ (127 meV) is much larger than that of $C_{60}H_4$-**1** (83 meV), and the average $\lambda_h$ of $C_{60}H_8$ (113 meV) is much larger than that of $C_{60}H_6$-**1** (71 meV). Highly hydrogenated $C_{60}H_n$ generally has large $\lambda_h$. The average $\lambda_h$ is $C_{60}H_{52}$ (254 meV), $C_{60}H_{54}$ (357 meV), $C_{60}H_{56}$ (294 meV), $C_{60}H_{58}$ (1074 meV), and $C_{60}H_{60}$ (140 meV). Only $C_{60}H_{60}$ has a smaller $\lambda_h$ than $C_{60}$.

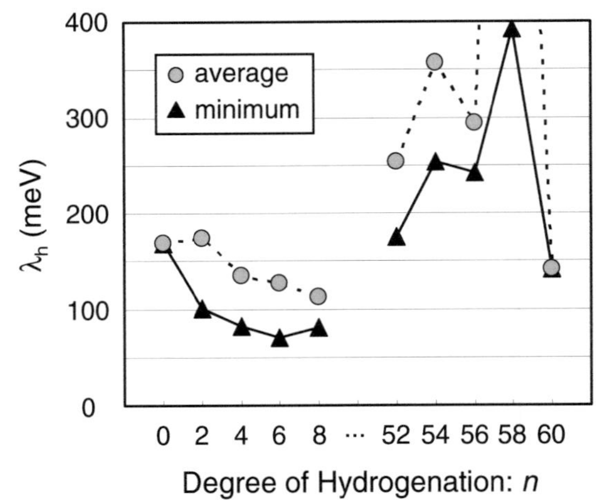


**Figure 6.** Dependence of minimum and average values of $\lambda_h$ on the degree of hydrogenation. The average value of $\lambda_h$ for $C_{60}H_{58}$ is very large (1074 meV), therefore it is not shown in the figure.

## 6. Analysis for molecular design

### 6.1. Geometrical relaxation

The reorganization energy is a stabilization energy by geometrical relaxation originating from the change in electronic structure [64–66]. The strong forces on the nuclei generally result in large $\lambda$ and geometrical relaxation. In our previous publications, we defined parameters $\Delta r$ that characterize the geometrical relaxation as

$$\Delta r = \sum_i |\Delta r_i|, \tag{7}$$

where $\Delta r_i$ is the difference in the $i$th bond length between neutral and cationic states. The summation is taken over all bonds. It has already been shown that there is an almost linear relationship between $\Delta r$ and $\lambda_h$ for $C_{60}H_n$ ($n$ = 2, 4, and 6) [11]. Figure 7 shows the relationship between $\lambda$ and $\Delta r$ for hole transport in $C_{60}H_8$, $C_{60}H_{52}$, $C_{60}H_{54}$, $C_{60}H_{56}$, $C_{60}H_{58}$, and $C_{60}H_{60}$. A linear relationship between $\Delta r$ and $\lambda_h$ is observed for $C_{60}H_8$ and $C_{60}H_{58}$. A clear linear relationship is not found for $C_{60}H_{52}$, $C_{60}H_{54}$, and $C_{60}H_{56}$ because the values of the reorganization energy for these species are almost the same. When the values of $\Delta r$ are the same, $C_{60}H_8$ has a smaller reorganization energy but $C_{60}H_{58}$ has a larger reorganization energy.

### 6.2. Molecular orbital pattern

It is well known that the electronic properties of the HOMO have a close relation to hole-transport properties. Therefore, we focus on the distribution of the HOMOs of $C_{60}H_n$. The HOMO of $C_{60}$, which is originally fivefold degenerate, splits because of the interaction between $C_{60}$ and the H atoms. The distribution of the HOMO easily changes depending on the interaction between $C_{60}$ and the H atoms.

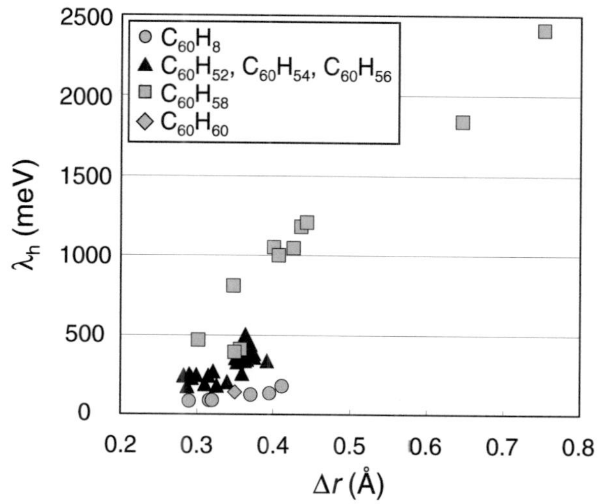


**Figure 7.** Relationship between reorganization energy ($\lambda$) and geometrical relaxation ($\Delta r$) for hole transport in $C_{60}H_n$ ($n$ = 8, 52, 54, 56, 58, and 60).

The HOMOs of isomer-**1** of $C_{60}H_n$ are shown in Figure 8. The HOMOs of $C_{60}H_8$-**1** and other lowly hydrogenated fullerenes [11, 13, 14] are distributed over the whole molecule. In

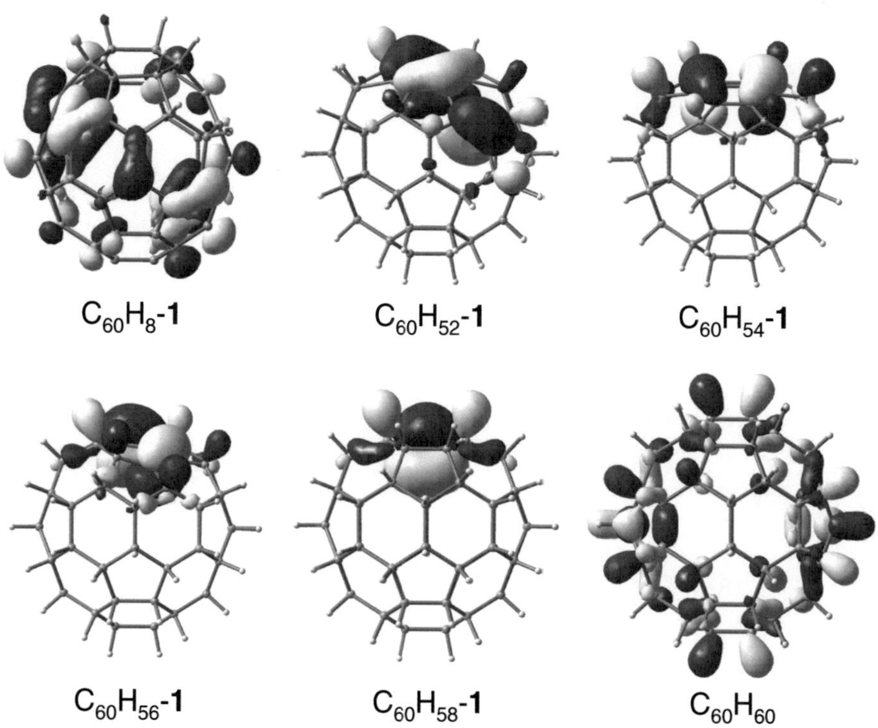


**Figure 8.** HOMOs of isomer-**1** of $C_{60}H_n$.

contrast, the HOMOs of $C_{60}H_{52}$-**1**, $C_{60}H_{54}$-**1**, $C_{60}H_{56}$-**1**, and $C_{60}H_{58}$-**1** are localized around the nonhydrogenated C atoms. Therefore, highly hydrogenated $C_{60}$ has a large reorganization energy. In particular, the HOMO of $C_{60}H_{58}$-**1** is localized on two carbon atoms, therefore the reorganization energy is very large. The HOMO of $C_{60}H_{60}$ is delocalized over the whole molecule because $C_{60}H_{60}$ has high symmetry, so that the value of $\lambda_h$ is smaller than that for other highly hydrogenated fullerenes.

## 7. Summary

Systematic analyses of reorganization energies of hydrogenated fullerenes $C_{60}H_n$ ($n = 2, 4, 6, 8, 52, 54, 56, 58$, and 60) give us very important knowledge for efficient design of useful $C_{60}$ materials:

- Considering only 1,2-addition, $C_{60}H_6$-**1** has the smallest reorganization energy (71 meV). Further hydrogenation does not reduce the reorganization energy.
- Highly hydrogenated fullerenes, especially $C_{60}H_{58}$, have very large reorganization energies because the HOMOs are localized around nonhydrogenated carbon atoms. Therefore, these $C_{60}$ derivatives are not suited for hole-transport materials.
- However, $C_{60}H_{60}$ has a smaller reorganization energy than other highly hydrogenated fullerenes and $C_{60}$ because the HOMO of $C_{60}H_{60}$ is distributed over the whole molecule.
- There is a linear relationship between the reorganization energy and the geometrical relaxation. When the values of $\Delta r$ are the same, lowly hydrogenated fullerenes have smaller reorganization energies than highly hydrogenated fullerenes.

## Acknowledgments

This work was supported by Grant-in-Aid for Scientific Research on Innovation Areas from the Ministry of Education, Culture, Sports, Science and Technology of Japan (No. 20118007) and Grant-in-Aid for Scientific Research (C) from Japan Society for Promotion of Sciences (No. 24560930). The author would like to thank Dr. Shigekazu Ohmori of the National Institute of Advanced Industrial Science and Technology (AIST) and Dr. Hiroshi Kawabata of Hiroshima University for collaboration in this research theme and helpful comments on this chapter. Theoretical calculations were mainly carried out using the computer facilities at the Research Institute for Information Technology, Kyushu University.

## Author details

Ken Tokunaga
*Division of Liberal Arts, Kogakuin University, Tokyo, Japan*

## 8. References

[1] Fong HH, Lun KC, So SK (2002) Chem. phys. lett. 407: 353.
[2] Tang CW, VanSlyke SA (1987) Appl. phys. lett. 51: 913.
[3] Brédas JC, Beljonne D, Coropceanu V, Cornil J (2004) Chem. rev. 104: 4971.
[4] Murphy AR, Fréchet JMJ (2007) Chem. rev. 107: 1066.

[5] Coropceanu V, Cornil J, da Silva Filho DA, Olivier Y, Silbey R, Brédas JL, Chem. rev. 107: 926.
[6] Mori T (2008) J. phys.: condens. matter 20: 184010.
[7] Sakanoue K, Motoda M, Sugimoto M, Sakaki S (1999) J. phys. chem. A 103: 5551.
[8] Sancho-García JC, Horowitz G, Brédas JL, Cornil J (2003) J. chem. phys. 119: 12563.
[9] Hutchison GR, Ratner MA, Marks TJ (2005) J. am. chem. soc. 127: 16866.
[10] Pan JH, Chiu HL, Chen L, Wang BC (2006) Comput. mater. sci. 38: 105.
[11] Tokunaga K (2011) In: Velasquez MA editor. Organic Semiconductors: Properties, Fabrication and Applications. New York: Nova Science Publishers.
[12] Tokunaga K (2012) In: Verner RF, Benvegnu C, editors. Handbook on Fullerene: Synthesis, Properties and Applications. New York: Nova Science Publishers.
[13] Tokunaga K, Ohmori S, Kawabata H, Matsushige K (2008) Jpn. j. appl. phys. 47: 1089.
[14] Tokunaga K, Kawabata H, Matsushige K (2008) Jpn. j. appl. phys. 47: 3638.
[15] Tokunaga K (2009) Chem. phys. lett. 476: 253.
[16] Tokunaga K, Ohmori S, Kawabata H (2009) Thin solid films 518: 477.
[17] Tokunaga K, Ohmori S, Kawabata H (2011) Jpn. j. appl. phys. 50: 01BA03.
[18] Tokunaga K, Ohmori S, Kawabata H (2011) Mol. cryst. liq. cryst. 539: 252(592).
[19] Tokunaga K, Ohmori S, Kawabata H (2012) Jpn. j. appl. phys. accepted.
[20] Könenkamp R, Priebe G, Pietzak B (1999) Phys. rev. B 60: 11804.
[21] Jahn HA, Teller E (1937) Proc. r. soc. London, ser. A 161: 220.
[22] Ceulemans A, Fowler PW (1990) J. chem. phys. 93: 1221.
[23] Singh TB, Marjanović M, Matt GJ, Günes S, Sariciftci NS, Ramil AM, Andreev A, Sitter H, Schwödiauer R, Bauer S (2005) Org. electron. 6: 105.
[24] Kitamura M, Kuzumoto Y, Kamura M, Aomori S, Arakawa Y (2007) Appl. phys. lett. 91: 183514.
[25] Singh TB, Marjanović M, Stadler P, Auinger M, Matt GJ, Günes S, Sariciftci NS, Schwödiauer R, Bauer S (2005) J. appl. phys. 97: 083714.
[26] Anthopoulos TD, de Leeuw DM, Cantatore E, van't Hof P, Alma J, Hummelen JC (2005) J. appl. phys. 98: 054503.
[27] Kubo M, Iketani K, Kaji T, Hiramoto M (2011) Appl. phys. lett. 98: 073311.
[28] Marcus RA (1956) J. chem. phys. 24: 966.
[29] Horowitz G (1998) Adv. mater. 10: 365.
[30] Nelsona SF, Lin YY, Gundlach DJ, Jackson TN (1998) Appl. phys. lett. 72: 1854.
[31] Dimitrakopoulos CD, Mascaro DJ (2001) IBM j. res. dev. 45: 11.
[32] Li XY, Tang XS, He FC (1999) Chem. phys. 248: 137.
[33] Wu Q, Van Voorhis T (2006) J. chem. phys. 125: 164105.
[34] Frisch, MJ *et al.* (2003) Gaussian 03 (Revision C. 02) Pittsburgh: Gaussian, Inc.
[35] Henderson CC, Cahill PA (1993) Science 259: 1885.
[36] Henderson CC, Rohlfing CM, Gillen KT, Cahill PA (1994) Science 264: 397.
[37] Ballenweg S, Gleiter R, Krätschmer W (1993) Tetrahedron lett. 34: 3737.
[38] Becker L, Evans TP, Bada J (1993) J. org. chem. 58: 7630.
[39] Avent AG, Darwish AD, Heimbach DK, Kroto HW, Meidine MF, Parsons JP, Remars C, Roers R, Ohashi O, Taylor R, Walton DRM (1994) J. chem. soc., perkin trans. 2: 15.
[40] Billups WE, Luo W, Gonzalez A, Arguello D, Alemany LB, Marriott T, Saunders M, Jiménes-Vázquez HA, Khong A (1997) Tetrahedron lett. 38: 171.
[41] Meier MS, Corbin PS, Vance VK, Clayton M, Mollman M, Poplawska M (1994) Tetrahedron lett. 35: 5789.

[42] Meier MS, Bergosh RG, Laske Cooke JA, Spielmann HP, Weedon BR (1997) J. org. chem. 62: 7667.
[43] Fukuzumi S, Suenobu T, Kawamura S, Ishida A, Mikami K (1997) Chem. commun. 291.
[44] Fukuzumi S, Suenobu T, Patz M, Hirasaka T, Itoh S, Fujitsuka M (1998) J. am. chem. soc. 120: 8060.
[45] Mandrus D, Kele M, Hettich RL, Guiochon G, Sales BC, Boatner LA (1997) J. phys. chem. B 101: 123.
[46] Wang GW, Li YJ, Li FB, Liu YC (2005) Lett. org. chem. 2: 595.
[47] Haufler RE, Conceicao J, Chibante LPF, Chai Y, Byrne NE, Flanagan S, Haley MM, O'Brien SC, Pan C, Xiao Z, Billups WE, Ciufolini MA, Hauge RH, Margrave JL, Wilson LJ, Curl RF, Smalley RE (1990) J. phys. chem. 94: 8634.
[48] Ruchardt C, Gerst M, Ebenhoch J, Beckhaus H, Campbell E, Tellgmann R, Schwarz H, Weiske T, Pitter S (1993) Angew. chem. 32: 584.
[49] Shigematsu K, Abe K, (1992) Chem. express 7: 905.
[50] Jin C, Hettich R, Compton R, Joyce D, Blencoe J, Burch T (1994) J. phys. chem. 98: 4215.
[51] Matsuzawa N, Dixon DA, Fukunaga T (1992) J. phys. chem. 96: 7594.
[52] Nossal J, Saini RK, Alemany LB, Meier M, Billups WE (2001) Eur. j. org. chem.: 4167.
[53] Shao Y, Jiang Y (1995) Chem. phys. lett. 242: 191.
[54] Cahill PA, Rohlfing CM (1996) Tetrahedron 52: 5247.
[55] Cross RJ, Jiménez-Vázquez HA, Li Q, Saunders M, Schuster DI, Wilson SR, Zhao H (1996) J. am. chem. soc. 118: 11454.
[56] Bergosh RG, Meiser MS, Cooke JAL, Spielmann HP, Weedon BR (1997) J. org. chem. 62: 7667.
[57] Alemany LB, Gonzalez A, Billups WE, Willcott MR, Ezell E, Gozansky E (1997) J. org. chem. 62: 5771.
[58] Canton SE, Yencha AJ, Kukk E, Bozek JD, Lopes MCA, Snell G, Berrah N (2002) Phys. rev. lett. 89: 045502.
[59] Manini N, Gattari P, Tosatti E (2003) Phys. rev. lett. 91: 196402.
[60] Bendale RD, Stanton JF, Zerner MC (1992) Chem. phys. lett. 194: 467.
[61] Boese AD, Scuseria GE (1998) Chem. phys. lett. 294: 233.
[62] Manini N, Corso AD, Fabrizio M, Tosatti E (2001) Philos. mag. B 81: 793.
[63] Devos A, Lannoo M (1998) Phys. rev. B 58: 8236.
[64] Sato T, Tokunaga K, Tanaka K (2006) J. chem. phys. 124: 024314.
[65] Tokunaga K, Sato T, Tanaka K (2006) J. chem. phys. 124: 154303.
[66] Sato T, Tokunaga K, Tanaka K (2008) J. phys. chem. A 112: 758.

For EU product safety concerns:
IN TECH d.o.o., Prolaz Marije Krucifikse Kozulić 3, 51000 Rijeka, Croatia,
info@intechopen.com or visit our website at intechopen.com.

# Meet the editor

Iyad Karamé is a professor at the Faculty of Sciences at the Lebanese University in Beirut. he is a director of the organometallic catalysis and coordination chemistry team, in the Department of Chemistry. He got his PhD degree from Claude Bernard-Lyon 1 University in France in January 2004. He was an assistant professor and a researcher (ATER) at the Ecole Normale Supérieure de Lyon, France, for one year (2004-2005). An Invited researcher at the Leibniz Institut für Katalyse in Rostock (2005-2006) and then a researcher at the Laboratory of Organometallic Chemistry of Surface, CPE-Lyon until 2008. His principal axis of research are organometallic catalysis and synthesis of chelating macrocyles for the complexation of hard metals.

# Contents

# Preface

The domain of catalytic hydrogenation continues to grow fast, reflecting the wide range of chemical applications that can be enhanced by the easy use of molecular hydrogen. The advances in characterization techniques and their application have improved our understanding of the catalytic processes and mechanisms occurring in both homogeneous and heterogeneous catalysis.

The aim of this volume, although not exhaustive, is to provide a general overview of new progress of the hydrogenation reactions.

This volume comprises a series of various contributions, as reviews or original articles, treating heterogeneously and homogeneously catalyzed hydrogenation reactions. It is composed of three parts: hydrogenation reactions in fine organic chemistry, hydrogenation reactions in environmental chemistry and renewable energy, and special topics in hydrogenation.

Among homogeneously catalyzed hydrogenation reactions, asymmetric hydrogenation is of major importance from an industrial viewpoint; the first two chapters of Part I review the asymmetric hydrogenation reactions.

Metal catalyzed hydrogenation belongs to the most important transformations in chemical industry and is a key technology for the manufacture of life science products. The third chapter introduces the role of homogeneous and heterogeneous catalytic hydrogenation reactions in processes for the production of vitamins, carotenoids, fragrance compounds and nutraceuticals. The fourth and fifth chapters discuses two examples of chemoselective hydrogenation: the 1,4-addition on conjugated C-C and C-O double bonds of arylidene tetramic acids and the selective hydrogenation of trifluoromethlphenyl diarizine derivatives.

Chapter six describes the partial hydrogenation of alkynes catalyzed by transitions metals catalysts. Kinetic study of this reaction catalyzed by supported Ni and Pd on alumina is discussed in chapter seven.

In the second part of this book, the hydroconversion of triglycerides into green liquid fuels is presented in chapter eight, while chapter nine reports transition metal sulfide catalyst for petroleum hydrodesulfurization reactions. Chapter ten focuses on recent advances in transition metal-catalyzed homogeneous hydrogenation of carbon dioxide in aqueous media. A work on hydrogenation of carbon oxides catalysed by nanoparticles constitutes chapter eleven.

The last and third part consists of two chapters. Chapter twelve treats photocatalytic hydrogenation on semiconducteur particles. In the last and thirteenth chapter, effect of hydrogenation on hole-transport properties of $C_{60}$ is theoretically estimated and is systematically discussed by the density functional theory (DFT) calculation, taking hydrogenated fullerenes $C_{60}H_n$ as examples.

This volume represents a good measure of high quality research; nevertheless, some subjects in the area of hydrogenation are not included and we look forward in the near future to complete this work by editing others volumes.

We hope you will find this book enjoyable and illuminating. Any comments you may have are of course very welcome.

We would like to take this opportunity to thank all contributors for their chapters. Their cooperation in adhering to the, at times tight, deadlines is very much appreciated. Finally, we wish to express our gratitude to the staff at INTECH for their kind assistance in bringing this book to fruition.

**Iyad Karamé**
Lebanese University
Beirut

Section 1

# Hydrogenation Reactions in Fine Organic Chemistry

Chapter 1

# Asymmetric Hydrogenation

Tsuneo Imamoto

Additional information is available at the end of the chapter

http://dx.doi.org/10.5772/48584

## 1. Introduction

The asymmetric hydrogenation of prochiral unsaturated compounds, such as alkenes, ketones, and imines, is one of the most efficient and straightforward methods for the preparation of optically active compounds. This method uses dihydrogen and small amounts of chiral transition metal complexes and is now recognized as economical, operationally simple, and environmentally friendly. It is frequently used in both academia and industry for the synthesis of chiral amino acids, amines, alcohols, and alkanes in an enantiopure or enantiomerically enriched form.

Asymmetric hydrogenation can basically be classified into two categories, homogeneous and heterogeneous hydrogenation. Heterogeneous hydrogenation is technically simple and has a longer history than homogeneous hydrogenation. In 1956, Akahori et al. reported the asymmetric hydrogenation of azalactones in the presence of silk-fibroin-supported palladium (Scheme 1) [1]. This pioneering work was later extended to the hydrogenation of prochiral ketones using a Raney nickel or platinum catalyst that was modified by chiral auxiliaries, such as tartaric acid or cinchona alkaloids. However, prepared heterogeneous catalysts have as yet provided moderate to good enantioselectivities but not very high selectivities, so the method is not useful in practice except in some limited cases. In sharp contrast, homogeneous hydrogenation has developed enormously in the past four decades, and has become the useful methodology in modern science and technology. Therefore, this chapter focuses on homogeneous asymmetric hydrogenation.

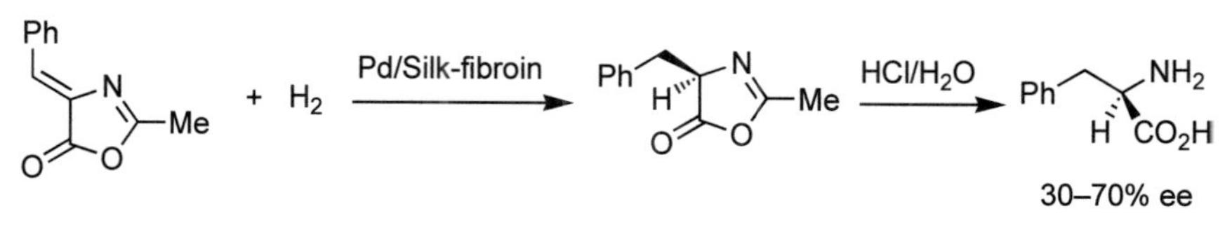


**Scheme 1.** Asymmetric hydrogenation of an azalactone catalyzed by silk-fibroin-supported palladium

INTECH
open science | open minds

Homogeneous asymmetric hydrogenation was first reported independently by Knowles and Horner in 1968 [2,3]. They replaced the triphenylphosphine of the Wilkinson catalyst ($RhCl(PPh_3)_3$) with optically active methylphenyl(*n*-propyl)phosphine and examined its catalytic performance in the hydrogenation of prochiral alkenes. The optical yields were low, but catalytic asymmetric hydrogenation was shown experimentally to have occurred unequivocally in the homogeneous system (Scheme 2).

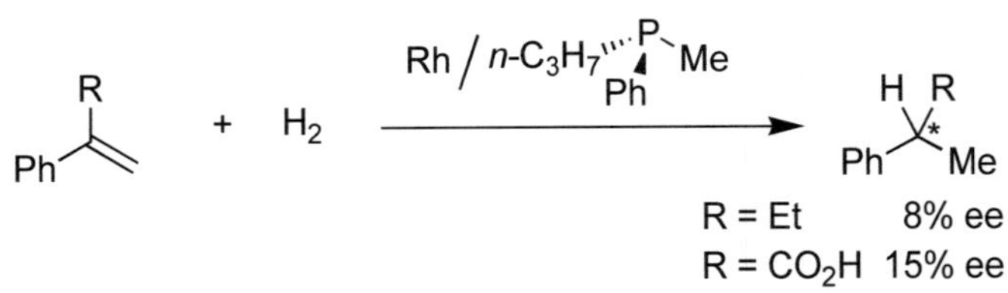


**Scheme 2.** First example of homogeneous asymmetric hydrogenation

In 1971, Kagan et al. synthesized a chelating diphosphine ligand with two phenyl groups on each of the two phosphorus atoms [4]. The ligand, 4,5-bis[(diphenylphosphino)methyl]-2,2-dimethyl-1,3-dioxolane (DIOP), is the first example of a $C_2$-symmetric phosphine ligand. Its high capacity for asymmetric induction, up to 88%, was demonstrated in the hydrogenation of α-dehydroamino acids and enamides [5], and these excellent results stimulated the design and synthesis of many other $C_2$-symmetric phosphine ligands. The most notable ligand reported in the period up to 1979 was 1,2-bis(*o*-anisylphenylphosphino)ethane (DIPAMP) developed by Knowles (Nobel laureate in 2001) et al. at Monsanto in 1975, which provided very high enantioselectivity values up to 96% in the hydrogenation of α-dehydroamino acids [6]. The methodology was used to produce (*S*)-3-(3,4-dihydroxyphenyl)alanine (L-DOPA), which is useful in the treatment of Parkinson's disease. This was the first example of asymmetric catalysis on an industrial scale (Scheme 3) [7].

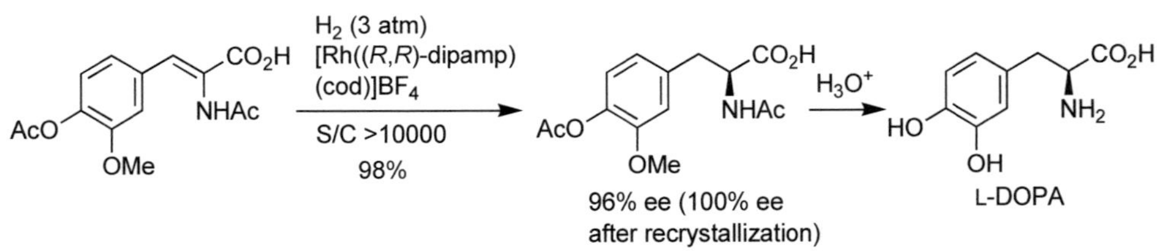


**Scheme 3.** The Monsanto process for the production of L-DOPA

Another landmark ligand was 2,2′-bis(diphenylphosphino)-1,1′-binaphthyl (BINAP), developed by Noyori (Nobel laureate in 2001) et al. in 1980 [8]. The appearance of BINAP heralded marked advances in asymmetric hydrogenation and other transition-metal-catalyzed asymmetric catalyses. The methodology developed by Noyori et al. using BINAP resolved longstanding problems, such as the limited applicability of the method, which was attributed to substrate specificity and unsatisfactory catalytic activity. Thus, a wide range of prochiral alkenes and carbonyl substrates, including simple ketones, were subjected to hydrogenation with much lower catalyst loadings, to generate the corresponding saturated compounds with exceedingly high enantioselectivity. The method based on the Ru-BINAP catalyst system has allowed the use of asymmetric hydrogenation in the industrial production of many useful optically active compounds such as pharmaceutical ingredients, agrochemicals, and flavors [9].

In 1993, the research groups of Pfaltz, Helmchen, and Williams independently reported a P,N-ligand phosphinooxazoline (PHOX) [10–12]. The utility of this ligand in asymmetric hydrogenation was demonstrated by Pfaltz et al. using its iridium complex. They showed that largely unfunctionalized alkenes were enantioselectively hydrogenated by Ir-PHOX and related catalysts [13,14]. Their studies significantly expanded the scope of asymmetric hydrogenation and offered a new tool for the efficient production of chiral building blocks.

In contrast, homogeneous asymmetric hydrogenation using chiral complexes of early transition metals or less-expensive late transition metals has also been investigated. Some success has been achieved in the hydrogenation of alkenes and imines with chiral catalysts containing titanium, zirconium, lanthanides, or iron. However, because of the length limitation on this chapter, rhodium-, ruthenium-, and iridium-catalyzed asymmetric hydrogenation will be described here.

Based on extensive experiments, computations, and theoretical considerations, asymmetric hydrogenation is now highly advanced, so any broad overview of this area is difficult. Fortunately, many exhaustive reviews have been published, together with excellent accounts of asymmetric hydrogenation. The author hopes that this chapter, together with the review articles [15–18], will provide good references for the process.

## 2. Chiral Phosphine Ligands for Asymmetric Hydrogenation

The design and synthesis of new chiral phosphine ligands are crucial for the development of transition-metal-catalyzed asymmetric catalysis. Over the past four decades, thousands of chiral phosphine ligands have been synthesized and their catalytic efficiencies evaluated [19–21]. Figure 1 illustrates representative phosphine ligands, including P,N-hybrid ligands, that have attracted much attention because of their novelty, conceptual importance, and/or practical utility.

Most of them are $C_2$-symmetric bidentate diphosphine ligands. In the hydrogenation process based on $C_2$-ligands, the number of structures that the catalyst–substrate complexes can adopt is reduced to half compared with those formed from $C_1$-symmetric catalysts, and consequently, $C_2$-symmetric ligands achieve higher enantioselectivity than $C_1$-symmetric ligands. Conversely, many $C_1$-symmetric ligands, including JosiPhos, Trichickenfootphos, and PHOX, display superior enantioselectivity, depending on the reaction.

DIPAMP is a typical $C_2$-symmetric and P-chiral (P-stereogenic) diphosphine ligand. This ligand played an outstanding role in the early stages of the history of asymmetric hydrogenation. Nevertheless, little attention had been paid to this class of P-chiral phosphine ligands for more than 15 years, mainly because of the difficulties inherent in their synthesis and apprehension about possible stereomutation at P-stereogenic centers. The author's

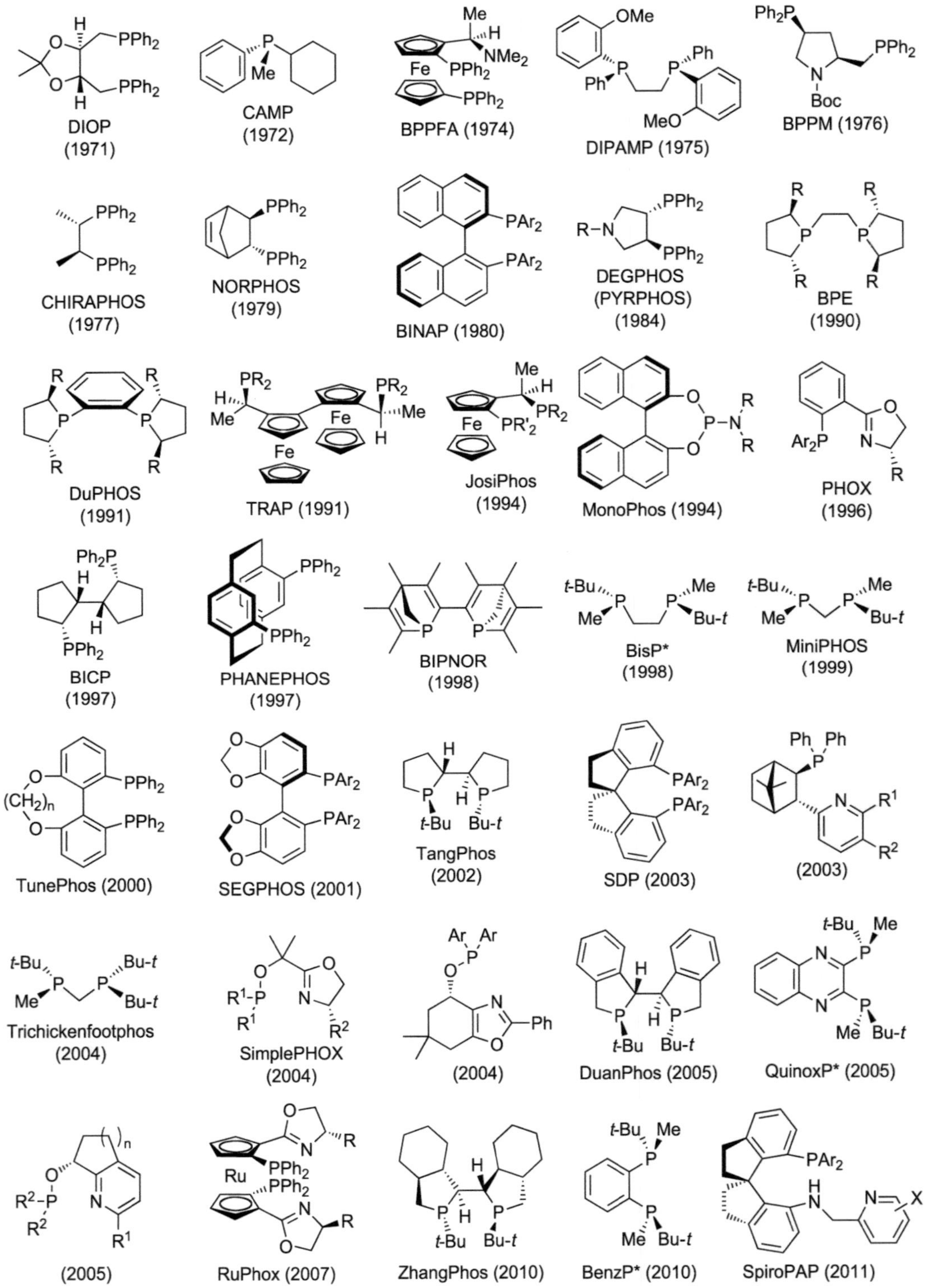


**Figure 1.** Representative chiral phosphine ligands

research group has developed efficient methods for the preparation of P-chiral phosphine ligands using phosphine–boranes as the key intermediates and prepared (*R,R*)-1,2-bis(*tert*-butylphenylphosphino)ethane in 1990, (*S,S*)-1,2-bis(*tert*-butylmethylphosphino)ethane (BisP*) in 1998, and (*R,R*)-bis(*tert*-butylmethylphosphino)methane (MiniPHOS) in 1999 [22–24]. Of these ligands, BisP* and MiniPHOS display enantioselectivities higher than those of DIPAMP in Rh-catalyzed asymmetric hydrogenation. These findings triggered the synthesis of structurally analogous but more rigid P-chiral phosphine ligands, and many highly efficient and practically useful ligands have since been reported (TangPhos, Trichickenfootphos, DuanPhos, QuinoxP*, ZhangPhos, BenzP*, etc.).

As mentioned above, many chiral phosphine ligands have been shown to exhibit excellent enantioselectivity and some outstanding ligands have been used in the industrial production of useful optically active compounds. However, there are no "omnipotent" ligands, and so the development of more efficient, operationally convenient, and widely applicable chiral phosphine ligands is still a vital research topic in the field of asymmetric catalysis.

## 3. Rhodium-catalyzed Asymmetric Hydrogenation

### 3.1. General scope

Rhodium-catalyzed hydrogenation is well suited to the enantioselective reduction of α- and β-dehydroamino acid derivatives and enamides. Thus, chiral α- and β-amino acids and secondary amine derivatives can be obtained in an enantiomerically pure or enriched form by the hydrogenation of amino-functionalized alkenes (Equations 1–3). The catalytic efficiency and enantioselectivity are largely dependent on the chiral ligands and substrates used. In general, electron-rich and structurally rigid ligands, such as DuPhos, DuanPhos, ZhangPhos, QuinoxP*, and BenzP*, provide the corresponding products in high to almost-perfect enantioselectivity. Di- or tri-substituted alkenes are readily hydrogenated, but tetrasubstituted alkenes require higher hydrogen pressure, higher catalyst loading, and/or a higher reaction temperature to facilitate the hydrogenation reaction.

Rhodium catalysts are also used for the hydrogenation of itaconic acid derivatives, enol esters, and ethenephosphonates (Equations 4–6). As in the hydrogenation of dehydroamino acids and enamides, the oxygen functional groups capable of coordination to the rhodium atom play an important role in accelerating the reaction, as well as in the enantioselection.

### 3.2. Reaction mechanism

Since the discovery of rhodium-catalyzed asymmetric hydrogenation, the reaction mechanism, including the catalytic cycle and the origin of the enantioselection process, has been studied extensively. Early studies using cationic rhodium complexes with $C_2$-symmetric diphosphine ligands with two diaryl substituents at each phosphorus atom led to the so-called "unsaturated mechanism". This mechanism, proposed by Halpern and Brown, is based on the following experimental facts and considerations [25–28].

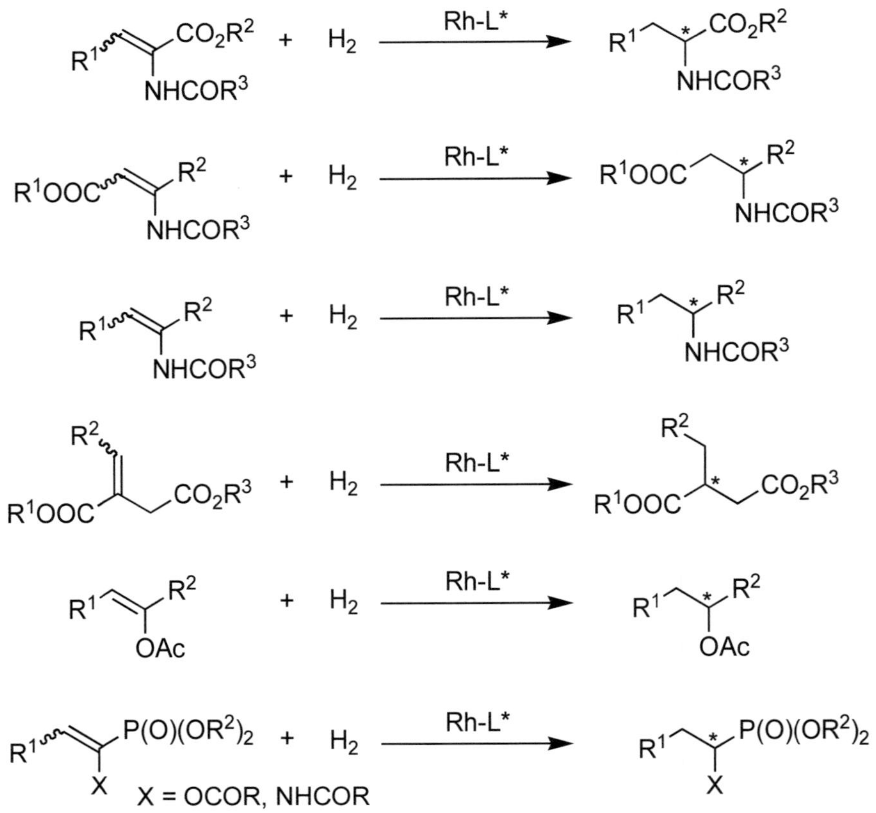


(1)
(2)
(3)
(4)
(5)
(6)

1. The solvate complex generated by the hydrogenation of a precatalyst reacts with a prochiral substrate, such as methyl (*Z*)-α-acetamidocinnamate (MAC), providing two diastereomeric catalyst–substrate complexes in a considerably high ratio. For example, the Rh-(*S,S*)-DIPAMP solvate complex binds to MAC to generate *Re*- and *Si*-coordinated adducts in a ratio of about 10:1.
2. The configuration of the major isomer does not correspond to the configuration of the product if it is assumed that the oxidative addition of $H_2$ occurs in an *endo*-manner and that the stereochemical integrity is maintained through to the final reductive elimination step.
3. At ambient temperatures, major and minor catalyst–substrate complexes are interconverted rapidly. The minor isomer is much more reactive with $H_2$ than the major isomer, and the reaction proceeds according to the Curtin–Hammett principle.
4. The oxidative addition of dihydrogen to the catalyst–substrate complex is rate-determining and irreversible, and enantioselection is determined at this step.
5. The kinetic and equilibration data are consistent with the stereochemical outcome (*R*:*S* = 98:2; 96% ee).
6. At low temperatures, enantioselectivity is significantly reduced. This fact is interpreted as reflecting that the interconversion between the major and minor isomers is very slow or almost in a frozen state at low temperatures. As a consequence, the major isomer competitively reacts with dihydrogen to generate the opposite enantiomeric product, resulting in lower enantioselectivity.

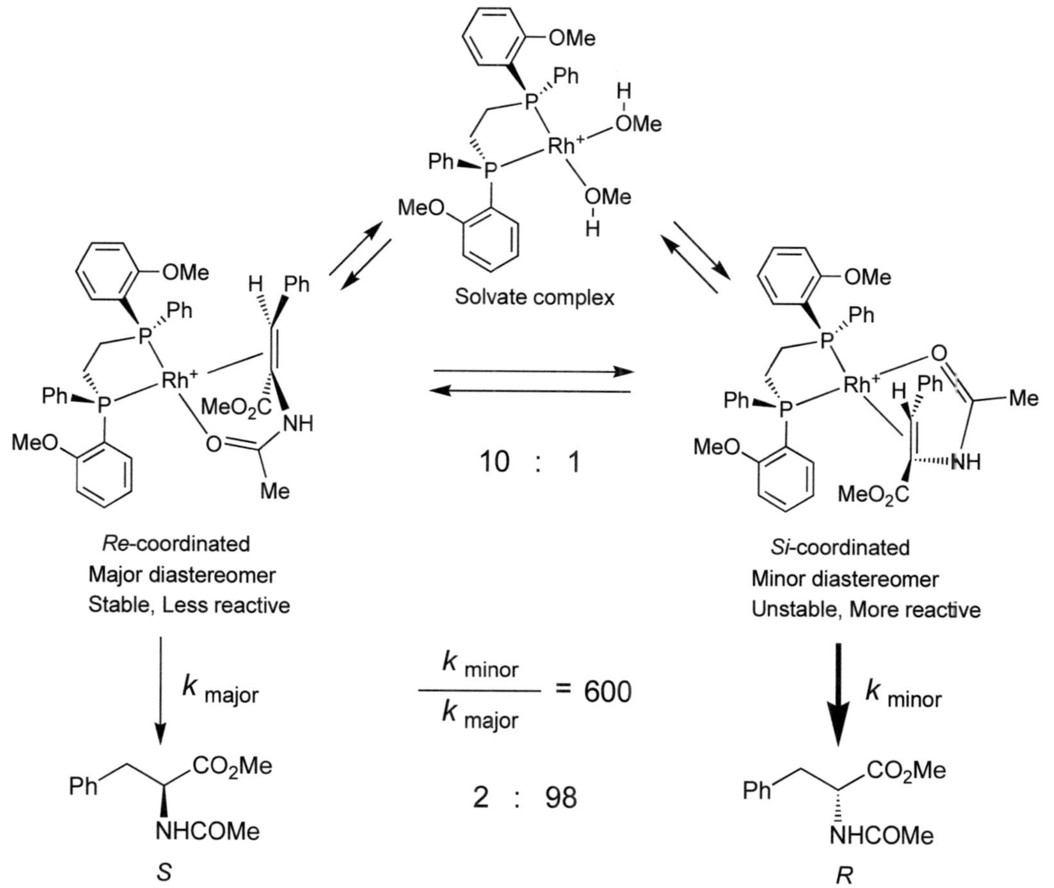


**Scheme 4.** Unsaturated mechanism: hydrogenation of MAC with Rh-(*S*,*S*)-DIPAMP leading tc (*R*)-phenylalanine methyl ester with 96% ee

7. A significant reduction in enantioselectivity is also observed when the reaction is performed under higher $H_2$ pressure. This fact is interpreted by considering that the reaction of the less-reactive major isomer with dihydrogen is facilitated under high $H_2$ pressure.

The key points in this mechanism are illustrated in Scheme 4. This enantioselection mechanism is quite unique, differing from those of other asymmetric catalyses. It should be noted that this mechanism does not correspond to the "lock and key" principle, which is widely invoked in stereoselective reactions catalyzed by enzymes.

In contrast, the development of electron-rich diphosphine ligands has revealed a new mechanistic aspect of rhodium-catalyzed asymmetric hydrogenation. It has been reported that rhodium catalysts with electron-rich phosphine ligands (DuPhos, BPE, BisP*, MiniPHOS, Trichickenfootphos, TangPhos, DuanPhos, ZhangPhos, QuinoxP*, BenzP*, etc.) display very high to almost-perfect enantioselectivity in the hydrogenation of many dehydroamino acids and enamides. The origin of this exceedingly high enantioselectvity

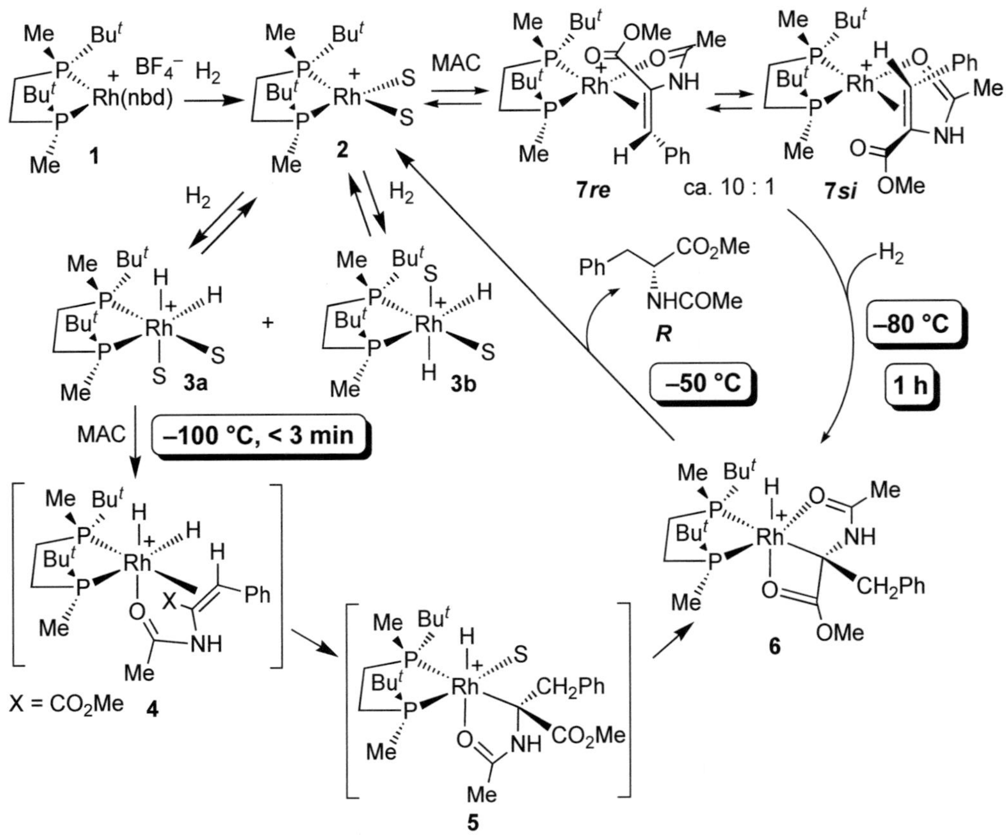

**Scheme 5.** Mechanism of the asymmetric hydrogenation of MAC with Rh-(*S*,*S*)-*t*-Bu-BisP*

cannot be explained well in terms of the "unsaturated mechanism" mentioned above. Gridnev and Imamoto et al. studied the hydrogenation mechanism using [Rh(*t*-Bu-BisP*)(nbd)]$BF_4$ (**1**) [29,30]. One of their notable findings was that the solvate complex [Rh(*t*-Bu-BisP*)($CD_3OD)_2$]$BF_4$ (**2**) reacted with $H_2$ at –90 °C to produce equilibrium amounts (ca. 20%) of rhodium dihydride complexes [$RhH_2$(*t*-Bu-BisP*)$BF_4$ (**3a** and **3b**; dihydride diastereomers). The dihydride complexes reacted with MAC, even at very low temperatures (–100 °C), and were rapidly (within 3 min) converted to the monohydride intermediate **6** (Scheme 5). The reaction is considered to proceed through the associated intermediate **4** and monohydride **5**.

On the contrary, the hydrogenation of the catalysts–substrate complexes (**7*re*** and **7*si*** = ca. 10:1) was relatively slow. It required about 1 h at –80 °C to generate the same concentration of monohydride **6**. The reaction is considered to proceed through the solvate complex **2**, which is generated by the reversible dissociation of **7*re*** and **7*si***, and to proceed via dihydrides **3a** and **3b**, **4**, and **5**. It is reasonable to infer that the enantioselection is determined at the migratory insertion step from **4** to **5**. There are eight possible diastereomers of **4**. Among them, complex **4** is energetically most stable, is preferentially formed, and undergoes migratory insertion via the lowest transition state, resulting in the formation of the (*R*)-hydrogenation product.

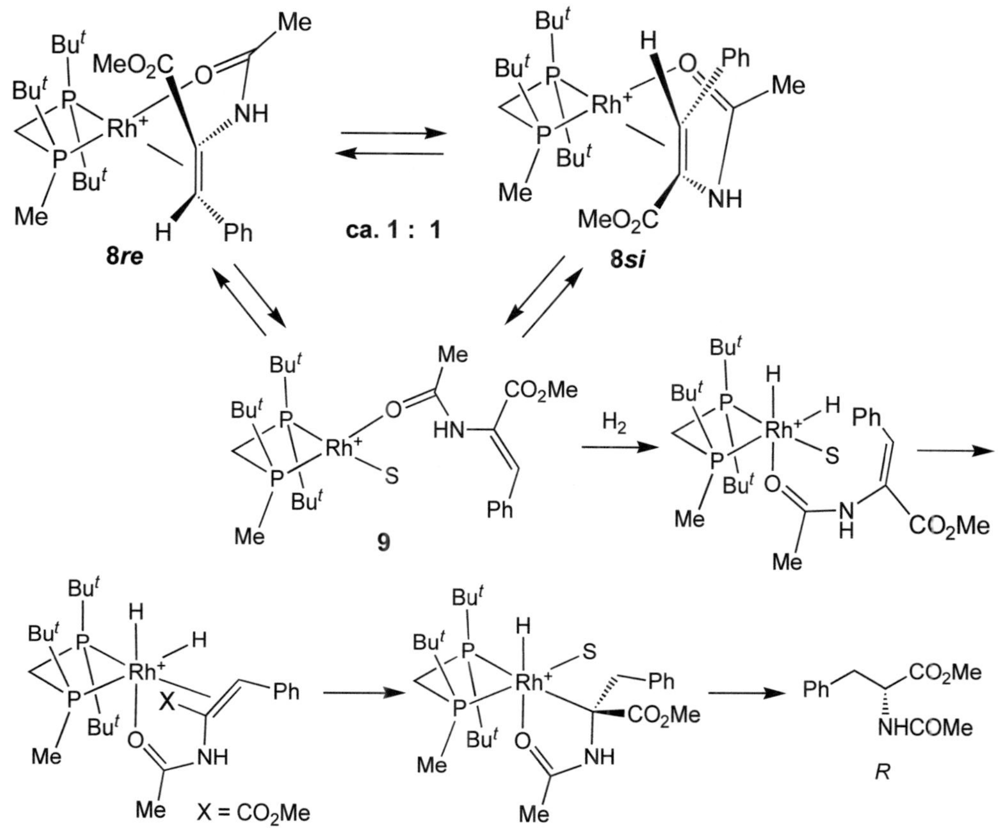


**Scheme 6.** Reaction pathway from catalyst–substrate complexes to (*R*)-*N*-acetylphenylalanine methyl ester

The origin of the enantioselection process has also been studied using MAC and Trichickenfootphos, a $C_1$-symmetric three-hindered phosphine ligand [31,32]. In this case, two of the four possible diastereomeric catalyst–substrate complexes are thermodynamically stable and exist in a ratio of about 1:1. Remarkably, the respective complexes reacted with dihydrogen to yield the same (*R*)-product. NMR and computational studies have demonstrated that the complexes (**8*re*** and **8*si***) dissociate the C=C double bond to generate nonchelating complex **9**, which in turn reacts with dihydrogen, with subsequent association and migratory insertion, to yield the (*R*)-product (Scheme 6).

Recently, the hydrogenation mechanism has also been studied using [Rh((*R*,*R*)-BenzP*)(nbd)]$BF_4$ [33]. Low-temperature NMR and density functional theory (DFT) calculations have revealed more detailed aspects of the mechanism. DFT calculations showed the relative stability of each intermediate and the transition state energies. Consequently, the most reasonable reaction pathway from the solvate complex **10** to the product is proposed to be as shown in Scheme 7. The solvate complex **10** is readily hydrogenated to dihydride **12** via **11**, followed by the reaction of **12** with MAC to produce the nonchelating dihydride intermediate **15**. The nonchelating catalyst–substrate complex **13** is also readily subjected to hydrogenation because dihydrogen is readily coordinated at the vacant site of the complex, leading to **15** via **14**. On the contrary, the hydrogenation of the chelating catalyst–substrate complex **16** requires a much higher activation energy, so the unsaturated pathway does not operate in this reaction system.

Enantioselection occurs at a later stage. The recoordination of the double bond of complex **15** to the rhodium atom occurs readily in the non-hindered quadrant to form the chelated dihydride intermediate **17**. This undergoes migratory insertion to produce monohydride **18**, followed by reductive elimination to generate a product with the correct absolute configuration.

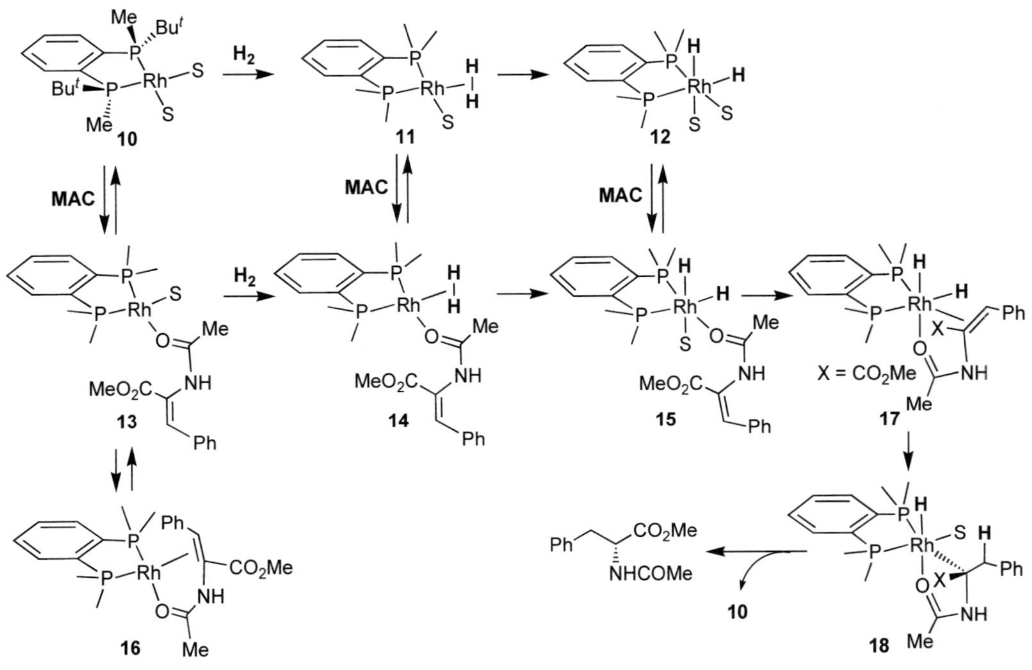

**Scheme 7.** The reaction pathway of the asymmetric hydrogenation of MAC catalyzed by the Rh-(*R*,*R*)-BenzP* complex

## 3.3. Application to the synthesis of useful optically active compounds

Rhodium complexes with chiral phosphine ligands have been widely used in academia and industry for the synthesis of the chiral building blocks of natural products, pharmaceuticals, and agrochemicals. Schemes 8–11 show representative examples.

Zhang et al. developed a new process for the production of ramipril, an angiotensin-converting enzyme inhibitor, used to treat high blood pressure and congestive heart failure (Scheme 8) [34]. The α-dehydroamino acid methyl ester **19** was efficiently hydrogenated under mild conditions with a rhodium–DuanPhos complex to yield compound **20** with 99% ee. The hydrolysis of the vinyl chloride moiety of compound **20**, followed by its cyclization, generated bicyclic amino acid **21**, which was converted to ramipril.

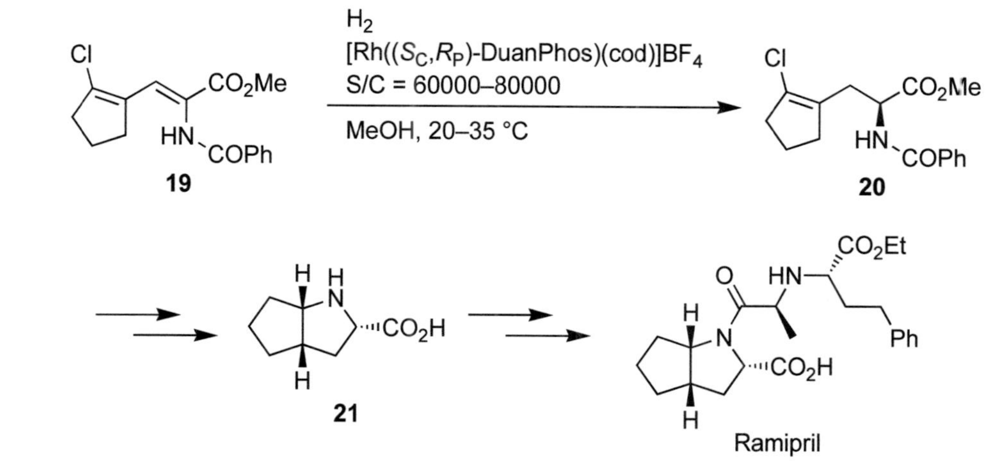


**Scheme 8.** Synthesis of ramipril via Rh-catalyzed asymmetric hydrogenation

Merck Research Laboratories identified taranabant, as a potential selective cannabinoid-1 receptor inverse agonist, for the treatment of obesity. One of the synthetic routes to taranabant is shown in Scheme 9, and involves the rhodium-catalyzed asymmetric hydrogenation of a tetrasubstituted enamide **22**. The hydrogenation reaction to introduce two stereogenic centers is achieved with a JosiPhos-type ligand and trifluoroethanol as the solvent, to produce compound **23** with 96% ee, and one recrystallization of the product increases the ee value to > 99.5%. The final dehydration of the primary amide with cyanuric chloride generates taranabant [35,36].

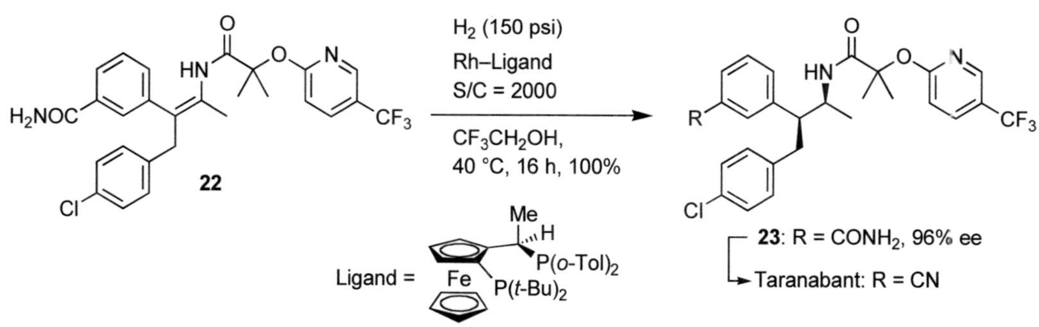


**Scheme 9.** Synthesis of taranabant via Rh-catalyzed asymmetric hydrogenation

Pregabalin, a kind of optically active γ-amino acid, is an anticonvulsant drug used for neuropathic pain and as an adjunct therapy for partial seizures. This drug is marketed by Pfizer under the trade name Lyrica. A chemical synthesis of pregabalin is shown in Scheme 10, where the key intermediate **25** is obtained by the asymmetric hydrogenation of *tert*-butylammonium (Z)-3-cyano-5-methyl-3-hexenoate (**24**) using a Rh-Trichickenfootphos catalyst. The very low catalyst loading (S/C =27,000), mild conditions (50 psi $H_2$ pressure, room temperature), and high enantioselectivity (98% ee) indicate the potential utility of this process in the large-scale production of pregabalin [37].

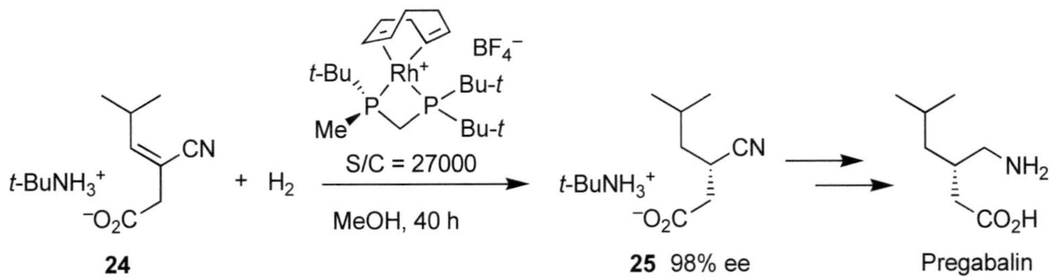


**Scheme 10.** Synthesis of a key intermediate in the production of pregabalin

Chiral β-amino acid derivatives are useful building blocks for the synthesis of β-peptides and β-lactam antibiotics. Asymmetric hydrogenation of β-dehydroamino acids with chiral rhodium catalysts is a useful method for the production of key chiral intermediates. An example of the preparation of a building block of the very late antigen-4 (VLA-4) antagonist S9059 is shown in Scheme 11. The hydrogenation of compound **26** in the presence of 0.1 mol % catalyst under 3 atm $H_2$ pressure proceeded rapidly, to produce the corresponding product **27** with 97.7% ee [33].

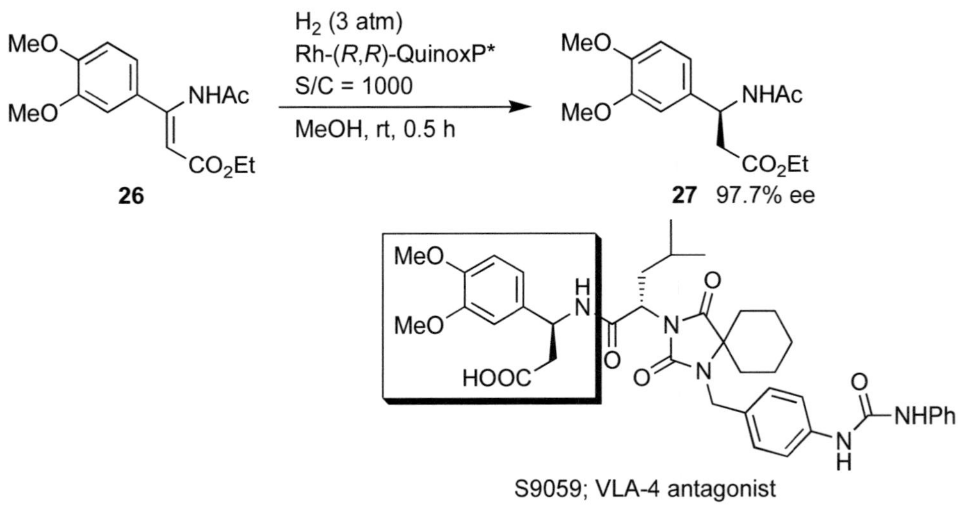


**Scheme 11.** Asymmetric hydrogenation of a *N*-acetyl-β-dehydroamino acid ester

# 4. Ruthenium-catalyzed Asymmetric Hydrogenation

## 4.1. Hydrogenation of functionalized alkenes

The discovery of chiral ruthenium catalysts significantly expanded the scope of asymmetric hydrogenation. Noyori et al. made the first breakthrough in this area using BINAP-Ru(II) dicarboxylate complexes. These complexes catalyze the highly enantioselective hydrogenation of the carbon–carbon double bonds of the substrates, the asymmetric hydrogenation of which had been difficult to achieve with the rhodium catalysts reported until then. For example, geraniol and its geometric isomer nerol, a kind of allyl alcohol, are subjected to hydrogenation with (*S*)-BINAP-Ru to produce (*R*)-citronellol and (*S*)-citronellol, respectively, and conversely, the use of (*R*)-BINAP-Ru produces the (*S*)- and (*R*)-products, respectively. Notably, the hydrogenation proceeds with a quite low catalyst loading (S/C = 50,000) to generate the products with a quantitative yield, with excellent enantioselectivities (96–99% ee) (Scheme 12) [38].

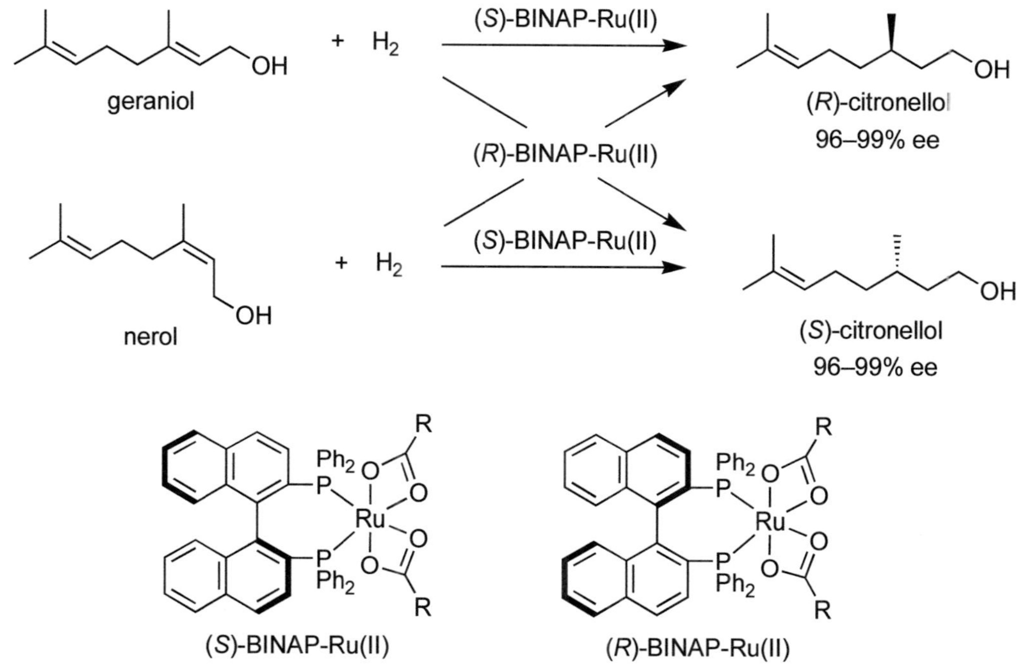


**Scheme 12.** Asymmetric hydrogenation of geraniol and nerol with BINAP-Ru(II) catalysts

The Ru(II) catalyst systems have been successfully applied to the enantioselective hydrogenation of α,β-unsaturated carboxylic acid esters, lactones, and ketones. Enamides are also efficiently hydrogenated with these catalysts. Using this catalyst system, isoquinoline alkaloids, morphine, and its artificial analogues can be prepared in an enantiopure form. A representative example, the synthesis of (*S*)-tetrahydropapaverine, is shown in Scheme 13 [39].

## 4.2. Hydrogenation of β-Keto esters and related substrates

Optically active β-hydroxy carboxylic esters are an important class of compounds in the synthesis of naturally occurring and biologically active compounds. Noyori et al. demonstrated a useful method for the catalytic asymmetric synthesis of this class of compounds using BINAP-Ru(II) complexes as the catalysts. The BINAP-Ru dicarboxylate complexes, which proved to be highly efficient for the enantioselective hydrogenation of various olefins, were not effective in this transformation. Instead, halogen-containing complexes $RuX_2$(binap) (X = Cl, Br, or I) were excellent catalyst precursors. The reactions with S/C > 1000 proceeded smoothly under 50–100 atm $H_2$ pressure, with excellent enantioselectivities, up to > 99% [40].

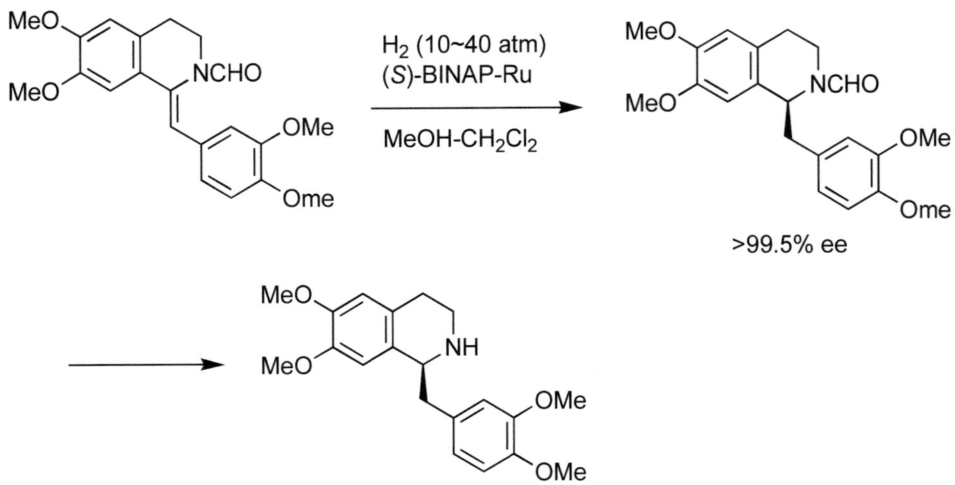


(*S*)-Tetrahydropapaverine

**Scheme 13.** Synthesis of (*S*)-tetrahydropapaverine via Ru-catalyzed asymmetric hydrogenation

The scope of this reaction was extensively expanded using various chiral phosphine ligands. As a result, a variety of β-keto esters, amides, and thiol esters with a functional group ($R^1$ = $ClCH_2$, alkoxymethyl, aryl, etc.) were hydrogenated in excellent enantioselectivities (Scheme 14). This method is currently used in academia and industry for the preparation of numerous chiral building blocks for the synthesis of biologically active compounds.

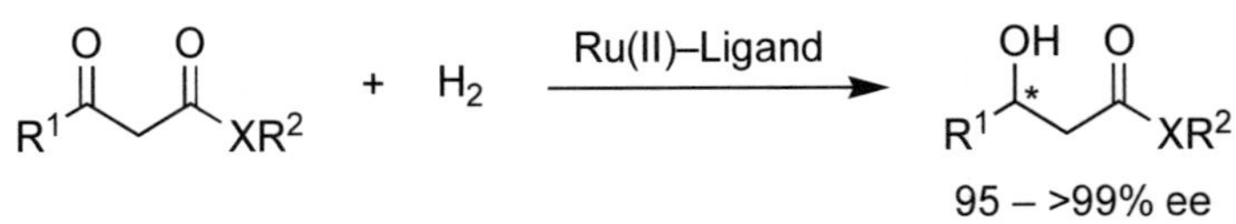


$R^1$ = Me, $ClCH_2$, Et, *i*-Pr, *n*-Bu, $PhCH_2OCH_2$, $PhCH_2OCH_2CH_2$, *i*-$Pr_3SiOCH_2$, *n*-$C_{11}H_{23}$, $(CH_3)_2CH(CH_2)_{11}$, $CF_3$, $PhCO_2CH_2$, $PhSO_2CH_2$, $CbzNHCH_2$, Aryl, etc
$XR^2$ = OMe, OEt, OPr-*i*, OBu-*t*, $NMe_2$, NHMe, SEt

**Scheme 14.** Ruthenium-catalyzed asymmetric hydrogenation of β-keto esters and related substrates

The hydrogenation of a β-keto ester bearing one substituent at the α-position provides four possible stereoisomeric β-hydroxy esters. Because stereomutation at the α-position of the β-keto ester occurs readily, it should be possible to selectively hydrogenate one of the β-keto ester enantiomers to yield only one stereoisomer, if the reaction conditions and the chiral ligand are selected appropriately. Noyori et al. established this dynamic kinetic resolution process using BINAP-Ru complexes [41,42]. The great utility of this method has been demonstrated in the production of many enantiopure building blocks. A representative example of the production of carbapenems by Takasago International Corporation is shown in Scheme 15 [43,44]. The hydrogenation of racemic **28** occurs with full conversion to yield the (2*S*,3*R*) product **29** with high diastereo- and enantioselectivity, and the product is further converted to the key intermediate, azetidinone **30**. The use of the DTBM-SEGPHOS-Ru(II)

(DTBM-SEGPHOS = 5,5′-bis[di(3,5-di-*tert*-butyl-4-methoxyphenyl)phosphino]-4,4′-bi-1,3-benzodioxole) complex for this reaction yields **29** almost exclusively (98.6% diastereomeric excess, 99.4% ee) [45].

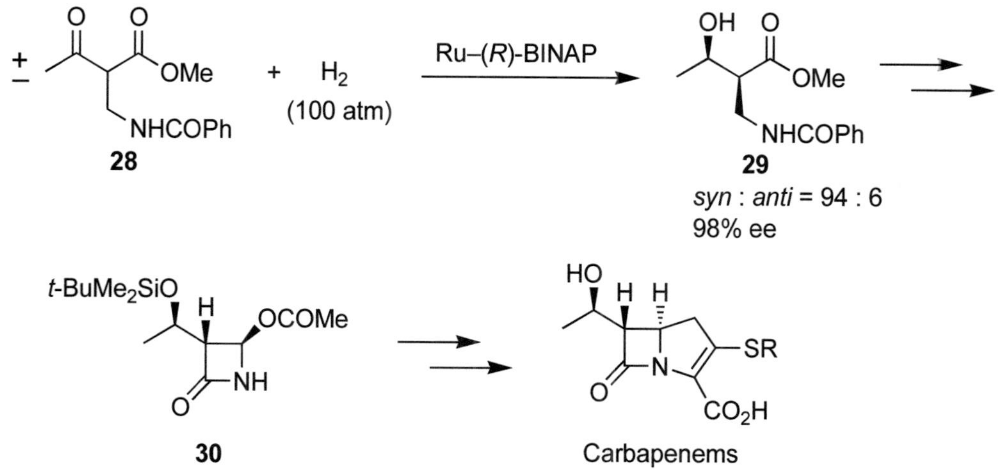


**Scheme 15.** Industrial synthesis of a carbapenem intermediate with Ru-BINAP-catalyzed hydrogenation

Another example is shown in Scheme 16. Racemic dimethyl 1-bromo-2-oxopropylphosphonate (**31**) is hydrogenated in the presence of the (S)-BINAP-Ru complex to yield (1*R*,2*S*)-1-bromo-2-hydroxypropylphosphonate (**32**) with 98% ee. The product is converted into fosfomycin, a clinically used antibiotic [46].

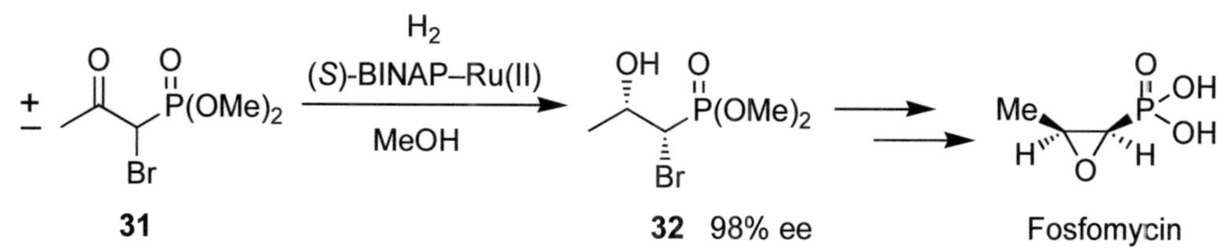


**Scheme 16.** Synthesis of fosfomycin via dynamic kinetic resolution

## 4.3. Hydrogenation of simple ketones

The development of ruthenium catalysts containing enantiopure diphosphines and diamines has allowed the asymmetric hydrogenation of simple ketones to optically active secondary alcohols. After examining numerous chiral diamines, Noyori, Ohkuma, and their co-workers found that the most effective catalyst systems were BINAP–DPEN (DPEN = 1,2-diphenylethylenediamine) (**33**) and BINAP–DAIPEN (DAIPEN = 1,1-di-4-anisyl-2-isopropyl-1,2-ethylenediamine) (**34**) (Fig. 2) [16,17,47]. In particular, the latter catalytic system (**34**), which has sterically more demanding 3,5-xylyl moieties on the phosphorus atoms exhibited exceedingly high catalytic activities and enantioselectivities in the hydrogenation of a wide range of ketone substrates.

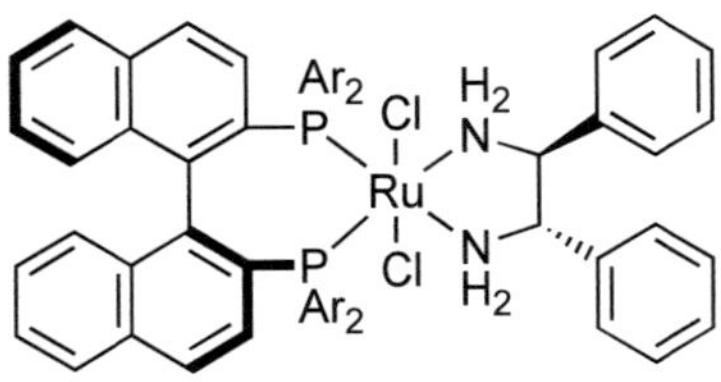


**33**

*trans*-$RuCl_2$[(*S*)-binap][(*S*,*S*)-dpen]

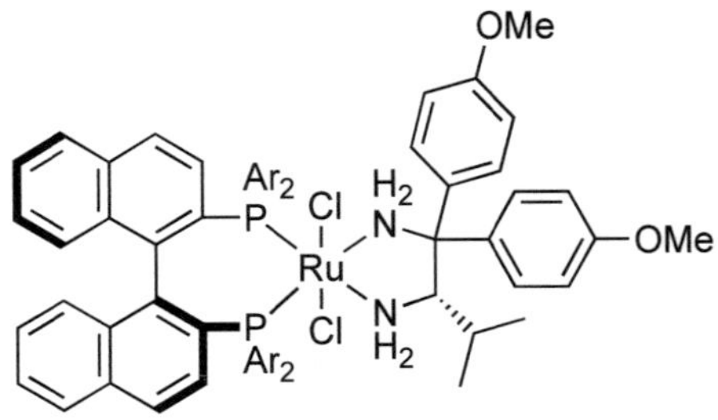


**34**

*trans*-$RuCl_2$[(*S*)-binap][(*S*)-daipen]

**Figure 2.** Ru(II) complexes with BINAP and chiral diamine

Representative examples of compounds obtained with these catalysts are shown in Figure 3. Alkyl aryl ketones, unsymmetric diaryl ketones, heteroaromatic ketones, unsymmetric dialkyl ketones, fluoro ketones, amino ketones, and $\alpha,\beta$-unsaturated ketones are hydrogenated with very high to almost-perfect enantioselectivities. High chemoselectivity is one of the characteristic features of this hydrogenation method. Therefore, only the carbonyl group is hydrogenated and the other functional groups, such as the carbon–carbon double bond and the nitro group, remain intact.

Recently, chiral ruthenabicylic complexes have been prepared and their exceedingly high catalytic performance has been demonstrated in the asymmetric hydrogenation of ketones [48]. Scheme 17 shows a typical example of the hydrogenation of acetophenone. The reaction under 50 atm $H_2$ pressure in the presence of 0.001 mol% catalyst proceeds very rapidly and is completed within 6 min, producing 1-phenylethanol with an essentially quantitative yield and more than 99% ee. The exceedingly high turnover frequency (> 600/s) and almost-perfect enantioselectivity are the best so far reported for ketone hydrogenation. The catalyst has been successfully applied to the asymmetric hydrogenation of several ketones, which are difficult substrates to reduce with high efficiency using existing catalysts. These facts, together with the easy preparation of these catalysts, strongly predict the promising results in the hydrogenation of a wide range of ketone substrates.

## 4.4. Mechanism of ketone hydrogenation catalyzed by ruthenium complexes of diphosphine and diamine

The mechanism of the Ru(II)-diphosphine/diamine-catalyzed asymmetric hydrogenation of ketones has been extensively studied by Noyori et al. [49]. The catalytic cycle demonstrated by them is shown in Scheme 18 [17,47,49].

The precatalyst **35** is converted via an induction process to the ruthenium hydride species **36**, which is equilibrated with other active species **37**, **38**, and **39**. The 18-electron Ru(II) hydride species **38** reacts with a ketone to produce a secondary alcohol and **39**. Complex **39** returns to **38** by the direct addition of $H_2$ or via **36** and **37**, and again reacts with the ketone. The marked catalytic activity and enantioselectivity originate from a nonclassical metal–ligand bifunctional mechanism. Therefore, the active species **38** involves the $H^{\delta-}$–$Ru^{\delta+}$–$N^{\delta-}$–$H^{\delta+}$ quadrupole, in

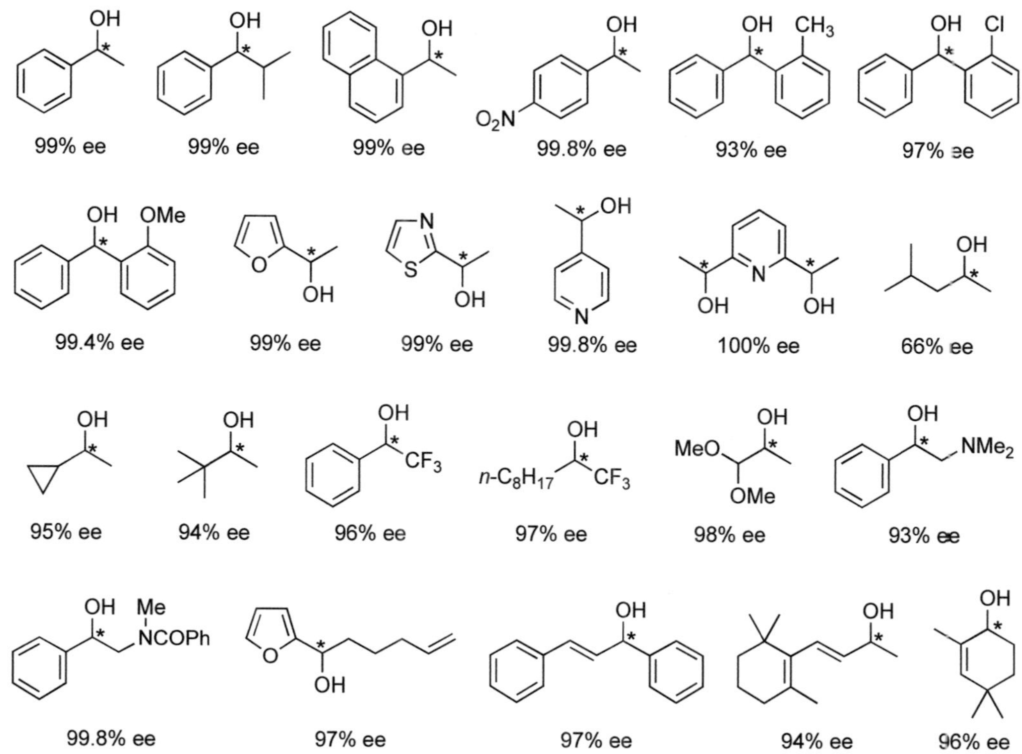


**Figure 3.** Representative examples of the ruthenium-catalyzed asymmetric hydrogenation of simple ketones

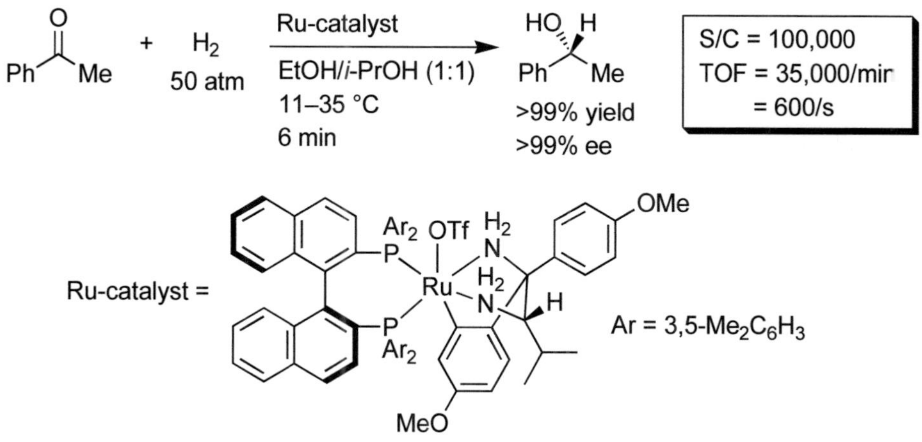


**Scheme 17.** Asymmetric hydrogenation of acetophenone catalyzed by a ruthenabicyclic complex

which two hydrogen atoms effectively interact with the $C^{\delta+}{=}O^{\delta-}$ dipole of the ketone, as shown in structure **40**. The reaction of the carbonyl group proceeds through a pericyclic six-membered transition state (**41**). It should be noted that the reduction of the carbonyl group occurs in an outer coordination sphere of 18-electron $Ru(H_2)$(diphosphine)(diamine), without any direct interaction with the metal center.

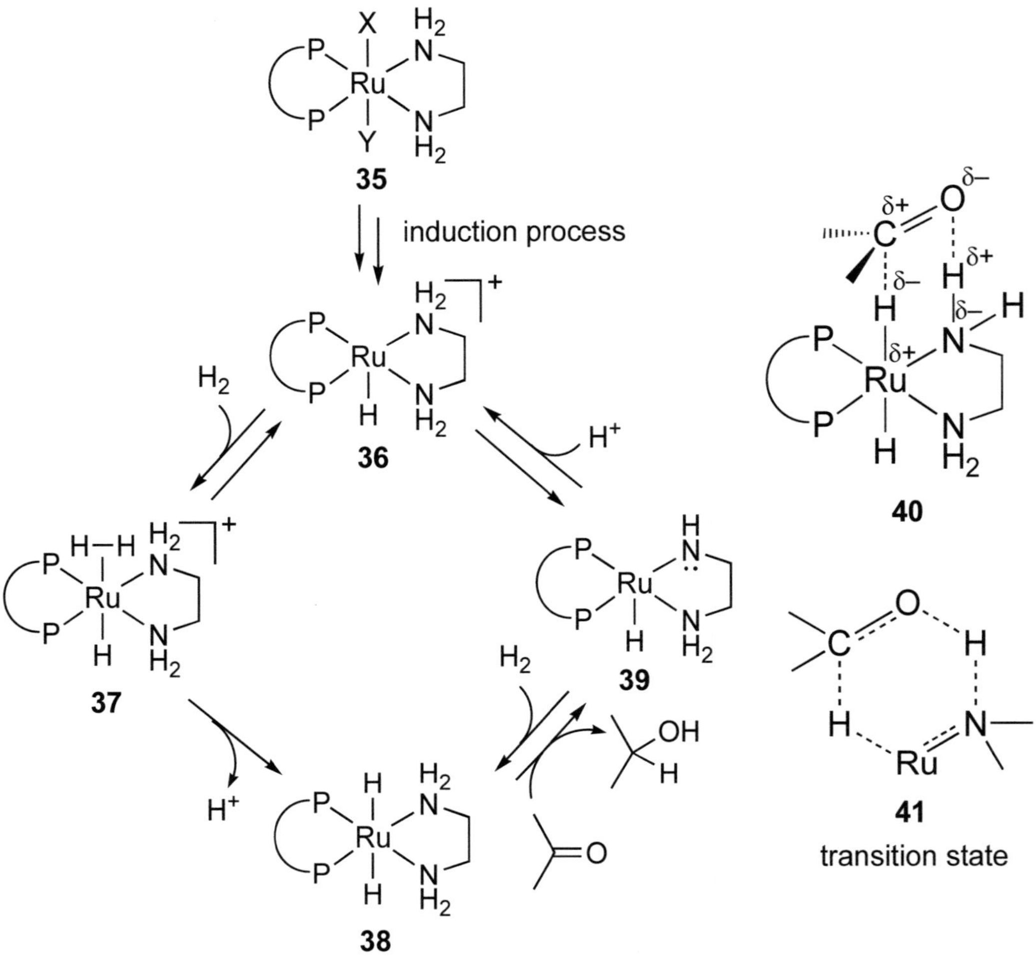


**Scheme 18.** Mechanism of ketone hydrogenation catalyzed by Ru(II)-diphosphine/diamine catalysts

# 5. Iridium-catalyzed Asymmetric Hydrogenation

## 5.1. Hydrogenation of unfunctionalized alkenes

Chiral rhodium and ruthenium catalysts are frequently used as the most versatile catalysts for the asymmetric hydrogenation of alkenes. However, the range of the substrates used is limited to alkenes with a coordinating functional group adjacent to the C=C double bond, except for several examples. The high enantioselectivities obtained by using rhodium or ruthenium catalysts are responsible for the coordination of the functional group to the metal center and the alkene π-bonding. In contrast, alkenes lacking coordinating groups have long been notoriously difficult to hydrogenate with high enantioselectivity. This difficulty was overcome by Pfaltz et al. in 1998 by using iridium complexes bearing chiral P,N-ligands [50]. Thus, they used Ir–PHOX complexes, which seemed to be the chiral analogues of Crabtree's catalyst $[Ir(cod)(PCy_3)(pyridine)]^+[PF_6]^-$ (Cy = cyclohexyl) [51,52]. Their initial study using

[Ir(phox)(cod)]$^+$[$PF_6$]$^-$ yielded high enantioselectivities of up to 98% ee in the hydrogenation of model substrates, but the turnover numbers were not large. The low activity of the catalysts was attributed to their deactivation during the hydrogenation reaction, and further experiments led them to the discovery of dramatic counterion effects. The replacement of the $PF_6^-$ anion with a bulky, apolar, and weakly coordinating anion BARF (tetrakis[3,5-bis(trifluoromethyl)phenyl]borate) ($BAr_F^-$) markedly improved the catalytic activity, allowing the use of catalyst loadings as low as 0.02 mol% (Scheme 19) [50,53].

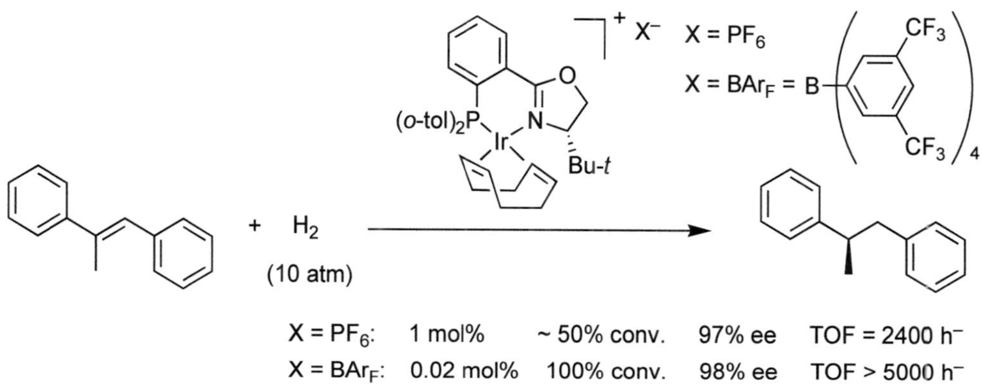


**Scheme 19.** Anion effect on the hydrogenation of (*E*)-α-methylstilbene

These successful results have significantly advanced this area of research with the development of numerous chiral P,N-ligands [13,54–58]. Representatives of the chiral iridium complexes so far reported are shown in Fig. 4. It should be noted that iridium complex **54**, with an *N*-heterocyclic carbene oxazoline ligand, is also effective in this kind of asymmetric hydrogenation [59].

Figure 5 shows some representative results for the asymmetric hydrogenation of unfunctionalized alkenes. Many rationally designed ligands display very high enantioselectivity (usually 99% ee) in the hydrogenation of a standard model substrate, (*E*)-α-methylstilbene. Purely alkyl-substituted alkenes are also reduced with high enantioselectivity. In the hydrogenation of 1,1-diarylethenes, two different aryl groups are effectively distinguished to produce the corresponding alkanes with good to excellent enantioselectivity. Notably, even tetrasubstituted alkenes are subject to hydrogenation, although the enantioselectivity depends largely on the substrate and the ligand structure.

Pfaltz et al. have demonstrated the practical utility of this methodology in the hydrogenation of γ-tocotrienyl acetate **55** to produce γ-tocopheryl acetate **56**, a precursor of γ-tocopherol, which is a component of vitamin E. The two prochiral (*E*)-configured C=C bonds of **55** are enantioselectively reduced under the conditions shown in Scheme 20 to generate the (*R*,*R*,*R*)-configuration product **56** with 98% purity [60]. This method provides a highly effective stereoselective route to this class of compounds and has great advantages over previous strategies, which used a stepwise approach to introduce the stereogenic centers into the side chain.

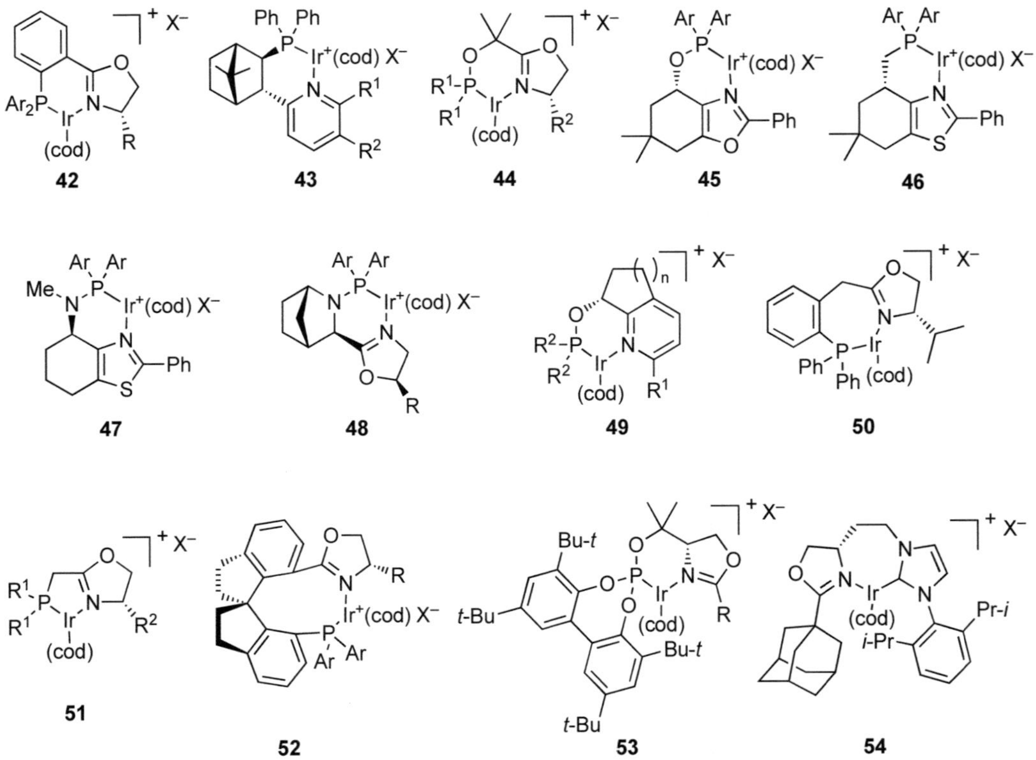


**Figure 4.** Representative chiral iridium complexes for asymmetric hydrogenation, X = BArF

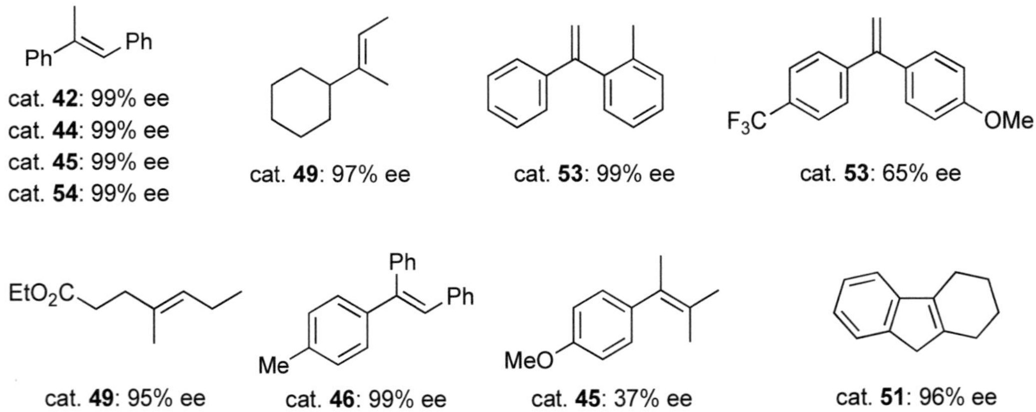


**Figure 5.** Representative examples of Ir-catalyzed largely unfunctionalized alkenes

## 5.2. Hydrogenation of functionalized alkenes

Recent studies of iridium-catalyzed asymmetric hydrogenation have significantly broadened its substrate spectrum. Therefore, not only unfuctionalized alkenes but also alkenes with functional groups connected to their C=C double bonds have been hydrogenated with high to excellent enantioselectivity. Figure 6 shows examples of the

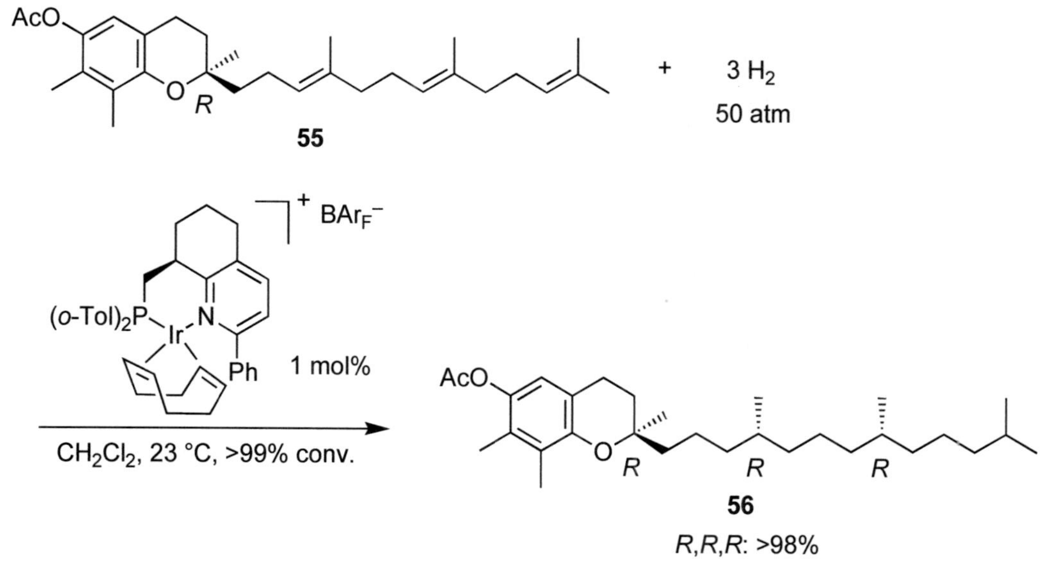


**Scheme 20.** Asymmetric hydrogenation of γ-tocotrienyl acetate

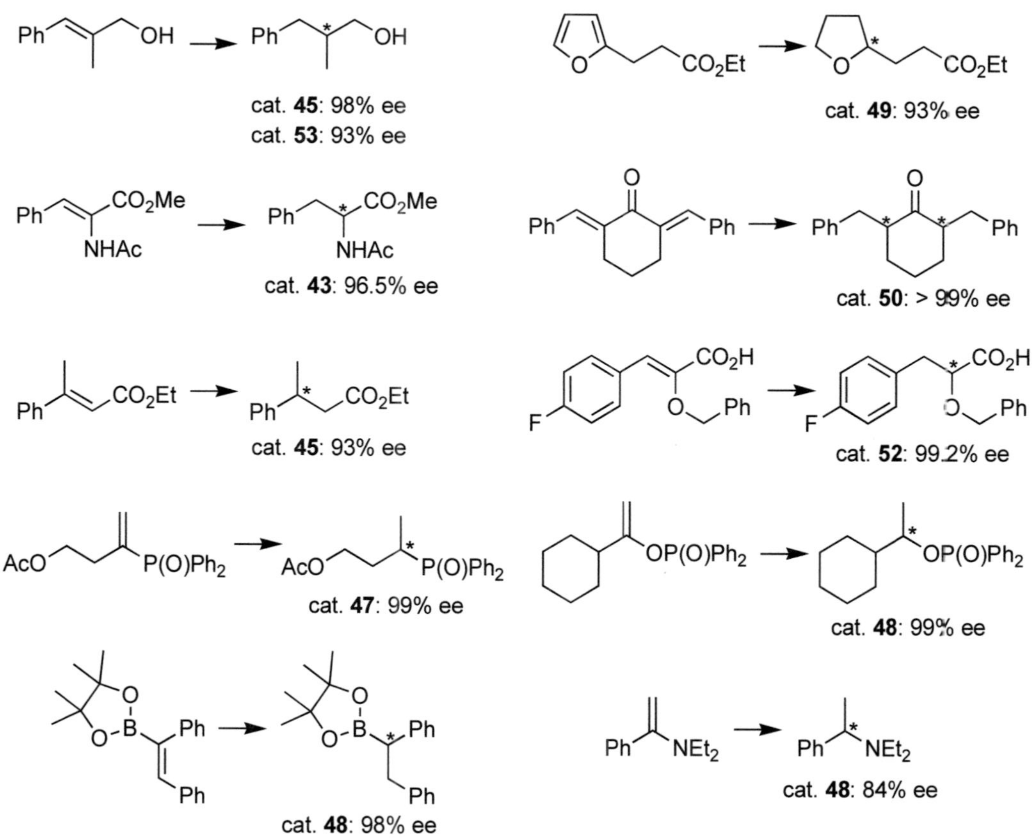


**Figure 6.** Representative examples of Ir-catalyzed asymmetric hydrogenation of functionalized alkenes

hydrogenation of allyl alcohols [61], furan rings [62], α-dehydroamino acid derivatives [63], α,β-unsaturated ketones [64], α,β-unsaturated carboxylic acid esters [61], α-alkoxy α,β-unsaturated acids [65], vinylphosphine oxides [66], enol phosphinates [67], vinyl boronates [68], and enamines [69,70]. Notably, substituted furans, vinyl boronates, and even enamines are hydrogenated with full conversion in high to excellent enantioselectivity.

### 5.3. Hydrogenation of simple ketones

It is well known that chiral iridium catalysts are applicable to the enantioselective hydrogenation of imines [71]. Recently, it has been shown that ketones, including α,β-unsaturated ketones, are also efficiently hydrogenated when iridium catalysts are used with P,N-ligands [72,73]. In contrast to the iridium complexes used with bidentate P,N-ligands, which tend to lose their activity under hydrogenation conditions, the complexes used with tridentate complexes resist deactivation and eventually exhibit high catalytic activity [73]. A typical example obtained by the use of catalyst **57** is shown in Scheme 20. The exceedingly high turnover number (TON), turnover frequency (TOF), and excellent enantioselectivity are comparable to those of chiral ruthenium complexes and indicate their great potential utility in the production of chiral secondary alcohols from ketones.

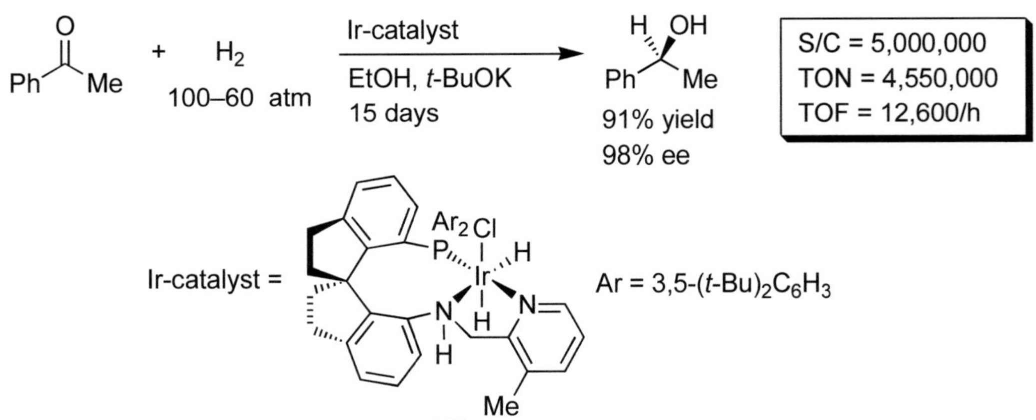


**57**

**Scheme 21.** Ir-catalyzed asymmetric hydrogenation of acetophenone

## 6. Conclusion

Since the discovery of homogeneous asymmetric hydrogenation, this area has progressed significantly over the past four decades. A variety of alkenes, including unfunctionalized alkenes, are hydrogenated enantioselectively using transition metal complexes with chiral ligands. Rhodium, ruthenium, and iridium are most frequently used as the center metals of these complexes, and the methods involving these complexes have become common processes in the efficient preparation of the chiral building blocks of natural products, pharmaceuticals, agrochemicals, and flavors.

Chiral complexes of titanium, zirconium, and lanthanides exhibit unique asymmetric hydrogenation properties, although at present, their practical use is limited to some special cases. Some late transition metals, such as palladium, cobalt, iron, and copper, are known to have potential utility in homogeneous asymmetric hydrogenation. The use of inexpensive metal complexes is clearly attractive for the manufacture of useful optically active compounds by asymmetric hydrogenation.

Asymmetric hydrogenation is a perfect atom-economic reaction, is usually carried out under mild conditions, and proceeds with an essentially quantitative yield. Undoubtedly, it is one of the most environmentally benign reactions and hence further investigations, using a variety of chiral metal catalysts, should allow the development of much more efficient and convenient methodologies for the preparation of optically active compounds.

## Author details

Tsuneo Imamoto
*Nippon Chemical Industrial Co., Ltd. and Chiba University, Japan*

## 7. References

[1] Akahori, S.; Sakurai, S.; Izumi, Y. & Fujii, Y. (1956), An Asymmetric Catalyst, *Nature*, Vol.178, p. 323

[2] Knowles, W.S. & Sabacky, M.J. (1968), Catalytic Asymmetric Hydrogenation Employing a Soluble Optically Active Rhodium Complex, *Chemical Comunications*, Vol.1968, pp. 1445-1446

[3] Horner, L.; Siegel, H. & Büthe, H. (1968), Asymmetric Catalytic Hydrogenation with an Optically Active Phosphinerhodium Complex in Homogeneous Solution, *Angewandte Chemie International Edition in English*, Vol.7, p. 942

[4] Dang, T.P. & Kagan, H.B. (1971), Asymmetric Synthesis of Hydratropic Acid and Amino Acids by Homogeneous Catalytic Hydrogenation, *Chemical Communications*, Vol.1971, p. 481

[5] Kagan, H.B. & Dang, T.P. (1972), Asymmetric Catalytic Reduction with Transition Metal Complexes. I. Catalytic System of Rhodium(I) with (–)-2,3-*O*-Isopropylidene-2,3-dihydroxy-1,4-bis-(diphenylphosphino)butane, a New Chiral Diphosphine, *Journal of the American Chemical Society*, Vol.94, pp. 6429-6433

[6] Knowles, W.S.; Sabacky, M.J.; Vineyard, B.D. & Weinkauff, D.J. (1975), Asymmetric Hydrogenation with a Complex of Rhodium and a Chiral Bisphosphine, *Journal of the American Chemical Society*, Vol.97, pp. 2567-2568

[7] Knowles, W.S. (2002), Asymmetric Hydrogenations, *Angewandte Chemie International Edition*, Vol.41, pp. 1998-2007.

[8] Miyashita, A.; Yasuda, A.; Takaya, H.; Toriumi, K.; Ito, T.; Souchi, T. & Noyori, R. (1980), Synthesis of 2,2'-Bis(diphenylphosphino)-1,1'-binaphthyl (BINAP), an Atropisomeric Chiral Bis(triaryl)phosphine, and Its Use in the Rhodium(I)-Catalyzed Asymmetric

Hydrogenation of α-(Acylamino)acrylic Acids, *Journal of the American Chemical Society*, Vol.102, pp. 7932-7934

[9] Noyori, R. (2002), Asymmetric Catalysis: Science and Opportunities, *Angewandte Chemie International Edition*, Vol.41, No.12, pp. 2008-2022.

[10] Matt, P.V. & Pfaltz, A. (1993), Chiral Phosphinoaryldihydrooxazoles as Ligands in Asymmetric Catalysis: Pd-Catalyzed Allylic Substitution, *Angewandte Chemie International Edition in English*, Vol.32, pp. 566-568

[11] Sprinz, J. & Helmchem, G. (1993), Phosphinoaryl- and Phosphinoalkyloxazolines as New Chiral Ligands for Enantioselective Catalysis: Very High Enantioselectivity in Palladium Catalyzed Allylic Substitutions, *Tetrahedron Letters*, Vol.34, pp. 1769-1772

[12] Dawson, G.J.; Frost, C.G. & Williams, J.M.J. (1993), Asymmetric Palladium Catalyzed Allylic Substitution Using Phosphorus Containing Oxazoline Ligands, *Tetrahedron Letter*, Vol.34, pp. 3149-3150

[13] Pfaltz, A. & Bell, S. (2007) Enantioselective Hydrogenation of Unfunctionalized Alkenes, In: *Handbook of Homogeneous Hydrogenation*, Vol.3, J.G. de Vries & C.J. Elsevier (Eds.), pp. 1049-1072, ISBN: 978-3-527-31161-3

[14] Woodmansee, D.H. & Pfaltz, A. (2011), Asymmetric Hydrogenation of Alkenes Lacking Coordinating Groups, *Chemical Communications*, Vol.47, pp. 7912-7916

[15] Noyori, R. (1994). *Asymmetric Catalysis in Organic Synthesis*, Wiley, ISBN 0-471-57267-5, New York, USA

[16] Noyori, R. & Ohkuma, T. (2001), Asymmetric Catalysis by Architectural and Functional Molecular Engineering: Practical Chemo- and Stereoselective Hydrogenation of Ketones, *Angewandte Chemie International Edition*, Vol.40, pp. 40-73

[17] Ohkuma, T. & Noyori, R. (2007), Enantioselective Ketone and β-Keto Ester Hydrogenations (Including Mechanisms), In: *Handbook of Homogeneous Hydrogenation*, Vol.3, J.G. de Vries & C.J. Elsevier (Eds.), pp. 1105-1163, ISBN: 978-3-527-31161-3, Wiley-VCH; Weinheim, Germany

[18] Shang, G.; Li, W. & Zhang, X. (2010). Transition Metal-Catalyzed Homogeneous Asymmetric Hydrogenation, In: *Catalytic Asymmetric Synthesis*, I. Ojima, (Ed.), pp. 343-436, Wiley, ISBN 978-0-470-17577-4

[19] Tang, W. & Zhang, X. (2003). New Chiral Phosphorus Ligands for Enantioselective Hydrogenation, *Chemical Reviews*, Vol.103, pp. 3029-3069

[20] Zhou, Q.-L. (Ed.) (2011) *Privileged Chiral Ligands and Catalysts*, Wiley-VCH, ISBN 978-3-527-32704-1, Weinheim, Germany

[21] Börner, A. Ed. (2008). *Phosphorus Ligands in Asymmetric Catalysis*, Wiley VCH, ISBN 978-3-527-31746-2, Weinheim, Germany

[22] Imamoto, T.; Oshiki, T.; Onozawa, T.; Kusumoto, T. & Sato, K. (1990), Synthesis and Reactions of Phosphine–Boranes. Synthesis of New Bidentate Ligands with Homochiral Phosphine Centers via Optically Pure Phosphine–Boranes, *Journal of the American Chemical Society*, Vol.112, pp. 5244-5252

[23] Imamoto, T.; Watanabe, J.; Wada, Y.; Masuda, H.; Yamada, H.; Tsuruta, H.; Matsukawa, S. & Yamaguchi, K. (1998), P-Chiral Bis(trialkylphosphine) Ligands and Their Use in

Highly Enantioselective Hydrogenation Reactions, *Journal of the American Chemical Society*, Vol.120, pp. 1635-1636

[24] Yamanoi, Y. & Imamoto, T. (1999), Methylene-Bridged P-Chiral Diphosphines in Highly Enantioselective Reactions, *Journal of Organic Chemistry*, Vol.64, pp. 2988-2989

[25] Halpern, J. (1982), Mechanism and Stereoselectivity of Asymmetric Hydrogenation, *Science*, Vol.217, pp. 401-407

[26] Halpern, J. (1985), Asymmetric Catalytic Hydrogenation: Mechanism and Origin of Enantioselection, In: *Asymmetric Synthesis*, J.D. Morrison (Ed.), Vol.5, Chapter 2, pp. 41-69, Academic Press, New York, USA

[27] Brown, J.M. (1999), Hydrogenation of Functionalized Carbon-Carbon Double Bonds, In: *Comprehensive Asymmetric Catalysis*, Vol. 2, E.N. Jacobsen, A. Pfaltz & H. Yamamoto (Eds.), pp. 121-182, ISBN 3-540-64336-2, Springer-Verlag, Berlin, Germany

[28] Brown, J.M. (2007), Mechanism of Enantioselective Hydrogenation, In: *Handbook of Homogeneous Hydrogenation*, Vol.3, J.G. de Vries & C.J. Elsevier (Eds.), pp. 1073-1103, ISBN: 978-3-527-31161-3, Wiley-VCH; Weinheim, Germany

[29] Gridnev, I.D.; Higashi, N.; Asakura, K. & Imamoto, T. (2000), Mechanism of Asymmetric Hydrogenation Catalyzed by a Rhodium Complex of (*S*,*S*)-1,2-Bis(*tert*-butylmethylphosphino)ethane. Dihydride Mechanism of Asymmetric Hydrogenation, *Journal of the American Chemical Society*, Vol.122, pp. 7183-7194

[30] Gridnev, I.D. & Imamoto, T. (2004), On the Mechanism of Stereoselection in Rh-Catalyzed Asymmetric Hydrogenation: A General Approach for Predicting the Sense of Enantioselectivity, *Accounts of Chemical Research*, Vol. 37, pp. 633-644

[31] Gridnev, I.D.; Imamoto, T.; Hoge, G.; Kouchi, M. & Takahashi, H. (2008), Asymmetric Hydrogenation Catalyzed by a Rhodium Complex of (*R*)-(*tert*-Butylmethylphosphino)(di-*tert*-butylphosphino)methane: Scope of Enantioselectivity and Mechanistic Study, *Journal of the American Chemical Society*, Vol.130, No.8, pp. 2560-2572

[32] Gridnev, I.D. & Imamoto, T. (2009), Mechanism of Enantioselection in Rh-Catalyzed Asymmetric Hydrogenation. The Origin of Utmost Catalytic Performance, *Chemical Communications*, No.48, pp. 7447-7464

[33] Imamoto, T.; Tamura, K.; Zhang, Z.; Horiuchi, Y.; Sugiya, M.; Yoshida, K.; Yanagisawa, A. & Gridnev, I.D. (2012). Rigid P-Chiral Phosphine Ligands with *tert*-Butylmethylphosphino Groups for Rhodium-Catalyzed Asymmetric Hydrogenation of Functionalized Alkenes, *Journal of the American Chemical Society*, Vol.134, pp. 1754-1769

[34] Liu, Z.; Lin, S.; Li, W.; Zhu, J.; Liu, X.; Zhang, X.; Lu, H.; Xiong, F. & Tian, Z. (2011), Enantioselective Synthesis of Cycloalkenyl-Substituted Alanines, U.S. Pat. Appl. Publ., US 20110257408 A1 20111020

[35] Wallace, D.J.; Campos, K.R.; Shultz, C.S.; Klapars, A.; Zewge, D.; Crump, B.R.; Phenix, B.D.; McWilliams, C.; Krska, S.; Sun, Y.; Chen, C. & Spindler, F. (2009), New Efficient Asymmetric Synthesis of Taranabant, a CB1R Inverse Agonist for the Treatment of Obesity, *Organic Process Research & Development*, Vol.13, pp. 84-90

[36] Sun, Y.; Krska, S.; Shultz, C.S.; & Tellers, D.M. (2010), Enabling Asymmetric Hydrogenation for the Design of Efficient Synthesis of Drug Substances, In: *Asymmetric*

*Catalysis on Industrial Scale*, Second Ed., H.-U. Blaser & H.-J. Federsel (Eds.), pp. 333-376, ISBN: 978-3-527-32489-7

[37] Hoge, G.; Wu, H.-P.; Kissel, W.S.; Pflum, D.A.; Greene, D.J. & Bao, J. (2004), Highly Selective Asymmetric Hydrogenation Using a Three Hindered Quadrant Bisphosphine Rhodium Catalyst, *Journal of the American Chemical Society*, Vol.126, pp. 5966-5967

[38] Takaya, H.; Ohta, T.; Sayo, N.; Kumobayashi, H.; Akutagawa, S.; Inoue, S.; Kasahara, I.; & Noyori, R. (1987), Enantioselective Hydrogenation of Allylic and Homoallylic Alcohols, *Journal of the American Chemical Society*, Vol.109, pp. 1596-1597

[39] Kitamura, M.; Hsiao, Y.; Ohta, M.; Tsukamoto, M.; Ohta, T.; Takaya, H. & Noyori, R. (1994), General Asymmetric Synthesis of Isoquinoline Alkaloids. Enantioselective Hydrogenation of Enamides Catalyzed by BINAP-Ruthenium(II) Complexes, *The Journal of Organic Chemistry*, Vol.59, pp. 297-310

[40] Noyori, R.; Ohkuma, T.; Kitamura, M.; Takaya, H.; Sayo, N.; Kumobayashi, H. & Akutagawa, S. (1987), Asymmetric Hydrogenation of β-Keto Carboxylic Esters. A Practical, Purely Chemical Access to β-Hydroxy Esters in High Enantiomeric Purity, *Journal of the American Chemical Society*, Vol.109, pp. 5856-5858

[41] Kitamura, M.; Tokunaga, M. & Noyori, R. (1993), Quantitative Expression of Dynamic Kinetic Resolution of Chirally Labile Enantiomers: Stereoselective Hydrogenation of 2-Substituted 3-Oxo Carboxylic Esters Catalyzed by BINAP-Ruthenium(II) Complexes, *Journal of the American Chemical Society*, Vol.115, pp. 144-152

[42] Noyori, R.; Tokunaga, M. & Kitamura, M. (1995), Stereoselective Organic Synthesis via Dynamic Kinetic Resolution, *Bulletin of the Chemical Society of Japan*, Vol.68, pp. 36-56

[43] Noyori, R.; Ikeda, T.; Ohkuma, T.; Widhalm, M.; Kitamura, M.; Takaya, H.; Akutagawa, S.; Sayo, N.; Saito, T.; Taketomi, T. & Kumobayashi, H. (1989), Stereoselective Hydrogenation via Dynamic Kinetic Resolution, *Journal of the American Chemical Society*, Vol.111, pp. 9134-9135

[44] Ohkuma, T.; Kitamura, M. & Noyori, R. (2000), Asymmetric Hydrogenation, in *Catalytic Asymmetric Synthesis*, 2$^{nd}$ edn (Ed. Ojima, I.), John Wiley & Sons, Inc., New York, pp. 1-110

[45] Shimizu, H.; Nagasaki, I.; Matsumura, K.; Sayo, N. & Saito, T. (2007), Developments in Asymmetric Hydrogenation from an Industrial Perspective, *Accounts of Chemical Research*, Vol.40, pp. 1385-1393

[46] Kitamura, M.; Tokunaga, M. & Noyori, R. (1995), Asymmetric Hydrogenation of β-Keto Phosphonates: A Practical Way to Fosfomycin, *Journal of the American Chemical Society*, Vol.117, pp. 2931-2932

[47] Ohkuma, T. (2010), Asymmetric Hydrogenation of Ketones: Tactics to Achieve High Reactivity, Enantioselectivity, and Wide Scope, *Proceedings of the Japan Academy, Ser. B*, Vol.86, pp. 202-219

[48] Matsumura, K.; Arai, N.; Hori, K.; Saito, T.; Sayo, N.; Ohkuma, T. (2011), Chiral Ruthenabicyclic Complexes: Precatalysts for Rapid, Enantioselective, and Wide-Scope Hydrogenation of Ketones, *Journal of the American Chemical Society*, Vol.133, pp. 10696-10699

[49] Sandoval, C.A.; Ohkuma, T.; Muniz, K. & Noyori, R. (2003), Mechanism of Asymmetric Hydrogenation of Ketones Catalyzed by BINAP/1,2-Diamine-Ruthenium(II) Complexes, *Journal of the American Chemical Society*, Vol.125, pp. 13490-13503

[50] Lightfoot, A.; Schnider, P. & Pfaltz, A. (1998), Enantioselective Hydrogenation of Olefins with Iridium–Phosphanodihydrooxazole Catalysts, *Angewandte Chemie International Edition*, Vol.37, pp. 2897-2899

[51] Crabtree, R.H.; Felkin, H. & Morris, G.E. (1977), Cationic Iridium Diolefin Complexes as Alkene Hydrogenation Catalysts and the Isolation of Some Related Hydrido Complexes, *Journal of Organometallic Chemistry*, Vol.141, pp. 205-215

[52] Crabtree, R.H. (1979), Iridium Compounds in Catalysis, *Accounts of Chemical Research*, Vol.12, pp. 331-337

[53] Smidt, S.P.; Zimmermann, N.; Studer, M. & Pfaltz, A. (2004), Enantioselective Hydrogenation of Alkenes with Iridium-PHOX Catalysts: A Kinetic Study of Anion Effects, *Chemistry: A European Journal*, Vol.10, pp. 4685-4693

[54] Roseblade, S.J. & Pfaltz, A. (2007), Iridium-Catalyzed Asymmetric Hydrogenation of Olefins, *Accounts of Chemical Research*, Vol.40, pp. 1402-1411

[55] Woodmansee, D. H. & Pfaltz, A. (2011), Asymmetric Hydrogenation of Alkenes Lacking Coordinating Groups, *Chemical Communications*, Vol.47, pp. 7912-7916

[56] Church, T.L. & Andersson, P.G. (2008), Iridium Catalysts for the Asymmetric Hydrogenation of Olefins with Nontraditional Functional Substituents, *Coordination Chemistry Reviews*, Vol.252, pp. 513-531

[57] Pamies, O.; Andersson, P.G. & Diéguez, M. (2010), Asymmetric Hydrogenation of Minimally Functionalised Terminal Olefins: An Alternative Sustainable and Direct Strategy for Preparing Enantioenriched Hydrocarbons, *Chemistry: A European Journal*, Vol.16, pp. 14232-14240

[58] Cui, X. & Burgess, K. (2005), Catalytic Homogeneous Asymmetric Hydrogenation of Largely Unfunctionalized Alkenes, *Chemical Reviews*, Vol.105, pp. 3272-3296

[59] Perry, M.C.; Cui, X.; Powell, M.T.; Hou, D.-R.; Reibenspies, J.H. & Burgess, K. (2003), Optically Active Iridium Imidazol-2-ylidene-oxazoline Complexes: Preparation and Use in Asymmetric Hydrogenation of Arylalkanes, *Journal of the American Chemical Society*, Vol.125, pp. 113-123

[60] Bell, S.; Wüstenberg, B.; Kaiser, S.; Menges, F.; Netscher, T. & Pfaltz, A. (2006), Asymmetric Hydrogenation of Unfunctionalized, Purely Alkyl-Substituted Olefins, *Science*, Vol.311, pp. 642-644

[61] Källström, K.; Hedberg, C.; Brandt, P.; Bayer, A. & Andersson, P.G. (2004), Rationally Designed Ligands for Asymmetric Iridium-Catalyzed Hydrogenation of Olefins, *Journal of the American Chemical Society*, Vol.126, pp. 14308-14309

[62] Kaiser, S.; Smidt, S.P. & Pfaltz, A. (2006), Iridium Catalysts with Bicyclic Pyridine-Phosphinite Ligands: Asymmetric Hydrogenation of Olefins and Furan Derivatives, *Angewandte Chemie International Edition*, Vol.45, pp. 5194-5197

[63] Bunlaksananusorn, T.; Polborn, K. & Knochel, P. (2003), New P,N Ligands for Asymmetric Ir-Catalyzed Reactions, *Angewandte Chemie International Edition*, Vol.42, pp. 3941-3943

[64] Lu, W.-J.; Chen, Y.-W. & Hou, X.-L. (2008), Iridium-Catalyzed Highly Enantioselective Hydrogenation of the C=C Bond of α,β-Unsaturated Ketones, *Angewandte Chemie International Edition*, Vol.48, pp. 10133-10136

[65] Li, S.; Zhu, S.-F.; Xie, J.-H.; Song, S.; Zhang, C.-M. & Zhou, Q.-L. (2010), Enantioselective Hydrogenation of α-Aryloxy and α-Alkoxy α,β-Unsaturated Carboxylic Acids Catalyzed by Chiral Spiro Iridium/Phosphino-Oxazoline Complexes, *Journal of the American Chemical Society*, Vol.132, pp. 1172-1179

[66] Cheruku, P.; Paptchikhine A.; Church, T.L. & Andersson, P.G. (2009), Iridium-N,P-Ligand-Catalyzed Enantioselective Hydrogenation of Diphenylvinylphosphine Oxides and Vinylphosphonates, *Journal of the American Chemical Society*, Vol.131, pp. 8285-8289

[67] Cheruku, P.; Gohil, S. & Andersson, P.G. (2007), Asymmetric Hydrogenation of Enol Phosphinates by Iridium Catalysts Having N,P Ligands, *Organic Letters*, Vol.9, pp. 1659-1661

[68] Paptchihine, A.; Cheruku, P.; Engman, M.; & Andersson, P.G. (2009), Iridium-Catalyzed Enantioselective Hydrogenation of Vinyl Boronates, *Chemical Communications*, pp.5996-5998

[69] Cheruku, P.; Church, T.L.; Trifonova, A.; Wartmann, T. & Andersson, P.G. (2008), Access to Chiral Tertiary Amines via the Iridium-Catalyzed Asymmetric Hydrogenation of Enamines, *Tetrahedron Letters*, Vol.49, pp. 7290-7293

[70] Hou, G.-H.; Xie, J.-H.; Yan, P.-C. & Zhou, Q.-L. (2009), Iridium-Catalyzed Asymmetric Hydrogenation of Cyclic Enamines, *Journal of the American Chemical Society*, Vol.131, pp. 1366-1367

[71] Xie, J.-H.; Zhu, S.-F. & Zhou, Q.-L. (2011), Transition Metal-Catalyzed Enantioselective Hydrogenation of Enamines and Imines, *Chemical Reviews*, Vol.111, pp. 1713-1760

[72] Xie, J.-B.; Xie, J.-H.; Liu, X.-Y.; Kong, W.-L.; Li, S. & Zhou, Q.-L. (2010), Highly Enantioselective Hydrogenation of α-Arylmethylene Cycloalkanones Catalyzed by Iridium Complexes of Chiral Spiro Aminophosphine Ligands, *Journal of the American Chemical Society*, Vol.132, pp. 4538-4539

[73] Xie, J.-H.; Liu, X.-Y.; Xie, J.-B.; Wang, L.-X. & Zhou, Q.-L. (2011), An Additional Coordination Group Leads to Extremely Efficient Chiral Iridium Catalysts for Asymmetric Hydrogenation of Ketones, *Angewandte Chemie International Edition*, Vol.50, pp. 7329-7332

Chapter 2

# Asymmetric Hydrogenation and Transfer Hydrogenation of Ketones

Bogdan Štefane and Franc Požgan

Additional information is available at the end of the chapter

http://dx.doi.org/10.5772/47752

## 1. Introduction

Optically active alcohols are important building blocks in the synthesis of fine chemicals, pharmaceuticals, agrochemicals, flavors and fragrances as well as functional materials (Arai & Ohkuma, 2011; Klingler, 2007). Furthermore, molecular hydrogen is without doubt the cleanest reducing agent, with complete atom efficiency. Therefore, the catalytic, asymmetric hydrogenation (AH) of prochiral ketones is the most practical and simplest method to access enantiomerically enriched secondary alcohols, on both the laboratory and industrial scales. Asymmetric transfer hydrogenation (ATH), on the other hand, represents an attractive alternative or complement to hydrogenation because it is easy to execute and a number of cheap chemicals can be used as hydrogen donors. For practical use and to address environmental issues a high catalyst activity (low loadings) and selectivity is preferable, as well as the employment of ''greener'' solvents, mild operating conditions and recyclable catalyst systems. High turnover numbers (TONs) and turnover frequencies (TOFs), and satisfactory stereo- and chemoselectivities are attainable only with a combination of well-defined metal catalysts and suitable reaction conditions. The reactivity and selectivity can be finely tuned by changing the bulkiness, chirality and electronic properties of the auxiliaries on the metal center of the catalyst.

## 2. Homogenous, asymmetric hydrogenation and transfer hydrogenation

Since the application of very efficient, chiral BINAP-derived ruthenium complexes in the AH of functionalized ketones (β-keto esters) at a high enantioselectivity level in the homogenous phase (Noyori et al., 1987), the development of more robust and reactive molecular catalysts is still highly desirable. Furthermore, because of the structural and functional diversity of organic substrates, no universal catalysts exist. Ruthenium complexes bearing chiral ligands are among the most commonly used catalysts for AH and ATH, following by rhodium and iridium, although in recent times other transition metals, like Fe, Cu, or Os have rapidly penetrated this field.

## 2.1. Ru-, Rh- and Ir-catalyzed hydrogenation and transfer hydrogenation

A major breakthrough in the wide-scope AHs of ketones was the discovery by Noyori and co-workers of the conceptually new and extremely efficient ruthenium bifunctional catalysts. They found that simple ketones like **1-5**, which lack anchoring heteroatoms capable of interacting with a metal center, can be reduced enatioselectively with $H_2$ (1-8 atm) in *i*-PrOH using a ternary catalyst system comprising a chiral BINAP-$RuCl_2$ precursor, a chiral 1,2-diamine ligand (**L1–L3**) and an alkaline base (*e.g.*, KOH) in a 1:1:2 molar ratio (Fig. 1) (Ohkuma et al., 1995a, 1995b). This catalyst system chemoselectively afforded the corresponding chiral alcohols in almost quantitative yields and up to 99% optical yields. Since then, a number of AHs catalyzed by Ru(II) complexes, like **C1** bearing chiral diphosphine, and diamine ligands for structurally diverse substrates, like alkyl-aryl ketones, heteroaromatic ketones, unsymmetrical benzophenones, aliphatic and α,β-unsaturated ketones, has been reported (Noyori & Ohkuma, 2001; Ohkuma, 2010). Furthermore, proper matching of a chiral ruthenium diphosphine with the correct enantiomer of diamine leads to exceptionally enantioselective catalysts, which are also highly chemoselective for C=O group *vs.* C=C and C≡C bonds, and tolerate many functionalities, like $NO_2$, $CF_3$, halogen, acetal, $CO_2R$, $NH_2$, NHCOR, etc.

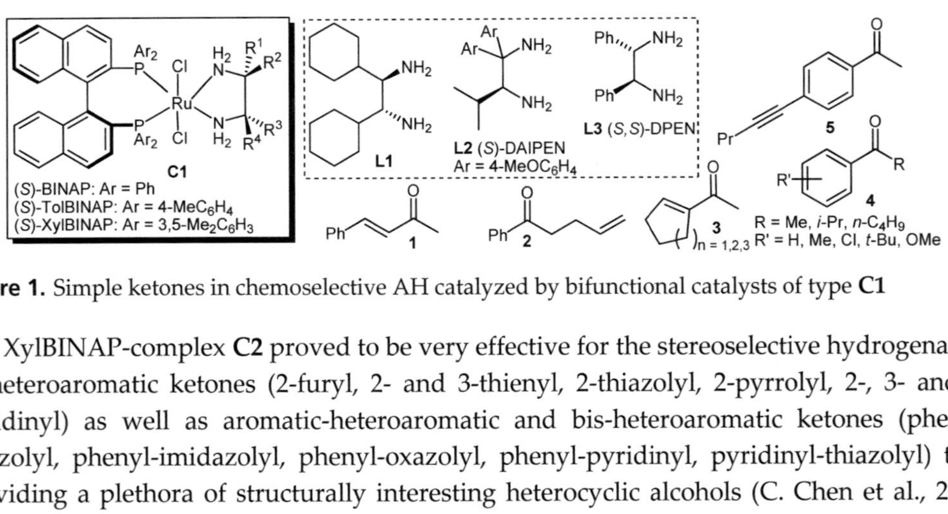


**Figure 1.** Simple ketones in chemoselective AH catalyzed by bifunctional catalysts of type **C1**

The XylBINAP-complex **C2** proved to be very effective for the stereoselective hydrogenation of heteroaromatic ketones (2-furyl, 2- and 3-thienyl, 2-thiazolyl, 2-pyrrolyl, 2-, 3- and 4-pyridinyl) as well as aromatic-heteroaromatic and bis-heteroaromatic ketones (phenyl-thiazolyl, phenyl-imidazolyl, phenyl-oxazolyl, phenyl-pyridinyl, pyridinyl-thiazolyl) thus providing a plethora of structurally interesting heterocyclic alcohols (C. Chen et al., 2003; Ohkuma et al., 2000). In fact, the complex **C2** has been established as one of the most efficient and selective pre-catalysts for the AH of a variety of ketones (Ohkuma et al., 1998) until the discovery of novel ruthenabicyclic complexes (Matsumura et al., 2011). The hydrogenation of acetophenone catalyzed by the ruthenabicyclic complex **C3** with a substrate-to-catalyst molar ratio (S/C) 10000 under 50 atm of $H_2$ in a *i*-PrOH/EtOH/*t*-BuOK mixture was completed in one minute to give (*R*)-1-phenylethanol in more than 99% *ee*, thus achieving a TOF of about $3.5 \cdot 10^4$ $min^{-1}$. For comparison, the pre-catalyst **C2** provided a similar outcome in four hours. This ruthenabicyclic pre-catalyst is better than all previous catalyst systems in terms of efficiency, enenatioselectivity and the scope of the ketone substrates (aromatic, aliphatic, cyclic and bicyclic ketones; **6-9**) (Fig. 2).

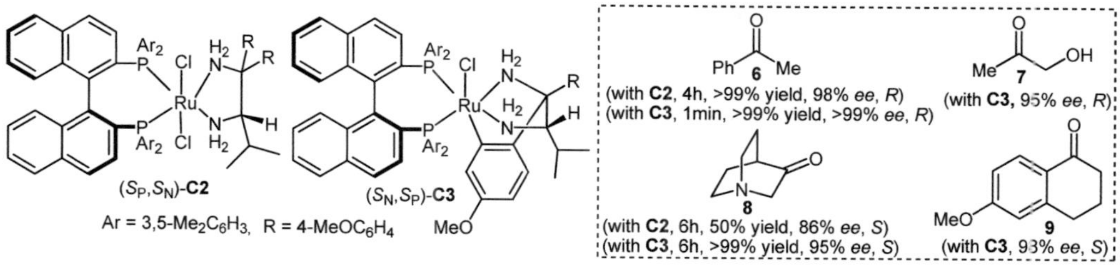


**Figure 2.** Ruthenabicyclic *vs.* standard Noyori catalyst for the AH of structurally different ketones

Since the Noyori's standard Ru(II) complexes of the type **C1** require the presence of a strong base as a co-catalyst to *in situ* generate an active catalyst, *i.e.*, $RuH_2$ species, some unwanted side reactions (*e.g.*, transesterification with an alcohol product in the case of **10**) may occur. Ohkuma et al. succeeded in preparing a relatively stable [RuH($\eta^1$-$BH_4$)(BINAP)(1,2-diamine)] catalyst **C4,** which allowed for the base-free AH of otherwise base-sensitive ketone substrates **10–13** in almost quantitative yields and excellent *ee* values (Fig. 3) (Ohkuma et al., 2002).

The extremely high reactivity and enantio-selectivity of [TunesPhos-Ru(II)-(1,2-diamine)] complexes combined with *t*-BuOK enabled the AH of ring-substituted acetophenones, 2-acetylthiophene, 2-acetylfuran, 1- and 2-acetylnaphthalen, and cyclopropyl methyl ketone with TONs up to 1000000 affording the corresponding chiral alcohols in *ee*'s up to >99% (W. Li et al., 2009). Among them, the catalyst precursor **C5** was found to be the most efficient, since decreasing the catalyst loading from 0.01 mol% to a ppm level had only a small impact on *ee* in the hydrogenation of acetophenone (99.8→98% *ee*), though high conversions necessitated longer reaction times (Fig. 3).

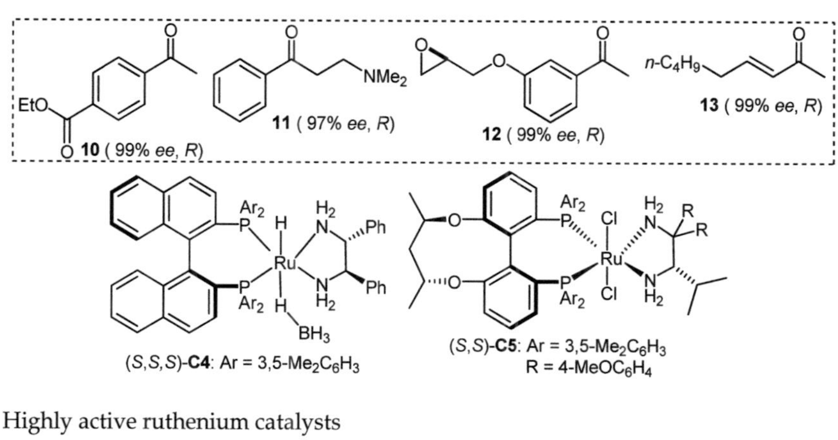


**Figure 3.** Highly active ruthenium catalysts

The discovery of new classes of hydrogenation catalysts that deviate from the Noyori-type **C1** may represent a good opportunity to reduce every type of ketone substrate with high reactivity and selectivity. Indeed, while the conventional [BINAP-Ru-(1,2-diamine)] catalysts have shown poor reactivity and enantio-selectivity in the hydrogenation of sterically congested *tert*-alkyl ketones, a reduction using the BINAP/(α-picolylamine)-based Ru complex **C6** in a ratio S/C as high as 100000 provided the corresponding *tert*-alkyl carbinols from ketones **14–18** in a high enantiomeric purity (Fig. 4) and proved to be chemoselective for enone **16** and also active for the highly hindered β-keto ester **18** (Ohkuma et al., 2005).

Interestingly, a combined amine-benzimidazole ligand in the complex **C7** influenced the reverse enantioselection from that typically observed in the AH of ring-substituted acetophenones and allowed the reduction to proceed in nonprotic solvents (toluene/*t*-BuOH 9:1) with S/C 1000 to 50000 giving (*S*)-alcohols in 82-99% *ee* (Fig. 4) (Y. Li et al., 2009).

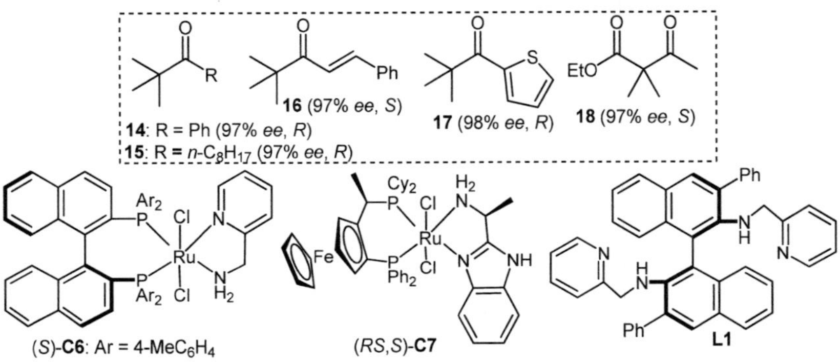


**Figure 4.** AH of sterically congested and poorly reactive ketones

AH using non-phosphine-based catalysts is attractive due to the toxicity of the catalyst precursors and the product contamination when Noyori-type catalysts are used. However, the efficiency of the π-allyl Ru precursor in combination with the phosphorous-free pyridyl-containing ligand **L1** did not exceed that of the original [BINAP-Ru-diamine] complexes (Fig. 4) (Huang et al., 2006). Interestingly, this new catalyst system catalyzes the hydrogenation of 1-indanone only in the absence of a base.

The most efficient AH catalysts tend to mimic that of Noyori as its excellent enantioselectivity is proposed to be a result of the synergistic effect of chiral phosphane and chiral amine ligands. Nevertheless, commercially available achiral diphosphanes (DPPF, DPEphos) in conjunction with rigid chiral biisoindoline-based diamines have been applied in the Ru-catalyzed AH of (hetero)aromatic ketones, affording excellent enantioselectivities (up to 99% *ee*) with an S/C up to 100000 (Zhu et al., 2011).

Since ketones coordinate more weakly to metals than olefins, many Rh-phosphane complexes show no activity for hydrogenation of simple ketones. However, the highly enantioselective direct hydrogenation of simple ketones **19–24** using an *in-situ*-prepared catalyst from simple precursors, $[Rh(COD)Cl]_2$ and the rigid chiral biphosphane ligand **L2** promoted by 2,6-lutidine (2,6-dimethylpyridine) and KBr has been reported (Fig. 5) (Q. Jiang et al., 1998). With this catalyst system, the hydrogenation of acetophenone was sluggish and gave only 57% *ee* of (*S*)-1-phenylethanol, whereas the presence of additives dramatically accelerated the reaction and enhanced the enantioselectivity (95% *ee*). While with aryl(heteroaryl) ketones (**19** and **20**) high *ee*'s were observed, more importantly, this hydrogenation procedure proved to be satisfactorily enantioselective for several alkyl-methyl ketones (**21–23**), even those bearing unbranched alkyl groups (**24**), which in principle represent the toughest problem for asymmetric reduction.

The complex prepared from $[Rh(COD)OCOCF_3)]_2$ and the amide-phosphine-phosphinite ligand **L3** catalyzed the AH of trifluoromethyl ketones **25** giving almost quantitative yields of the corresponding alcohols in 83-97% *ee* (Kuroki et al., 2001). Interestingly, this Rh-catalyst showed preferential activity and stereoselectivity for fluorinated ketone substrates since acetophenone gave only a 2% yield of 1-phenylethanol in 8% *ee*.

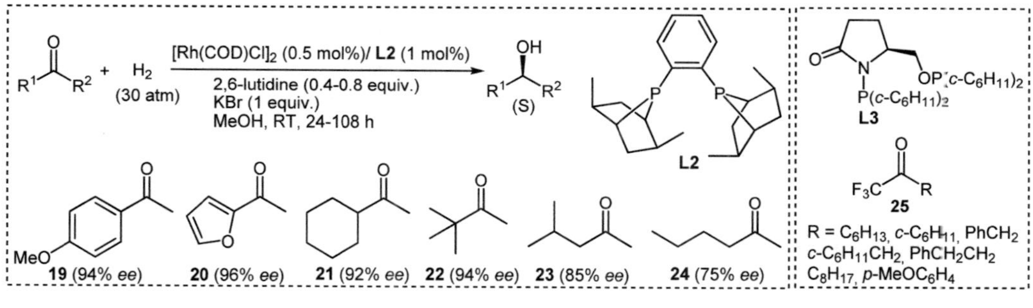


**Figure 5.** Rh-catalyzed AH of simple and fluorinated ketones

The hydrogenation of ketones catalyzed by chiral iridium complexes has been well studied and developed because iridium is less expensive than rhodium (Malacea et al., 2010). Generally, Ir(I) or Ir(III) complexes with chiral diamines, diphosphines or a combination of both, very similar to those in Ru-catalyzed hydrogenation, have been successfully employed in the AH of various aromatic ketones and β-keto esters. On the other hand, chiral Ir(I) complexes bearing N-heterocyclic carbenes as ligands proved to be far less efficient (Diez & Nagel, 2009). Although complexes of $[Ir(COD)Cl]_2$ and planar-chiral ferrocenyl phosphine-thioethers (*e.g*, **L4**) (Le Roux et al., 2007) or spiro aminophosphine ligands (*e.g.*, **L5**) (J.-B. Xie et al., 2010) efficiently catalyze the AH of acetophenone-type substrates and more importantly *exo*-cyclic α,β-unsaturated ketones **26**, chiral Ir-complexes with phosphorous-nitrogen ligands tend to lose their activity under hydrogenation conditions. The introduction of an additional coordination group in the bidentate spiro aminophosphine ligand **L6** led to a very stable and efficient catalyst for the AH of simple ketones **27**, affording the chiral alcohols **28** in up to 99.9% *ee* (Fig. 6) (J.-H. Xie et al., 2011). For example, acetophenone was reduced with a $2 \cdot 10^{-5}$ mol% catalyst loading to give (*S*)-1-phenylethanol in 98% *ee*, reaching a TON of $4.55 \cdot 10^6$ and a TOF of $1.26 \cdot 10^4$ $h^{-1}$.

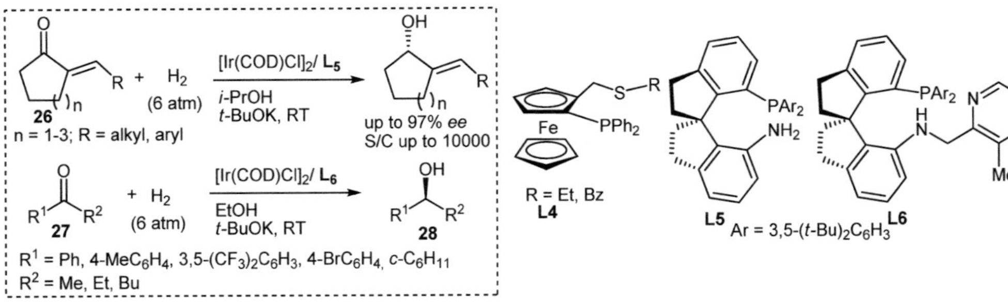


**Figure 6.** Ir-catalyzed AH

With its origin in Meerwein-Pondorf-Verley reduction, and later developed in its asymmetric version, the transfer hydrogenation of ketones has emerged as an operationally simpler and significantly safer alternative to catalytic $H_2$-hydrogenation as there is no need for special vessels and high pressures (Ikariya & Blacker, 2007; Palmer & Wills 1999). Moreover, chemo-, regio- and stereoselectivity can often be different from that of AH. In the ATH process, the transition-metal catalyst is able to abstract a hydride and a proton from the hydrogen donor and deliver them to the carbonyl moiety of the ketone. Suitable catalysts for ATH are typically complexes of homochiral ligands with Ru, Rh or Ir, whilst *i*-PrOH/base (hydroxide or alkoxyide) or formic acid/triethylamine (FA/TEA, 5:2 azeotrope) are the most common hydrogen donors usually being the solvents at the same time. A major drawback of using *i*-PrOH is the reaction reversibility, giving limited conversions and affecting the enantiomeric purity of the products after long reaction times. The use of formic acid can overcome these drawbacks, although only a narrow range of catalysts that tolerate formic acid is available.

In parallel with the discovery of efficient ruthenium catalysts for AH, Noyori and co-workers found a prototype of chiral (arene)Ru(II) catalysts of type **C8** bearing *N*-sulfonated 1,2-diamines (*e.g.*, TsDPEN = *N*-(*p*-toluenesulfonyl)-1,2-diphenyl-ethylenediamine) or amino alcohols such as chiral ligands for the highly enantio-selective ATH of (hetero)aromatic ketones in *i*-PrOH/KOH or in FA/TEA (Fig. 7) (Fujii et al., 1996; Hashiguchi et al., 1995; Takehara et al., 1996). After this milestone discovery a large number of related or novel ligands and catalysts for ATH have been developed that display a broad substrate scope and provide optically active alcohols in a high enantiomeric purity (Baratta & Rigo, 2008; Everaere et al., 2003; Gladiali et al., 2006).

The stereochemically rigid β-amino alcohols **L7** or **L8** work very well as ligands for Ru-catalyzed ATH in basic *i*-PrOH, outperforming *N*-(*p*-toluenesulfonyl)-1,2-diamines in some cases, but in general these types of ligands appear to be incompatible with a FA/TEA reduction system (Fig. 7) (Palmer et al., 1997; Alonso et al., 1998).

An *in-situ*-prepared complex from $[RuCl_2(benzene)]_2$ and "roofed" *cis*-diamine ligand **L9**, which is both conformationally rigid and sterically congested, functions as an excellent catalyst for ATH with the FA/TEA of aryl ketones, including sterically bulky ketones (Matsunaga et al., 2005).

It was first disclosed by Noyori, that a N-H moiety is necessary for an efficient transfer of hydrogen from the metal hydride. However, the Ru complex with the oxazolyl-pyridyl-benzimidazole-based NNN ligand **L10** featuring no N-H functionality exhibited a high catalytic activity in the ATH of different acetophenones (Fig. 7) (Ye et al., 2011).

Another type of ligands lacking a basic NH group like **L11** are based on a combination of *N*-boc-protected α-amino acids and a sugar amino alcohol unit and have shown a high enantioselectivity (typically >99 *ee*) in the Ru-catalyzed ATH of aryl ketones, where the enantioselectivity is exclusively controlled by the sugar moiety (Coll et al., 2011). It was found that the addition of LiCl for the ATH in a *i*-PrOH/THF mixture catalyzed by Ru complexes bearing *N*-boc-protected α-amino acid hydroxyamide **L12** significantly enhanced the activity and selectivity, hence suggesting a non-classical bimetallic hydrogen-transfer mechanism (Fig. 7) (Wettergren et al., 2009).

The combination of $[RuCl_2(p\text{-cymene})]_2$ and the chiral BINOL-derived diphosphonite ligand **L13** constitutes yet another Ru catalyst system solely composed of P-ligands for the efficient ATH (*i*-PrOH/*t*-BuOK) of alkyl-aryl and alkyl-alkyl ketones, although the *ee*'s were lower for the latter (Fig. 7) (Reetz & Li, 2006). In contrast, $H_2$-hydrogenation is less successful when using this system.

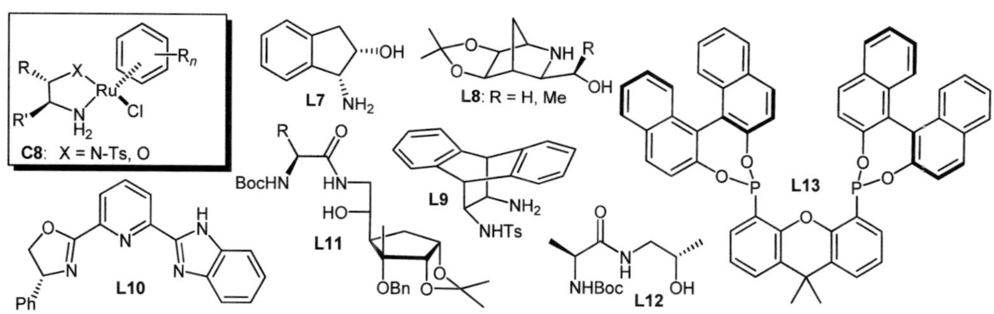


**Figure 7.** Selected ligands for ATH

There is a continuing search for stable catalysts that would not degrade easily during the hydrogenation process, thus making it possible to execute as many as possible catalytic cycles. In this respect, the covalent linkage from the diamine to the $\eta^6$-arene unit in the "tethered" catalysts **C9** provide extra stability and a significant increase in rate relative to the "unthetered" catalyst in some cases (Fig. 8) (Cheung et al., 2007). With these catalysts, ring-substituted acetophenones, α-chloroacetopehones, dialkyl ketones and ketopyridines were converted to the corresponding chiral alcohols in FA/TEA, mostly near to room temperature.

It has been shown that the Rh complex with the "achiral" but tropos benzophenone-derived ligand **L14** and a chiral diamine activator (*e.g.*, **L3**) affords higher enantioselectivities in the ATH of acetophenones and 1-acetylnaphthalene than those obtained by the enantiopure BINAP counterpart (Fig. 8) (Mikami et al., 2006). Cyclometalated Ru(II), Rh(III) and Ir(III) complexes **C10–C12** being easily prepared from commercial ligands, have shown a satisfactory catalytic activity and a high-to-very high enantioselectivity (*ee*'s up to 98%) in the ATH of different ketones (cyclic ketone, aryl-alkyl ketone, 2-acetylfuran, cyclopropyl-phenyl ketone) (Fig. 8) (Pannetier et al., 2011). The complexes **C11** and **C12** were not isolated but used *in situ*.

The unique phenomenon of an enhancement of the enantioselectivity by using the chiral bulky alcohol (*S*)-1-(9-anthracenyl)ethanol as an additive in the ATH of 4'-phenylacetophenone as well as in the $H_2$-hydrogenation of several acetophenone derivatives with the catalyst **C13** was recently demonstrated (Fig. 8) (Ito et al., 2012).

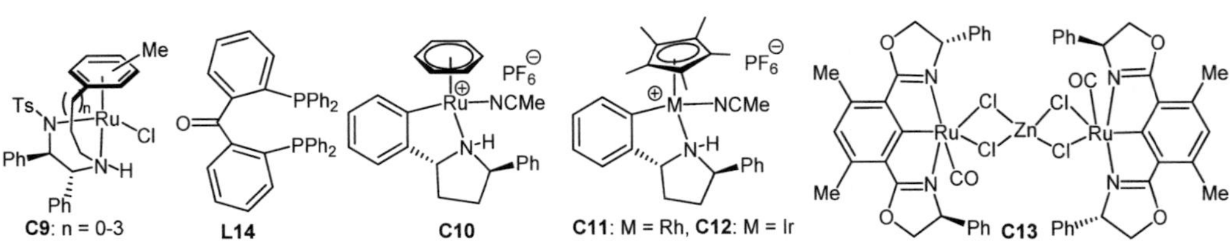


**Figure 8.** ATH catalyst systems

## 2.2. Hydrogenation and transfer hydrogenation employing other transition metals

Although Ru(II) complexes have enzyme-like properties reaching high TONs and TOFs, many times near to room temperature, and deliver the secondary alcohols in near-quantitative *ee*'s, the limited availability of precious metals, their high price and their toxicity reduce their attractiveness for future use. In this respect the development of catalysts with similar properties to replace platinum-group metals is very desirable from both the economic and environmental points of view. In fact, iron is cheap and ubiquitous, and its traces in final products are not as serious a problem as traces of ruthenium, for example (Morris, 2009).

The first hydrogenation of ketones catalyzed by a well-defined iron catalyst was effected with an iron hydride Shvo-type complex **C14** (Casey & Guan, 2007), while later on Morris and co-workers succeeded in the ATH of simple ketones catalyzed by iron complexes containing chiral PNNP tetradentate ligands, attaining *ee* values up to 99% in the best cases (Mikhailine et al., 2009; Sues et al., 2011). For example, acetophenone was reduced to (*S*)-1-phenylethanol in 82% *ee* and a TOF as high as $3.6 \cdot 10^3$ $h^{-1}$ with the pre-catalyst **C15**, while installing the sterically more hindered P-ligand in the complex **C16** even increased the activity ($2.6 \cdot 10^4$ $h^{-1}$) and enantioselectivity (90% *ee*) at the beginning of the reaction (Fig. 9).

An asymmetric Shvo-type iron complex **C17** was found to be a very poor catalyst for the transfer hydrogenation of acetophenone with FA/TEA, since after 48 hours only a 40% conversion and a 25% *ee* were observed (Hopewell et al., 2012).

Enantioselective, copper-catalyzed homogenous $H_2$-hydrogenation was introduced by Shimizu and co-workers, who used a catalyst system based on $[Cu(NO_3)\{P(3,5\text{-Xylyl})_3\}_2]$,

(*R*)-SEGPHOS (**L15**) or (*S,S*)-BDPP (**L16**), and *t*-BuONa for the reduction of (hetero)aryl ketones, affording good yields and *ee*'s up to 92% (Shimizu et al., 2007, 2009). A range of aryl, alkyl, cyclic, heterocyclic, and aliphatic ketones were hydrogenated under 50 bar of $H_2$ with a combination of inexpensive $Cu(OAc)_2$ and monodentate binaphthophosphepine ligand **L17** (Junge et al., 2011). On the other hand, $Cu(OTf)_2$ with the bisoxazoline ligand **L18** mimics alcohol dehydrogenase and catalyzes the ATH of α-ketoesters with Hantzsch esters as hydrogen donors (Fig. 10) (J. W. Yang & List, 2006).

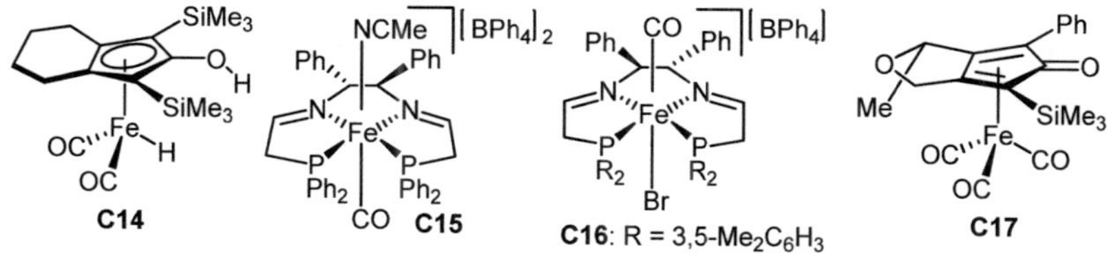


**Figure 9.** Selected iron catalysts

Owing to the stronger bonding of Os compared to Ru, robust and thermally stable complexes can be obtained, which is important for achieving highly productive catalysts. Os(II) CNN pincer complexes **C18** exhibited a high catalytic activity and productivity in both the AH (5 atm $H_2$/*t*-BuOK) and ATH (*i*-PrOH/*i*-PrONa) of ketones (Baratta et al., 2008). Enantioselectivities up to 98% *ee* are possible with a remarkably low catalyst loading (0.005-0.02 mol%). More active and productive [$OsCl_2$(diphosphane)(diamine)] complexes like **C19**, resembling those of Noyori, catalyzed the AH of alkyl-aryl, *tert*-butyl and cyclic ketones with S/C ratios of 10000–100000 and TOFs up to $10^4$ $h^{-1}$ (Baratta et al., 2010) (Fig. 10).

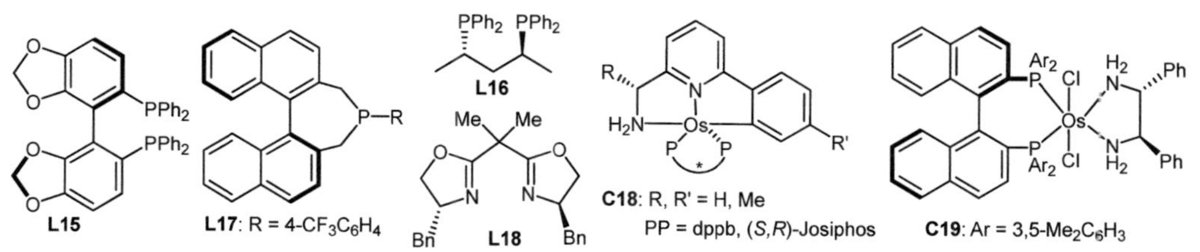


**Figure 10.** Ligands for Cu-mediated hydrogenation and Os-complexes

## 2.3. Hydrogenation and transfer hydrogenation in water and ionic liquids

As a consequence of the increasing demand for "greener" laboratory and industrial applications, the development of water-operating catalytic systems for the asymmetric hydrogenation of ketones has been of great interest (Wu & Xiao, 2007). The main disadvantage, however, is the low solubility of the homogenous metal catalysts and most of the organic substrates when going from organic to aqueous media, which may be reflected in a reduced activity and selectivity. To circumvent this, either hydrophilic, often charged, functionalities can be introduced to ligands to render the catalysts water-soluble, or different surfactants can be added in order to solvate the reaction partners, although in some cases water-insoluble catalysts can deliver a superior activity and selectivity.

Water-soluble Ru, Ir or Rh catalysts were prepared *in situ* using modified Noyori-type ligands **L19** and enabled the ATH in *i*-PrOH in the presence of water (Bubert et al., 2001, Thorpe et al., 2001), while Chung and co-workers communicated the first examples of the ATH of aromatic ketones with $HCO_2Na$ in neat water catalyzed by $[RuCl_2(p\text{-cymene})]_2$ together with the (S)-proline amide ligand **L20** attaining *ee*'s comparable with those in a homogenous solution (Rhyoo et al., 2001). The latter catalyst system appeared to be quite stable, since it could be recycled six times with little loss of performance. Similarly, an *in-situ*-prepared catalytic complex from the proline-functionalized ligand **L21** and $[RuCl_2(p\text{-cymene})]_2$ in a 1:1 ratio showed good activity for the aqueous ATH of acetophenone-type ketones as well as bicyclic ketones (Manville et al., 2011). Due to its difficult purification, the ligand **L22** was replaced by another water-soluble ligand **L23**, and its complex with $[C_5Me_5RhCl_2]_2$ was active for the ATH of α-bromomethylaromatic ketones, besides ring-substituted acetophenones, and bicyclic ketones (L. Li et al., 2007). The tethered Rh complex **C20** reported by Wills acts as a very productive catalyst for aqueous-reduction as it continues to turnover a reaction at low loadings, even at 0.01 mol%, typically associated with the best $H_2$-hydrogenation catalysts, without any decrease in the enantioselectivity (Matharu et al., 2006). The chiral aqua Ir(III)-complex **C21** bearing non-sulfonated diamine was shown to be very flexible in the ATH of α-cyano- and α-nitroacetophenones as the reaction can be conducted at pH 2 (formic acid) as well as at pH 5.5 ($HCO_2Na$) in a water-methanol system without affecting the selectivity (Vázquez-Villa et al., 2011) (Fig. 11).

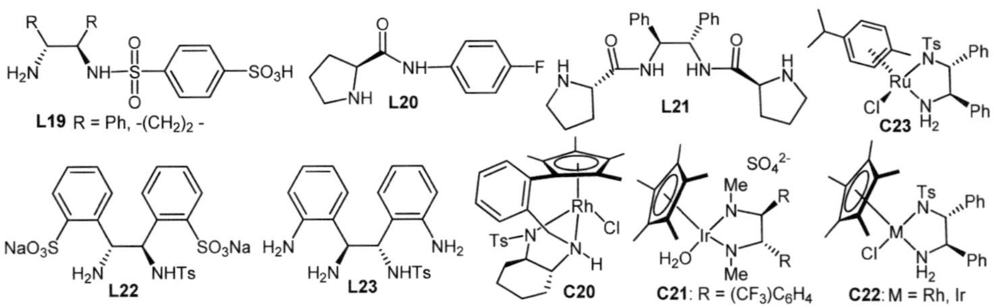


**Figure 11.** Selected ligands and complexes for aqueous hydrogenation

Surfactants are often added as co-solvents to obtain a sufficient solubility of the reactants, products and metal catalysts, thus retaining the activity and selectivity of the hydrogenation process. The ATH of ketones, particularly α-bromomethyl aromatic ketones, was successfully performed with $HCO_2Na$ by employing the unmodified and hydrophobic Ru-, Rh- and Ir-TsDPEN complexes **C22** and **C23** in the presence of single-tailed, cationic and anionic surfactants and to form micelles and vesicles (Fig. 11) (Wang et al., 2005). It is notable that catalysts embedded in these micro-reactors can be separated from the organic phase and reused for at least six times without any loss of activity and enantioselectivity.

In recent years ionic liquids (ILs) have attracted an increasing interest because of their non-volatility, non-flammability and low toxicity. Additionally, ILs are capable of immobilizing homogenous catalysts and facilitating the recycling of catalysts. Ideally, organic products can be easily separated by extraction with a less polar solvent and the IL phase containing catalyst can be reused. Such an immobilization of catalysts also promises to prevent the leaching of toxic metals into the organic products, which is especially desirable in the production of pharmaceutical intermediates.

Various aromatic ketones were reduced with FA/TEA in an ionic liquid **L25** at 40 °C, catalyzed by an *in-situ*-generated catalyst from $[RuCl_2(p\text{-cymene})]_2$ and the ionic chiral aminosulfonamide ligand **L24**, affording good-to-excellent conversions and *ee* values (Fig. 12) (Zhou et al., 2011). The catalytic system could be recovered and reused three times with a slight loss of enantioselectivity from 97% to 94% *ee* for the reduction of acetophenone. In contrast, the catalyst activity showed a remarkable drop with each cycle, and therefore the reaction times had to be prolonged for high conversions.

While for the AH of β-alkyl β-ketoesters high enantioselectivities can be attained by using the Ru-BINAP system, for the analogous β-aryl ketoesters much more inferior *ee* values were obtained (Noyori et al., 1987). However, the highly enantioselective hydrogenation of a wide range of β-aryl ketoesters **29** in the homogenous ionic liquid **L26**/methanol system was possible with Ru catalysts bearing 4,4′-substituted BINAP ligands **L27** (Fig. 12) (Hu et al., 2004a). The catalysts were recycled and reused four times, but there was a remarkable deterioration in the conversion rates and *ee* values, which were more pronounced with the ligand R = $SiMe_3$.

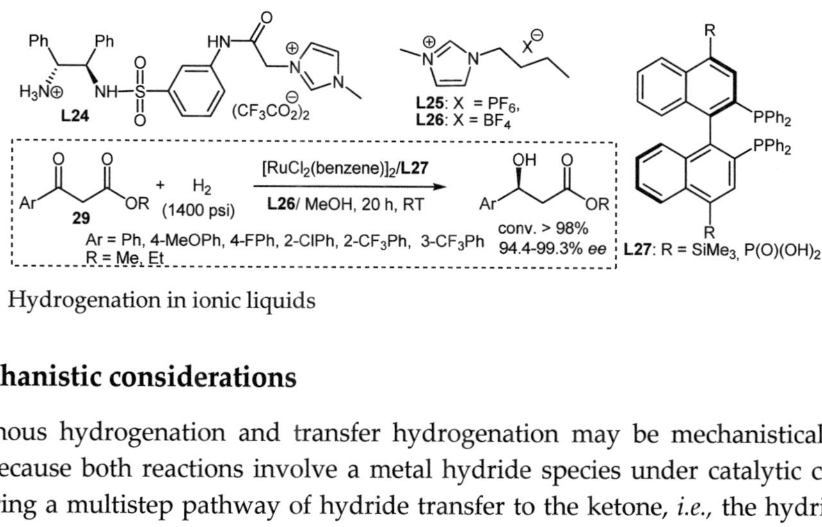


**Figure 12.** Hydrogenation in ionic liquids

## 2.4. Mechanistic considerations

Homogenous hydrogenation and transfer hydrogenation may be mechanistically closely related because both reactions involve a metal hydride species under catalytic conditions, thus sharing a multistep pathway of hydride transfer to the ketone, *i.e.*, the hydridic route, which can operate in the inner or outer coordination sphere of the catalyst metal center (Clapham et al., 2004). Applied only to the transfer hydrogenation, direct hydrogen transfer (Meerwein-Ponndorf-Verly reaction) from the metal alkoxyide to the ketone without the involvement of metal hydrides proceeding through a six-membered transition state has also been proposed, and is typical for non-transition metals (*e.g.*, Al) (deGraauw et al., 1994).

Noyori and co-workers proposed metal-ligand bifunctional catalysis for their Ru catalysts containing chiral phosphine-amine ligands and for (arene)Ru-diamine catalysts, which consequently resulted in a widely accepted mechanism to be responsible for the highly enantio-selective hydrogenation and transfer hydrogenation of prochiral ketones (Noyori et al., 2001, 2005). The actual catalysts, Ru-hydrides **31** or **34,** are usually created in a basic alcoholic solution (under $H_2$ or not) at the beginning of the catalytic reaction from the Ru precursors **30** or **33**. Note that only the *trans*-Ru$H_2$ **31** is a very active catalyst. A key feature of bifunctional catalysts is that the N-H unit of a diamine ligand forms a hydrogen bond with carbonyl oxygen, thus stabilizing the six-membered pericyclic transition state (**TS1** or **TS1′**) and hence facilitating the hydride transfer from Ru-H, which adds to the carbonyl carbon concurrently with a transfer of the acidic proton from N-H to the carbonyl oxygen. This concerted process results in the formation of an alcohol product and Ru-amido species (**32** or **35**). The hydride intermediate (**31** or **34**) is then regenerated either by the addition of molecular hydrogen or by the reverse hydrogen transfer from a dihydrogen source (*e.g.*, *i*-PrOH) to the formal 16-electron Ru-amido intermediate (**32** or **35**). The latter step is considered to be a rate-limiting step. The overall process is occurring outside the coordination sphere of the metal without the interacting of the ketone or alcohol with the metal center. This is known as an outer-sphere mechanism. It is depicted in Fig. 13 for the hydrogenation with molecular hydrogen catalyzed by the diphosphine-Ru-diamine system (a) and for transfer hydrogenation catalyzed by the (arene)Ru-diamine complex (b) in its simplified representation.

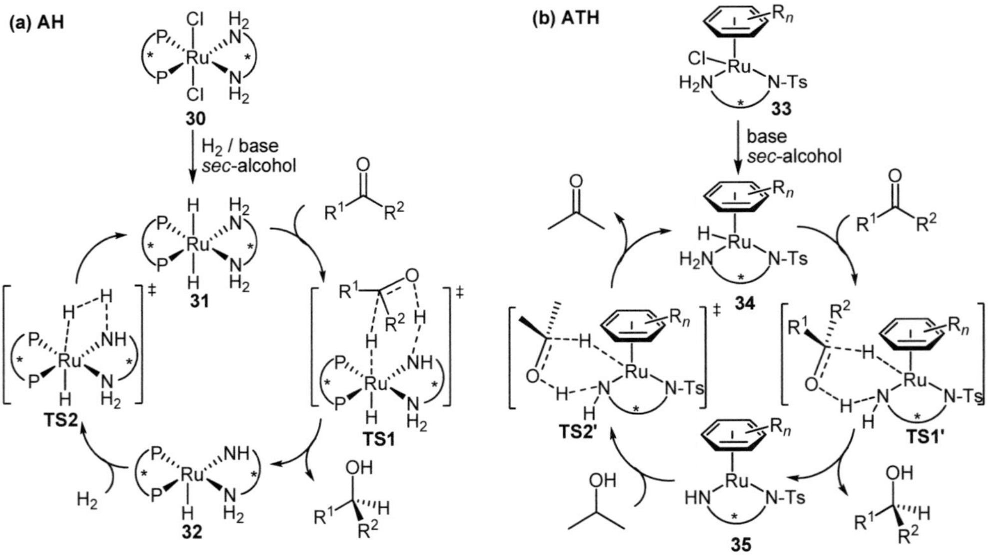


**Figure 13.** Outer-sphere hydridic route for bifunctional catalysts

Depending on transition-metal catalysts, an ionic mechanism has also been proposed where the proton and hydride transfer occur in separate steps (Bullock, 2004).

The active species in catalytic cycles, Ru-hydride (**31** or **34**) and Ru-amido complexes (**32** or **35**), have not only been detected but also isolated in some cases (Abdur-Rashid et al., 2001, 2002; Haack et al 1997).

The absolute configuration of the alcohol product in AH is determined in the six-membered transition state resulting from the reaction of a chiral diphosphine-diamine-$RuH_2$ complex with a prochiral ketone (Noyori et al., 2005). Because the enantiofaces of the ketone are differentiated on the molecular surface of the saturated $RuH_2$ complex, a suitable combination of the catalyst and substrate is necessary for high efficiency. The prochiral ketone (*e.g.*, acetophenone) approaches in such a ways as to minimize the non-bonded repulsion between the phosphine Ar group and the phenyl ring of the ketone, and to maximize the electronic NH/π attraction (Fig 14 (a)).

The stereoselectivity in the hydrogenation of prochiral aryl ketones catalyzed by (arene)Ru(II) complexes (mostly in ATH) has been ascribed not only to the chiral environment originating from the amine ligand, but also to the contribution of the arene ligand to the stabilization of the transition state through the CH/π interaction (Fig 14 (b)) (Yamakawa et al., 2001). This interaction as well as the NH/π interaction occurring in the transition states with diphosphine-Ru-(1,2-diamine) systems may explain why aryl ketones usually give better *ee* values than simple unfunctionalized alkyl-alkyl ketones.

Depending on the ligands attached to the metal center (M = transition metal) the inner-sphere mechanisms, in which monohydride or dihydride species are involved, can operate in $H_2$-hydrogenation and transfer hydrogenation (Clapham et al., 2004, Samec et al., 2006; Wylie et al., 2011). In contrast to the outer-sphere mechanism, here the ketone and alcohol interact with the metal center.

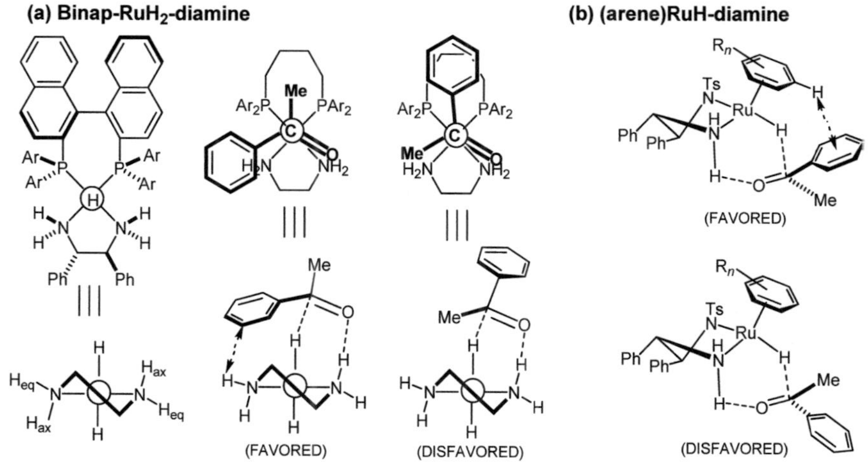


**Figure 14.** Enantiodifferentiation in the bifunctional-catalyzed hydrogenation of acetophenone

## 3. Heterogeneous hydrogenation

For the heterogeneous, asymmetric, catalytic reduction of the C=O functionality, there are two types of heterogeneous catalysts. One is chirally modified supported metals, and the other is the immobilized homogeneous catalyst on a variety of organic and inorganic polymeric materials. There are also two major reasons for preparing and studying heterogeneous catalysts: firstly, and most importantly, the better and advanced separation and handling properties, and, secondly, the potential to create catalytic positions with an improved catalytic performance. The ultimate heterogeneous catalyst can easily be renewed, reused without of loss of activity and selectivity, which are at least as good or even better than those of the homogeneous analogue.

## 3.1. Immobilized chiral complexes

The immobilization of a homogeneous metal coordination complex is a useful strategy in the preparation of new hydrogenation catalysts. Much effort has been devoted to the preparation of such heterogenized complexes over the past decade due to their ease of separation from the reaction mixture and the desired minimal product contamination caused by metal leaching, as well as to their efficient recyclability without any significant loss of activity. Preferably, Rh, Ir, and Ru complexes have been employed in the hydrogenations of carbonyl functionality (Corma et al., 2006). Chemically different supports have been used for the immobilization of various homogeneous complexes, including polymeric organic and inorganic supports (Saluzzo et al., 2002; Bergbreiter, 2002; Fan et al., 2002). Due to their chemical nature, organic polymeric supports have some drawbacks concerning reduced stability that affects the reusability of the catalysts, mainly due to their swelling and deformation (Bräse et al., 2003; Dickerson et al., 2002). Supports of an inorganic nature are more suitable owing to their physical properties, chemical inertness and stability (with respect to swelling and deformation) in organic solvents. The above-mentioned properties of the inorganic supports will facilitate the applications of the materials in reactions carried at higher temperatures and their use in continuous-flow reactions. In the past decade a lot of research effort has been devoted to the development of adequate procedures to attach homogenous catalysts onto inorganic supports (Merckle & Blümel, 2005; Crosman et al., 2005; Corma et al., 2005; Jones et al., 2005; Melero et al., 2007). Immobilization via covalent bonds is undoubtedly the most convenient, but on the other hand, it is the most challenging method for immobilization to perform on such supports (Jones et al., 2005; Steiner et al., 2004; Pugin et al., 2002; Sandree et al., 2001). For example, micelle templated silicas (MTS) featuring a unique porous distribution and high thermal and mechanical stabilities can be easily functionalized by the direct grafting of the functional organo-silane groups on their surfaces (McMorn & Hutchings, 2004; Heckel & Seebach, 2002; Bigi et al., 2002, Clark & Macquarrie, 1998; Tada & Iwasawa, 2006). On the other hand, polar solvents such as water or alcohols and high temperatures during the catalytic procedure can promote the hydrolysis of the grafted moieties.

The heterogenized catalysts can potentially combine the advantages of both homogenous and heterogeneous systems. In 2003, Hu and coworkers developed a novel chiral porous solid catalyst based on zirconium phosphonates for the practically useful enantio-selective hydrogenation of unfunctionalized aromatic ketones (Fig. 15) (Hu et al., 2003a).

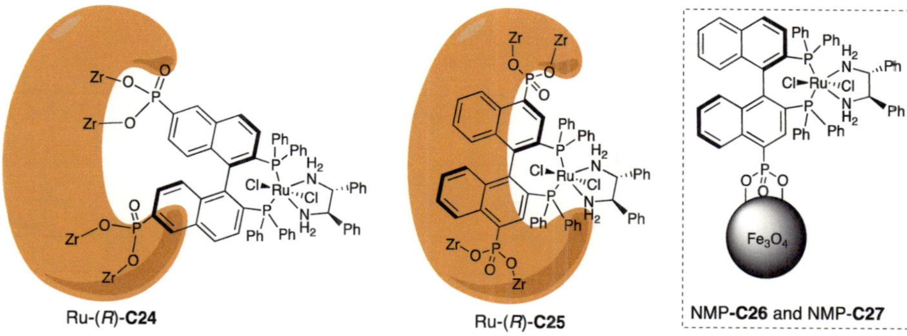


**Figure 15.** Schematic presentation of chiral porous Zr-phosphonate-Ru-(*R*)-**C24** in Ru-(*R*)-**C25** heterogeneous catalysts

With the built-in Ru-BINAP-DPEN moieties, porous solids of Ru-(*R*)-**C24** and Ru-(*R*)-**C25** exhibited high activity and enantioselectivity in the hydrogenation of aromatic ketones (Table 1). Acetophenone was hydrogenated, producing 1-phenylethanol with a complete conversion and 96.3% *ee* in *i*-PrOH with a 0.1 mol% loading of Ru-(*R*)-**C24**. This level of enantioselectivity is higher than that observed for the parent Ru-BINAP-DPEN homogeneous catalyst, which gives ~80% *ee* for the hydrogenation of acetophenone (Ohkuma et al., 1995a; Doucet et al., 1998). As indicated in table 1, the Ru-(*R*)-**C24** immobilized catalyst has also been tested to catalyze the hydrogenation of other aromatic ketones resulting in the formation of the corresponding alcohols with the same high enantioselectivity (90.6-99.0% *ee*) and complete consumption of the starting ketone. Although the Ru-(*R*)-**C25** catalyst is also highly active for the hydrogenation of aromatic ketones, the enantoselectivity is modest and similar to that of the parent Ru-BINAP-DPEN homogeneous catalyst. The authors believe that the modest enantioselectivities observed for the Ru-(*R*)-**C25** catalyst originate in the substituent effects on the BINAP ligand. Furthermore, the catalysts were successfully reused without any deterioration of the enantioselectivity in eight cycles. The activities did not decrease for the first six cycles, but began to drop during the seventh run (95% conversion), reaching 85% of conversion in the eighth cycle. Furthermore, the Ru(II) catalysts of type Ru-(*R*)-**C24** and Ru-(*R*)-**C25** having dimethylformamide as a ligand instead of 1,2-diphenylethylenediamine were developed and used for the heterogeneous AH of β-keto esters with *ee* values from 91.7 up to 95.0 % with the same enantio enrichment as is the case in the parent homogenous BINAP-Ru catalyst. The substrates, β-aryl-substituted β-keto esters, are hydrogenated with the same modest *ee* values (69.6 % *ee*) as observed when using the homogenous BINAP-Ru analogue (Noyori & Takaya, 1990). The introduced catalysts can be readily recycled and reused (Hu et al., 2003b). Structurally similar Ru(II) catalysts with phosphonic-acid-substituted BINAP were prepared and afterwards immobilized on magnetite nanoparticles prepared by the thermal decomposition method (MNP-**C26**, Fig. 15) or by the coprecipitation method (NMP-**C27**, Fig. 15) (Hu et al., 2005). The catalysts were tested for the heterogeneous asymmetric hydrogenation of aromatic ketones showing a remarkably high activity and enantioselectivity (Table 1).

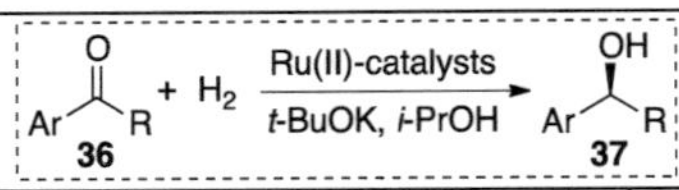

| Substrate **36** | Ru-(*R*)-**C24**; *ee* (%) | Ru-(*R*)-**C25**; *ee* (%) | MNP-**C26**; *ee* (%) | MNP-**C27**; *ee* (%) |
|---|---|---|---|---|
| Ar = Ph, R = Me | 96.3 | 79.0 | 87.6 | 81.7 |
| Ar = 2-naphtyl, R = Me | 97.1 | 82.1 | 87.6 | 82.0 |
| Ar = 4-*t*Bu-Ph, R = Me | 99.2 | 91.5 | 95.1 | 91.1 |
| Ar = 4-MeO-Ph, R = Me | 96.0 | 79.9 | 87.6 | 77.7 |
| Ar = 4-Cl-Ph, R = Me | 94.9 | 59.3 | 76.6 | 70.6 |
| Ar = 4-Me-Ph, R = Me | 97.0 | 79.5 | 87.9 | 80.5 |
| Ar = Ph, R = Et | 93.1 | 83.9 | 88.9 | 86.3 |
| Ar = Ph, R = *cyclo*-Pr | 90.6 | – | – | – |
| Ar = 1-naphtyl, R = Me | 99.2 | 95.8 | – | – |

**Table 1.** Heterogeneous hydrogenation of the aromatic ketones using Ru(II) catalyst

Heterogeneous chiral Ru(II)-TsDPEN-derived catalysts based on Noyori's (1*S*,2*S*)- or (1*R*,2*R*)-*N*-p-tosylsulfonyl)-1,2-diphenylethylenediamine (TsDPEN) were successfully immobilized onto amorphous silica gel and silica mesopores of MCM-41 and SBA-15 using an easily accessible approach (P.-N. Liu et al., 2004a, 2004b, 2005). The immobilized catalysts demonstrated high catalytic activities and enantioselectivities (up to >99% *ee*, **38a-38l**) (Fig. 16) for the heterogeneous ATH of different ketones. In particular, the catalyst could be recovered and reused in multiple consecutive runs (up to 10 uses) with a completely maintained enantioselectivity.

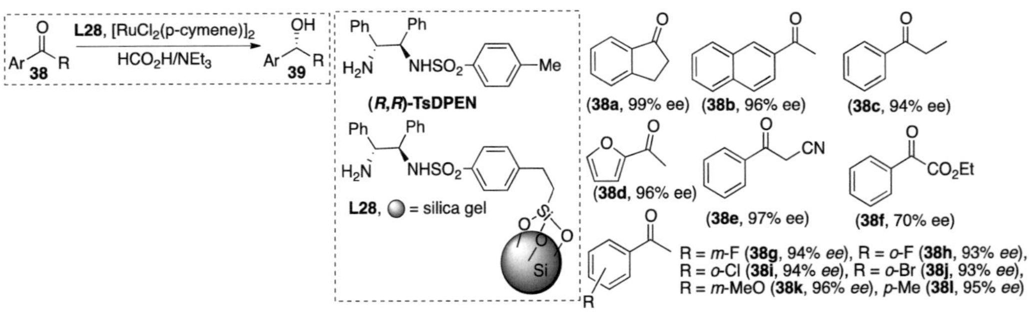


**Figure 16.** Heterogeneous Ru$^{II}$ mesoporous silica-supported catalysts

Additionally, Li and coworkers (J. Li et al., 2009) developed a Ru(II)-TsDPEN-derived catalyst that was immobilized in a magnetic siliceous mesocellular foam material. The heterogeneous catalyst showed comparable activities and enantioselectivities (*ee* 89-97%) with the parent catalyst Ru(II)-TsDPEN in the ATH of imines and simple aromatic ketones. Polymer-supported-TsDPEN ligands combined with [RuCl$_2$(*p*-cymene)]$_2$ have been shown to exhibit high activities (93-98%) and enantioselectivities (86-97% *ee*) for the heterogeneous ATH of aromatic ketones, which are suitable intermediates for the synthesis of (*S*)-fluoxetine with a 75% yield and a 97% *ee* (Y. Li et al., 2005).

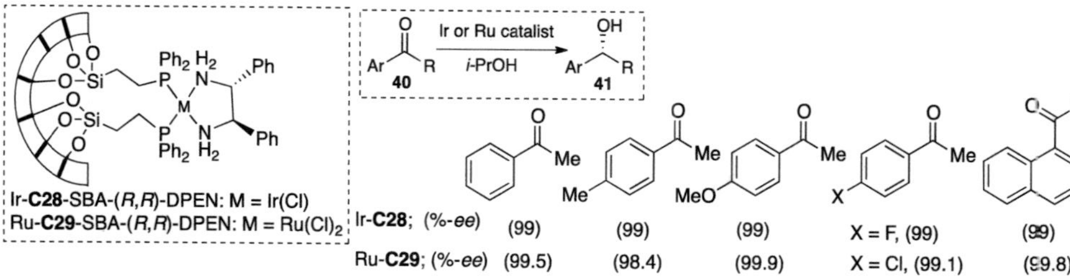


**Figure 17.** Ir and Ru mesoporous silica-supported catalysts

Chiral Ru and Ir, mesoporous, silica-supported catalysts were introduced by Liu and coworkers (G. Liu et al., 2008a, 2008b). The Ir-**C28**-SBA-(*R*,*R*)-DPEN catalyst was investigated using a series of aromatic ketones as substrates (Fig. 17). In general, high conversions (95-99 %) and an excellent enantioselectivity, producing the corresponding *R* enantiomers, were observed by applying 40 atm of $H_2$ at 50 °C and 0.4 mol% of catalyst loading. The catalyst was recovered and reused several times without considerably affecting the *ee* values. The analogous Ru catalyst, Ru-**C29**-SBA-(*R*,*R*)-DPEN, also displays a high catalytic activity and enantioselectivity under similar reaction conditions (Fig. 17) for the ATH of aromatic ketones.

Two magnetic chiral Ir and Rh catalysts were prepared *via* directly post-grafting 1,2-diphenylethylenediamine and 1,2-cyclohexanediamine-derived organic silica onto silica-coated iron oxide nanoparticles (G. Liu et al., 2011). The synthesis was followed by a complexation with Ir(III) or Rh(III) complexes. High catalytic activities (up to 99% conversion) and enantioselectivities (up to 92% *ee*) were obtained in the ATH reaction, reducing the aromatic ketones in an aqueous medium (Fig. 18). Both catalysts could be recovered by magnetic separation and be reused ten times without significantly affecting their catalytic activities and enantioselectivities.

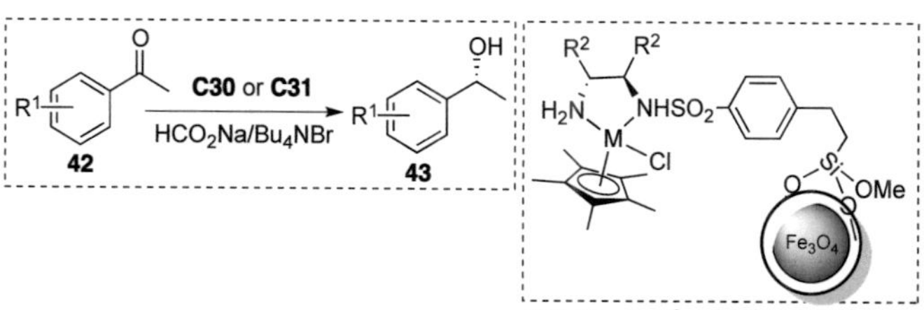


| C30 (R¹, Conv. %, [*ee* %]) | C31 (R¹, Conv. %, [*ee* %]) |
|---|---|
| (H, 98.6, [89.6]) | (H, 99.9, [87.6]) |
| (4-Cl, 99.6, [88.7]) | (4-Cl, 98.2, [84.0]) |
| (4-F, 98.4, [92.5]) | (4-F, 99.5, [86.7]) |
| (4-Me, 96.7, [87.2]) | (4-Me, 99.4, [85.9]) |
| (4-MeO, 96.4, [81.6]) | (4-MeO, 99.9, [85.0]) |

**C30**: M = Ir; R² = (*S*,*S*)-1,2-diphenylethylenediamine
**C31**: M = Rh; R² = (*R*,*R*)-1,2-cyclohexanediamine

**Figure 18.** Magnetic Ir and Rh chiral catalysts

The mesoporous SBA-15 anchored 9-amino *epi*-cinchonine-[Ir(COD)Cl]$_2$ complex shows good activity and moderate enantio-selectivity (45-78% *ee*) in the ATH reaction of substituted acetophenones (Shen et al., 2010).

The chiral RuCl$_2$-diphosphine-diamine complex with siloxy functionality was successfully immobilized on mesoporous silica nanospheres with three-dimensional channels (Fig. 19) (Mihalcik & Lin, 2008). Upon activation with *t*-BuOK, the catalysts **C32-C36** can be used for the AH of aromatic ketones; however, **C32-C36** exhibit lower enantioselectivities than their parent homogeneous catalysts. The highest *ee* value of 82% was observed for the hydrogenation of 2-acetonaphthone using **C33** as a catalyst. A similar drop in enantio-selectivities has been noticed for many asymmetric catalysts immobilized on bulk mesoporous silica (Song & Lee, 2002). Catalysts of the type **C32-C34** were also examined in a dynamic kinetic resolution of α-branched aryl aldehydes. The highest *ee* value of 97% was obtained using 0.1 mol% of the **C33** catalyst and 700 psi of $H_2$ pressure on 3-methyl-2-phenylbutanal as a substrate.

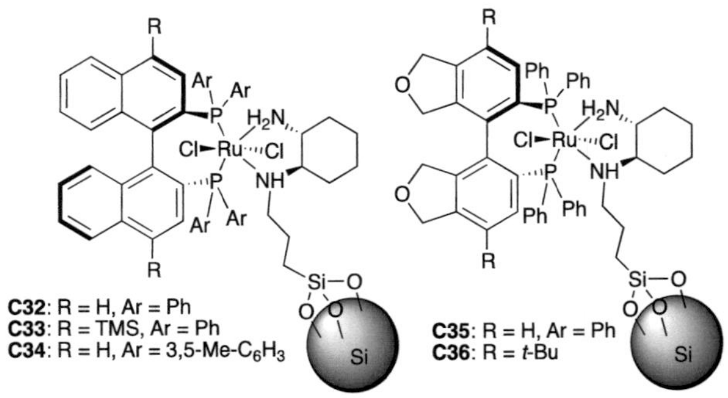


**Figure 19.** Chiral $RuCl_2$-diphosphine-diamine complexes immobilized on mesoporous silica nanospheres

Differently substituted Rh complexes were anchored on an $Al_2O_3$ support and applied for the enantioselective C=O hydrogenation with reasonable activity and enantioselectivities with *ee*s up to 80% (Zsigmond et al., 2008). Due to the fact that an immobilized catalyst did not show a superior enantio-selectivity compared to its homogenous counterparts, the major advantage of the catalyst's immobilization is the possibility to recycle the catalysts.

The immobilization of the rhodium complexes [Rh((*R*)-BINAP)(COD)]$CF_3SO_3$, [Rh((*S*)-BINAP)(COD)]$ClO_4$·thf, and [Rh((*S,S*)-chiraphos)(NOR)]$ClO_4$, and the ruthenium complexes [Ru((*R*)-BINAP)($PPh_3$)$Cl_2$] and [Ru((*R*)-BINAP)$Cl_3$] in a thin film of silica-supported ionic liquid enhanced the enantioselectivity of the parent catalyst. As the model reaction, the stereo-selective hydrogenation of acetophenone as a non-chelating prochiral ketone was studied. The enantioselectivities in a moderate range (up to 74%) were observed (Fow et al., 2008). Furthermore, a mesoporous material-supported ionic liquid phase was used as a carrier medium to immobilize the chiral ruthenium complex composed of a chiral 1,2-diamine and an achiral monophosphine (Lou et al., 2010). All the prepared catalysts were active in the hydrogenation of simple aromatic ketones enabling an enantioselectivity from 45 up to 78% *ee*.

Furthermore, a series of polystyrene-supported TsDPEN ligands were prepared in one step and converted to the corresponding Ru(II) complexes by a treatment with [$RuCl_2$(*p*-cymene)]$_2$ in dichloromethane at 40 °C for an hour (Marcos et al., 2011). The so-prepared polystyrene-based Ru(II)-catalytic resins showed a low conversion (37%, 48 h, 40 °C) of acetophenone to the corresponding (*R*)-alcohol (85% *ee*) in the ATH ($HCO_2H/Et_3N$ = 5/2) in water. The more promising results were obtained in dichloromethane, where (*R*)-1-phenylethanol was produced in 99% conversion and with 97% *ee*. The catalytic resin could be recycled three times without any significant loss of conversion and enantioselectivity, but further recycling shows a major drop in performance of the catalytic resin. A modified tethered Rh(III)-*p*-toluenesulfonyl-1,2-diphenylethylenediamine (Rh-TsDPEN) complex immobilized on polymeric supports (amino-functionalized polyethylene microparticles) was used in kinetic and up-scaling experiments on the ATH of acetophenone in water. A second-order model describes the enantioselective conversion of acetophenone to phenylethanol and mainly the solution pH was found to play a pivotal role for the activity and reusability of the catalyst (Dimroth et al., 2011). Polyethylene glycol (PEG) supported chiral ligands have also been developed and examined in the Ru-catalyzed ATH of prochiral aromatic ketones in water using $HCO_2Na$ as the hydrogen source. Xiao et al. introduced a PTsDPEN ligand that has two PEG chains (PEG-2000) on the *meta*-position of the TsDPEN's phenyl groups. Comparing the results of the Ru-TsDPEN catalyst in water, the PEG Ru(II) catalyst in the ATH of various aromatic ketones by $HCO_2Na$ in water gave faster rates and a good reusability (X. Li et al., 2004a, 2004b). As an alternative for attaching a PEG chain onto the TsDPEN-tipe ligands, a medium-length PEG chain (PEG-750) at the *para*-position of the aryl sulfonate group was introduced (J. Liu et al., 2008). The corresponding Ru-PEG-BsDPEN catalyst displays a high activity, reusability and enantioselectivity (up to 99% *ee*) in the ATH in water.

A series of dendrimers and hybrid dendrimers based on Noyori-Ikariya's TsDPEN ligand were prepared and the application of their Ru(II) complexes in the ATH of acetophenones was studied. A high catalytic activity and completely maintained enantio-selectivity (acetophenone, 93.4-98.2% *ee*; 4-bromoacetophenone, 90.1-92.7% *ee*; 1-(naphthalen-2-yl)ethanone, 92.8-95.1% *ee*; 1,2-diphenylethanone, 93.9% *ee*) were observed. Higher-generation core-functionalized dendritic catalysts could be recovered through solvent precipitation and reused several times without any major loss of activity and enantio-selectivity (Y.-C. Chen et al., 2001, 2002, 2005; W. Liu et al., 2004). Hydrophobic Fréchet-type dendritic chiral 1,2-diaminocyclohexane-Rh(III) complexes have also been tested for ATH in water (Jiang et al., 2006). Excellent conversions (70-99%) and enantioselectivity, acetophenone (96% *ee*), 4-chloroacetophenone (93% *ee*), 4-methoxyacetophenone (94% *ee*), 1-tetralone (97% *ee*), 2-acetylpyridine (91% *ee*), 2-acetyltiophene (96% *ee*), ethyl 2-oxo-2-phenylacetate (72% *ee*), and (*E*)-4-phenylbut-3-en-2-one (52% *ee*) were obtained.

## 3.2. "Self-supported" and solid-supported heterogeneous catalysts

Among various approaches for homogeneous catalyst immobilization, the "self-supported" strategy exhibits some relevant characteristics, such us easy preparation, good stability, high density of catalytically active sites, and high stereocontrol performance, as well as simple recovery (Dai, 2004; Ding et al., 2007). Self-supported Noyori-type catalysts **C37-C40** for the AH of ketones by the programmed assembly of bridged diphosphine and diamine ligands with Ru(II) ions were developed (Fig. 20) (Liang et al., 2005; Liang et al., 2006). The enantioselectivity of the hydrogenation of the aromatic ketones under the catalysis of the self-supported catalyst **C40** was in some cases significantly higher than the *ee* values obtained in the homogeneous catalysis. However, it is expected that the enantioselectivities achieved in the hydrogenation of ketones with the catalysts **C37** and **C38** composed of chirally flexible biphenylphosphine ligands are lower than those of the **C39** and **C40** constructed with chiral BINAP-containing ligands. This might be explained using Mikami's mechanistic considerations obtained by an $^1H$ NMR study of the monomeric complex of DM-BIPHEP/$RuCl_2$/(*S,S*)-DPEN (Mikami et al., 1999). Furthermore, this type of catalyst can be readily recovered and reused with the retention of enantioselectivity and reactivity.

A very interesting example is the asymmetric synthesis of the chiral alcohol function that makes use of the strength of ion pairing in ionic liquids (Schulz et al., 2007). The hydrogenation of substrate **46** using $H_2$ (60 bar) at 60 °C in the presence of the heterogeneous, achiral catalyst Ru/C in an ethanolic solution, gave the corresponding hydroxyl-functionalized ionic liquid in a quantitative yield and up to 80% *ee* (Fig. 21). The degree of enantioselectivity is dependent on the concentration of the substrate **46** in ethanol during the transformation. The higher the concentration of **46**, the higher the *ee* value of the hydrogenated cation that was observed. This behavior can be explained by considering the ion-pair separating effect of the ethanol solvent.

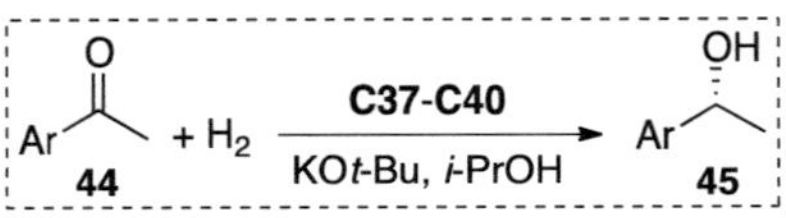


| Catalyst | $R^1$ | ee (%), [*R*] |
|---|---|---|
| **C37** | Ph | 46.2 |
| **C38** | Ph | 83.2 |
| **C39** | Ph | 78.2 |
| **C40** | Ph | 97.4 |
| **C38** | 2-Me-$C_6H_4$ | 81.4 |
| **C38** | 3-Me-$C_6H_4$ | 86.9 |
| **C38** | 4-Me-$C_6H_4$ | 74.9 |
| **C40** | 4-Me-$C_6H_4$ | 97.5 |
| **C38** | 4-Cl-$C_6H_4$ | 71.9 |
| **C40** | 4-F-$C_6H_4$ | 96.2 |
| **C40** | 4-Cl-$C_6H_4$ | 96.9 |
| **C40** | 4-Br-$C_6H_4$ | 97.2 |
| **C40** | 4-MeO-$C_6H_4$ | 96.2 |

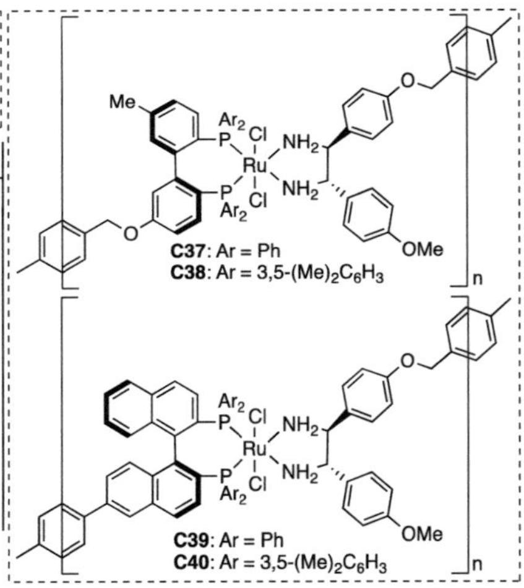


**Figure 20.** Self-supported Noyori-type catalysts **C37-C40** for the AH of ketones

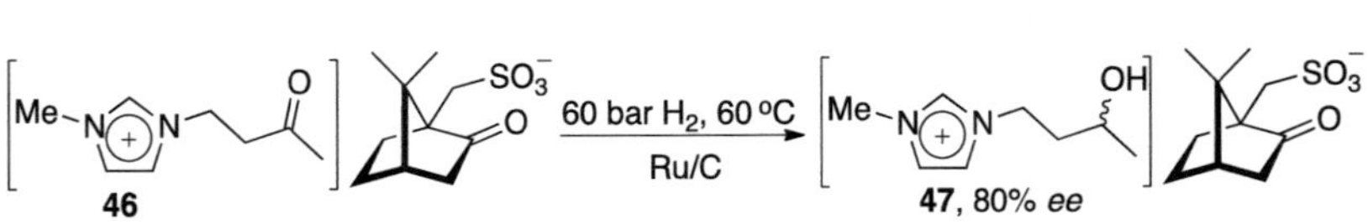


**Figure 21.** Enantio-selective hydrogenation of a keto-functionalized ionic liquid

Importing chirality to a catalytic active metal surface by the adsorption of a chiral organic molecule (often referred to as a chiral modifier) seems to be one of the promising strategies to obtain new chiral heterogeneous catalytic systems. In the hydrogenation of C=O function, chirality-modified supported metal catalysts represent a promising approach with synthetic potential. Orito et al. introduced the strategy of a cinchona-alkaloid-modified platinum catalyst system in 1979 (Orito et al., 1979). Following the early work of Blaser et al. (Studer et al., 1999, 2000, 2003; Blaser et al., 2000), Baiker et al. (Heinz et al., 1995, von Arx et al., 2002), and others, the methodology developed in the sense of the substrate scope, and on the other hand, extensive efforts were carried out to get more insight into understanding the mechanistic aspects of the transformation. The method was found to have excellent performance in the hydrogenation of activated ketones (Fig. 22).

The modifiers derived from CD and quinine (QN) lead to an excess of (*R*)-ethyl lactate, whereas the CN and QD derivatives preferentially lead to the *S* enantiomer. It has been shown that substituted aliphatic and aromatic α-keto ethers are suitable substrates for the enantioselective hydrogenation catalyzed by cinchona-modified Pt catalysts and both kinetic and dynamic kinetic resolution is possible (Studer et al., 2002). For conversions less than 50%, *ee*'s of up to 98% were observed when starting with racemic substrates (kinetic resolution). Strong acceleration of the reaction was noticed in the presence of KOH, but without of the enantiomeric excess. In order to get dynamic kinetic resolution the $OH^-$ ions had to be immobilized on a solid ion-exchange resin enabling *ee*'s of more than 80%.

A systematic structure-selectivity study of the hydrogenation of activated ketones catalyzed by a modified Pt-catalyst revealed a high substrate specificity of the catalytic system. Relatively small structural changes in the substrate or modifier can strongly affect the enantio-selectivity and often in the opposite manner, especially when comparing reactions in toluene and AcOH (Exner et al., 2003). Fluorinated β-diketones can be enantioselectively hydrogenated on cinchona-alkaloids-modified $Pt/Al_2O_3$ catalysts. Methyl, ethyl, and isopropyl 4,4,4-trifluoroacetoacetates were hydrogenated in the presence of MeOCD-modified $Pt/Al_2O_3$ catalysts, producing the corresponding alcohols in 93-96% *ee* (van Arx et al., 2002).

Synthetically obtained (*R,R*)-pantoyl-naphtylethylamine ((*R,R*)-PNEA) provides 93% *ee* in the hydrogenation of 1,1,1-trifluoro-2,4-pentanedione and 85% *ee* in the case of 1,1,1-trifluoro-5,5-dimethyl-2,4-hexanedione (Diezi et al., 2005a, 2005b, Hess et al., 2004). A thorough investigation concerning the origin of the chemo- and enantioselectivity in the hydrogenation of diketones on platinum revealed that the structures of ammonium ion-enolate-type ion pairs formed between the modifier and 1,3-diketones are different in solution and on the surface of the metal. The chemoselectivity is attributed to the selective interaction of the protonated amine group of the modifier to the absorbed activated keto-carbonyl function and prevention of the interaction of the non-activated carbonyl group with the metal surface (Diezi et al., 2006). Results on the enantioselective hydrogenation of α-fluoroketones, a group of activated ketones on chiral platinum-alumina surface have shown that the Orito reaction is also suitable for the preparation of the corresponding chiral α-fluoroalcohols. The enantioselectivity of 92% was achieved in the hydrogenation of 2,2,2-trifluoroacetophenone under optimized reaction conditions using a CD-modified Pt catalyst (von Arx et al., 2001a). However, the enantioselectivities obtained on other α-fluorinated ketones were only moderate (Varga et al., 2004; Felföldi et al., 2004; Szöri et al., 2009).

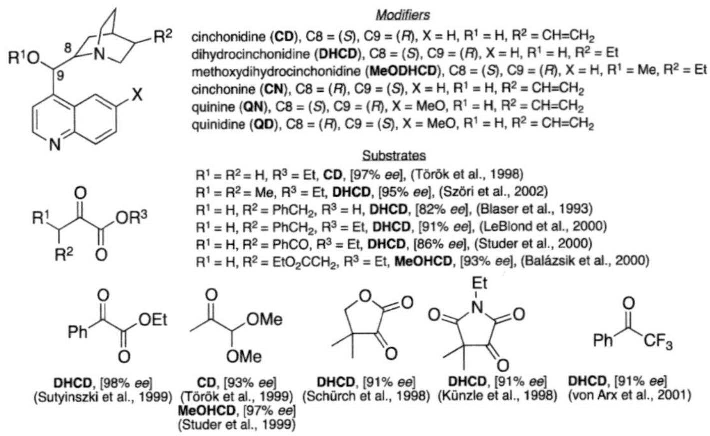


**Figure 22.** Enantio-selective hydrogenation of activated ketones.

A supported ($SiO_2$) iridium catalyst, which is stabilized by $PPh_3$ and modified by a chiral diamine, derived from cinchona alkaloids, exhibits a high activity and high enantioselectivity for the hydrogenation for the simple aromatic ketones (Fig. 23). The addition of different bases (*t*-BuOK, LiOH, NaOH, or KOH) improves both the activity and the enantioselectivity of the reaction (Jiang et al., 2008). A similar ruthenium catalyst (Ru/$\gamma$-$Al_2O_3$) was also developed and a broad range of aromatic ketones over this catalyst can be hydrogenated (Jiang et al. 2010).

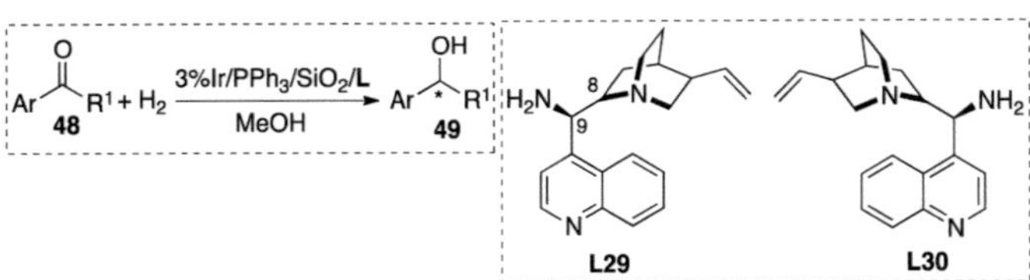


| R¹ | Ar | L29; *ee* (%) | L30; *ee* (%) |
|---|---|---|---|
| Me | Ph | 88 | 74 |
| Me | 2-Cl-$C_6H_4$ | 96 | 86 |
| Me | 2-MeO-$C_6H_4$ | 94 | 86 |
| Me | 3-Cl-$C_6H_4$ | 75 | — |
| Et | Ph | 86 | 75 |
| Me | 2-F-$C_6H_4$ | 86 | 77 |
| Me | 2-Br-$C_6H_4$ | 92 | 90 |
| Me | 2-$CF_3$-$C_6H_4$ | 81 | 83 |
| Me | 4-MeO-$C_6H_4$ | 87 | 76 |
| *i*-Pr | Ph | 52 | — |

**Figure 23.** Enantioselective hydrogenation of activated ketones.

A series of silica ($SiO_2$) supported iridium catalysts stabilized by cinchona alkaloids were also prepared and applied in the heterogeneous asymmetric hydrogenation of acetophenone. Cinchona alkaloids display a substantial capability to stabilize and disperse the Ir particles. A synergistic effect between the (*1S,2S*)-DPEN (modifier) and the CD (stabilizer) significantly accelerates the activity as well as the enantioselectivity (up to 79% *ee*) on acetophenone (Yang et al. 2009).

Besides improving the cinchonidine-platinum catalyst system, extensive efforts have been made in developing a reliable mechanistic interpretation. To understand the adsorption behavior of the modifier and reactant, their conformation, and their intra-molecular interactions at solid-liquid interface, an *in-situ* attenuated, total-reflection, infrared study has been performed. The adsorption of cinchonidine on the Pt/$Al_2O_3$ in the presence of a solvent and $H_2$ is strongly concentration dependent. The quinolone moiety of the modifier is responsible for the absorption on the Pt surface (Ferri & Bürgi, 2001).

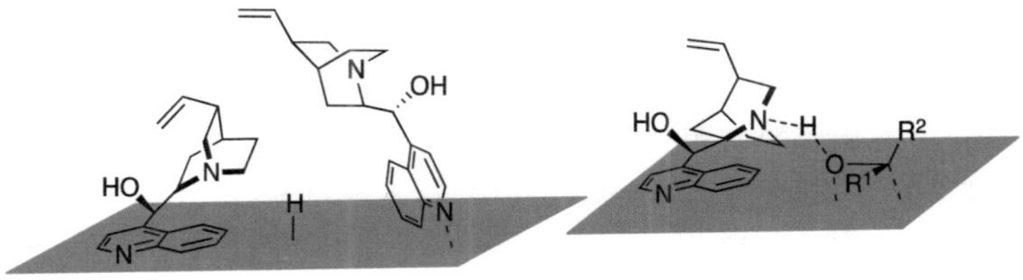


**Figure 24.** Schematic representation of the adsorption mode of CD on the metal surface and interactions between the half-hydrogenated state of the activated ketone and the basic quinuclidine-N atom of the chiral modifier.

An inversion of the enantioselectivity occurs in the asymmetric hydrogenation of the activated ketones by changing the solvent composition, including water and acid additives (von Arx et al., 2001b; Bartók et al., 2002). Hydrogenation of the ethyl pyruvate over Pt/$Al_2O_3$ (Huck et al., 2003a) and 4-methoxy-6-methyl-2-pyrone over Pd/$TiO_2$ (Huck et al., 2003b), an equimolar mixture of cinchona alkaloids CD and QD resulted in *ee*'s similar to those obtained with CD alone, while QD gave a high *ee* of the opposite enantiomers. This was explained by different adsorption strengths and absorption modes of the modifier (Fig. 24). Furthermore, cinchona ether homologues can give opposite enantiomers through maintaining the same absolute configuration of the parent alkaloid. In the hydrogenation of ketopantolactone the CD alkaloid produced (*R*)-pantolactone in 79% *ee*, whereas *O*-phenylcinchonidine (PhOCD) gave *S*-enantiomere in 52% *ee*. It seems that the OH group of CD is not involved in the substrate-modifier interaction during the hydrogenation process, which is also confirmed by the fact that *O*-methyl-CD and *O*-ethyl-CD gave the same enantiomer in excess than CD. The inversion of enantioselectivity is explained by the change in the chiral pocket experienced by the incoming reactant and the change is related to the conformational behavior of the absorbed alkaloid and the steric effects of the ether group. PhOCD can generate conformations whose adsorption energy is decreased with respect to the parent CD. An equally important change is also the alteration of the chiral pocket obtained upon absorption of the modifier (Fig. 24). These changes are enough to induce the inversion of enantio-selectivity (Bonalumi et al., 2005; Vargas et al., 2006, 2007). The aspects of the interaction of different modifiers, MeOCD, *t*-MeSiOCD (Bonalumi et al., 2007), (*R*)-iCN (Schmidt et al., 2008), and tryptophan and tryptophan-based di end tripeptides (Mondelli et al., 2009) with a metal surface have also been studied experimentally (using TEM, XPS, and ATR-IR spectroscopy) and theoretically (DFT calculations). Furthermore, it has been shown that the rate of hydrogenation and enantioselectivity outcome depends on the shape and terrace sites (Pt{100}or {111}) of the nanoparticles. Both the rate and the *ee* increased in the hydrogenation of ethyl pyruvate and ketopantolactone when Pt {111} nanoparticles were modified using CD or QN as the chiral modifiers (Schmidt et al., 2009).

## 4. Conclusions

This chapter discusses the transition-metal-catalyzed, asymmetric, homogenous and heterogeneous hydrogenation of prochiral ketones, not so much focusing on the reactions providing valuable chiral alcohols, but rather it gives prominent and interesting examples of the ketone substrates and catalyst systems that are found in the recent literature. Despite the tremendous effort being made in the catalytic, asymmetric hydrogenation of prochiral ketones, approaching the enzymatic performance in some cases, there is still much potential for the continued development of these reactions. Concerning the environmental and economic issues, the introduction of non-toxic, cheap, and at the same time efficient and universal catalyst systems, being able to operate under mild conditions in a highly selective manner and for a broad range of substrates, remains a challenge for future research. Additionally, more rational catalyst designs are possible with better mechanistic understandings of the catalytic cycles in catalytic AH and ATH reactions.

## Author details

Bogdan Štefane and Franc Požgan
*Faculty of Chemistry and Chemical Technology, University of Ljubljana,*
*EN-FIST Centre of Excellence, Slovenia*

## Acknowledgement

The Ministry of Higher Education, Science and Technology of the Republic of Slovenia, the Slovenian Research Agency (P1-0230-0103), EN→FIST Centre of Excellence, and Krka, Pharmaceutical company, d.d. are gratefully acknowledged for their financial support.

## 5. References

Abdur-Rashid, K.; Faatz, M.; Lough, A. J. & Morris, R. H. (2001). Catalytic Cycle for the Asymmetric Hydrogenation of Prochiral Ketones to Chiral Alcohols: Direct Hydride and Proton Transfer from Chiral Catalysts *trans*-Ru(H)$_2$(diphosphine)(diamine) to Ketones and Direct Addition of Dihydrogen to the Resulting Hydridoamido Complexes. *Journal of the American Chemical Society*, Vol.123, No.30, (August 2001), pp. 7473-7474, ISSN 0002-7863

Abdur-Rashid, K.; Clapham, S. E.; Hadzovic, A.; Harvey, J. N.; Lough, A. J. & Morris, R. H. (2002). Mechanism of the Hydrogenation of Ketones Catalyzed by *trans*-Dihydrido(diamine)ruthenium(II) Complexes. *Journal of the American Chemical Society*, Vol.124, No.50, (December 2002), pp. 15104-15118, ISSN 0002-7863

Alonso, D. A.; Guijarro, D.; Pinho, P.; Temme, O. & Andersson, P. G. (1998). (1*S*,3*R*,4*R*)-2-Azanorbornylmethanol, an Efficient Ligand for Ruthenium-Catalyzed Asymmetric Transfer Hydrogenation of Ketones. *The Journal of Organic Chemistry*, Vol.63, No.8, (April 1998), pp. 2749-2751, ISSN 0022-3263

Arai, N. & Ohkuma, T. (2010). Reduction of Carbonyl Groups: Hydrogenation, In: *Science of Synthesis: Stereoselective Synthesis* 2, G.A Molander, (Ed.), 9-57, Thieme, ISBN 978-313-1541-21-5, Stuttgart, Germany

Balázsik, K.; Szöri, K.; Felföldi, K.; Török, B. & Bartók, M. (2000). Asymmetric synthesis of alkyl 5-oxotetrahydrofuran-2-carboxylates by enantioselective hydrogenation of dialkyl 2-oxoglutarates over cinchona modified $Pt/Al_2O_3$ catalysts. *Chemical Communications*, No.17, pp. 555-556, ISSN 1359 7345

Baratta, W.; Ballico, M.; Chelucci, G.; Siega, K. & Rigo P. (2008). Osmium(II) CNN Pincer Complexes as Efficient Catalysts for Both Asymmetric Transfer and $H_2$ Hydrogenation of Ketones. *Angewandte Chemie International Edition*, Vol.47, No.23, (May 2008), pp. 4362-4365, ISSN 1433-7851

Baratta, W. & Rigo, P. (2008). 1-(Pyridin-2-yl)methanamine-Based Ruthenium Catalysts for Fast Transfer Hydrogenation of Carbonyl Compounds in 2-Propanol. *European Journal of Inorganic Chemistry*, No.26, (September 2008), pp. 4041-4053, ISSN 1099-0682

Baratta, W.; Barbato, C.; Magnolia, S.; Siega, K. & Rigo, P. (2010). Chiral and Nonchiral [$OsX_2$(diphosphane)(diamine)] (X: Cl, $OCH_2CF_3$) Complexes for Fast Hydrogenation of Carbonyl Compounds, *Chemistry-A European Journ*al, Vol.16, No.10, (March 2010), pp. 3201-3206, ISSN 1521-3765

Bartók, M.; Sutyinszki, M.; Felföldi, K. & Szöllősi, Gy. (2002). Unexpected change of the sense of the enantioselective hydrogenation of ethyl pyruvate catalyzed by a Pt–alumina-cinchona alkaloid system. *Chemical Communications*, No.10, pp. 1130- 1131, ISSN 1359 7345

Bergbreiter, D. E. (2002). Using Soluble Polymers To Recover Catalysts and Ligands. *Chemical Reviews*, Vol.102, No.10, (August 2002), pp. 3345-3384, ISSN 0009-2665

Bigi, F.; Moroni, L.; Maggi, R. & Sartori, G. (2002) Heterogeneous Enantioselective Epoxidation of Olefins Catalysed by Unsymmetrical (salen)Mn(III) Complexes Supported on Amorphous or MCM-41 Silica Through a New Triazine-Based Linker. *Chemical Communications*, No.7, pp. 716-717, ISSN 1359 7345

Blaser, H. U. & Jalett, H. P. (1993). Enantioselective Hydrogenation of $\alpha$-Ketoacids Using Platinum Catalysts Modified With Cinchona Alkaloids. *Studies in Surface Science and Catalysis*, Vol.78, pp. 139-146, ISBN: 978-0-444-89063-4

Blaser, H. U.; Jalett, H. P.; Lottenbach, W. & Studer, M. (2000). Heterogeneous Enantioselective Hydrogenation of Ethyl Pyruvate Catalyzed by Cinchona- Modified Pt Catalysts: Effect of Modifier Structure. *Journal of the American Chemical Society*, Vol.122, No.51, (December 2000), pp. 12675-12682, ISSN 0002-7863

Bonalumi, N.; Vargas, A.; Ferri, D.; Bürgi, T.; Mallat, T. & Baiker, A. (2005). Competition at Chiral Metal Surfaces: Fundamental Aspects of the Inversion of Enantioselectivity in Hydrogenations on Platinum. *Journal of the American Chemical Society*, Vol.127, No.23, (June 2005), pp. 8467–8477, ISSN 0002-7863

Bonalumi, N.; Vargas, A.; Ferri, D.; & Baiker, A. (2007). Chirally Modified Platinum Generated by Adsorption of Cinchonidine Ether Derivatives: Towards Uncovering the Chiral Sites. *Chemistry - A European Journal*, Vol.13, No.33, (November 2007), pp. 9236–9244, ISSN 1434-193X

Bräse, S.; Lauterwasser, F. & Ziegert, R. E. (2003). Recent Advances in Asymmetric C-C and C- Heteroatom Bond Forming Reactions using Polymer-Bound Catalysts. *Advanced Synthesis & Catalysis*, Vol.345, No.8, (August 2003), pp. 869-929, ISSN 1615-4169

Bubert, C.; Blacker, J.; Brown, S. M.; Crosby, J.; Fitzjohn, S.; Muxworthy, J. P.; Thorpe, T. & Williams, J. M. J. (2001). Synthesis of water-soluble aminosulfonamide ligands and their application in enantioselective transfer hydrogenation. *Tetrahedron Letters* Vol.42, No.24, (June 2001), pp. 4037-4039, ISSN 0040-4039

Bullock, R. M. (2004). Catalytic Ionic Hydrogenations. *Chemistry-A European Journal,* Vol.10, No.10, (May 2004), pp. 2366-2374, ISSN 1521-3765

Bürgi, T. & Baiker, A. (2004). Heterogeneous Enantioselective Hydrogenation over Cinchona Alkaloid Modified Platinum: Mechanistic Insights into a Complex Reaction. *Accounts of Chemical Research,* 2004, Vol.*37, No.11* (November 2004), pp. 909–917, ISSN 0001-4842

Casey, C. P. & Guan, H. (2007). An Efficient and Chemoselective Iron Catalyst for the Hydrogenation of Ketones. *Journal of the American Chemical Society*, Vol.129, No.18, (May 2007), pp. 5816-5817, ISSN 0002-7863

Chen, Y.-C.; Wu, T.-F.; Deng, J.-G.; Liu, H.; Jiang, J.-Z.; Choi, M. C. K. & Chan, A. S. C. (2001). Dendriticcatalysts for asymmetric transfer hydrogenation.*Chemical Communications*, No.16, pp. 1488-1489, ISSN 1359 7345

Chen, Y.-C.; Wu, T.-F; Deng, J.-G.; Liu, H.; Cui, X.; Zhu, J.; Jiang, Y.-Z.; Choi, M. C. K. & Chan, A. S. C. (2002). Multiple Dendritic Catalysts for Asymmetric Transfer Hydrogenation. *The Journal of Organic Chemistry*, Vol.67, No.15 (Juliy 2002), pp. 5301-5306, ISSN 0022-3263

Chen, C.; Reamer, R. A.; Chilenski, J. R. & McWilliams, C. J. (2003). Highly Enantioselective Hydrogenation of Aromatic-Heteroaromatic Ketones. *Organic Letters,* Vol.5, No.26, (December 2003), pp. 5039-5042, ISSN 1523-7060

Chen, Y.-C.; Wu, T.-F.; Jiang, L.; Deng, J.-G.; Liu, H.; Zhu, J. & Jiang, Y.-Z. (2005). Synthesis of Dendritic Catalysts and Application in Asymmetric Transfer Hydrogenation. The *Journal of Organic Chemistry,* Vol.70, No.3 (February 2005), pp. 1006-1010, ISSN 0022-3263

Cheung, F. K.; Lin, C.; Minissi, F.; Crivillé A. L.; Graham, M. A.; Fox, D. J. & Wills, M. (2007). An Investigation into the Tether Length and Substitution Pattern of Arene-Substituted Complexes for Asymmetric Transfer Hydrogenation of Ketones. *Organic Letters*, Vol.9, No.22, (October 2007), pp. 4659-4662, ISSN 1523-7060

Clapham, S. E.; Hadzovic, A. & Morris, R. H. (2004). Mechanism of the $H_2$-hydrogenation and transfer hydrogenation of polar bonds catalyzed by ruthenium hydride complexes. *Coordination Chemistry Reviews*, Vol.248, No.21-24, (December 2004), pp. 2201-2237, ISSN 0010-8545

Clark, J. H. & Macquarrie, D. J. (1998). Catalysis of Liquid Phase Organic Reactions Using Chemically Modified Mesoporous Inorganic Solids. *Chemical Communications*, No.8, pp. 853-860, ISSN 1359 7345

Coll, M., Pàmies, O.; Adolfsson, H. & Diéguez, M. (2011). Carbohydrate-based pseudo-dipeptides: new ligands for the highly enantioselective Ru-catalyzed transfer hydrogenation reaction. *Chemical Communications*, Vol.47, No.44, pp. 12188-12190, ISSN 1359 7345

Corma, A.; Das, D.; García, H. & Leyva, A. (2005). A Periodic Mesoporous Organosilica Containing a Carbapalladacycle Complex as Heterogeneous Catalyst for Suzuki Cross-Coupling. *Journal of Catalysis*, Vol.229, No.2, (January 2005), pp. 322-331, ISSN 0021-9517

Corma, A. & Garcia, H. (2006). Silica-Bond Homogenous Catalysts as Recoverable and Reusable Catalysts in organic Synthesis. *Advanced Synthesis & Catalysis*, Vol.348, No.12-13, (August 2006), pp. 1391-1412, ISSN 1615-4169

Crosman, A. & Hoelderich, W. F. (2005). Enantioselective Hydrogenation Over Immobilized Rhodium Diphosphine Complexes on Aluminated SBA-15. *Journal of Catalysis*, Vol.232, No.1, (May 2005), pp. 43-50, ISSN 0021-9517

Dai, L.-X. (2004). Chiral Metal–Organic Assemblies–A New Approach to Immobilizing Homogeneous Asymmetric Catalysts. *Angewandte Chemie International Edition*, Vol.43, No.43, (November 2004), pp. 5726–5729, ISSN 1433-7851

de Graauw, C. F.; Peters, J. A.; van Bekkum, H. & Huskens, J. (1994). Meerwein-Ponndorf-Verley Reductions and Oppenauer Oxidations: An Integrated Approach. *Synthesis*, No.10, (September 1994), pp. 1007-1017, ISSN 0039-7881

Dickerson, T. J.; Reed, N. N. & Janda, K. D. (2002). Soluble Polymers as Scaffolds for Recoverable Catalysts and Reagents. *Chemical Reviews*, Vol.102, No.10, (September 2002), pp. 3325-3344, ISSN 0009-2665

Diez, C. & Nagel, U. (2010). Chiral iridium(I) bis(NHC) complexes as catalysts for asymmetric transfer hydrogenation. *Applied Organometallic Chemistry*, Vol.24, No.7, (July 2010), pp. 509-516, ISSN: 1099-0739

Diezi, S.; Hess, M.; Orglmeister, E.; Mallat, T. & Baiker, A. (2005). Chemo and enantioselective hydrogenation of fluorinated ketones on platinum modified with (R)-1-(1-naphthyl)ethylamine derivatives. *Journal of Molecular Catalysis A: Chemical*, Vol.239, No.1-2, (September 2005), pp. 49-56, ISSN 1381-1169

Diezi, S.; Hess, M.; Orglmeister, E.; Mallat, T. & Baiker, A. (2005). An efficient synthetic chiral modifier for platinum. *Catalysis Letters*, Vol.102, No.3-4, (August 2005), pp. 121-125, ISSN 1011-372X

Diezi, S.; Ferri, D., Vargas, A.; Mallat, T. & Baiker, A. (2006). The Origin of Chemo- and Enantioselectivity in the Hydrogenation of Diketones on Platinum. *Journal of the American Chemical Society*, Vol.128, No.12, (March 2006), pp. 4048-4057, ISSN 0002-7863

Dimroth, J.; Schedler, U.; Keilitz, J.; Haag, R. & Schomäcker, R. (2011). New Polymer-Supported Catalysts for the Asymmetric Transfer Hydrogenation of Acetophenone in Water–Kinetic and Mechanistic Investigations. *Advanced Synthesis & Catalysis*, Vol.353, No.8 (May 2011), pp. 1335-1344, ISSN 1615-4169

Ding, K.; Wang, Z. & Shi, L. (2007). Self-supported chiral catalysts for heterogeneous enantioselective reactions. *Pure and Applied Chemistry*, Vol.79, No.9, pp. 1531-1540, ISSN 0033-4545

Doucet, H.; Ohkuma, T.; Murata, K.; Yokozawa, T.; Kozawa, M.; Katayama, E.; England, A. F.; Ikariya, T. & Noyori, R. (1998). *trans*-[$RuCl_2$(phosphane)$_2$(1,2-diamine)] and Chiral *trans*-[$RuCl_2$(diphosphane)(1,2-diamine)]: Shelf-Stable Precatalysts for the Rapid, Productive, and Stereoselective Hydrogenation of Ketones. *Angewandte Chemie International Edition*, Vol.37, No.12, (July 1998), pp. 1703-1707, ISSN 1433- 7851

Exner, C.; Pfaltz, A.; Studer, M. & Blaser, H.-U. (2003). Heterogeneous Enantioselective Hydrogenation of Activated Ketones Catalyzed by Modified Pt-Catalysts: A Systematic Structure-Selectivity Study. *Advanced Synthesis & Catalysis*, Vol.345, (November 2003), No.11, pp. 1253–1260, ISSN 1615-4169

Everaere, K.; Mortreux, A. & Carpentier, J.-F. (2003). Ruthenium(II)-Catalyzed Asymmetric Transfer Hydrogenation of Carbonyl Compounds with 2-Propanol and Ephedrine-Type Ligands. *Advanced Synthesis & Catalysis*, Vol.345, No.1-2, (January 2003), pp. 67-77, ISSN: 1615-4169

Fan, Q.-H.; Li, Y.-M. & Chan, A. S. C. (2002). Recoverable Catalysts for Asymmetric Organic Synthesis. *Chemical Reviews*, Vol.102, No.10, (September 2002), pp. 3385- 3465, ISSN 0009-2665

Felföldi, K.; Varga, T.; Forgó P. & Bartók, M. (2004). Enantioselective Hydrogenation of Trifluoromethylcyclohexyl Ketone on Cinchona Alkaloid Modified Pt-Alumina Catalyst. *Catalysis Letters*, Vol.97, No.1-2, (August 2004), pp. 65-70, ISSN: 1011-372X

Ferri, D. & Bürgi, T. (2001). An in Situ Attenuated Total Reflection Infrared Study of a Chiral Catalytic Solid−Liquid Interface: Cinchonidine Adsorption on Pt. *Journal of the American Chemical Society*, Vol.123, No.48, (December 2001), pp. 12074–12084, ISSN 0002-7863

Fow, K. L.; Jaenicke, S.; Müller, T. E. & Sievers, C. (2008). Enhanced enantioselectivity of chiral hydrogenation catalysts after immobilisation in thin films of ionic liquid. *Journal of Molecular Catalysis A: Chemical*, Vol.279, No.2, (January 2008), pp. 239-247, ISSN 1381-1169

Fujii, A.; Hashiguchi, S.; Uematsu, N.; Ikariya, T. & Noyori, R. (1996). Ruthenium(II)-Catalyzed Asymmetric Transfer Hydrogenation of Ketones Using a Formic Acid−Triethylamine Mixture. *Journal of the American Chemical Society*, Vol.118, No.10, (March 1996), pp. 2521-2522, ISSN 0002-7863

Gladiali, S. & Alberico, E. (2006). Asymmetric transfer hydrogenation: chiral ligands and applications. *Chemical Society Reviews*, Vol.35, No.3, (March 2006), pp. 226-236, ISSN 0306-0012

Haack, K.-J.; Hashiguchi, S.; Fujii, A.; Ikariya, T. & Noyori, R. (1997). The Catalyst Precursor, Catalyst, and Intermediate in the $Ru^{II}$-Promoted Asymmetric Hydrogen Transfer between Alcohols and Ketones. *Angewandte Chemie International Edition in English*, Vol.36, No.3, (February 1997), pp. 285-288, ISSN 1433-7851

Hashiguchi, S.; Fujii, A.; Takehara, J.; Ikariya, T. & Noyori, R. (1995). Asymmetric Transfer Hydrogenation of Aromatic Ketones Catalyzed by Chiral Ruthenium(II) Complexes. *Journal of the American Chemical Society*, Vol.117, No.28, (July 1995), pp. 7562-7563, ISSN 0002-7863

Heckel, A. & Seebach, D.; (2002). Preparation and Characterization of TADDOLs Immobilized on Hydrophobic Controlled-Pore-Glass Silica Gel and Their Use in Enantioselective Heterogeneous Catalysis. *Chemistry - A European Journal*, Vol.8, No.3, (January 2002), pp. 559-572, ISSN 1521-3765

Heinz, T.; Wang, G.; Pfaltz, A.; Minder, B.; Schürch, M.; Mallat, T. & Baiker, A. (1995). 1-(1-Naphtyl)ethylamine and Derivatives Thereof as Chiral Modifiers in the Enantioselective

Hydrogenation of Ethyl Pyruvate over Pt-Alumina. *Journal of the Chemical Society, Chemical Communications*, No. pp. 1421-1422, ISSN 1359 7345

Hess, R.; Diezi, S.; Mallat, T. & Baiker, A. (2004). Chemo- and enantioselective hydrogenation of the activated keto group of fluorinated β-diketones. Tetrahedron: Asymmetry, Vol.15, No.2, (January 2004), pp. 251-257, ISSN 0957-4166

Hopewell, J. P.; Martins, J. E. D.; Johnson, T. C.; Godfrey, J. & Wills, M. (2012). Developing asymmetric iron and ruthenium-based cyclone complexes; complex factors influence the asymmetric induction in the transfer hydrogenation of ketones. *Organic & Biomolecular Chemistry*, Vol.10, No.1, (January 2012), pp. 134-145, ISSN

Hu, A.; Ngo, H. L. & Lin, W. (2003). Chiral Porous Hybrid for Practical Heterogeneous Asymmetric Hydrogenation of aromatic ketones. *Journal of the American Chemical Society*, Vol.125, No.38, (September 2003) pp. 11490-11491, ISSN 0002-7863

Hu, A.; Ngo, H. L. & Lin, W. (2003). Chiral, Porous, Hybrid Solids for Highly Enantioselective Heterogeneous Asymmetric Hydrogenation of β-Keto Esters. *Angewandte Chemie International Edition*, Vol.43, No.9, (December 2003), pp. 6000-6003, ISSN 1433-7851

Hu, A.; Ngo, H. L. & Lin, W. (2004). Remarkable 4,4′-Substituent Effects on Binap: Highly Enantioselective Ru Catalysts for Asymmetric Hydrogenation of β-Aryl Ketoesters and Their Immobilization in Room-Temperature Ionic Liquids. *Angewandte Chemie International Edition*, Vol.43, No.19, (May 2004), pp. 2501-2504, ISSN 1433-7851

Hu, A.; Yee, T. G. & Lin, W. (2005). Magnetically Recoverable Chiral Catalysts Immobilized on Magnetite Nanoparticles for Asymmetric Hydrogenation of Aromatic Ketones. *Journal of the American Chemical Society*, Vol.127, No.36, (September 2005), pp. 12486-12487, ISSN 0002-7863

Huck, W.-R.; Mallat, T. & Baiker, A. (2003). Non-Linear Effect of Modifier Composition on Enantioselectivity in Asymmetric Hydrogenation over Platinum Metals. *Advanced Synthesis & Catalysis*, Vol.345, No.1-2, (January 2003), pp. 255-260, ISSN 1615-4169

Huck, W.-R.; Bürgi, T.; Mallat, T. & Baike, A. (2003). Asymmetric hydrogenation on platinum:nonlinear effect of coadsorbed cinchona alkaloids on enantiodifferentiation. *Journal of Catalysis*, Vol.216, No.1–2, (May–June 2003), pp. 276–287, ISSN 0021-9517

Huang, H.; Okuno, T.; Tsuda, K.; Yoshimura, M. & Kitamura, M. (2006). Enantioselective Hydrogenation of Aromatic Ketones Catalyzed by Ru Complexes of Goodwin–Lions-type $sp^2N/sp^3N$ Hybrid Ligands R-BINAN-R′-Py. *Journal of the American Chemical Society*, Vol.128, No.27, (July 2006), pp. 8716-8717, ISSN 0002-7863

Ikariya, T. & Blacker, A. J. (2007). Asymmetric Transfer Hydrogenation of Ketones with Bifunctional Transition Metal-Based Molecular Catalysts. *Accounts of Chemical Research*, Vol.40, No.12, (December 2007), pp. 1300-1308, ISSN 0001-4842

Ito, J.; Teshima, T. & Nishiyama, H. (2012). Enhancement of enantioselectivity by alcohol additives in asymmetric hydrogenation with bis(oxazolinyl)phenyl ruthenium catalysts. *Chemical Communications*, Vol.48, No.8, pp. 1105-1107, ISSN 1359 7345

Jiang, Q.; Jiang, Y.; Xiao, D.; Cao, P. & Zhang, X. (1998). Highly Enantioselective Hydrogenation of Simple Ketones Catalyzed by a Rh-PennPhos Complex. *Angewandte Chemie International Edition*, Vol.37, No.8, (May 1998), pp. 1100-1103, ISSN 1433-7851

Jiang, L.; Wu, T.-F.; Chen, Y.-C.; Zhu, J. & Deng, J.-G. (2006). Asymmetric transfer hydrogenation catalysed by hydrophobic dendritic DACH–rhodium complex in water. *Organic & Biomolecular Chemistry*, Vol.4, No.17, pp. 3319-3324, ISSN 1477-0520

Jiang, H.; Yang, C.; Li, C.; Fu, H.; Chen, H.; Li, R. & Li, X. (2008). Heterogeneous Enantioselective Hydrogenation of Aromatic Ketones Catalyzed by Cinchona- and Phosphine-Modified Iridium Catalysts. *Angewandte Chemie International Edition*, Vol.47, No.48, (November 2008), pp. 9240-9244, ISSN 1433-7851

Jiang, H.; Chen, H. & Li, R. (2010). Cinchona-modified Ru catalysts for enantioselective heterogeneous hydrogenation of aromatic ketones. *Catalysis Communications*, Vol.11, No.7, (March 2010), pp. 584-587, ISSN 1566-7367

Jones, C. W.; McKittrick, M. W.; Nguyen, J. V. & Yu, K. (2005). Design of silica-tethered metal complexes for polymerization catalysis. *Topics in Catalysis*, Vol.34, No.1-4, pp. 67-76, ISSN 1022-5528

Junge, K.; Wendt, B.; Addis, D.; Zhou, S.; Das, S.; Fleischer, S. & Beller, M. (2011). Copper-Catalyzed Enantioselective Hydrogenation of Ketones. *Chemistry - A European Journal*, Vol.17, No.1, (January 2011), pp. 101-105, ISSN 1521-3765

Klingler, F. D. (2007). Asymmetric Hydrogenation of Prochiral Amino Ketones to Amino Alcohols for Pharmaceutical Use. *Accounts of Chemical Research*, Vol.40, No.12, (December 2007), pp. 1367-1376, ISSN 0001-4842

Kuroki, Y.; Sakamaki, Y.; & Iseki, K. (2001). Enantioselective Rhodium(I)-Catalyzed Hydrogenation of Trifluoromethyl Ketones. *Organic Letters*, Vol.3, No.3, (January 2001), pp. 457-459, ISSN 1523-7060

Künzle, N.; Szabo, A.; Schürch, M. & Wang, G. (1998). Enantioselective hydrogenation of a cyclic imidoketone over chirally modified Pt/Al2O3. *Chemical Communications*, No.13, pp. 1377-1378, ISSN 1359 7345

LeBlond, C.; Wang, J.; Liu, J.; Andrews, A. T. & Sun, Y.-K. (1999). Highly Enantioselective Heterogeneously Catalyzed Hydrogenation of $\alpha$-Ketoesters under Mild Conditions. *Journal of the American Chemical Society*, Vol.121, No.20, (May 1999), pp. 4920-4921, ISSN 0002-7863

Le Roux, E.; Malacea, R.; Manoury, E.; Poli, R.; Gonsalvi, L. & Peruzzini, M. (2007). Highly Efficient Asymmetric Hydrogenation of Alkyl Aryl Ketones Catalyzed by Iridium Complexes with Chiral Planar Ferrocenyl Phosphino-Thioether. *Advanced Synthesis & Catalysis*, Vol.349, No.3, (February 2007), pp. 309–313, ISSN: 1615-4169

Li, X.; Chen, W.; Hems, W.; King, F. & Xiao, J. (2004). Asymmetric transfer hydrogenation of ketones with a polymer-supported chiral diamine. *Tetrahedron Letters*, Vol.45, No.5, (January 2004), pp. 951–953, ISSN 0040-4039

Li, X.; Wu, X.; Chen, W.; Hancock, F. E.; King, K. & Xiao, J. (2004)- Asymmetric Transfer Hydrogenation in Water with a Supported Noyori-Ikariya Catalyst. *Organic Letters*, 2004, Vol.6, No.19, (September 2004), pp. 3321–3324, ISSN 1523-7060

Li, Y.; Li, Z.; Li, F.; Wang, Q. & Tao, F. (2005). Preparation of polymer-supported Ru-TsDPEN catalysts and use for enantioselective synthesis of (*S*)-fluoxetine. *Organic & Biomolecular Chemistry*, Vol.3, No.14, pp. 2513-2518, ISSN 1477-0520

Li, Y.; Ding, K. & Sandoval, C. A. (2009). Hybrid $NH_2$-Benzimidazole Ligands for Efficient Ru-Catalyzed Asymmetric Hydrogenation of Aryl Ketones. *Organic Letters*, Vol.11, No.4, (Februar 2009), pp. 907-910, ISSN 1523-7060

Li, L.; Wu, J.; Wang, F.; Liao, J.; Zhang, H.; Lian, C.; Zhu, J. & Deng, J. (2007). Asymmetric transfer hydrogenation of ketones and imines with novel water-soluble chiral diamine as ligand in neat water. *Green Chemistry*, 2007, Vol.9, No.1, pp. 23-25, ISSN 1463-9262

Li, J.; Zhang, Y.; Han, D.; Gao, Q. & Li, C. (2009). Asymmetric transfer hydrogenation using recoverable ruthenium catalyst immobilized into magnetic mesoporous silica. *Journal of Molecular Catalysis A: Chemical*, Vol. 298, (February 2009), pp. 31-35, ISSN 1381-1169

Li, W.; Sun, X.; Zhou, L.; Hou, G.; Yu, S. & Zhang, X. (2009). Highly Efficient and Highly Enantioselective Asymmetric Hydrogenation of Ketones with TunesPhos/1,2-Diamine-Ruthenium(II) Complexes. *The Journal of Organic Chemistry*, Vol.74, No.3, (Februar 2009), pp. 1397-1399, ISSN 0022-3263

Liang, Y.; Jing, Q.; Li, X.; Shi, L. & Ding, K. (2005). Programmed Assembly of Two Different Ligands with Metallic Ions: Generation of Self-Supported Noyori-type Catalysts for Heterogeneous Asymmetric Hydrogenation of Ketones. *Journal of the American Chemical Society*, Vol.127, No.21, pp. 7694–7695, ISSN 0002-7863

Liang, Y.; Wang, Z. & Ding, K. (2006). Generation of Self-Supported Noyori-Type Catalysts Using Achiral Bridged-BIPHEP for Heterogeneous Asymmetric Hydrogenation of Ketones. *Advanced Synthesis & Catalysis*, Vol.348, No.12-13, (August 2006), pp. 1533–1538, ISSN 1615-4169

Liu, P.-N.; Gu, P.-M.; Wang, F. & Tu, Y.-Q. (2004). Efficient Heterogeneous Asymmetric Transfer Hydrogenation of Ketones Using Highly Recyclable and Accessible Silica-Immobilized Ru-TsDPEN Catalysts. *Organic Letters*, Vol.6, No.2, (January 2004), pp. 169-172, ISSN 1523-7060

Liu, P.-N.; Gu, P.-M.; Deng, J. G.; Tu, Y.-Q. & Ma, Y.-P. (2005). Efficient Heterogeneous Asymmetric Transfer Hydrogenation Catalyzed by Recyclable Silica-Supported Ruthenium Complexes. *European Journal of Organic Chemistry*, No.15 (August 2005), pp. 3221-3227, ISSN 1434-193X

Liu, P.-N.; Deng, J. G.; Tu, Y.-Q. & Wang, S. H. (2004). Highly efficient and recyclable heterogeneous asymmetric transfer hydrogenation of ketones in water. *Chemical Communications*, No.18, pp. 2070-2071, ISSN 1359 7345

Liu, W.; Cui, X.; Cun, L.; Zhu, J. & Deng, J. (2004). Tunable dendritic ligands of chiral 1,2-diamine and their application in asymmetric transfer hydrogenation. *Tetrahedron: Asymmetry*, Vol.16, No.15, (August 2005), pp. 2525–2530, ISSN 0957-4166

Liu, J.; Zhou, Y.; Wu, Y.; Li, X. & Chan, A. S. C. (2008). Asymmetric transfer hydrogenation of ketones with a polyethylene glycol bound Ru catalyst in water. *Tetrahedron: Asymmetry*, Vol.19, No.7, (April 2008), pp. 832–837, ISSN 0957-4166

Liu, G.; Yao, M.; Zhang, F.; Gao, Y. & Li, H. (2008). Facile synthesis of a mesoporous silica-supported catalyst for Ru-catalyzed transfer hydrogenation of ketones. *Chemical Communications*, No.3, pp. 347-349, ISSN 1359 7345

Liu, G.; Yao, M.; Wang, J.; Lu, X.; Liu, M.; Zhang F. & Li, H. (2008). Enantioselective Hydrogenation of Aromatic Ketones Catalyzed by a Mesoporous Silica-Supported

Iridium Catalyst. *Advanced Synthesis & Catalysis*, Vol.350, No.10 (July 2008), pp. 1464-1468, ISSN 1615-4169

Liu, G.; Gu, H.; Sun, Y.; Long, J.; Xu, Y. & Li, H. (2011). Magnetically Recoverable Nanoparticles: Highly Efficient Catalysts for Asymmetric Transfer Hydrogenation of Aromatic Ketones in Aqueous Medium. *Advanced Synthesis & Catalysis*, Vol.353, No.8, (May 2011) pp. 1317–1324, ISSN 1615-4169

Lou, L.-L.; Dong, Y.; Yu, K.; Jiang, S.; Song, Y.; Cao, S. & Liu, S. (2010). Chiral Ru complex immobilized on mesoporous materials by ionic liquids as heterogeneous catalysts for hydrogenation of aromatic ketones. *Journal of Molecular Catalysis A: Chemical*, Vol.333, No.1-2, (December 2010), pp. 20-27, ISSN 1381-1169

Malacea, R.; Poli, R. & Manoury, E. (2010). Asymmetric hydrosilylation, transfer hydrogenation and hydrogenation of ketones catalyzed by iridium complexes. *Coordination Chemistry Reviews*, Vol.254, No.5-6, (March 2010), pp. 729-752, ISSN 0010-8545

Manville, C. V.; Docherty, G.; Padda, R. & Wills, M. (2011). Application of Proline-Functionalised 1,2-Diphenylethane-1,2-diamine (DPEN) in Asymmetric Transfer Hydrogenation of Ketones. *European Journal of Organic Chemistry*, No.34, (December 2011), pp. 6893–6901), ISSN 1099-0690

Marcos, R.; Jimeno, C. & Pericàs, M. A. (2011). Polystyrene-Supported Enantiopure 1,2-Diamines: Development of a Most Practical Catalyst for the Asymmetric Transfer Hydrogenation of Ketones. *Advanced Synthesis & Catalysis*, Vol.353, No.8, (May 2011) pp. 1345–1352, ISSN 1615-4169

Matharu, D. S.; Morris, D. J.; Clarkson, G. J. & Wills, M. (2006). An outstanding catalyst for asymmetric transfer hydrogenation in aqueous solution and formic acid/triethylamine. *Chemical Communications*, No.30, pp. 3232-3234, ISSN 1359 7345

Matsumura, K.; Arai, N.; Hori, K.; Saito, T.; Sayo, N. & Ohkuma, T. (2011). Chiral Ruthenabicyclic Complexes: Precatalysts for Rapid, Enantioselective, and Wide-Scope Hydrogenation of Ketones. *Journal of the American Chemical Society*, Vol.133, No.28, (July 2011), pp. 10696-10699, ISSN 0002-7863

Matsunaga, H.; Ishizuka, T. & Kunieda, T. (2005). Highly efficient asymmetric transfer hydrogenation of ketones catalyzed by 'roofed' *cis*-diamine-Ru(II) complex. *Tetrahedron Letters*, Vol.46, No.21, (May 2005), pp. 3645-3648, ISSN 0040-4039

McMorn, P. & Hutchings, G. J. (2004). Heterogeneous Enantioselective Catalysts: Strategies for the Immobilisation of Homogeneous Catalysts. *Chemical Society Reviews*, Vol.33. No.2, pp. 108-122, ISSN 0306-0012

Merckle, C. & Blümel, J. (2005). Improved Rhodium Hydrogenation Catalysts Immobilized on Silica. *Topics in Catalysis*, Vol.34, No.1-4, pp. 5-15, ISSN 1022-5528

Melero, J. A.; Iglesias, J.; Arsuaga, J. M.; Sainz-Pardo, J.; Frutos, P. & Blazquez, S. (2007). Synthesis and Catalytic Activity of Organic–Inorganic Hybrid Ti-SBA-15 Materials. *Journal of Materials Chemistry*, Vol.17, (November 2006), pp. 337-385, ISSN 0959-9428

Mihalcik, D. J. & Lin, W. (2008). Mesoporous Silica Nanosphere Supported Ruthenium Catalysts for Asymmetric Hydrogenation. *Angewandte Chemie International Edition*, Vol.47, No.33, pp. 6229-6232, ISSN 1521-3773

Mikami, K.; Korenaga, T.; Terada, M.; Ohkuma, T.; Pham, T. & Noyori, R. (1999). Conformationally Flexible Biphenyl-phosphane Ligands for Ru-Catalyzed Enantioselective Hydrogenation. *Angewandte Chemie International Edition*, Vol.38, No.4, (February, 1999), pp. 495–497, ISSN 1521-3773

Mikami, K.; Wakabayashi, K.; Yusa, Y. & Aikawa, K. (2006). Achiral benzophenone ligand–rhodium complex with chiral diamine activator for high enantiocontrol in asymmetric transfer hydrogenation. *Chemical Communications*, No.22, (June 2006), pp. 2365-2367, ISSN 1359 7345

Mikhailine, A.; Lough, A. J. & Morris, R. H. (2009). Efficient Asymmetric Transfer Hydrogenation of Ketones Catalyzed by an Iron Complex Containing a P–N–N–P Tetradentate Ligand Formed by Template Synthesis. *Journal of the American Chemical Society*, Vol.131, No.4, (February 2009), pp. 1394-1395, ISSN 0002-7863

Mondelli, C.; Vargas, A.; Santarossa, G. & Baiker, A. (2008). Fundamental Aspects of the Chiral Modification of Platinum with Peptides: Asymmetric Induction in Hydrogenation of Activated Ketones. *The Journal of Physical Chemistry C*, Vol.113, No.34, (Avgust 2009), pp. 15246-15259, ISSN 1932-7447

Morris, R. H. (2009). Asymmetric hydrogenation, transfer hydrogenation and hydrosilylation of ketones catalyzed by iron complexes. *Chemical Society Reviews*, Vol.38, No.8, (August 2009), pp. 2282-2291, ISSN 0306-0012

Noyori, R.; Ohkuma, T.; Kitamura, M.; Takaya, H.; Sayo, N.; Kumobayashi, H. & Akutagawa, S. (1987). Asymmetric hydrogenation of beta-keto carboxylic esters. A practical, purely chemical access to beta-hydroxy esters in high enantiomeric purity. *Journal of the American Chemical Society*, Vol.109, No.19, (September 1987), pp. 5856-5858, ISSN 0002-7863

Noyori, R. & Takaya, H. (1990). BINAP: an efficient chiral element for asymmetric catalysis. *Accounts of Chemical Research*, Vol.23, No.10 (October 1990), pp. 345–350, ISSN 0001-4842

Noyori, R. & Ohkuma, T. (2001). Asymmetric Catalysis by Architectural and Functional Molecular Engineering: Practical Chemo- and Stereoselective Hydrogenation of Ketones. *Angewandte Chemie, International Edition*, Vol.40, No.1, (January 2001), pp. ISSN 1433-7851

Noyori, R.; Yamakawa, M. & Hashiguchi, S. (2001). Metal-Ligand Bifunctional Catalysis: A Nonclassical Mechanism for Asymmetric Hydrogen Transfer between Alcohols and Carbpnyl Compounds. *The Journal of Organic Chemistry*, Vol.66, No.24, (November 2001), pp. 7931-7944, ISSN 0022-3263

Noyori, R.; Sandoval, C. A.; Muñiz, K. & Ohkuma, T. (2005). Metal-ligand bifunctional catalysis for asymmetric hydrogenation. *Philosophical Transactions of The Royal Society A*, Vol.363, No.1829, (April 2005), pp. 901-912, ISSN 1364-503X

Ohkuma, T.; Ooka, H.; Ikariya, T. & Noyori, R. (1995). Preferential hydrogenation of aldehydes and ketones. *Journal of the American Chemical Society*, Vol.117, No.41, (October 1995), pp. 10417-10418, ISSN 0002-7863

Ohkuma, T.; Ooka, H.; Hashiguchi, S.; Ikariya, T. & Noyori, R. (1995). Practical Enantioselective Hydrogenation of Aromatic Ketones. *Journal of the American Chemical Society*, Vol.117, No.9, (March 1995), pp. 2675-2676, ISSN 0002-7863

Ohkuma, T.; Koizumi, M.; Doucet, H.; Pham, T.; Kozawa, M.; Murata, K.; Katayama, E.; Yokozawa, T.; Ikariya, T. & Noyori, R. (1998), Asymmetric Hydrogenation of Alkenyl, Cyclopropyl, and Aryl Ketones. $RuCl_2$(xylbinap)(1,2-diamine) as a Precatalyst Exhibiting a Wide Scope. *Journal of the American Chemical Society*, Vol.120, No.54, (December 1998), pp. 13529-13530, ISSN 0002-7863

Ohkuma, T.; Koizumi, M.; Yoshida, M., & Noyori, R. (2000). General Asymmetric Hydrogenation of Hetero-aromatic Ketones. *Organic Letters*, Vol.2, No.12, (May 2000), pp. 1749-1751, ISSN 1523-7060

Ohkuma, T.; Koizumi, M.; Muñiz, K.; Hilt, G.; Kabuto, C. & Noyori, R. (2002). *trans*-RuH($\eta^1$-$BH_4$)(binap)(1,2-diamine): A Catalyst for Asymmetric Hydrogenation of Simple Ketones under Base-Free Conditions. *Journal of the American Chemical Society*, Vol.124, No.23, (June 2002), pp. 6508-6509, ISSN 0002-7863

Ohkuma, T.; Sandoval, C. A.; Srinivasan, R.; Lin, Q.; Wei, Y.; Muñiz, K. & Noyori, R. (2005). Asymmetric Hydrogenation of *tert*-Alkyl Ketones. *Journal of the American Chemical Society*, Vol.127, No.23, (June 2005), pp. 8288-8289, ISSN 0002-7863

Ohkuma, T. (2010). Asymmetric hydrogenation of ketones: Tactics to achieve high reactivity, enantioselectivity, and wide scope. *Proceedings of the Japan Academy, Series B*, Vol.86, No.3, (March 2010), pp. 202-219, ISSN 0386-2208

Orito, Y.; Imai, S. & Niwa, S. (1979). Asymmetric hydrogenation of methyl pyruvate using a platinum-carbon catalyst modified with cinchonidine. *Nippon Kagaku Kaishi*, No.8, pp. 1118-1120, ISSN 0369-4577

Palmer, M. J. & Wills, M. (1990). Asymmetric transfer hydrogenation of C=O and C=N bonds. *Tetrahedron: Asymmetry*, Vol.10, No.11, (June 1999), pp. 2045–2061, ISSN: 0957-4166

Palmer, M.; Walsgrove, T. & Wills, M. (1997). (1*R*,2*S*)-(+)-*cis*-1-Amino-2-indanol: An Effective Ligand for Asymmetric Catalysis of Transfer Hydrogenations of Ketones. *The Journal of Organic Chemistry*, Vol.62, No.15, (July 1997), pp. 5226-5228, ISSN 0022-3263

Pannetier, N.; Sortais, J.-B.; Issenhuth, J.-T.; Barloy, L.; Sirlin, C.; Holuigue, A.; Lefort, L.; Panella, L.; de Vries, J. G. & Pfeffer, M. (2011). Cyclometalated Complexes of Ruthenium, Rhodium and Iridium as Catalysts for Transfer Hydrogenation of Ketones and Imines. *Advanced Synthesis & Catalysis*, Vol.353, No.14-15, (October 2011), pp. 2844-2852, ISSN: 1615-4169

Pugin, B.; Landert, H.; Spindler, F. & Blaser, H. R. (2002). More than 100,000 Turnovers with Immobilized Ir-Diphosphine Catalysts in an Enantioselective Imine Hydrogenation. *Advanced Synthesis & Catalysis*, Vol.344, No.9, (October 2002), pp. 974-979, ISSN 1615-4169

Reetz, M. T. & Li, X. (2006). An Efficient Catalyst System for the Asymmetric Transfer Hydrogenation of Ketones: Remarkably Broad Substrate Scope. *Journal of the American Chemical Society*, Vol.128, No.4, (February 2006), pp. 1044-1045, ISSN 0002-7863

Rhyoo, H. Y.; Park, H.-J. & Chung, Y. K. (2001). The first Ru(II)-catalysed asymmetric hydrogen transfer reduction of aromatic ketones in aqueous media. *Chemical Communications*, No.20, pp. 2064-2065, ISSN 1359-7345

Saluzzo, C. & Lemaire, M. (2002). Homogeneous-Supported Catalysts for Enantioselective Hydrogenation and Hydrogen Transfer Reduction. *Advanced Synthesis & Catalysis*, Vol.344, No.9, (October 2002), pp. 915-928, ISSN 1615-4169

Samec, J. S. M.; Bäckvall, J.-E.; Andersson, P. G. & Brandt, P. (2006). Mechanistic aspects of transition metal-catalyzed hydrogen transfer reactions. *Chemical Society Reviews*, Vol.35, No.3, (March 2006), pp. 237-248, ISSN 0306-0012

Sandree, A. J.; Reek, J. N. H; Kamer, P. C. J. & van Leeuwen, P. W. N. M. (2001). A Silica-Supported, Switchable, and Recyclable Hydroformylation–Hydrogenation Catalyst. *Journal of the American Chemical Society*, Vol.123, No.36, (September 2011) pp. 8468-8476, ISSN 0002-7863

Schulz, P. S.; Müller, N.; Bösmann, A. & Wasserscheid, P. (2007). Effective Chirality Transfer in Ionic Liquids through Ion-Pairing Effects. *Angewandte Chemie International Edition*, Vol.46, No.8, (February 2007), pp. 1293–1295, ISSN 1433-7851.

Schürch, M.; Künzle, N.; Mallat, T. & Baiker, A. (1998). Enantioselective Hydrogenation of Ketopantolactone: Effect of Stereospecific Product Crystallization during Reaction. *Journal of Catalysis*, Vol.176, No.2, (June 1998), pp. 569-571, ISSN 0021-9517

Schmidt, E.; Ferri, D.; Vargas, A. & Baiker, A. (2008). Chiral Modification of Rh and Pt Surfaces: Effect of Rotational Flexibility of Cinchona-Type Modifiers on Their Adsorption Behavior. *The Journal of Physical Chemistry C*, Vol.112, No.10, (March 2008), pp. 3453-4018, ISSN 1932-7447

Schmidt, E.; Ferri, D.; Vargas, A.; Mallat, T. & Baiker, A. (2009). Shape-Selective Enantioselective Hydrogenation on Pt Nanoparticles. *Journal of the American Chemical Society*, Vol.131, No.34, (September 2009), pp. 12358-12367, ISSN 0002-7863

Shen, Y.; Chen, Q.; Lou, L.-L.; Yu, K.; Ding, F. & Liu, S. (2010). Asymmetric Transfer Hydrogenation of Aromatic Ketones Catalyzed by SBA-15 Supported Ir(I) Complex Under Mild Conditions. *Catalysis Letters*, Vol.137, No.1-2, (June 2010), pp. 104-109, ISSN 1011-372X

Shimizu, H.; Igarashi, D.; Kuriyama, W.; Yusa, Y.; Sayo, N. & Saito, T. (2007). Asymmetric Hydrogenation of Aryl Ketones Mediated by a Copper Catalyst. *Organic Letters*, 2007, Vol.9, No.9, (April 2007), pp. 1655–1657, ISSN 1523-7060

Shimizu, H.; Nagano, T.; Sayo, N.; Saito, T.; Ohshima, T. & Mashima, K. (2009). Asymmetric Hydrogenation of Heteroaromatic Ketones and Cyclic and Acyclic Enones Mediated by Cu(I)-Chiral Diphosphine Catalysts. *Synlett*, 2009, No.19, (December 2009), pp. 3143-3146, ISSN 0936-5214

Shiomi, T. & Nishiyama, H. (2007). Intermolecular Asymmetric Reductive Aldol Reaction of Ketones as Acceptors Promoted by Chiral Rh(Phebox) Catalyst. *Organic Letters*, Vol.9, No.9, (April 2007), pp. 1655-1657, ISSN 1523-7060

Song C. E. & Lee, S. (2002). Supported Chiral Catalysts on Inorganic Materials. *Chemical Reviews*, Vol.102, No.10, (August 2002), pp. 3495-3524, ISSN 0009-2665

Steiner, I.; Aufdenblatten, R.; Togni, A.; Blaser, H. U. & Pugin, B. (2004). Novel silica gel supported chiral biaryl-diphosphine ligands for enantioselective hydrogenation *Tetrahedron: Asymmetry*, Vol.15, No.14, (July 2004) pp. 2307-2311, ISSN 0957-4166

Studer, M.; Burkhardt, S. & Blaser, H.-U. (1999). Enantioselective hydrogenation of α-keto acetals with cinchona modified Pt catalyst. *Chemical Communications,* No.17, pp. 1727-1728, ISSN 1359 7345

Studer, M.; Burkhardt, S.; Indolese, A. F. & Blaser, H.-U. (2000). Enantio- and chemoselective reduction of 2,4-diketo acid derivatives with cinchona modified Pt-catalyst-Synthesis of (*R*)-2-hydroxy-4-phenylbutyric acid ethyl ester. *Chemical Communications,* No.14, pp. 1327-1328, ISSN 1359 7345

Studer, M.; Blaser, H.-U. & Burkhardt, S. (2002). Hydrogenation of α-Keto Ethers: Dynamic Kinetic Resolution with a Heterogeneous Modified Catalyst and a Heterogeneous Base. *Advanced Synthesis & Catalysis*, Vol.344, No.5, (July 2002), pp. 511-515, ISSN 1615-4169

Studer, M.; Blaser, H.-U. & Exner, C. (2003). Enantioselective Hydrogenation Using Heterogeneous Modified Catalysts: An Update. *Advanced Synthesis & Catalysis*, Vol.345, No.1-2, (January 2003), pp. 45-65, ISSN 1615-4169

Sues, P. E.; Lough, A. J. & Morris, R. H. (2011). Stereoelectronic Factors in Iron Catalysis: Synthesis and Characterization of Aryl-Substituted Iron(II) Carbonyl P–N–N–P Complexes and Their Use in the Asymmetric Transfer Hydrogenation of Ketones. *Organometallics*, Vol.30, No.16, (August 2011), pp. 4418–4431, ISSN 0276-7333

Sutyinszki, M.; Szöri, K.; Felföldi, K. & Bartók, M. (2002). 98% Enantioselectivity in the asymmetric synthesis of a useful chiral building block by heterogeneous method: Enantioselective hydrogenation of ethyl-benzoylformate over cinchona modified $Pt/Al_2O_3$ catalysts in the acetic acid. *Catalysis Communications*, Vol.3, No.3, (March 2002), pp. 125-127, ISSN 1566-7367

Szőri, K. Sutyinszki, M. Felföldi, K. & Bartók, M. (2002). Heterogeneous asymmetric reactions: Part 28. Efficient and practical method for the preparation of (*R*)- and (*S*)- α-hydroxy esters by the enantioselective heterogeneous catalytic hydrogenation of α-ketoesters. *Applied Catalysis A: General*, Vol.237, No.1-2, (November 2002), pp. 275-280, ISSN 0926-860X

Szőri, K.; Balázsik, K.; Cserényi, S.; Szőllősi, G. & Bartók, M. (2009). Inversion of enantioselectivity in the 2,2,2-trifluoroacetophenone hydrogenation over Pt- alumina catalyst modified by cinchona alkaloids. *Applied Catalysis A: General*, Vol.362, No.1-2, (June 2009), pp. 178-184, ISSN 0926-860X

Tada, M. & Iwasawa, Y. (2006). Advanced chemical design with supported metal complexes for selective catalysis. *Chemical Communications*, No.27, pp. 2833-2844, ISSN 1359 7345

Takehara, J.; Hashiguchi, S.; Fujii, A.; Inoue, S.; Ikariya, T. & Noyori, R. (1996). Amino alcohol effects on the ruthenium(II)-catalysed asymmetric transfer hydrogenation of ketones in propan-2-ol. *Chemical Communications*, No.3, (Februar 1996), pp. 233-234, ISSN 1359 7345

Thorpe, T.; Blacker, J.; Brown, S. M.; Bubert, C.; Crosby, J.; Fitzjohn, S.; Muxworthy, J. P. & Williams, J. M. J. (2001). Efficient rhodium and iridium-catalysed asymmetric transfer hydrogenation using water-soluble aminosulfonamide ligands. *Tetrahedron Letters*, Vol.42, No.24, (June 2001), pp. 4041-4043, ISSN 0040-4039

Török, B.; Balázsik, K. Bartók, M.; Felföldi, K. & Bartók, M. (1999). New synthesis of a useful C3 chiral building block by a heterogeneous method: enantioselective hydrogenation of pyruvaldehyde dimethyl acetal over cinchona modified Pt/$Al_2O_3$ catalysts. *Chemical Communications*, No.17, pp. 1725-1726, ISSN 1359 7345

Török, B.; Felföldi, K.; Szakonyi, G.; Balázsik, K. & Bartók, M. (1998). Enantiodifferentiation in asymmetric sonochemical hydrogenations. *Catalysis Letters*, Vol.52, No.1-2 (June 1998), pp. 81-84, ISSN 1011-372X

Varga, T.; Felföldi, K.; Forgó, P. & Bartók, M. (2004). Heterogeneous asymmetric reactions: Part 38. Enantioselective hydrogenation of fluoroketones on Pt–alumina catalyst. *Journal of Molecular Catalysis A: Chemical*, Vol.216, No.2, (July 2004), pp. 181-187, ISSN 1381-1169

Vargas, A.; Bonalumi, N.; Ferri, D. & Baiker, A. (2006). Solvent-Induced Conformational Changes of *O*-Phenyl-cinchonidine: A Theoretical and VCD Spectroscopy Study. *The Journal of Physical Chemistry A*, Vol.110, No.3, (January 2006), pp. 1106–1117, ISSN 1089-5639

Vargas, A.; Ferri, D.; Bonalumi, N.; Mallat T. & Baiker, A. (2007). Controlling the Sense of Enantioselection on Surfaces by Conformational Changes of Adsorbed Modifiers. *Angewandte Chemie International Edition*, Vol.46, No.21, (May 2007), pp. 3905-3908, ISSN 1521-3773

Vázquez-Villa, H.; Reber, S.; Ariger, M. A. & Carreira, E. M. (2011). Iridium Diamine Catalyst for the Asymmetric Transfer Hydrogenation of Ketones. *Angewandte Chemie International Edition*, Vol.50, No.38, (September 2011), pp. 8979-8981, ISSN 1433-7851

von Arx, M.; Mallat, T. & Baiker, A. (2001). Platinum-catalyzed enantioselective hydrogenation of aryl-substituted trifluoroacetophenones. *Tetrahedron: Asymmetry*, Vol.12, No.22, (December 2001), pp. 3089-3094, ISSN 0957-4166

von Arx, M.; Mallat, T. & Baiker, A. (2002). Asymmetric Hydrogenation of Activated Ketones on Platinum: Relevant and Spectator Species. *Topics in Catalysis*, Vol. 19, No.1 (March 2002), pp. 75-87, ISSN 1022-5528

von Arx, M., Mallat, T. & Baiker, A (2001). Inversion of Enantioselectivity during the Platinum-Catalyzed Hydrogenation of an Activated Ketone. *Angewandte Chemie International Edition*, Vol.40, No.12, (June 2001), pp. 2302-2305, ISSN 1521-3773

von Arx, M. Mallat T. & Baiker, A. (2002). Highly Efficient Platinum-Catalyzed Enantioselective Hydrogenation of Trifluoroacetoacetates in Acidic Solvents. *Catalysis Letters*, Vol.78, No.1-4, (March 2002), pp. 267-271, ISSN 1011-372X

Wang, F.; Liu, H.; Cun, L.; Zhu, J.; Deng, J. & Jiang, Y. (2005). Asymmetric Transfer Hydrogenation of Ketones Catalyzed by Hydrophobic Metal–Amido Complexes in Aqueous Micelles and Vesicles. *The Journal of Organic Chemistry*, Vol.70, No.23, (November 2005), pp. 9424-9429, ISSN: 0022-3263

Wu, X. & Xiao, J. (2007). Aqueous-phase asymmetric transfer hydrogenation of ketones – a greener approach to chiral alcohols. *Chemical Communications*, No.24, (June 2007), pp. 2449-2466, ISSN 1359-7345

Wylie, W. N. O; Lough, A. J. & Morris, R. H. (2011). Mechanistic Investigation of the Hydrogenation of Ketones Catalyzed by a Ruthenium(II) Complex Featuring an N-Heterocyclic Carbene with a Tethered Primary Amine Donor: Evidence for an Inner Sphere Mechanism. *Organometallics*, Vol.30, No.5, (March 2011), pp. 1236-1252, ISSN 0276-7333

Wettergren, J.; Buitrago, E.; Ryberg, P. & Adolfsson, H. (2009). Mechanistic Investigations into the Asymmetric Transfer Hydrogenation of Ketones Catalyzed by Pseudo-Dipeptide Ruthenium Complexes. *Chemistry - A European Journal*, Vol.15, No.23, (June 2009), pp. 5709-5718, ISSN 1521-3765

Xie, J.-B.; Xie, J.-H.; Liu, X.-Y.; Kong, W.-L.; Li S. & Zhou, Q.-L. (2010). Highly Enantioselective Hydrogenation of $\alpha$-Arylmethylene Cycloalkanones Catalyzed by Iridium Complexes of Chiral Spiro Aminophosphine Ligands. *Journal of the American Chemical Society*, Vol.132, No.13, (April 2010), pp. 4538-4539, ISSN 0002-7863

Xie, J.-H.; Liu, X.-Y.; Xie, J.-B.; Wang, L.-X. & Zhou, Q.-L. (2011). An Additional Coordination Group Leads to Extremely Efficient Chiral Iridium Catalysts for Asymmetric Hydrogenation of Ketones. *Angewandte Chemie International Edition*, Vol.50, No.32, (August 2011), pp. 7329-7332, ISSN 1433-7851

Yamakawa, M.; Yamada, I. & Noyori, R. (2001). CH/$\pi$ Attraction: The Origin of Enantioselectivity in Transfer Hydrogenation of Aromatic Carbonyl Compounds Catalyzed by Chiral $\eta^6$-Arene-Ruthenium(II) Complexes. *Angewandte Chemie International Edition*, Vol.40, No.15, (August 2001), pp. 2818-2821, ISSN 1433-7851

Yang, J. W. & List, B. (2006). Catalytic Asymmetric Transfer Hydrogenation of $\alpha$-Ketoesters with Hantzsch Esters. *Organic Letters*, Vol.8, No.24, (November 2006), pp. 5653-5655, ISSN 1523-7060

Yang, C.; Jiang, H.; Feng, J., Fu, H.; Li, R.; Chen, H. & Li, X. (2009). Asymmetric hydrogenation of acetophenone catalyzed by cinchonidine stabilized Ir/$SiO_2$. *Journal of Molecular Catalysis A: Chemical*, Vol.300, No.1-2, (March 2009), pp. 98-102, ISSN 1381-1169

Ye, W.; Zhao, M.; Du, W.; Jiang, Q.; Wu, K.; Wu, P. & Yu, Z. (2011), Highly Active Ruthenium(II) Complex Catalysts Bearing an Unsymmetrical NNN Ligand in the (Asymmetric) Transfer Hydrogenation of Ketones. *Chemistry - A European Journal*, Vol.17, No.17, (April 2011), pp. 4737-4741, ISSN 1521-3765

Zhou, Z.; Sun Y. & Zhang, A. (2011), Asymmetric transfer hydrogenation of prochiral ketones catalyzed by aminosulfonamide-ruthenium complexes in ionic liquid. *Central European Journal of Chemistry*, Vol.9, No.1, (February 2011), pp. 175-179, ISSN 1895-1066

Zhu, Q.; Shi, D.; Xia, C. & Huang, H. (2011). Ruthenium Catalysts Containing Rigid Chiral Diamines and Achiral Diphosphanes for Highly Enantioselective Hydrogenation of Aromatic Ketones. *Chemistry - A European Journal*, Vol.17, No.28, (July 2011), pp. 7760-7763, ISSN 1521-3765

Zsigmond, Á.; Undrala, S.; Notheisz, F.; Szöllősy Á. & Bakos J. (2008). The effect of substituents of immobilized Rh complexes on the asymmetric hydrogenation of acetophenone derivatives. *Central European Journal of Chemistry*, Vol.6, No.4, (December 2008), pp. 549-554, ISSN 1895-1066

Chapter 3

# Hydrogenation in the Vitamins and Fine Chemicals Industry – An Overview

Werner Bonrath, Jonathan Medlock, Jan Schütz, Bettina Wüstenberg and Thomas Netscher

Additional information is available at the end of the chapter

http://dx.doi.org/10.5772/48751

## 1. Introduction

In the pharmaceutical and partly also in the fine chemicals industry many chemical conversions require stoichiometric amounts of reagents, and thus generate large amounts of waste [1, 2]. This is in contrast to the production of bulk chemicals which mostly relies on catalysis. This difference can be explained by the higher complexity of pharmaceuticals and fine chemicals which makes catalysis more demanding and process development more expensive.

According to Sheldon's classification [3], most vitamins are typical fine chemicals with production volumes of about 100 to 10'000 tons per year. Some vitamins can be placed in the class of bulk chemicals. Typically these compounds have been produced industrially for decades in multi-step syntheses with high overall yields. The application of catalytic methods in the highly competitive field of vitamins has increased significantly in recent years because of price pressure on these products. Research and development is thus driven by the necessity to reduce waste, use less toxic reagents and solvents, improve energy efficiency, recycle catalysts and reagents, and combine unit operations to reduce costs and achieve more sustainable processes. These goals are mostly in accordance with the twelve principles of "green chemistry"[4,5,6].

Catalytic hydrogenation is certainly the most widely applicable method for the reduction of organic compounds and belongs to the most important transformations in chemical industry. Catalytic hydrogenations in the fine chemicals industry are usually carried out with heterogeneous catalysts. Homogeneous catalysts are typically applied for highly selective transformations, particularly enantioselective reductions. In the case of full recycling of heterogeneous or homogeneous catalysts, hydrogenation with molecular hydrogen is an atom economic transformation and undoubtedly the cleanest possible method for reducing a compound. Alternatively, hydrogen donors such as isopropanol or formic acid can be applied in transfer hydrogenations. In the field of catalytic hydrogenation reactions several significant inventions have been reported in the last 150 years. Sabatier and co-worker investigated the application of highly dispersed metals, e.g. nickel, in the hydrogenation of organic compounds [7,8]. The elective semi-hydrogenation of C≡C-bonds in presence of lead-doped palladium on calcium carbonate catalysts found by Lindlar was a further milestone in the field of catalytic hydrogenation reactions [9-11]. During the last decades asymmetric hydrogenations, pioneered by W.S. Knowles and R. Noyori, were a further highlight in the field of hydrogenation applied in organic synthesis [12,13].

Catalytic hydrogenations can be carried out in a variety of ways, either in the liquid or gas phase and in batch-wise or continuous mode. In continuous processes usually fixed-bed reactors or fluidized-bed reactors are used. The suitable choice of a reactor system depends on various factors such as, in particular, the choice of catalyst, the reaction conditions, heat formation, space-time-yield, residence time, hydrogen pressure, mass-transport phenomena, temperature, solvent and economic reasons.

In this contribution we will focus on industrially important catalytic hydrogenation reactions which are of interest for DSM Nutritional Products concerning the manufacture of vitamins, carotenoids and nutraceuticals. The various hydrogenations are organized by reaction types, rather than by the products prepared: hydrogenations of C=C double bonds, selective semi-hydrogenations of C≡C triple bonds, hydrogenations of C=X/C≡X multiple bonds (X = oxygen or nitrogen), and stereoselective hydrogenations.

## 2. Hydrogenation of C=C double bonds

Probably the most common hydrogenation reaction performed in industry is the hydrogenation of carbon-carbon double bonds. A wide variety of catalysts are available from commercial suppliers and this transformation is considered a robust and atom-economical reaction. Even so, careful optimisation of the reaction conditions can be required to obtain full conversion and reduce or eliminate by-products. This is also important in the synthesis of vitamins and fine chemicals.

(all-*rac*)-α-Tocopherol (**3**) is the economically most important member of the group of vitamin E compounds, due to its biological and antioxidant properties. This fat-soluble vitamin is produced on a scale of >30’000 tonnes per year (mostly in form of its acetate derivative **4**) for applications in human and animal nutrition. One of the key building blocks for the chemical production of synthetic vitamin E is trimethylhydroquinone (TMHQ, **1**), which is converted into (all-*rac*)-α-tocopherol (**3**) by condensation with (all-*rac*)-isophytol (**2**, Scheme 1) [14-16].

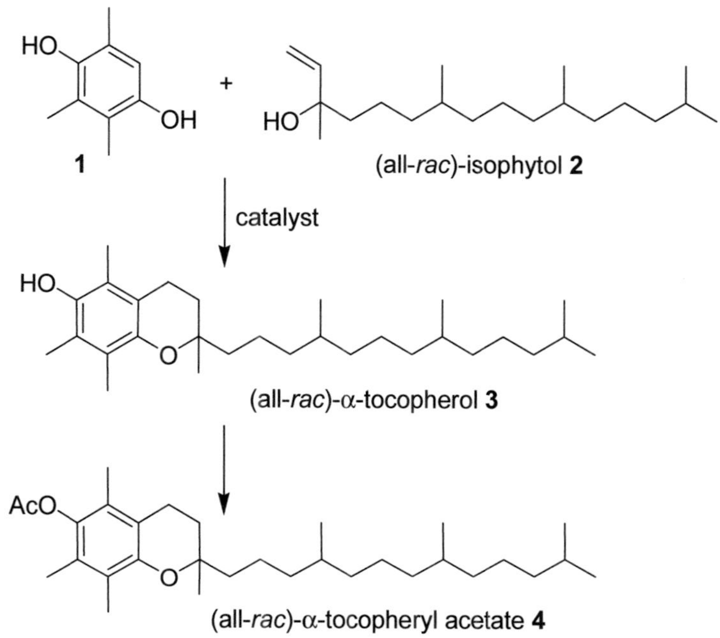


**Scheme 1.** Final steps of the chemical production of vitamin E.

The synthesis of isophytol can be carried out starting from acetone (**5**) and building up the isoprenoic side chain by a sequence of $C_2$ and $C_3$ elongations (*via* **8** and **10**, Scheme 2). Another approach starts from citral (**9**) followed by reaction to linalool (**11**) and elongation to geranylacetone (**13**) [14,15]. Routes to **9** based on prenyl chloride and myrcene (**6**) are described in literature but are not competitive. The use of cheap isobutene (**7**) is preferred which represents also an access to other isoprenoic building blocks, e.g. **14**. The $C_3$-elongation can be carried out by Saucy-Marbet or Carroll reactions and the $C_2$ elongation by ethynylation or vinylation reactions (vinyl Grignard addition) [1]. In such reaction sequences several hydrogenations of C=C bonds are necessary. A key intermediate in the synthesis of isophytol (**2**) is hexahydropseudoionone (**12**), which can be produced from pseudoionone (**10**) or geranylacetone (**13**). In the past the hydrogenation reactions were carried out batch-wise in presence of a Pd/C catalyst below 80 °C and <10 bar pressure [17]. This hydrogenation was also investigated using Pd- or Rh-containing polymers on $Al_2O_3$ [18]. The liquid phase hydrogenation of pseudoionone (**10**), geranylacetone (**13**) or dihydrogeranylacetone in presence of a suspended catalyst in a special reactor which allows an easy hold-up of the catalyst has been described in [19]. The synthesis of these saturated ketones in a continuous fixed-bed mode applying a Pd catalyst supported on $SiO_2$ results in excellent yield under nearly full conversion [20].

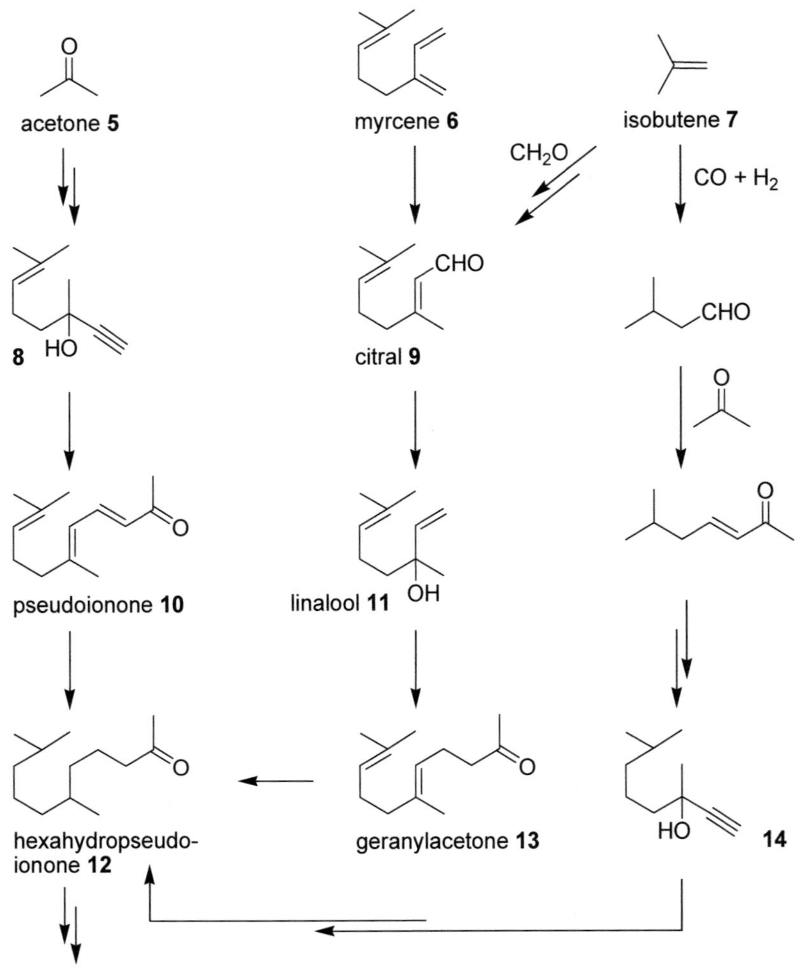


**Scheme 2.** Synthesis of isophytol (*E*/*Z* isomerism of olefins is omitted here).

The other key intermediate for the synthesis of vitamin E, TMHQ (**1**), is accessible *via* catalytic hydrogenation of trimethylbenzoquinone (TMQ, **15**) using a palladium on carbon catalyst (Scheme 3).

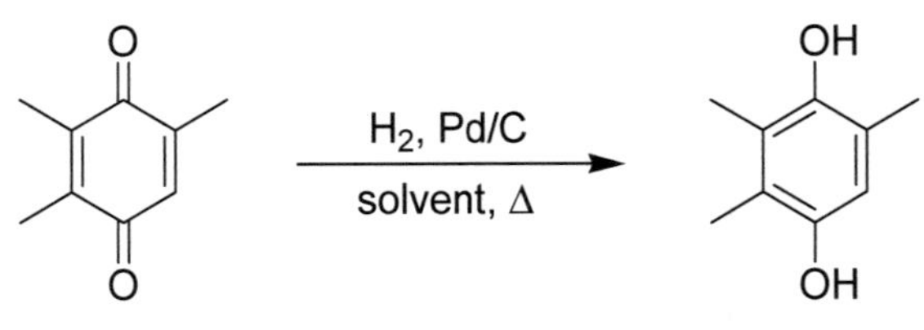


TMQ **15** **1**

**Scheme 3.** Catalytic hydrogenation of TMQ to TMHQ.

The hydrogenation is usually carried out in a continuous mode at medium to low pressure and elevated temperature to prevent crystallization of the product from the reaction mixture. Yields are generally quantitative.

Alternatively, a range of different Pd-catalyzed hydrogenation conditions can be found in the literature, e.g. using Pd/C in solvents like carboxylic esters [21] or acetone [22], or palladium on acidic [23,24] or basic oxides [25] in lower alcohols. Also a Pt-catalyzed hydrogenation has been described using platinum nitrate and aluminium oxide in *i*-butanol [25].

An alternative access to TMHQ (**1**) starts from less methylated 1,4-benzoquinones, e.g. 2,6-dimethylbenzoquinone (2,6-DMQ, **16**), which is first hydrogenated to 2,6-dimethylhydroquinone (**17**) and methylated later, e.g. by aminomethylation *via* **18** followed by hydrogenolysis to TMHQ (**1**) (Scheme 4) [26].

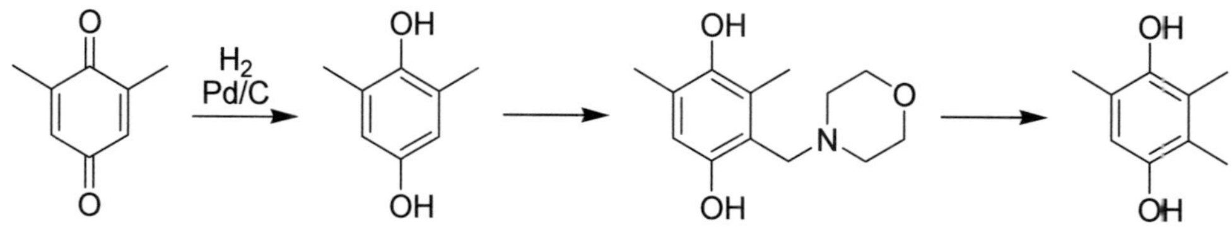


**Scheme 4.** Synthesis of TMHQ from 2,6-DMQ.

The K-vitamins are a group of substituted 2-methyl-1,4-naphthoquinones with (or without) a prenyl chain of different length in C-3 position (Figure 1). The core unit of all K-vitamins is menadione (vitamin $K_3$, **21**) [27,28]. The standard synthesis of vitamins $K_1$ and $K_2$ (**19**,**20**) is the coupling of the aromatic unit with the (poly)prenyl side chain, similar to vitamin E. The direct coupling of the side chain to menadione (**21**) is not possible; therefore a key step is the hydrogenation of menadione to menadiol (**22**) so that alkylation can take place. The hydrogenation is usually carried out batch-wise using a palladium on carbon catalyst.

After esterification, condensation with isophytol (**2**) according to the methods of the group of Isler [29,30] or Hirschmann *et al.* [31], and subsequent saponification the hydroquinone is then re-oxidized to the corresponding naphthoquinone, e.g. in the industrial preparation of synthetic vitamin $K_1$ (phylloquinone, Konakion®, **23**) (Scheme 5).

A remarkable stereoselective hydrogenation of a trisubstituted olefinic C=C double bond was used in several total synthesis routes to the water-soluble vitamin (+)-biotin (**26**, Scheme 6) [32]. This product is produced on a scale of about 100 tonnes per year, and only the (3a*S*,4*S*,6a*R*)-stereoisomer exhibits full biological activity. Stereocenter C-4 of the thiophane ring can be introduced by catalytic hydrogenation of the exocyclic olefin **24** with undefined double-bond stereochemistry on Pd/C or other heterogeneous catalysts, yielding **25** with the desired all-*cis* relative configuration at centers C-4, C-3a, and C-6a. The N-benzyl protecting groups are stable under those conditions. This strategy, originally developed by Goldberg and Sternbach [33,34], was later on used by other groups in various syntheses of racemic and optically active intermediates [32].

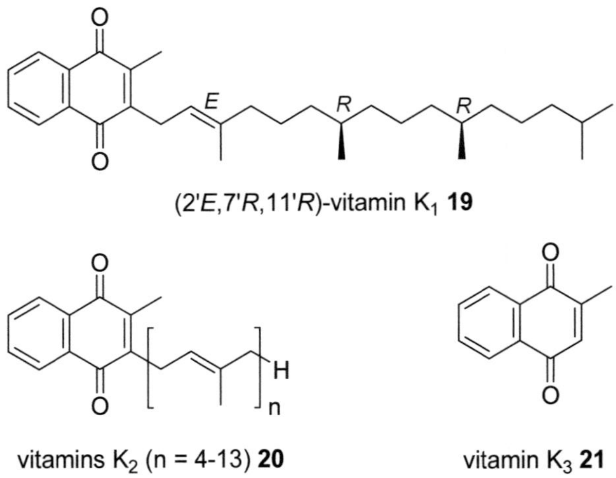


**Figure 1.** Molecular structures of the different K vitamins.

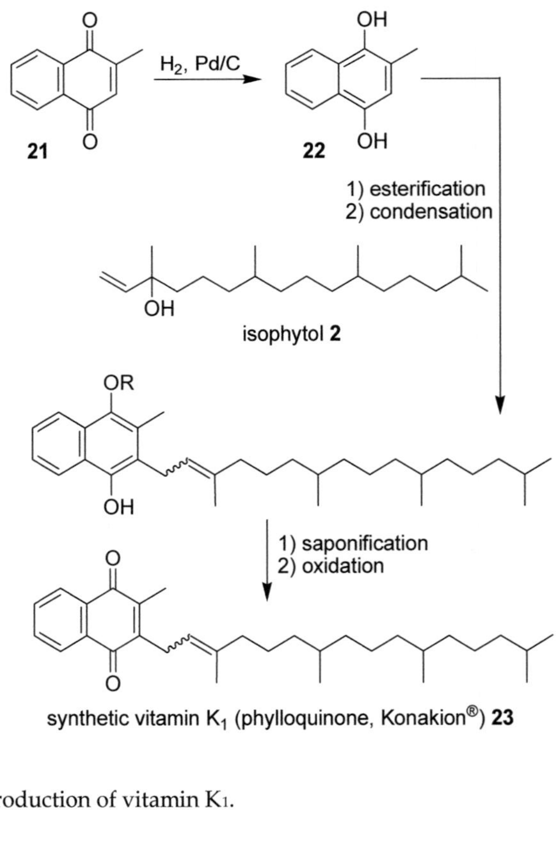


**Scheme 5.** Chemical production of vitamin $K_1$.

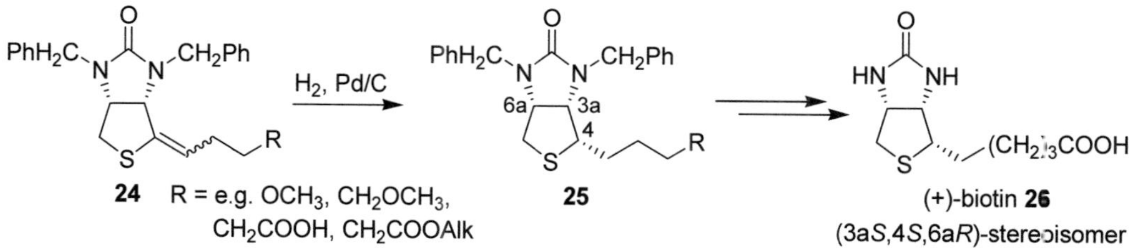


**Scheme 6.** Stereoselective heterogeneous hydrogenation in routes towards (+)-biotin.

## 3. Semi-hydrogenation of C≡C triple bonds (Lindlar type)

The semi-hydrogenation of carbon-carbon triple bonds to alkenes is one of the most useful hydrogenations for the production of vitamins, however careful choice of catalyst and reaction conditions are required to obtain high selectivity. In general, hydrogenation of acetylenes with a metal catalyst results in the formation of the fully saturated alkane product, since the second hydrogenation (alkene to alkane) is generally faster than the first (alkyne to alkene). However, as long as some of the starting alkyne remains in the reaction mixture, selectivity can be high since the alkynes bind more strongly to the metal surface. The selectivity can be enhanced by the use of suitable catalyst poisons which modify the activity of the metal catalyst.

One of the most widely used and most selective catalysts is the one originally developed by Lindlar [9,11]. With this catalyst, the palladium supported on calcium carbonate is doped with a lead acetate solution during manufacture. This catalyst can then be used either directly in the hydrogenation or modified further by an organic compound such as an amine. Since the hydrogen is delivered from the metal surface to the alkyne, usually high selectivity is obtained for the *Z*-(*cis*)-alkene product. "Lindlar catalysts" generally have 5% palladium loading and 2-5% lead loading, depending on the application.

One of the earliest uses of the catalyst developed by Lindlar was the semi-hydrogenation of a vitamin A key intermediate (**27**, Scheme 7) to give tetraene **28**. Whilst this could be achieved with poisoned palladium on charcoal or palladium on calcium carbonate [35], selectivities were significantly higher with the lead-doped catalyst and the reaction could easily be stopped after the uptake of just one equivalent of hydrogen gas.

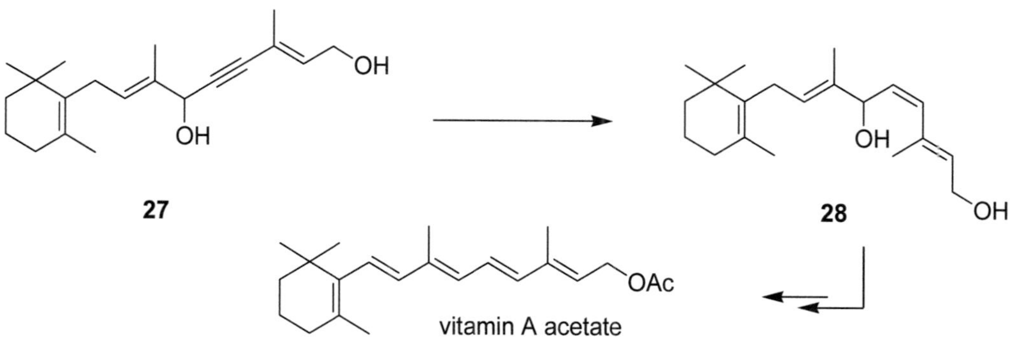


**Scheme 7.** Semi-hydrogenation of a vitamin A intermediate.

Since this early success the "Lindlar catalyst" (as it has become known) has been used in many different production processes. It is of particular importance in the synthesis of vitamins A & E and also intermediates for the fragrance industry. An important starting material in DSM Nutritional Products's production of such compounds is methylbutenol (MBE, **30**, Scheme 8). MBE is synthesised by the partial hydrogenation of the corresponding alkyne MBY (**29**) in a batch-wise process. Selectivity is very high (>98%) and the catalyst can be recycled multiple times.

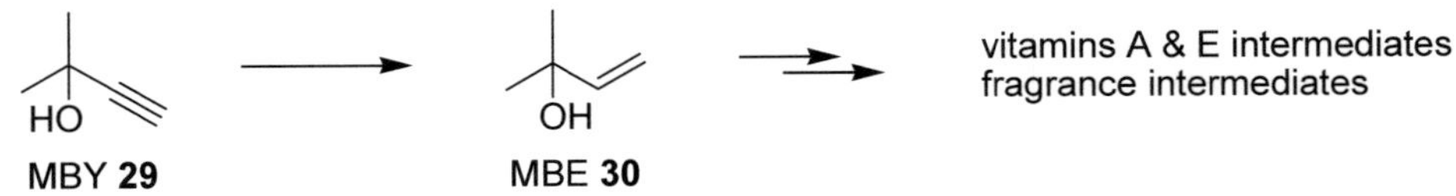


**Scheme 8.** Semi-hydrogenation of MBY.

From MBE, the chain is extended in a sequential manner to obtain dehydroisophytol (**31**, Scheme 9). This is then reduced in another semi-hydrogenation to give isophytol (**2**, cf. Scheme 2) [15]. Isophytol can then be coupled with TMHQ (**1**), as described previously, to form $\alpha$-tocopherol (cf. Scheme 1). As with MBY-MBE, the hydrogenation is carried out in a batch-wise process at 2-5 bar hydrogen pressure.

**Scheme 9.** Preparation of isophytol by semi-hydrogenation.

Two compounds of interest to the fragrance industry are linalool (**11**) and linalyl acetate (**33**, Scheme 10). Both have pleasant floral and spicy odours and are found in a wide range of natural flowers and spice plants. Their main uses are as perfume components in soaps, shampoos and lotions. They can both be synthesised by semi-hydrogenation of **8** and **32** using Lindlar catalysts, however the reaction conditions had to be optimised independently since even minor changes to the substrate structure can significantly affect the hydrogenation selectivity.

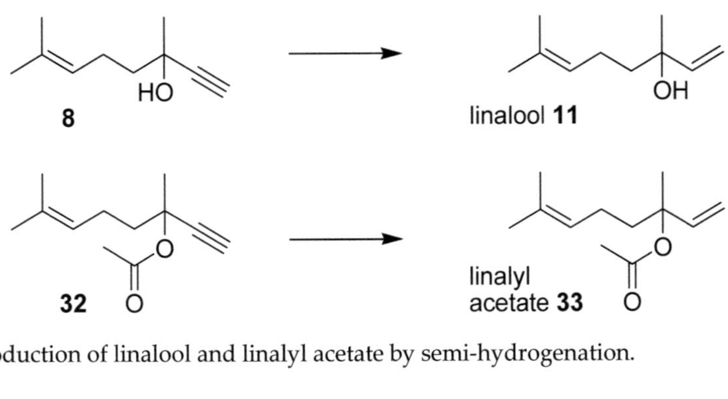


**Scheme 10.** Production of linalool and linalyl acetate by semi-hydrogenation.

As an extension to the work above, another fragrance compound, dimethyloctenol (DMOE, **36**), can be prepared by the combination of a hydrogenation of a C=C double bond and a selective semi-hydrogenation of a C≡C triple bond (Scheme 11). Methylheptenone (**34**) can be hydrogenated with a range of Pd/C or Pd/$Al_2O_3$ catalysts to give methylheptanone (**35**). The reaction was run without solvent at a range of temperatures (30-80 °C) and pressures (1-10 bar hydrogen). The latter compound was ethynylated to give the alkyne **14**, which can undergo semi-hydrogenation with high selectivity (up to 95%) to give the desired tertiary allyl alcohol **36**. Lindlar catalysts with varying amounts of lead-doping were successful at moderate temperatures (20-40 °C) and pressures (1-10 bar hydrogen) [36].

New approaches for the application of Lindlar-type catalysts are the use of supported palladium nanoparticles. By carefully controlling their preparation, a narrow range of diameters can be obtained and deposited on a carbon support [37,38]. These catalysts allow the hydrogenation of C≡C bonds with low metal loadings and in several cases with a high selectivity, e.g. hex-3-yne can be hydrogenated to hex-3-ene in high selectivity at full conversion, however to the best of our knowledge, these have not yet been applied on an industrial scale for the production of vitamins and fine chemicals.

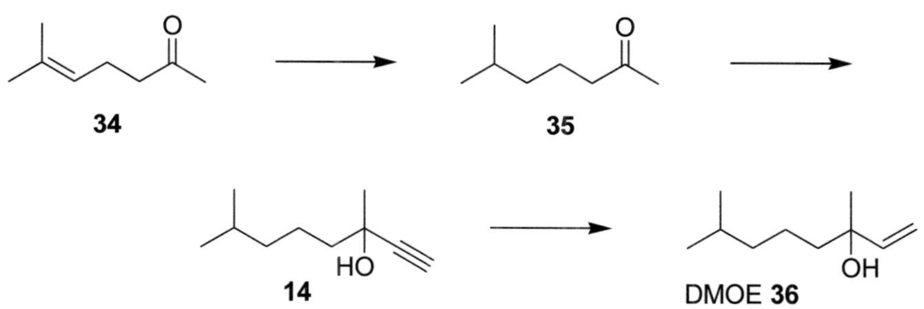


**Scheme 11.** Synthesis of DMOE by combination of two different kinds of hydrogenation.

Further trends in the research on Lindlar hydrogenations are focusing on the addition of $FeCl_2$ and tetramethylammonium chloride to the catalyst, and the use of palladium on metal sintered fibers. This allows the hydrogenation of triple bonds in a continuous reaction mode, also with low Pd loadings, in which Pb doping is not necessary [39-42].

Resveratrol (**40**, Scheme 12) is a polyphenol that can be isolated from natural sources such as red grapes or giant knotweed. It has attracted significant attention due to its possible health benefits in areas such as anti-cancer, anti-aging, anti-inflammatory and cardiovascular protection. The synthesis of resveratrol has been reported a number of times and can be industrially performed using Mizoroki-Heck coupling of 3,5-diacetoxy-styrene and 4-acetoxybromobenzene as key step [43,44].

An alternative approach to resveratrol uses the selective hydrogenation of tolan derivatives **37** in presence of a Lindlar-type catalyst to form intermediate **38** [45]. The precursor **37** can be synthesized by Sonogashira coupling (Scheme 12). In general this procedure can be applied to the synthesis of electron-rich stilbene derivatives as in the preparation of natural compounds like combretastatin (**39**) [46].

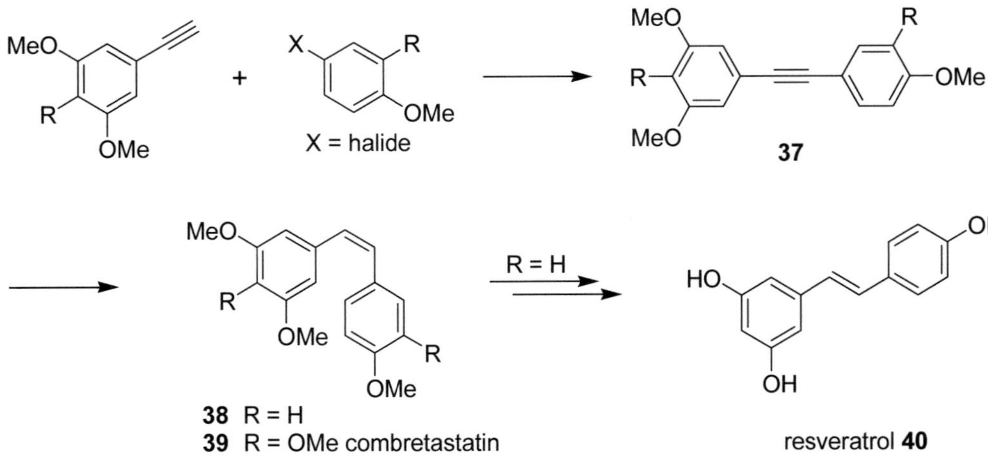


**Scheme 12.** Synthesis of combretastatin and resveratrol via Sonagashira coupling and semi-hydrogenation.

Z-Butene-1,4-diol (**42**) is an intermediate in the synthesis of vitamin $B_6$ (**43**), and is synthesized by selective hydrogenation of butynediol (**41**) in presence of a Pd/$CaCO_3$ catalyst (Scheme 13) [47,48]. The hydrogenation also proceeds well in water. The application of Pd/C or Pd/$Al_2O_3$ has been described under similar conditions [49]. The hydrogenation can also be carried out in a monolith bubble column reactor. Full conversion and high selectivity can be achieved in presence of a Pd-catalyst [50].

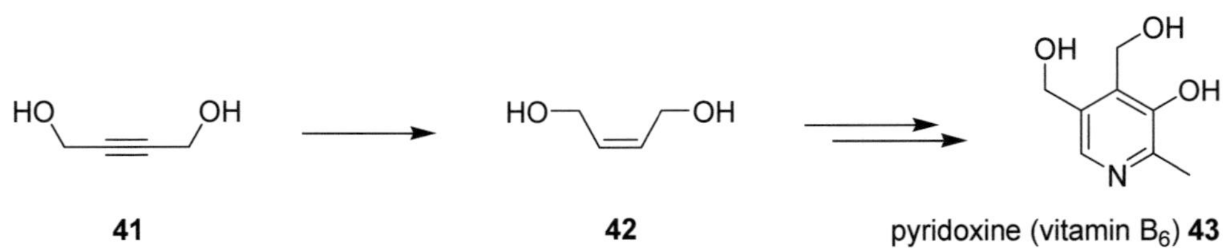


**Scheme 13.** Semi-hydrogenation of butynediol for vitamin $B_6$.

As described above, Lindlar hydrogenations are usually carried out in a batch-wise mode in presence of a Pb-doped Pd catalyst on a calcium carbonate carrier. However, new trends in the selective semi-hydrogenation of C≡C bonds are continuous processing and the application of environmentally friendly solvents. The application of supercritical fluids (sc-fluids) in fine chemical processes, e.g. extractions or as process solvent is well documented [51].

Lindlar-type hydrogenations in supercritical fluids, e.g. sc-$CO_2$, can be carried out in a continuous manner applying a plug-flow reactor set-up. The set-up allows the usage of a new type of Pd-catalyst, amorphous $Pd_{81}Si_{19}$ in a Pb-free system [52].

## 4. Hydrogenation of C=O and C≡N functional groups

As well as the hydrogenation of C-C multiple bonds, the hydrogenation of C-X multiple bonds is used extensively in the production of vitamins and fine chemicals.

L-Ascorbic acid (vitamin C, **46**) is the vitamin that is produced on the largest volume worldwide (approximately 100,000 tonnes per year). The most important method for its industrial manufacture is the Reichstein process [53], which transforms D-glucose (**44**) into L-ascorbic acid *via* a mixed series of high-yielding chemical and microbiological steps. In the first step of the synthesis sequence D-glucose is hydrogenated to D-sorbitol (**45**) in presence of a Ni-alloy catalyst (Scheme 14). The reaction is usually carried out at high pressure and elevated temperature in a batch-mode or continuous process. Under these conditions the product is obtained in high selectivity and almost quantitative yield with only minor amounts of D-mannitol and L-iditol as by-products.

Alternatively to the catalytic hydrogenation, microbiological [54] and electrochemical [55] methods also are known for the reduction of D-glucose to D-sorbitol.

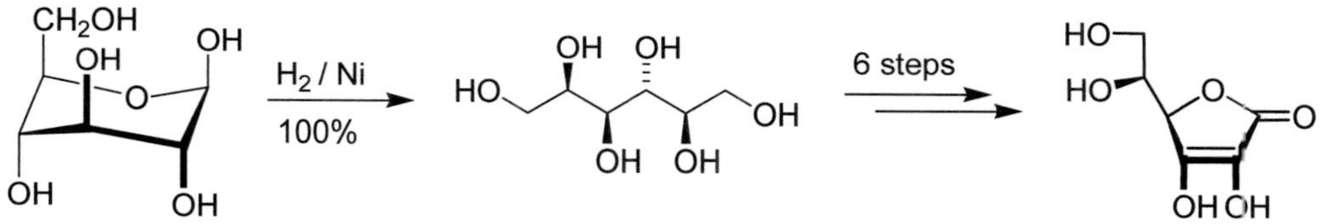


**Scheme 14.** Reduction of D-glucose to D-sorbitol in the Reichstein process for the synthesis of vitamin C.

Vitamin $B_1$ (thiamine chloride, **49**, Scheme 15) contains two heterocyclic rings linked by a methylene unit. In all industrial syntheses, the key intermediate is Grewe diamine (**48**), onto which the thiazole ring is constructed. In general, the diamine is prepared by hydrogenation of a pyrimidino nitrile (**47**) using a nickel-alloy catalyst. This was first reported in 1944 by Huber [56] who found that the use of palladium and platinum supported catalysts gave significant amounts of secondary amine **50** as by-product. The use of nickel catalysts also gave significant quantities of secondary amines, but these could be reduced to less than 5% by the addition of ammonia to the reaction mixture.

Modern industrial syntheses still use the same basic process, although optimisation has improved the reaction significantly, further reducing the unwanted by-products. Reactions proceed in a batch-wise process with a solvent saturated with ammonia at moderate (>10 bar) hydrogen pressure to ensure high activity and throughput. The catalyst can be recycled multiple times and usually remains in the reaction after batch. The choice of nickel catalyst is also important and recently nickel-alloy catalysts (Centoprime®) have been developed that reduce amine by-products further [57-60].

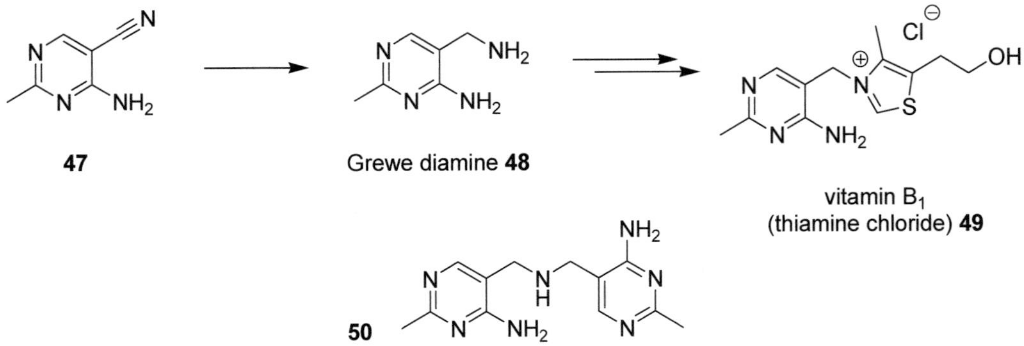


**Scheme 15.** Production of Grewe diamine for vitamin B1.

Resveratrol (**40**) can be produced industrially by an alternative route to that described earlier (cf. Scheme 12). A key intermediate in this production is the benzylic alcohol **52** [43]. This is prepared by the hydrogenation of acetophenone derivative **51** using a transition metal catalyst [61]. A number of catalysts can successfully perform this transformation, but the suppression of by-products is essential. Use of palladium or platinum on carbon resulted in yields of up to 90% (depending on the solvent); however levels of critical by-products were too high to be applied on large scale. The problems were solved by the use of nickel-alloy catalysts. Using these catalysts in ethyl acetate (2-10 bar hydrogen) allowed the production of the required alcohol **52** in greater than 95% yield. In addition the catalyst can remain in the reactor and be recycled multiple times resulting in a cost effective process.

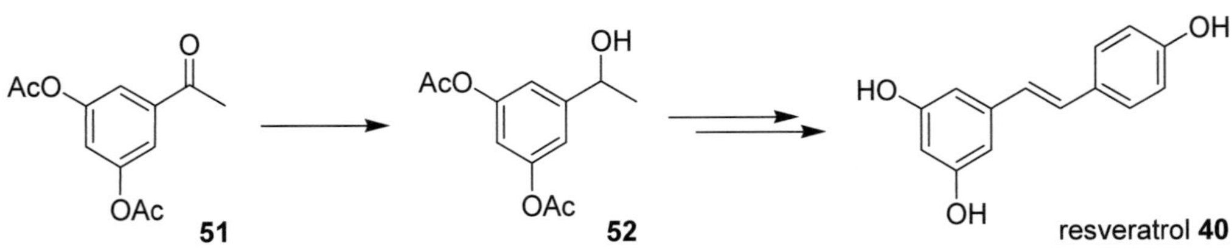


**Scheme 16.** Hydrogenation of a ketone for the synthesis of resveratrol.

## 5. Stereoselective hydrogenation of C=C double bonds

Stereoselective processes, in particular the asymmetric hydrogenation of C=C double bonds, play an increasingly important role in the total synthesis of isomerically pure biologically active products. The naturally occurring fat soluble antioxidant (2*R*,4′*R*,8′*R*)-α-tocopherol (**53**, Scheme 17) is not available in sufficient amounts from natural source starting material. In the course of the considerable efforts towards an economic total synthesis of **53** during the last decades [16], exceptionally efficient new asymmetric hydrogenation processes were developed for the introduction of chirality to the aliphatic tocopherol side chain.

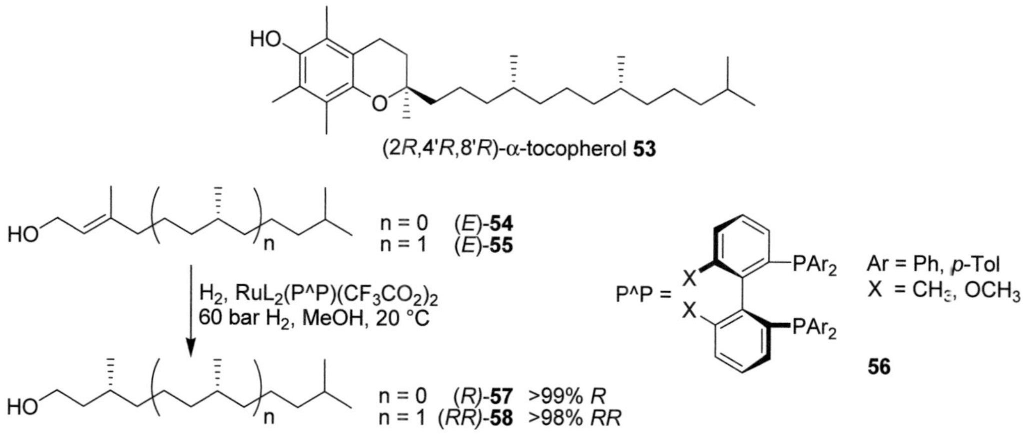


**Scheme 17.** Asymmetric hydrogenation reactions of allylic alcohols in isoprenoid chemistry.

The homogeneous asymmetric hydrogenation of allylic alcohols catalyzed by ruthenium complexes could be performed on pilot scale with substrate-to-catalyst ratios of up to 150'000 (Scheme 17). The $C_{10}$-building block (*E*)-**54** was transformed into (*R*)-**57** with >99% selectivity by using (*S*)-MeOBIPHEP (**56**, Ar = *p*-Tol, X = $OCH_3$) as ligand. Under similar conditions, hydrogenation of (*E*)-**55** gave (*R*,*R*)-**58** (>98% *RR*) with the catalyst derived from (*S*)-*p*-Tol-BIPHEMP (**56**, Ar = *p*-Tol, X = $CH_3$) [62].

Even two chiral centers can be introduced by the one-pot reduction of unfunctionalized trialkyl substituted olefins in the presence of Ir-$BAr_F$ complexes containing chiral P,N-ligands (Scheme 18). By applying this novel retrosynthetic concept, an (all-*R*)-tocopherol derivative could be obtained for the first time from the corresponding (all-*E*)-tocotrienol derivative in a collaboration of the Pfaltz group with DSM Nutritional Products. Asymmetric hydrogenation of γ-tocotrienyl acetate (*R*,*E*,*E*)-**59** with pyridyl phosphinite **60** yielded acetate **61** with complete conversion and excellent stereoselectivity (>98% *R*,*R*,*R*) [63,64].

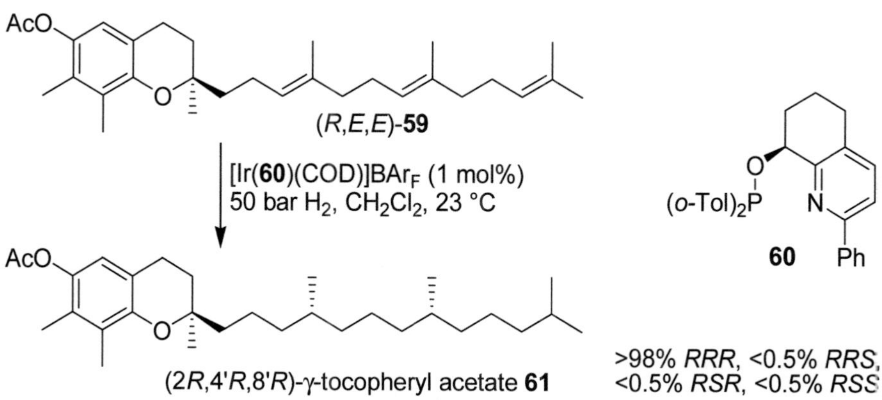


**Scheme 18.** Asymmetric hydrogenation reactions of unfunctionalized trisubstituted olefinic double bonds.

A Rh(I)-catalysed highly diastereoselective hydrogenation is the basis for a very short synthesis of (+)-biotin (**26**) developed by Lonza and applied on a technical scale (Scheme 19). The hydrogenative key transformation was the result of a cooperation with the catalysis group of former Ciba-Geigy [65]. After ligand screening and optimisation, the conversion of substrate **63**, easily accessible from diketene *via* cheap tetronic acid (**62**), could be improved to high selectivity (**65a:65b** >99:1) with ferrocene derived josiphos2 (**64**) as ligand. Although the synthesis *via* thiolactone **66** and subsequent stereoselective heterogeneous hydrogenation of olefin **67** (cf. Chapt. 2, Scheme 6, **23**→**24**) was operated on tonne scale, production had to be terminated due to high production costs. The final debenzylation step to yield **26** led to destruction of the chirality in the expensive (*R*)-methylbenzylamine auxiliary used for introduction of one of the two nitrogen functionalities.

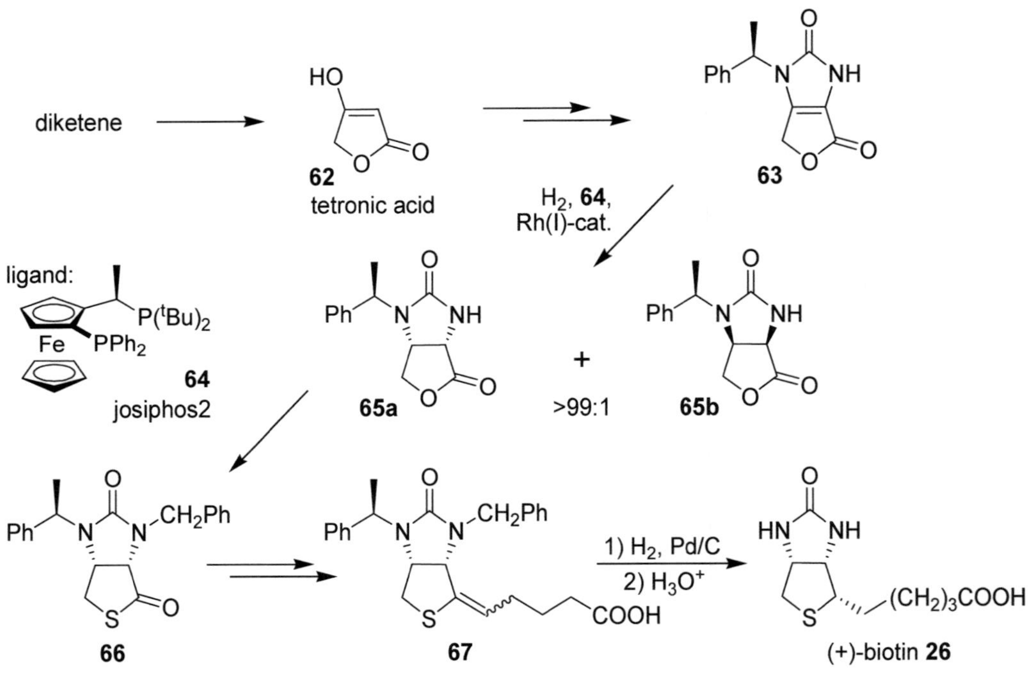


**Scheme 19.** Diastereoselective hydrogenation in the Lonza procedure to (+)-biotin.

The chiral D-lactone **70** is the key intermediate in all other commercially interesting biotin syntheses (Scheme 20) [66,67]. In former times, diastereomeric acetals derived from hydroxylactone **69** were used. *Meso*-compound **72** (obtained from fumaric acid **68** *via* diacid **71**) was recognised as an easily available starting material for introduction of the optical activity. Classical optical resolution to form diastereomeric imides or ammonium salts, diastereoselective ring opening with chiral alcohols (e.g. → **74**), as well as enzyme catalyzed transesterification reactions were used. All those processes, however, are elaborative and need chiral auxiliaries in stoichiometric amounts which have to be recycled. Often, additional protective group transformations are necessary. Until 2006, the direct desymmetrisation of anhydride **72** to D-lactone **70** and further elaboration to thiolactone **73** was only possible by reduction with expensive (*R*)-BINAL-H in over-stoichiometric amounts under low-temperature conditions [68], which is not applicable on larger scale.

A breakthrough was achieved in an inter-company cooperation, between DSM Nutritional Products and the catalysis group of Solvias [66,67]. The asymmetric hydrogenation of anhydride **72** to D-lactone **70** with Ir- and Rh-complexes with atropisomeric ligands [69] yielded best results. After a short series of screening and optimisation experiments, full conversion and *ee* values of >95% could be reached, resulting in successful production trials on tonne scale (Scheme 21). The transformation of thioanhydride **75** to thiolactone **73** needed some more drastic reaction conditions, delivering lower selectivities and moderate yield

(*R*)-Pantothenic acid (*R*-**77**) is naturally occurring as a component of coenzyme A and is essential for various biochemical processes. *R*-**77** and its commercial form calcium (*R*)-pantothenate are industrially produced by optical resolution from racemic pantolactone (*RS*-**76**, Scheme 22) as the key intermediate. The latter is prepared from isobutyraldehyde, formaldehyde, and hydrogen cyanide, and subsequent acidic hydrolysis [70]. Alternative processes rely on the enantioselective hydrogenation of 2-oxopantolactone (**78)** [71]. With a rhodium catalyst containing *m*TolPOPPM as a ligand, high turn-over numbers (TON) of 200'000 and high optical purity (91% *ee*) could be obtained. The heterogeneous version of this transformation using a Pt/alumina catalyst in the presence of a chiral modifier, e.g. cinchonidine, delivered very similar *ee* values [72].

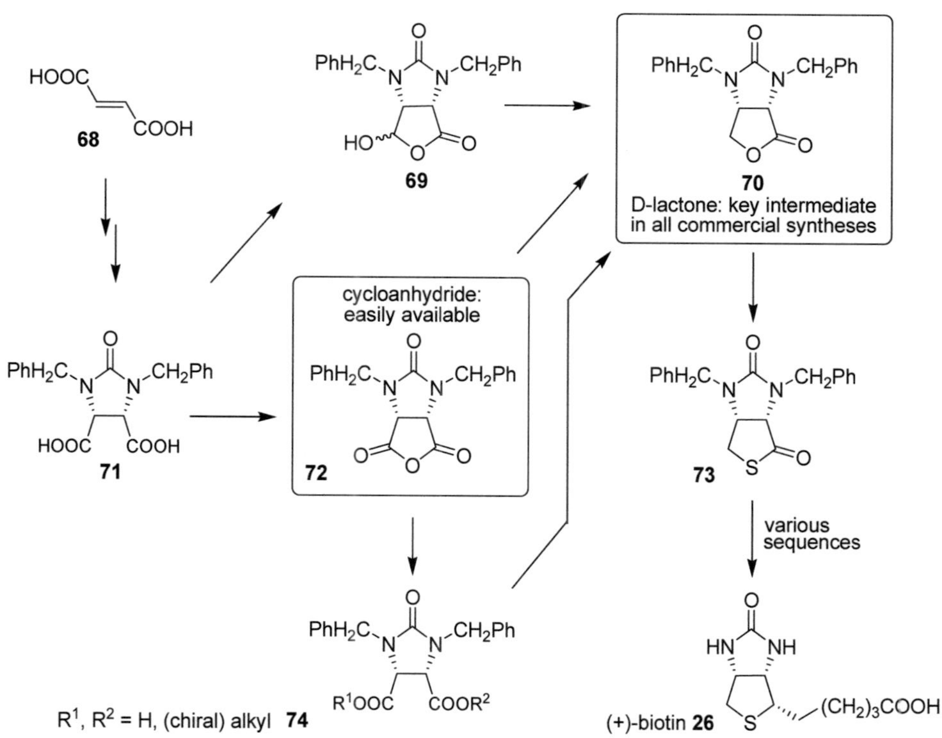


**Scheme 20.** Synthetic strategies for the introduction of optical activity into (+)-biotin.

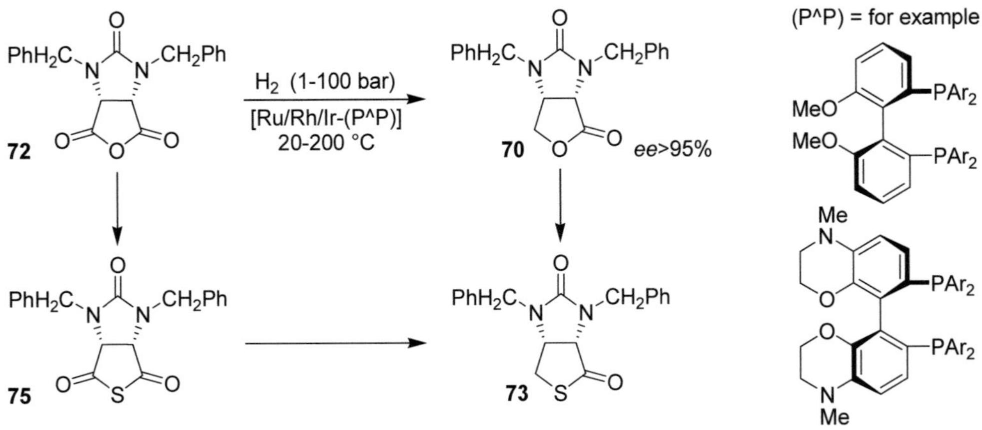


**Scheme 21.** Preparation of an optically active lactone by catalytic asymmetric hydrogenation.

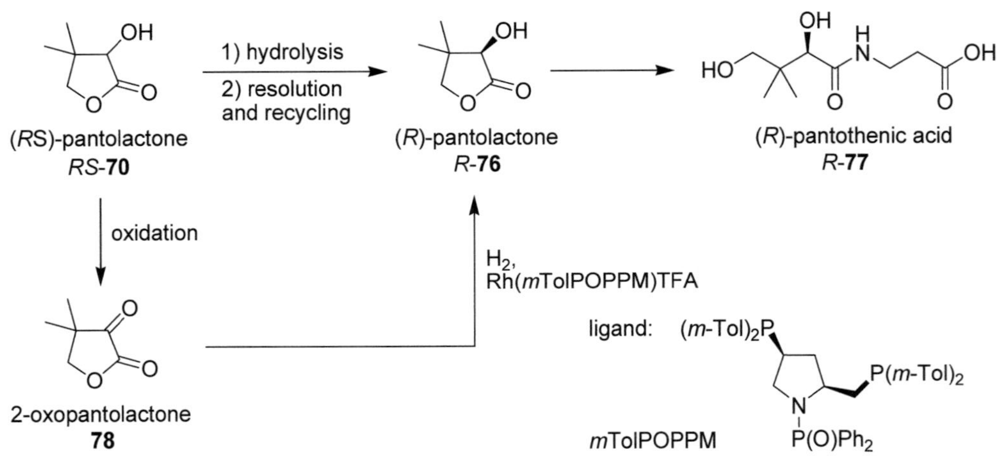


**Scheme 22.** Enantioselective hydrogenation of 2-oxopantolactone in the synthesis of pantothenic acid.

(3*R*,3'*R*)-Zeaxanthin [(3*R*,3'*R*)-**83**] is found in the human eye and is of interest for the treatment of age-related macular degeneration (AMD) [73] which is a major cause of blindness in the Western World. (3*R*,3'*R*)-**83** occurs in nature in corn (maize) and egg yolk. Its total synthesis starts from ketoisophorone (**79**, Scheme 23) which is efficiently transformed by yeast fermentation into the chiral intermediate (*R*)-levodione [(*R*)-**80**]. The overall reaction scheme consists of various reaction steps (*via* (*R,R*)-**81**, **82**). Stereoselective hydrogenation of (*R*)-levodione [(*R*)-**80**] to *trans*-actinol, however, was difficult to achieve [74].

While the heterogeneous nickel catalyzed hydrogenation of (*R*)-**80** yielded a 80:20 mixture of (*R,R*)-*trans*-actinol [(*R,R*)-**81**] and the corresponding (4*S*,6*R*)-isomer, the homogeneous Ru-BINAP catalyzed asymmetric hydrogenation gave only poor chemo- and enantioselectivities. The breakthrough could be achieved by a ruthenium catalyzed asymmetric transfer hydrogenation with isopropanol using a dianion complex [Ru(N-*p*Ts-ethylenediamine){-2H}($\eta^6$-arene)] [75]. At a substrate-to-catalyst ratio of around 1000, high *de* values were obtained when performing the reaction at room temperature and reaction times below 24 h. This was the basis for up-scaling and transfer to technical implementation.

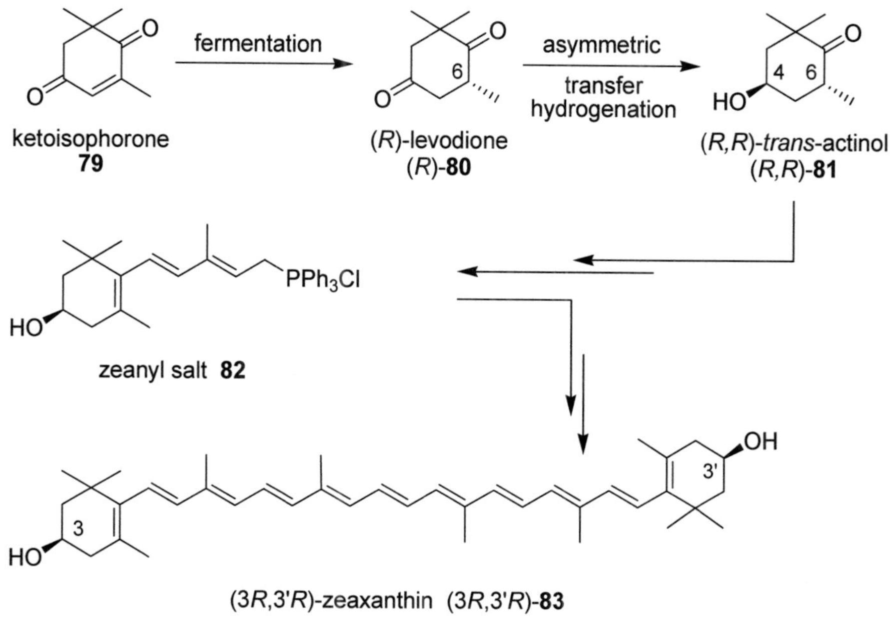


**Scheme 23.** Asymmetric transfer hydrogenation of (*R*)-levodione in the synthesis of (3*R*,3'*R*)-zeaxanthin.

## 6. Conclusions

Various types of hydrogenation reactions are indispensable parts of economically and ecologically beneficial manufacturing processes towards valuable products in the vitamins and fine chemicals industry. The increasing importance of environmentally benign production methods is addressed by developing concepts for improving the efficiency of transformations, continuous processing and recycling, and achieving high chemo- and stereoselectivities, thus avoiding laborious separation protocols and waste formation. The examples presented and discussed in detail show many achievements in this field during recent decades, but also the necessity to further search for alternative solutions.

## Author details

Werner Bonrath, Jonathan Medlock, Jan Schütz,
Bettina Wüstenberg and Thomas Netscher*
*DSM Nutritional Products, Research and Development, Basel, Switzerland*

* Corresponding Author

## 7. References

[1] Bonrath W, Netscher T (2005) Catalytic processes in vitamins synthesis and production. Appl. Catal. A: General 280: 55-73.

[2] Sheldon R.A, van Bekkum H (2001) Fine Chemicals through Heterogeneous Catalysis. Weinheim, Germany: Wiley-VCH.

[3] Sheldon R.A (2000) Atom efficiency and catalysis in organic synthesis. Pure Appl. Chem. 72: 1233-1246.

[4] Angrick M, Kümmerer K, Meizner L (2006) Nachhaltige Chemie. Erfahrungen und Perspektiven, Reihe Ökologie und Wirtschaftsforschung, Vol. 66. Marburg, Germany: Metropolis.

[5] United States Environmental Protection Agency (2010), 'Green Chemistry'. Available: http://www.epa.gov/greenchemistry. Accessed 2012 Mar 08.

[6] Anastas P.T, Eghbali N (2010) Green Chemistry: Principles and Practice. Chem. Soc. Rev. 39: 301-312.

[7] Sabatier P, Sendrens J.B (1900) Hydrogenation of ethylene in presence of different metals. Comptes Rendus Hebdomadais des Seances de l'Academie des science 130: 1781.

[8] Sabatier P, Sendrens J.B (1899) Action of hydrogen on acetylene in presence of nickel. Comptes Rendus Hebdomadais des Seances de l'Academie des science 128: 1173.

[9] Lindlar H (1952) Ein neuer Katalysator für selektive Hydrierungen, Helv. Chim. Acta 35: 446-450.

[10] Lindlar H (1954) Hydrogenation of acetylenic bond utilizing a palladium-lead catalyst. US2681938.

[11] Lindlar H, Dubuis R (1966) Palladium Catalyst for Partial Reduction of Acetylenes. Organic Syntheses 45: 89-92.

[12] Knowles W.S (2002) Asymmetric hydrogenations. Angew. Chem. Int. Ed. 41: 1998-2007.

[13] Noyori R (2002) Asymmetric catalysis: science and opportunities. Angew. Chem. Int. Ed. 41: 2008-2022.

[14] Baldenius K.-U, von dem Bussche-Hünnefeld L, Hilgemann E, Hoppe P, Stürmer R (1996) Chapter 4: Vitamin E (Tocopherols, Tocotrienols). In: Ullmann's Encyclopedia of Industrial Chemistry, Vol. A 27. Weinheim, Germany: VCH, pp. 478-488.

[15] Bonrath W, Eggersdorfer M, Netscher T (2007) Catalysis in the industrial preparation of vitamins and nutraceuticals. Catal. Today 121: 45-57.

[16] Netscher T (2007) Synthesis of Vitamin E. Vitamins and Hormones 76: 155-202.

[17] Surmatis J.D, Weber J (1957) Hexahydropseudoionone. US 2783257.

[18] Mirzoeva E.S, Bronstein L.M, Valetska P.M, Sulman, E.M (1995) Catalytic hydrogenation properties of Pd- and Rh-containing polymers immobilized on $Al_2O_3$. Reactive Polymers 24: 243-250.

[19] Goebbel H.-G, Kaibel G, Miller C, Dobler W, Dirnsteiner T, Hahn T, Breuer K, Aquila W (2004) Manufacture of tetrahydrogeranylacetone by hydrogenation of pseudoionone mixture with geranylacetone and/or dihydrogeranylacetone. WO2004007413.

[20] Bonrath W, Kircher T, Kuenzi R, Tschumi J (2006) Process for the preparation of saturated aliphatic ketones. WO2006029737.

[21] Kawaguchi T, Nishida T, Ohmura Y, Tanomura M, Nakao K, Takagi T, Ninagawa Y, Itoi K (1973) 2,3,5-Trimethylhydroquinone. DE2250066.

[22] Tamai Y, Itoi K (1973) 2,3,5-Trimethylhydroquinone. DE2309051.

[23] Schuster L (1971) Trimethylhydroquinone. DE1940386.

[24] Yui T, Ito A, Takata T (1986) Preparation of 2,3,5-trimethylhydroquinone by hydrogenation of 2,3,4-trimethylbenzoquinone in the presence of zeolite-supported platinum group metal catalysts. EP198476.

[25] Broecker F.J, Tavs P, Laas H, Aders W.K (1985) Trimethylhydroquinone. EP152842.

[26] Bonrath W, Netscher T, Schütz J, Wüstenberg B (2012) Process for manufacture of TMHQ. WO2012025587.

[27] Weber F, Rüttimann A (2005) Chapter 5: Vitamin K. In: Ullmann's Encyclopedia of Industrial Chemistry, 7th ed., Vitamins, Vol. A 27. Weinheim, Germany: Wiley-VCH, pp. 47-67.

[28] Rüttimann A (1986) Recent advances in the synthesis of K-vitamins. Chimia 40: 290-306.

[29] Isler O, Doebel K (1954) Syntheses in the vitamin K series. I. The total synthesis of vitamin $K_1$. Helv. Chim. Acta 37: 225-233.

[30] Lindlar H (1957) Verfahren zur Herstellung von Kondensationsprodukten. CH320582.

[31] Hirschmann R, Miller R, Wendler N.L (1954) The synthesis of vitamin $K_1$. J. Am. Chem. Soc. 76: 4592-4594.

[32] De Clercq P.J (1997) Biotin: A Timeless Challenge for Total Synthesis. Chem. Rev. 97: 1755-1792.

[33] Goldberg M.W, Sternbach L.H (1949) Synthesis of biotin. US 2489232.

[34] Goldberg M.W, Sternbach L.H (1949) Synthesis of biotin. US 2489235.

[35] Isler O, Huber W, Ronco A, Kofler M (1947) Synthese des Vitamin A, Helv. Chim. Acta 30: 1911-1927.

[36] Bonrath W, Tschumi J, Medlock J (2012) Process for the manufacture of 3,7-dimethyl-1-octen-3-ol. WO2012025559.

[37] Witte P.T (2009) Process for the preparation of an aqueous colloid precious metal suspension.WO2009096783.

[38] Witte P.T, de Groen M, de Rooij R.M, Donkervoort P, Bakermans H.G, Geus J.W (2010) Highly active and selective precious metal catalysts by use of the reduction-deposition method. Stud. Surf. Sci. Catal. 175: 135-144.

[39] Bonrath W, Kiwi-Minsker L, Renken A, Semagina N (2008) Platinum metal group catalysts and process for the preparation thereof. WO2008101602.

[40] Bonrath W, Grasemann M, Renken A, N. Semagina, Kiwi-Minsker L (2008) Novel catalysts and related hydrogenations. WO2008101603.

[41] Bonrath W, Müller T, Kiwi-Minsker L, Renken A, Iourov I (2011) Hydrogenation process of alkynols to alkenols in the presence of structured catalysts based on sintered metal fibers. WO2011092280.

[42] Bonrath W, Kiwi-Minsker L, Iouranov I (2012) Sintered metal fiber-based catalyst. WO2012001166.

[43] Haerter R, Lemke U, Radspieler A (2005) Process for the Preparation of Stilbene Derivatives. WO2005023740.

[44] Wüstenberg B, Stemmler R.T, Letinois U, Bonrath W, Hugentobler M, Netscher T, (2011) Large-Scale Production of Bioactive Ingredients as Supplements for Healthy Human and Animal Nutrition. Chimia 65: 420-428.

[45] Letinois U, Bonrath W (2011) Preparation of substituted electron rich diphenylacetylens and stilbene derivatives. WO201109888.

[46] Letinois U, Bonrath W, Roth F (2009) Heterogeneously catalyzed access to combretastatin. Poster presentation, Regio Symposium. Rheinfelden, Germany, 23.09.-25.09.2009.

[47] Pauling H, Weimann B.J (1996) Chapter 8: Vitamin B6. In: Ullmann's Enzyclopedia of Industrial Chemistry, Vol. A 27. Weinheim, Germany: VCH, pp. 530-540.

[48] Fukuda T, Kusama T (1958) Partial hydrogenation of 1,4-butynediol. Bull. Chem. Soc. Jpn 31: 339-342.

[49] Hort E.V (1978) Butenediol. DE2818260.

[50] Winterbottom, J.M, Marwan H, Stitt E.H, Natividad R (2003) The palladium catalysed hydrogenation of 2-butyne-1,4-diol in a monolith bubble column reactor. Catalysis Today 79-80: 391-399.

[51] Bonrath W, Karge R (2002) Application of Supercritical Fluids in the Fine Chemical Industry. In: High Pressure Chemistry, Klärner F.G, van Eldick R, editors. Weinheim, Germany: Wiley-VCH, pp. 398-421.

[52] Tschan R, Schubert M.M, Baiker A, Bonrath W, Lansink-Rotgerink H (2001) Catal. Lett. 75: 31-36.

[53] Oster B, Fechtel U (1996) Vitamin C (L-Ascorbic Acid). In: Ullmann's Encyclopedia of Industrial Chemistry, Vol. A 27. Weinheim, Germany: VCH, pp. 552-554.

[54] Agency of Industrial Sciences and Technology (1985) Production of sorbitol by Fusarium. JP60083588.

[55] Kassim A Bin, Rice C.L, Kuhn A.T (1981) Formation of sorbitol by cathodic reduction of glucose. J. Appl. Electrochem. 11: 261-267.

[56] Huber W (1944) Hydrogenation of Basic Nitriles with Raney Nickel. J. Am. Chem. Soc. 66: 876-879.

[57] Ostgard D.J, Roessler F, Karge R, Tacke T (2007) The treatment of activated nickel catalysts for the selective hydrogenation of pynitrile. Chem. Ind. Catalysis of Organic Reactions 115: 227-234.

[58] Degischer O.G, Roessler F (2001) Modification of catalysts for hydrogenation of nitriles to primary amines. EP1108469.

[59] Ostgard D, Berweiler M, Roeder S (2002) Production of primary and secondary amines by hydrogenation of nitriles and imines using Raney-type catalysts in the hollow bodies. WO2002051791.

[60] Ostgard D, Duprez V, Olindo R, Roeder S, Berweiler M (2006) Process for modifying Ni, Cu, Co catalysts by deposition of carbon compounds and the use of the catalysts. WO2006060749.

[61] Bonrath W, Letinois U, Hugentobler M, Karge R, Lehmann H (2010) Process for Resveratrol Intermediate. WO2010079123.

[62] Netscher T, Scalone M, Schmid R (2004) Enantioselective hydrogenation: Towards a large-scale total synthesis of (R,R,R)-$\alpha$-tocopherol. In: Blaser H.-U, Schmidt E, editors. Asymmetric Catalysis on Industrial Scale. Weinheim, Germany: Wiley-VCH, pp. 71–89.

[63] Bell S, Wüstenberg B, Kaiser S, Menges F, Netscher T, Pfaltz A (2006) Asymmetric Hydrogenation of Unfunctionalized, Purely Alkyl-Substituted Olefins. Science 311: 642-644.

[64] Bonrath W, Menges F, Netscher T, Pfaltz A, Wüstenberg B (2006) Asymmetric hydrogenation of alkenes using chiral iridium complexes. WO2006066863.

[65] Imwinkelried R (1997) Catalytic Asymmetric Hydrogenation in the Manufacture of d-Biotin and Dextrometorphan. Chimia 51: 300-302.

[66] Bonrath W, Karge R, Netscher T, Roessler F, Spindler F (2009) Biotin – The Chiral Challenge. Chimia 63: 265-269.

[67] Bonrath W, Karge R, Netscher T, Roessler F, Spindler F (2010) Chiral Lactones by Asymmetric Hydrogenation – a Step Forward in (+)-Biotin Synthesis. In: Blaser H.-U, Federsel H.-J, editors. Asymmetric Catalysis on Industrial Scale: Challenges, Approaches, and Solutions, 2nd ed.. Weinheim, Germany: Wiley-VCH, pp. 27-39.

[68] Matsuki K, Inoue H, Takeda M (1993) Highly Enantioselective Reduction of *Meso*-1,2-Dicarboxylic Anhydrides. Tetrahedron Lett. 34: 1167-1170.

[69] Bonrath W, Karge R, Roessler F (2006) Manufacture of lactones. WO2006108562.

[70] Kaiser K, de Potzolli B (1996) Chapter 11: Pantothenic Acid. In: Ullmann's Enzyclopedia of Industrial Chemistry, Vol. A 27. Weinheim, Germany: VCH, pp. 559-566.

[71] Schmid R (1996) Homogeneous catalysis with metal complexes in a pharmaceuticals' and vitamins' company: Why, what for, and where to go? Chimia 50: 110-113.

[72] Schürch M, Künzle N, Mallat T, Baiker A (1998) Enantioselective Hydrogenation of Ketopantolactone: Effect of Stereospecific Product Crystallization during Reaction. J. Catal. 176: 569–571.

[73] Britton G, Liaaen-Jensen S, Pfander H, editors (2009) Carotenoids, Vol. 5: Nutrition and Health. Basel, Switzerland: Birkhäuser.

[74] Püntener K, Scalone M (2010) Enantioselective Hydrogenation: Applications in Process R&D of Pharmaceuticals. In: Blaser H.-U, Federsel H.-J, editors. Asymmetric Catalysis on Industrial Scale: Challenges, Approaches, and Solutions, 2nd ed.. Weinheim, Germany: Wiley-VCH, pp. 13-25.

[75] Crameri Y, Puentener K, Scalone M (1999) Verfahren zur Herstellung von trans-(R,R)-Actinol. EP915076.

# Chapter 4

# Homogeneous Chemoselective Hydrogenation of Heterocyclic Compounds – The Case of 1,4 Addition on Conjugated C-C and C-O Double Bonds of Arylidene Tetramic Acids

Christos S. Karaiskos, Dimitris Matiadis,
John Markopoulos and Olga Igglessi-Markopoulou

Additional information is available at the end of the chapter

http://dx.doi.org/10.5772/48739

## 1. Introduction

Homogeneous hydrogenation constitutes an important synthetic procedure and is one of the most extensively studied reactions of homogeneous catalysis. The impressive development of coordination and organometallic chemistry has allowed for the preparation of a wide variety of soluble metal complexes active as homogeneous hydrogenation catalysts under mild conditions. [1,2]

Early advances in chemoselective olefin hydrogenation were dominated by the introduction of homogeneous transition metal complexes. [3-5] Many of them allow for the preferential reduction of carbon-carbon double bonds over a coexisting C=O functionality. [6-11]

Catalysis is a multidisciplinary scientific concept that serves a broad range of industries covering specialty, fine, intermediate, commodity and life science chemicals. Catalysts are commonly used for the hydrogenation of alkenes, alkynes, aromatics, aldehydes, ketones, esters, carboxylic acids, nitro groups, nitriles and imines. These materials may be in the form of bio-, homogeneous, heterogeneous and heterogenised homogeneous catalysts where each type has its own special properties that can be adjusted for their optimal use. The trend is towards selective hydrogenation of specific groups of a fine chemical, leaving all other structural groups and activated cites intact. This is of great importance for the medicinal industry. Although the enantioselective hydrogenation of C=C and C=hetero-atom double bonds lays mostly in the field of homogeneous catalysis, intensive research has led to the use of heterogeneous reactions for those applications. There have been some examples identified, where heterogeneous catalysts show promising results. [12]

The catalytic hydrogenation of $\alpha,\beta$-unsaturated ketones in particular has been widely investigated. This unsaturated system is encountered in many organic structures linear or cyclic. Literature reports the use of heterogeneous catalysts, in addition to homogeneous catalysts. Therefore, it is essential to investigate all relative findings in the area of heterogeneous catalysis, along with homogeneous catalysis to reveal the possibilities for chemoselective and regioselective hydrogenation of conjugated and non-conjugated unsaturated systems.

The implementation of the $H_4Ru_4(CO)_9$[(*S*)-BINAP] complex in the process allows for wide range of possibilities. This versatile four-Ru-nuclei complex is a potential hydrogenation agent, but could as well catalyze a number of undesired side-reactions that need to be avoided. Past reviews have indicated ruthenium as a transition metal with many possibilities depending on the environment provided. [13] Transformations including additions, redox isomerizations, coupling reactions and cycloadditions can be achieved using the appropriate ligands and reaction conditions. On the other hand, a large number of ligands has been developed and tested for their performance in asymmetric hydrogenation. [14] Ligands of particular interest are phosphorus ligands and more specifically phosphane and phosphine ligands (Figure 1), not neglecting the value of phosphonate and phosphinite ligands. [15] In any case, we need to focus on selectivity (enantioselectivity, chemoselectivity, regioselectivity, etc.) for reasons of economy. The word economy does not only refer to substrate or catalyst economy, but also to atom economy. What is demanded today is a process as clean as possible, from which we derive the desired product only, limiting or even extinguishing all byproducts.

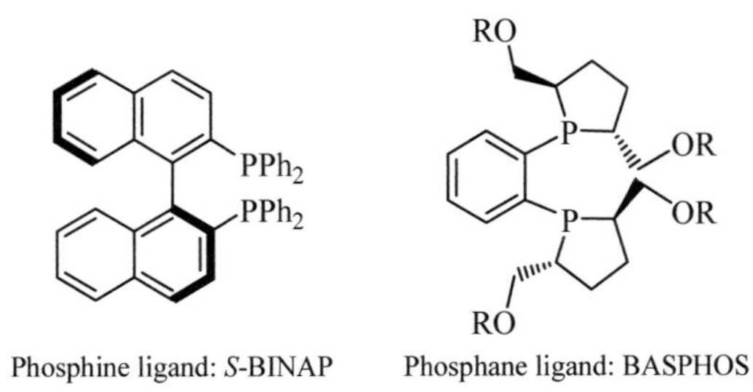


**Figure 1.** Examples of phosphorus ligands for enantioselective hydrogenation.

Arylidene tetramic acids present high structural versatility. The pyrrolidine-2,4-dione nucleus ring can be substituted by a variety of functional groups, incorporating either electron donating or withdrawing characteristics. It is found that 3,5-bisarylidene tetramic acids can be synthesized in high yields, providing a very good and versatile substrate for the catalytic hydrogenation. [16] This multi-conjugated structure includes an $\alpha,\beta$-unsaturated ketone built in the heterocyclic nucleus, that is possible to be subjected to hydrogenation, as it is depicted in Figure 2.

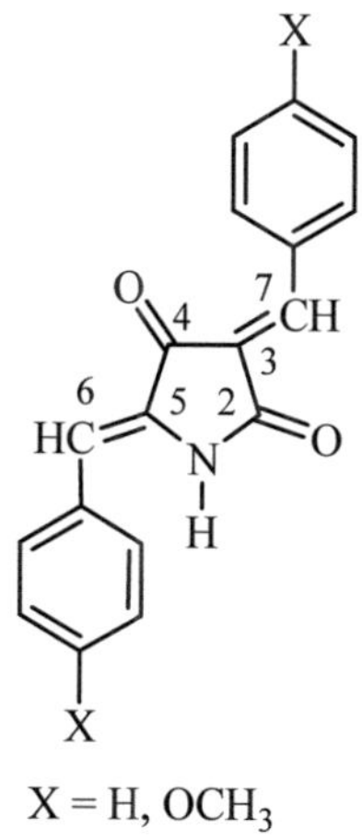


**Figure 2.** 3,5-Bisarylidene tetramic acids.

This unsaturated carbon backbone structure can be divided in sections that exhibit interpretable characteristics. So, we can distinguish two aromatic monoenes (C7-arylidene and C6-arylidene groups) and three *α,β*-unsaturated ketones (C2=O-C3=C7, C4=O-C3=C7 and C4=O-C5=C6 groups). Of course, the extended conjugated structure through the aromatic rings cannot be neglected; however, for reasons of simplicity we will regard this system as an addition of several simple antagonistic unsaturated systems. So, it is clear that we need to investigate the behavior of different unsaturated systems and their antagonistic action in conditions of catalytic hydrogenation if we are to discover the processes involved in the hydrogenation of 3,5-bisarylidene tetramic acids.

## 2. On the hydrogenation of conjugated and non-conjugated olefins

The most common form of a conjugated system is that of conjugated dienes. Indeed, the most common conjugated system, 1,3-butadiene, was discovered in the early 20th Century and since then it has been widely investigated. Aromatic monoenes, and furthermore monoenes are also very important, as they comprise a similar to the conjugated dienes substrate activity towards catalytic hydrogenation. The case of *α,β*-unsaturated ketones is more complex and will be dealt with later in the text. Economics of the hydrogenation process lead to the preferential employment of heterogeneous metal catalysts, since the benefits against the homogeneous catalysts are multiple:

- Cheaper transition metals.
- Cheaper supporting materials.
- Easier to recycle and regenerate.

On the other hand homogeneous catalysts are usually more selective and can be specifically modified to meet the needs for a particular substrate application. This will be further discussed later in the chapter. It is also possible to heterogenize an homogeneous catalyst, compromising with a partial loss of activity.

The competitive addition of hydrogen in the case of non-conjugated double and triple C-C bonds has been studied in the past. [17] It has been found that the hydrogenation of triple bonds, especially terminal triple bonds, is thermodynamically favorable compared to double bonds. In this case, protection of the ethynyl group by silylation of the 1-(4-Ethynylphenyl)-4-propyl-2,6,7-tioxabicyclo[2.2.2]octane, allowed for the selective hydrogenation of the olefinic bond, in moderate rates (Scheme 1). To achieve even higher selectivity immobilized platinum on carbon was employed, since platinum is known to favor the hydrogenation of olefins.

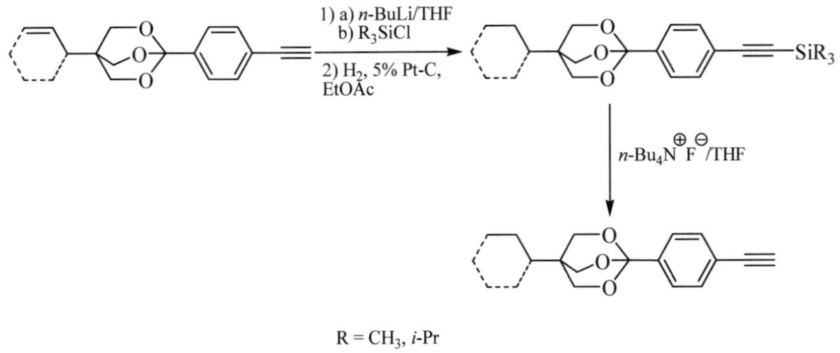


**Scheme 1.** Silylation and selective hydrogenation of the olefinic bond of 1-(4-Ethynylphenyl)-4-propyl-2,6,7-tioxabicyclo[2.2.2]octane.

Studies on the homogeneous catalytic hydrogenation of conjugated dienes and cyclic conjugated dienes point out the steric effects that can potentially inhibit the addition of hydrogen. [18] In particular, conjugated dienes where 1,4-hydrogenated using (Naphthalene)$Cr(CO)_3$ in atmospheric hydrogen pressure, room temperature and polar coordinated solvents. The observations include the following:

- Naphthalene is substituted by the solvent in the Cr complex.
- The hydrogenation follows 1$^{st}$ order kinetics at low concentration of substrate and catalyst.
- 1,4-trans substituted cyclic dienes experience steric effects which results in slow reaction rates.
- The rate determining step is the oxidative addition of hydrogen (isotope kinetic studies).
- Addition of free naphthalene in the solvent decreases the rate of hydrogenation (displaces solvent in the complex formation).

In Scheme 2, the proposed mechanism for the 1,4-hydrogenation is presented.

On the other hand, as Cho and Alper suggest, the application of oxygen-preactivated $[(Bu^t_2PH)PdPBu^t_2]_2$, a binuclear palladium complex, in the homogeneous hydrogenation of various olefin and unsaturated ketone dienes, leads to selective 1,2-hydrogenation. The process proceeds in good yields, under mild conditions and is effective in several cyclic compounds as well. [19]

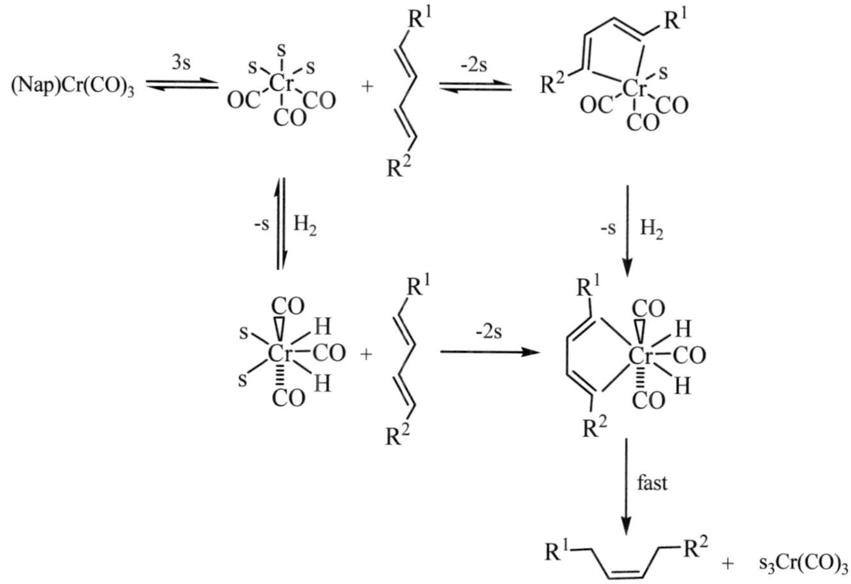


**Scheme 2.** Hydrogenation mechanism for the 1,4-hydrogen addition on conjugated dienes (s = solvent).

Similar results were observed during the study of the catalytic behavior of a dendrimer-bound $PdCl_2$ complex. [20] In this case, cyclopentadiene is hydrogenated in atmospheric oxygen, in a polar solvent (ethanol). This study clearly states the superior performance of polar solvents. The addition of hydrogen occurs only at one of the two conjugated double bonds, leading to a monoene (cyclopentene). No 1,4-hydrogenation is observed. Moreover, the hydrogenation rate is dramatically reduced after the cyclopentadiene is consumed and there is no cyclopentane formation observed.

Dahlén and Hilmersson proved that the employment of THF, a moderate polarity solvent along with the catalytic system of $SmI_2/H_2O$/amine, delivers high versatility and selectivity for the hydrogenation of C-C double and triple bonds. The process proceeds on a high rate and with mild conditions. Depending on the amine employed, different results can be obtained. [21] The main conclusions can be summarized as follows:

- Conjugated dienes are preferably hydrogenated, rather than non-conjugated dienes.
- Cyclic conjugated dienes are 1,2-hydorgenated (no 1,4-hydrogenation occurs).
- Linear or branched conjugated dienes afford a mixture of 1,2 and 1,4-hydrogenated products (still non-conjugated double bonds are not hydrogenated).

The possibility of the selective hydrogenation of non-conjugated dienes to monoenes was demonstrated by more recent studies. [22] The catalytic system of $NiCl_2$-Li-DTBB (4,4'-di-*tert*-butylbiphenyl) in a polar ROH solvent, generates Ni(0) nanoparticles and molecular hydrogen in situ. This enables the selective hydrogenation of cyclic, bi-cyclic and linear non-conjugated dienes to monoenes in mild, atmospheric conditions. Moreover, linear alkenes are completely saturated under these conditions. The system of $NiCl_2$-Li-DTBB works efficiently with THF as well.

In the field of homogeneous catalysis a wide range of Ru and Rh catalysts has been developed. The most common type of ligand in these complexes is the phosphorus-based ligand. The application of these catalysts in enantioselective hydrogenation of olefins has proven to be successful throughout the last decades. The successful hydrogenation results cover many types of olefins, starting from $\alpha$-Dehydroamino acid derivatives, enamides and ($\beta$-Acylamino) acrylates, to Enol esters (Scheme 3). Hydrogenation results are characterized mainly by the mild conditions applied, good enantiomeric excess and high yield. The solvents of choice are mainly MeOH, DCM, EtOH and THF. [15]

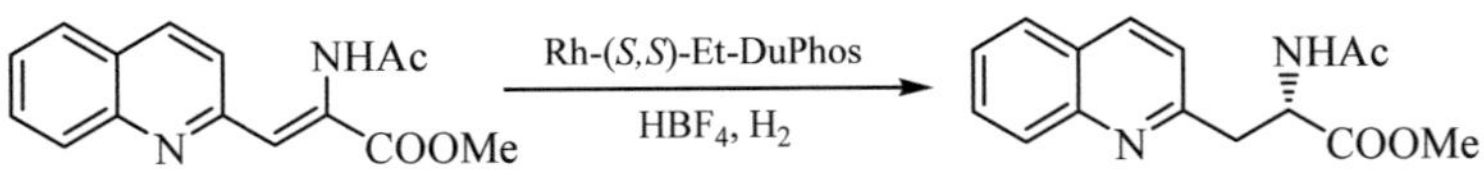


**Scheme 3.** DuPhos homogeneous catalytic hydrogenation of an $\alpha$-Dehydroamino acid derivative.

## 3. The case of ketones

Ketones and unsaturated ketones are an important class of organic compounds since they are biologically active substances or precursors for the synthesis of biologically active substances. In many cases, the required process for the transformation of the precursor to the active derivative is the catalytic hydrogenation process. That is the reason why many researchers focus on the investigation of this process. To pronounce the importance of the catalytic hydrogenation of ketones we will refer to examples for both functionalized and unfunctionalized ketones, analyzing the hydrogenation conditions involved and the catalytic complexes employed. One interesting case is the Ru catalytic hydrogenation of aryl-pyridyl ketones to afford Carbinoxamine precursors. Carbinoxamine is an important histamine H1 antagonist (Figure 3). The Ru complex employed (Figure 3) is very effective especially when the solvent of the reaction is polar (MeOH, EtOH), but best results are derived with 2-propanol in terms of high yield and enantioselectivity (up to 98% e.e.). The reaction conditions are mild and the reaction scheme is presented in Scheme 4. [23]

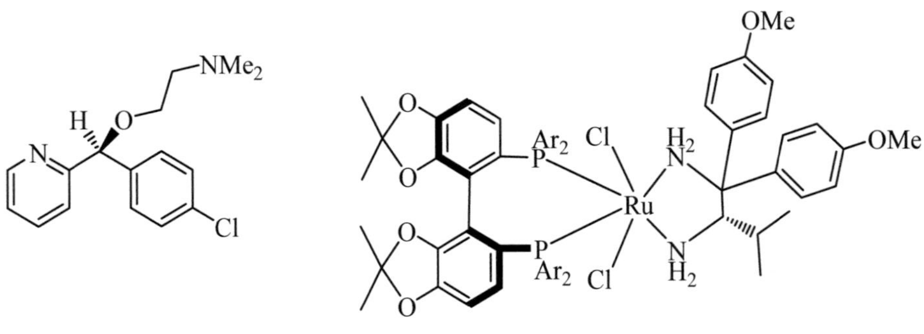


Ar = $C_6H_5$ or 4-$CH_3C_6H_4$ or 3,5-$(CH_3)_2C_6H_3$

Carbinoxamine

Ru(II)-SunPhos/DAIPEN

**Figure 3.** Carbinoxamine, an histamine H1 antagonist and the catalyst Ru(II)-SunPhos/DAIPEN.

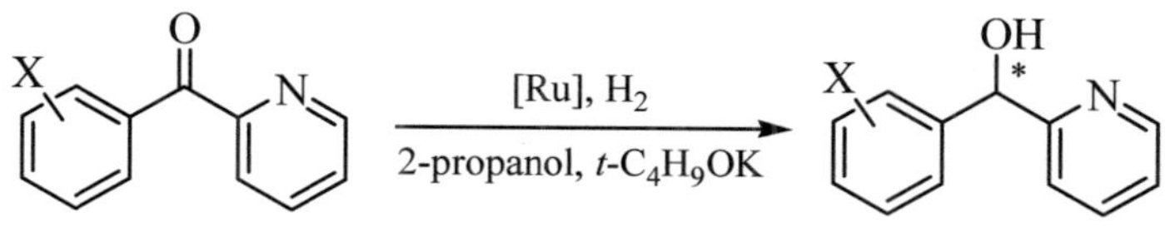


X = $CH_3$, $CH_3O$, F, Cl, Br

Aryl-Pyridyl Ketone

**Scheme 4.** Catalytic hydrogenation of a substituted aryl-pyridyl ketone with Ru(II)-SunPhos/DAIPEN.

The electron density of the central metal of the catalytic system employed has a major contribution in the overall hydrogenation process, since it enables the intermediate hydride formation. [24] Spogliarich et al. showed that Ir complexes present high selectivity in the hydrogenation of conjugated enones. The selectivity is even higher if there is an olefinic bond substitution by an aromatic group. It seems that electron withdrawing groups favor the reduction of the carbonyl group, as it happens when employing Ru catalysts, in hydrogen transfer hydrogenation reactions. In a similar study Spogliarich presented an analogous behavior in the homogeneous catalytic hydrogenation of cyclic conjugated enones with Ir complexes. [25]

V. Ponec investigated in a series of studies the behavior of carbonyl compounds subjected to hydrogenation. In the original study the competitive hydrogenation of acetone and propanal is investigated. [26] The system $Pt/SiO_2$ is employed, promoter-free or promoted by Ga, Ge or Fe compounds. The conditions involve a seldom encountered continuous flow glass reactor, atmospheric temperature and pressure and ethanol as solvent. It is clear by this study that pure Pt preferably catalyzes acetone hydrogenation, whereas doped Pt works better with propanal, with high selectivity. Pd provides very poor results when employed for the hydrogenation of carbonyl compounds, when at the same time presents very low self poisoning effects. When a mixture of acetone and propanal is subjected to hydrogenation certain observations occur. Propanal is much more strongly adsorbed than acetone, affording very low hydrogenation rates. On the other hand, in homogeneous catalysis propanal is more readily hydrogenated. [27] In a more extended work, Ponec denotes the role of ionic admixture to metal/support catalysts for the selective hydrogenation of the carbonyl group of $\alpha,\beta$-unsaturated aldehydes. Focusing on the mechanism of catalysis on such conjugated systems, Ponec suggests an 1,4 adsorption of the unsaturated aldehyde over Pd, that ultimately lowers the selectivity towards C=O hydrogenation. Promoters may induce ensemble effects that minimize the electronic (ligand) effect and favor the hydrogenation of the C=O group. Other mechanism suggestions for the promoter effect involve the modification of the surface composition of the support, the particle size morphology and 'chemical' promotion, like the electrostatic field promotion which changes the field on and occupation of the orbitals mediating the metal-O or metal-C bonds. Regarding the selectively in reducing carbonyl groups, the platinum-group metals can be ranked according to their selectivities as follows:

$$Ir > Pt > Ru > Rh$$

It is established that application of Sn, Ga, Ge or Fe promoters may enhance the effectiveness of these metals by activating the oxygen of the carbonyl group. The activation involves the formation of a chemical bond between the oxygen and the cation of the promoter. The material used as a catalyst support is another important parameter. $TiO_2$ is identified as an excellent support promoting the C=O hydrogenation. [28]

A different approach by Kazuyuki H. et al. showed that the system Pd/C(en), that is expected to be a poor hydrogenation catalyst for single or conjugated ketones, functions rather well with aromatic ketones. [29] This is a mild hydrogenation reaction that takes place in atmospheric conditions. A very important result is derived from this study. The key factor to control the selectivity of the hydrogenation and avoid any undesirable hydrogenolysis on the substrate is to employ THF as a solvent rather than a polar solvent, such as MeOH. This is most important if bis-aromatic ketones are subjected to hydrogenation.

A highly chemoselective and regioselective homogeneous catalytic hydrogenation of unsaturated aldehydes and ketones to unsaturated alcohols is achieved by Jian-Xin Chen et al. employing a binuclear-bidendate Phosphine-Copper(I) hydride complex (Figure 4). [30] The mechanism involves the initial fragmentation of the Cu dimmer to functional monomers. THF, or the system benzene/*t*-butanol are the preferred solvents and a major observation is that a higher hydrogen pressure (above 70psi) inhibits the catalyst's functionality. High chemoselectivity is achieved for cyclic and aromatic ketones. However, 1,2 hydrogenation of $\alpha,\beta$-unsaturated aldehydes and ketones requires a higher hydrogen pressure (above 1atm), and benzene/*t*-butanol as a solvent system. Also, it is not sufficiently regioselective, affording a mixture of hydrogenated products.

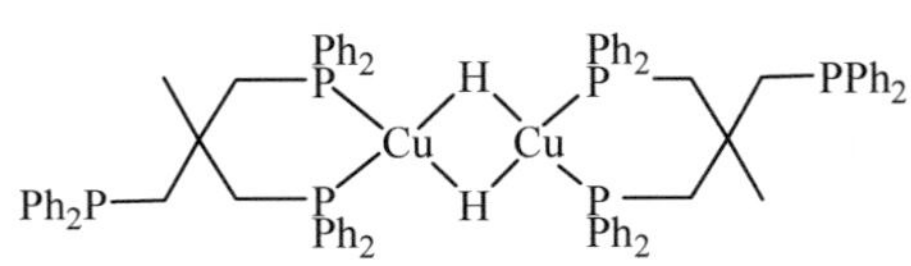


**Figure 4.** Binuclear-bidendate Phosphine-Copper(I) hydride complex: $[(\eta^2\text{-tripod})CuH]_2$, tripod = 1,1,1-tris (diphenylphosphinomethyl)ethane).

In another application of a phosphine monodendate (based on 4,5-dihydro-3*H*-dinaphthophosphepines), with Rh nucleus this time, very good enantioselectivity was derived for the homogeneous hydrogenation of enol carbamates, as presented by S. Enthaler et al (Scheme 5). [31] Very high enantioselectivities are achieved, up to 96% e.e. The optimization study reveals a moderate effect of the temperature and hydrogen pressure to the hydrogenation and enantioselectivity result. The solvents with the best behavior are the polar ones (methanol and ethanol). However, a rather high value of hydrogen pressure (25bar) is derived as optimal. Furthermore, due to ligand dissociation the temperature cannot be well elevated above 90°C.

An alternative to copper hydrides is presented by N. Ravasio et al. [32] This alternative involves the system $Cu/SiO_2$ and is employed for the hydrogenation of $\alpha,\beta$-unsaturated ketones. The reaction conditions are quite mild and the overall performance shows good selectivity. In fact, this catalyst affords only saturated ketones, in both cyclic and linear structures. The only handicap is the use of toluene as a solvent.

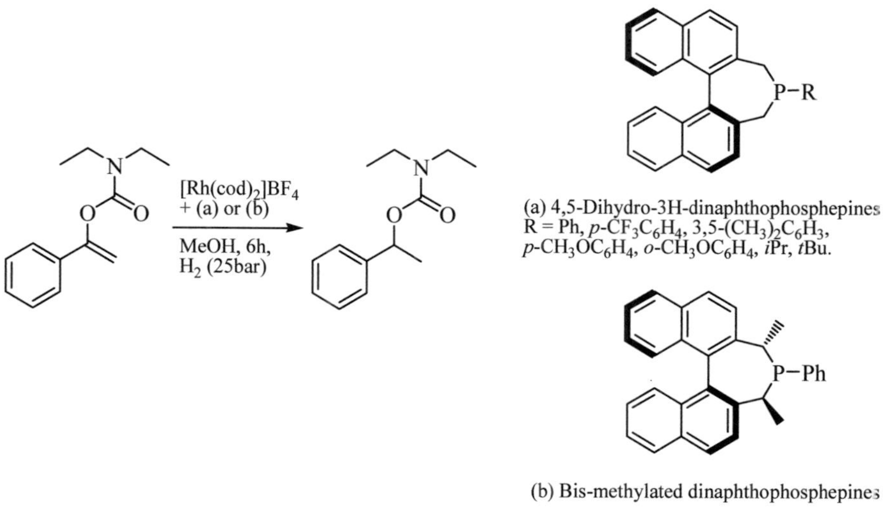


**Scheme 5.** Substituted dinaphthophosphepines homogeneous hydrogenation of enol carbamates.

Two separate studies investigate the application of the more expensive gold/support systems. [33,34] Gold is supported either on $Fe_2O_3$ or $Al_2O_3$, in different weight percentage and different particle size. Depending on the catalyst there is a differentiation in the selectivity during the hydrogenation of *α,β*-unsaturated ketones. Usually, a mixture of a saturated ketone and unsaturated alcohol is derived, with only a small percentage of saturated alcohol. XRD diffraction pattern also reveals low temperature reduction of amorphous iron by the attached gold, to form magnetite ($Fe_3O_4$). These studies confirm that the major role in the hydrogenation selectivity is played by the support, and more particularly the reducibility of the support. Reduced $Fe_2O_3$ increases the electron density of the gold particles, increasing the selectivity towards C=O reduction.

So far, we mostly encounter high selectivities for the hydrogenation of the C=O group of *α,β*-unsaturated ketones. The reverse selectivity may be obtained with the application of a Ruthenium heterogeneous catalyst. [35] The Ru catalyst is employed in the form of Ru nanoparticles immobilized on hectorite. The use of ethanol as solvent and moderate hydrogen pressure (1-10bar) at room temperature ultimately affords saturated ketones, with a selectivity over 99%. Common substrates are 3-buten-2-one and 3-penten-2-one. The highest benefit is the possibility to recycle and reuse the hectorite-supported Ru nanoparticles.

Phosphorus ligand homogeneous catalysts present very poor enantioselectivity results when it comes to *α,β*-unsaturated ketones, lactames, amides and carboxylic acid esters. However, a successful attempt on the hydrogenation of 3-ethoxy pyrrolidinone with a dicationic (*S*)-di-*t*-Bu-MeO-BIPHEP-Ru complex in 2-propanol is reported (Scheme 6).

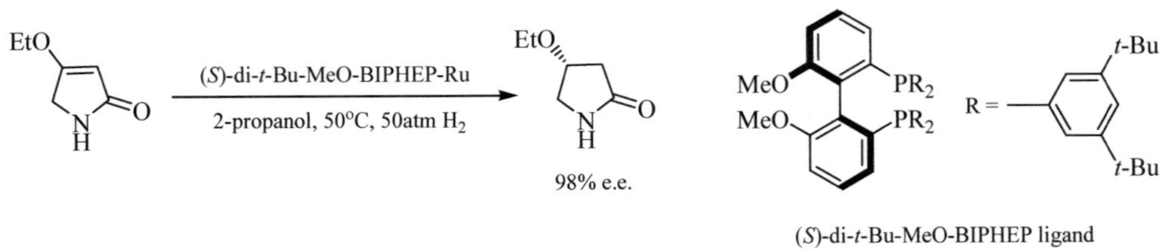


**Scheme 6.** [Ru] homogeneous enantioselective catalytic hydrogenation of 3-ethoxy pyrrolidinone.

In the case of $\alpha$-,$\beta$- and $\gamma$-ketoesters as well as amino ketones, homogeneous chiral Ru catalysts are quite effective, even though the reaction times are prolonged. In the case of unfunctionalized ketones (aromatic ketones, aliphatic ketones, unsaturated ketones) homogeneous Ru catalysts presented low selectivity towards the C=O bond. It was Noyori's catalyst *trans*-[RuCl2(diphosphane)(1,2-diamine)] that first enabled the enantioselective hydrogenation of C=O double bond of $\alpha$,$\beta$-unsaturated ketones and cyclic enones. [15]

Unfuctionalized ketones, and more specifically aromatic ketones, were investigated by Xu et al. [36] The research group discovered that a Ru(II) catalyst containing the BINOL backbone (Figure 5a) is able to successfully hydrogenate a wide range of aromatic ketones, with high yields and up to 99% e.e. All reactions are conducted in 2-propanol. Later, a variation of this amino-phenyl catalyst (Figure 5b) was adopted by Arai et al. for the catalytic hydrogenation of $\alpha$-branched aromatic ketones. High enantionselectivities and diastereoselectivities are achieved in mild reaction conditions even for the hydrogenation of racemic $\alpha$-amido ketones. The solvent of choice is 2-propanol. [37]

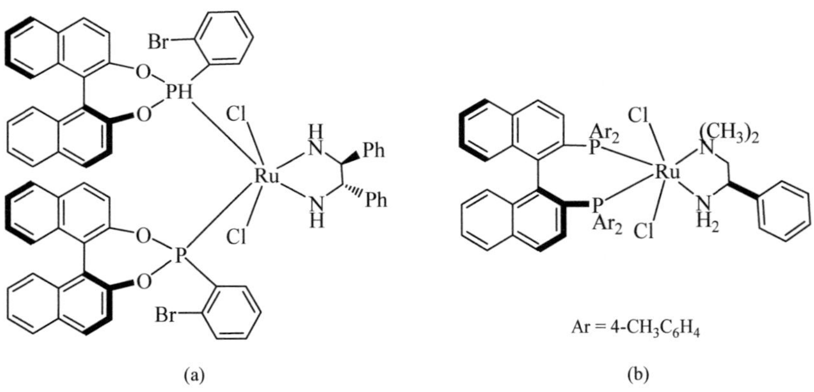


**Figure 5.** (a) Ru(II)-BINOL and (b) Ru(II)-TolBINAP/DMAPEN catalysts for the hydrogenation of aromatic ketones.

Another study on the hydrogenation of aromatic ketones employs complexes Ru(II)-TunePhos/DPEN and Ru(II)-TunePhos/DAIPEN which are similar to Ru(II)-TolBINAP/DMAPEN. [38] These catalytic complexes are proven very efficient for the hydrogenation of a wide range of unfunctionalized ketones. Yields and enantioselectivities are always over 80% and the conditions are quite mild, with the exception of the elevated hydrogen pressure (10-50bar). The solvent of choice is 2-propanol.

Another ketone hydrogenation application for a R(II)-DAIPEN complex is for the synthesis of 3,6,7,8-tetrahydrochromeno[7,8-*d*]imidazoles. The full structure of the catalytic complex this time is: $RuCl_2$[Xyl-P-Phos][DAIPEN]. The hydrogenation reaction is depicted in Scheme 7. This process exhibits high selectivity. However, it requires elevated temperature and pressure conditions to afford good yields (50-80°C and 30-80bar). Rh and Ru catalysts are tested, with various ligands. Best results are derived with the Noyori-type complex $RuCl_2$[Xyl-P-Phos][DAIPEN], that affords 100% yield with over 90% e.e. The solvent of choice is 2-propanol with 1M solution of *t*-BuOK in *t*-BuOH. [39]

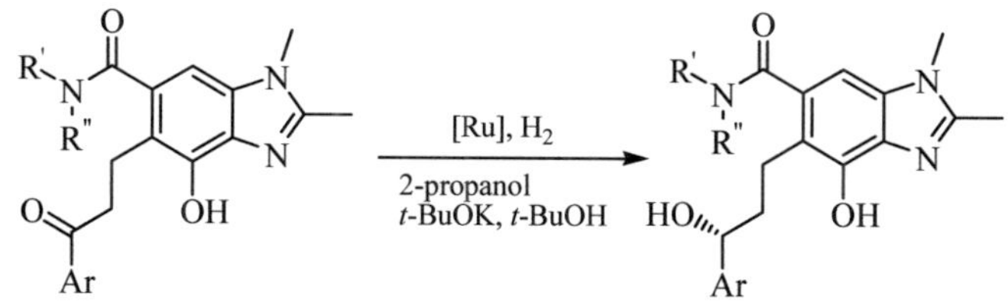


**Scheme 7.** $RuCl_2$[Xyl-P-Phos][DAIPEN] catalytic hydrogenation of ketones for the synthesis of imidazole derivatives.

Another interesting result from the same research is the keto-enol tautomerism that is observed in one of the hydrogenation byproducts (Scheme 8). The conclusions derived are that keto-enol tautomers are very common in the case of exocyclic ketones, and that the formation of significant byproducts cannot be avoided even under thoroughly controlled reaction conditions.

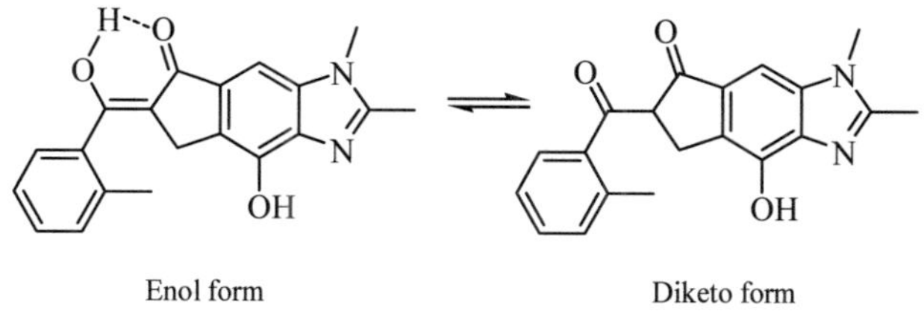


**Scheme 8.** Keto-enol tautomerism of the byproduct formed during the Scheme's 5 hydrogenation.

Exocyclic ketones can be catalytically hydrogenated to afford chiral cyclic alcohols which are important intermediates for the synthesis of a variety of biologically active molecules. In particular, the catalytic hydrogenation of exocyclic $\alpha,\beta$-unsaturated ketones enables the synthesis of exocyclic allylic alcohols (Scheme 9). Xie et al. employed an Ir catalyst (Scheme 9) to achieve this hydrogenation in 2-propanol/*t*-BuOK. The results are remarkable, with a yield over 95% and e.e. over 90%. [40] Earlier, Fogassy et al. attempted to achieve the hydrogenation of differentiated exocyclic $\alpha,\beta$-unsaturated ketones (Scheme 10) using the heterogeneous catalytic system Pd/$TiO_2$/cinchonidine and various polar and non-polar solvents. Pd is recognized as a poor C=O hydrogenation catalyst and the activity is directed towards the C=C double bond. Additionally, the presence of $TiO_2$ substrate and the cinchonidine promoter enhances the functionality and enantioselectivity of the catalyst. Still, the enantiomeric excess of the product reported by Fogassy et al. is not higher than 54%. [41]

In total, we have described two different hydrogenation methods for exocyclic $\alpha,\beta$-unsaturated ketones, which selectively hydrogenate different unsaturated centers.

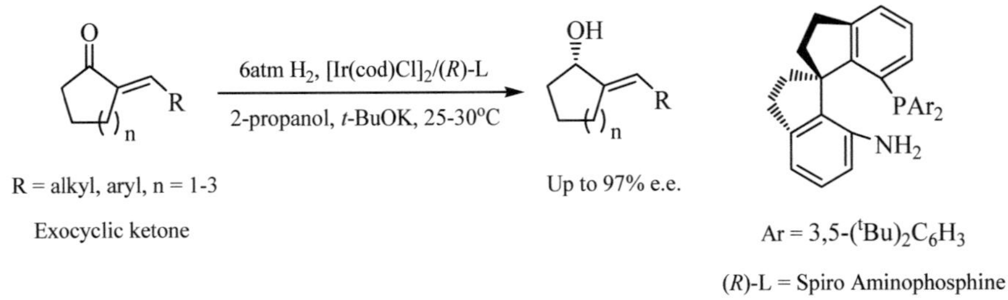


**Scheme 9.** Exocyclic ketone catalytic hydrogenation with the homogeneous $[Ir(cod)Cl]_2$/(*R*)-L catalyst.

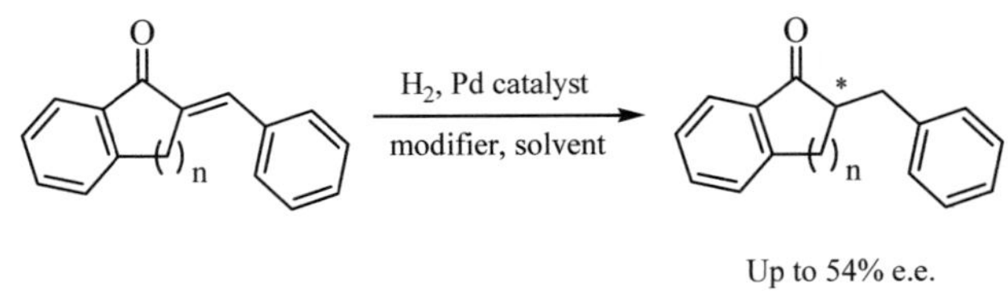


**Scheme 10.** Catalytic heterogeneous hydrogenation of the C=C double bond of exocyclic $\alpha,\beta$-unsaturated ketones with Pd/supported modified catalysts.

After this in depth investigation of the hydrogenation of ketones, given the fact that is a field with the most important advances over the last years, we have to pay attention to the heterocyclic compounds. Indeed, the presence of a carbonylic group in the ring of an heterocyclic compound changes completely the behavior of the C=O double bond. Few advances were made in that field and an insight of this process will reveal the complexity that is hidden in the structures of heterocyclic compounds.

## 4. Hydrogenation of heterocyclic compounds

Heterocyclic compounds are in general more versatile than cyclic or linear organic structures. Electron distribution is largely unbalanced and there are strong intramolecular interactions because of the heteroatom presence in the backbone structure. Usually, it is hard to theoretically disassemble these structures into smaller functional groups to investigate their behavior, but sometimes it is the only way to interpret the experimentally observed results. In this paragraph we will refer to some characteristic examples of catalytic hydrogenation of heterocyclic compounds to emphasize the versatility of these structures.

He et al. successfully attempted to hydrogenate Spiroindene Dimethyl Acetic Acid to (*S*)-Spiroindane Dimethyl Acetic Acid (Scheme 11) using Rh and Ru asymmetric catalysts. Spiroindane and its analogues are potent MC4R antagonists and could be used as a treatment for obesity as well as erectile dysfunction. The hydrogenation reaction proceeds in mild conditions, and Ru is proven to be much more efficient compared to Rh when the same ligands are employed. The solvent of choice is ethanol and conversion reaches 100% in the case of Ru, with good enantiomeric excess. Observing Scheme 11 we can understand that this is an aromatic-conjugated system with an attached heterocyclic ring.

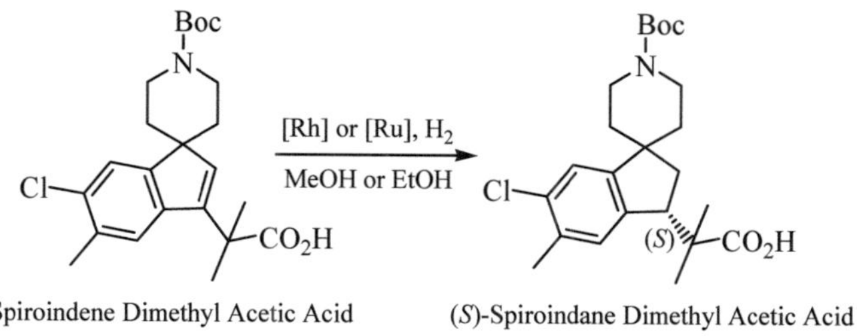


**Scheme 11.** Catalytic hydrogenation of Spiroindene Dimethyl Acetic Acid to (*S*)-Spiroindane Dimethyl Acetic Acid.

Still, the selectivity is high and the extended unsaturated system is not an antagonist to the hydrogenation of the conjugated double bond. [42]

The system of substituted pyridines and quinolines (Figure 6) was investigated by Solladié-Cavallo et al. This is a complex heterocyclic and polyaromatic system, bearing a number of potential sites for catalytic hydrogenation. As the authors report, they achieve partial hydrogenation of the substrates using $PtO_2$ and either HCl or $CF_3CO_2H$, or mixture of the two. Depending on the conditions applied and the substrate employed different products are derived. Quinolyl and pyridyl compounds (2-substituted with a carbonyl group) provide clean and total formation of the desired amino alcohol with the catalytic system $PtO_2$/0.5 equiv. HCl. All other hydrogenation conditions lead to complex mixtures of products. When the heterocyclic ring is substituted by an alkyl group, then only the system $PtO_2/CF_3CO_2H$ affords a single clean product with the complete hydrogenation of the aromatic ring. [43]

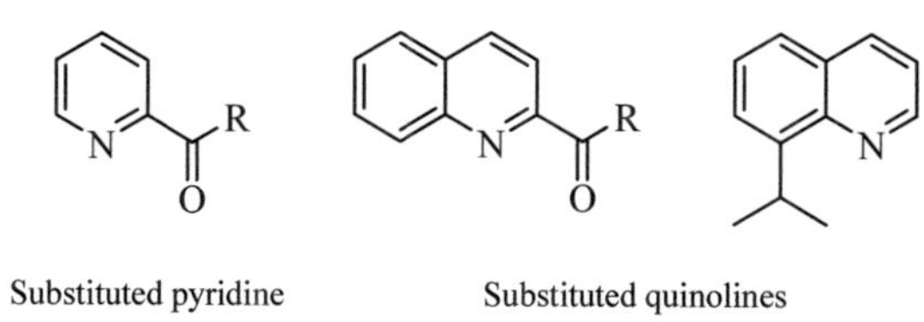


**Figure 6.** Substituted pyridine and quinolines as substrates for catalytic hydrogenation.

It is, therefore, clear that choosing the right conditions' combination for a given substrate is very important for achieving the hydrogenation, on the one hand, and for eliminating all byproducts and deriving a single pure product, on the other hand.

Amides constitute a very interesting moiety of heterocyclic compounds. The hydrogenation of the carbonyl of the heterocyclic ring can lead to important precursors for the synthesis of biologically active compounds. A very interesting point is that the carbonyl hydrogenation leads to a simultaneous ring opening. Ito et al. have successfully hydrogenated enantioselectively various amides and prochiral glutarimides as the first stage of a synthetic process for the production of (-)-Paroxetine, an antidepressant. The hydrogenation requires high temperatures (80°C) and proceeds smoothly in relatively low $H_2$ pressure, in the presence of ($\eta^5$-$C_5Me_5$)Ru(Aminophosphine), 2-propanol and *t*-BuOK. The reaction is presented in Scheme 12. Here, it is important to note the functionality of 2-propanol in the hydrogenation mechanism. Ito et al. suggest that 2-propanol mainly promotes the reaction by participating in the heterolytic cleavage of $H_2$ possibly through a hydrogen-bonding network and it hardly serves as a hydrogen source in the present reaction conditions. [44]

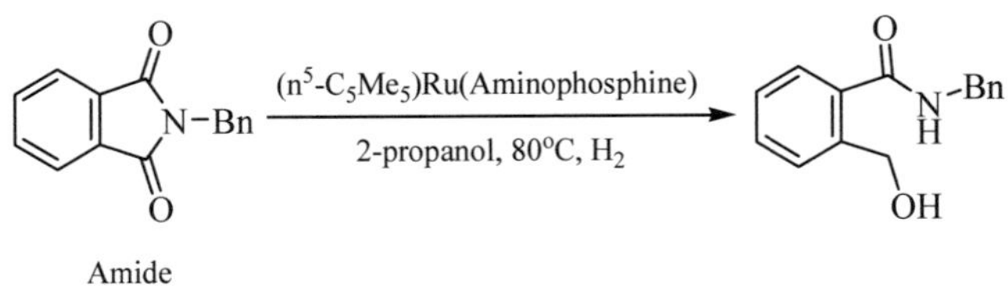


**Scheme 12.** Catalytic hydrogenation of amides by homogeneous Ru catalysts.

## 5. Selectivity in the hydrogenation reactions

We have already encountered the selectivity problem in almost every aspect of the catalytic hydrogenation. Indeed, it is such an important feature of the catalyst and the reaction conditions that cannot be ignored. It defines the refined hydrogenation method that focuses in a particular unsaturated center and which delivers a clean and high quality result. Additionally, we can refer to a few more examples. A very elegant work in the area of hydrogenation selectivity was made by Á. Molnár et al. [45] Covering the catalytic action of heterogeneous palladium catalysts, this could also be viewed under the scope of heterogeneous catalysis. Palladium is recognized as the best metal to achieve high regioselectivity with respect to the hydrogenation of dienes. Also, it exhibits high chemoselectivity in semihydrogenations. Considering conjugated dienes, the main selectivity principle is the same as in all multiple-bond compounds; the terminal double bond presents higher activity and is hydrogenated preferentially. After a monoene is formed, there is competition between the unreacted diene and the formed monoene. As a general principle, conjugated dienes are more reactive than non-conjugated dienes and monoenes. This is attributed to the fact that the entire $\pi$-system of dienes is involved in adsorption through di-$\pi$-coordination, which is more favorable than the d-$\sigma$ mode of adsorption of a single double bond. Versatility in the hydrogenation of dienes is achieved via various supports for Pd: Pd-on-$Al_2O_3$, Pd-on-$SiO_2$, Pd-on-C, Pd-on-graphite, bimetallic Pd samples and so on. Investigation of the hydrogenation of butadiene and isoprene reveals a hydrogenation mechanism that involves 1,4 addition via $\pi$-allyl intermediates. Palladium favors the formation of 1-butene and *trans*-2-butene, in contrast to other metals such as gold and copper. The Pd particles dispersion has a maximum of activity at about 25-30%, and then the activity decreases with increasing dispersion. Carbon deposits partially poison the catalyst, allowing for the adsorption of butadiene that selectively leads to the formation of *n*-butenes. Pd-Cu-on-$Al_2O_3$ hydrogenates butadiene at 99% selectively to butenes, and the system Pd-Au-on-$SiO_2$ retains this selectivity. In the case of 1,5-hexadiene, a non-conjugated diene, the high activity of Pd induces a double bond migration to form 2,4-hexadiene which then is reduced to 2-hexenes. Improvement of the selectivity of Pd towards *n*-hexenes is achieved by employment of $TiO_2$ support. Bimetallic catalysts show the same selectivity in 1,5-hexadiene, as in butadiene. An important application, catalytic hydrogenation of *trans,trans*-2,4-hexadienic acid using Pd-on-C leads selectively to the formation of semihydrogenated methyl *trans*-2-hexenoate, which is the desired product. The conclusions above are visualized in Table 1.

Another confirmation about the Pd selectivity for the hydrogenation of olefin double bonds in conjugated carbonyl systems is presented by B. C. Ranu and A Sarkar. [46] Using the catalytic system ammonium formate/Pd-C they successfully hydrogenated the olefin double bond that is conjugated to a carbonyl group, for a variety of substrates, without affecting the isolated double bonds. High efficiencies and short reaction times are the main characteristics of the method.

| Catalysts | Substrates | Hydrogenation intermediates | Hydrogenation products |
|---|---|---|---|
| Pd-on-$Al_2O_3$<br>Pd-on-$SiO_2$<br>Pd-on-C<br>Pd-on-graphite<br>Pd-Ag-on-$\alpha$-$Al_2O_3$ | 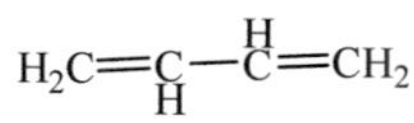<br>Butadiene | 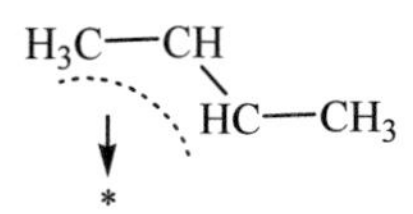<br>π-Allyl intermediate | 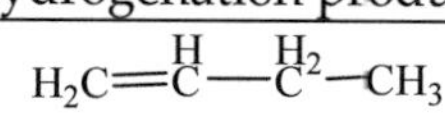<br>1-Butene<br>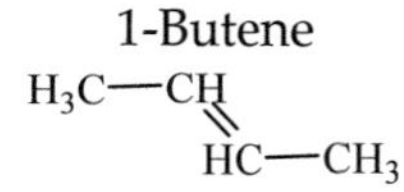<br>*Trans*-2-butene |
| Pd-Cu-on-$Al_2O_3$<br>Pd-Au-on-$Al_2O_3$<br>Pd-Ag-on-$SiO_2$<br>Pd-on-$TiO_2$ | 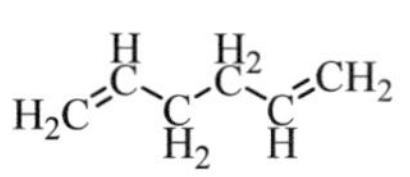<br>1,5-Hexadiene | 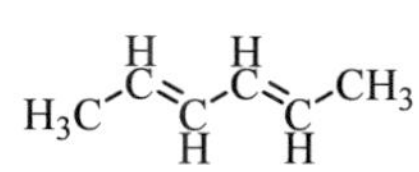<br>2,4-Hexadiene | 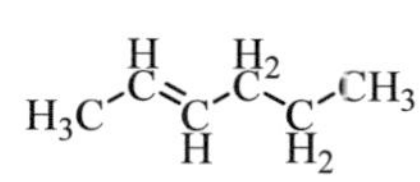<br>2-Hexene |

**Table 1.** Supported Pd heterogeneous catalysis summary.

Fibroin-palladium catalyses the hydrogenation of olefinic double bonds, conjugated to aromatic ketones (Scheme 13). The ketone double bond remains intact. This, however, is not the case when there is an ester group in $\alpha$ position to the ketone. In both cases atmospheric conditions were employed and MeOH was the solvent of choice. A very important observation is that when MeOH is employed as solvent for the hydrogenation of benzyl ester derivatives, a partial hydrogenolysis of the substrate occurs. This can be avoided if THF replaces MeOH as the reaction solvent. [47]

Heterogeneous or immobilized homogeneous catalysts are the systems of choice in order to achieve selective hydrogenation with high purity products. Sahoo et al. employed an immobilized Ru triphenylphosphine complex over mesoporous silica SBA-15 and successfully hydrogenated, chemoselectively, prochiral and $\alpha,\beta$-unsaturated ketones. [48] The turnover frequencies are very high and the enantioselectivity achieved for the prochiral ketones is also at very good levels. The chemoselectivities in the conversion of $\alpha,\beta$-unsaturated ketones to the respective allyl alcohols are greater than 90%.

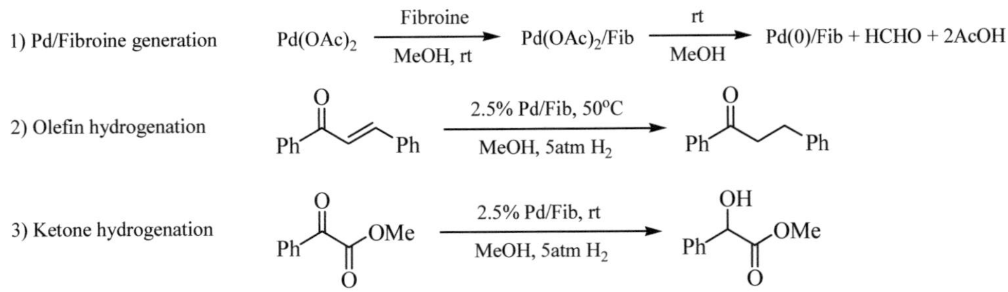


**Scheme 13.** Fibroin-palladium hydrogenation of olefinic and carbonylic confugated double bonds.

A Cp*Ir (pentamethylcyclopentadienyl) complex is employed for the regio- and chemoselective transfer hydrogenation of quinolines. The reaction solvent also serves as the hydrogen source and in this case is 2-propanol. The quinolines bearing an electron donating or withdrawing group are readily hydrogenated, but for isoquinolines or pyridines this methodology is unsuccessful. The proposed mechanism involves the protonation of the quinoline and the simultaneous formation of an iridium-hydride, which is followed by the adsorption of the carbon-carbon double bond on the iridium-hydride complex. The final stage is a protonolysis to afford 1,2,3,4-tetrahydroquinoline. [49]

# 6. The case of 3,5-bisarylidene tetramic acids

## 6.1. Chemoselective hydrogenation of C3=C7 double bond

The various oxidation states of Ru along with the tremendously rich list of available ligands for homogeneous Ru complex formation provide a strong background for the creation of specific catalysts to perform a certain function. An appropriately structured Ru complex can catalyze very difficult and hard to achieve transformations. [13]

The possibilities provided by the $H_4Ru_4(CO)_9$[(*S*)-BINAP] complex employed here, have not yet been sufficiently investigated. Since the purpose of application is the hydrogenation of 3,5-bisarylidene tetramic acids the conditions of the reaction must be controlled absolutely in order to avoid side reactions. Also, the purity of raw materials must be high, since a small impurity could initiate various uncontrolled and unidentified side reactions.

Judging from the cases we have encountered so far, there are two possibilities arising: A C3=C7 monoene selective hydrogenation, or a C7-C4=O 1,4-hydrogen addition (Scheme 14). To analyze this, first we have to outline the parameters affecting the mechanism, provided the catalyst is the $H_4Ru_4(CO)_9$[(*S*)-BINAP] complex:

- Reaction temperature.
- Gas $H_2$ reaction pressure.
- Reaction solvent.
- Substrate.

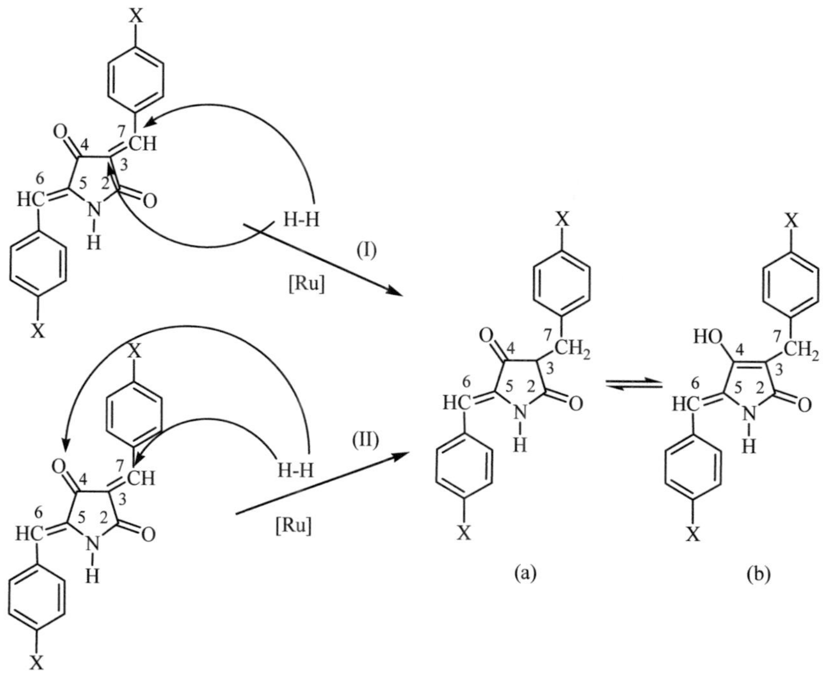


X = H, $OCH_3$, [Ru] = $H_4Ru_4(CO)_9[(S)\text{-BINAP}]$

**Scheme 14.** Selective homogeneous catalytic hydrogenation of 3,5-arylidene tetramic acid, at C3-C7.

The $-OCH_3$ substituted aromatic groups afford slightly higher yields than the benzylidene derivatives. This means that the electron donating group of $-OCH_3$ enhances the catalytic hydrogenation. Additionally, the employment of EtOH as the reaction solvent leads to a selective hydrogenation at C3-C7, regardless the temperature and pressure applied. The later two parameters affect only the kinetics of the reaction at this point. However, if the reaction solvent is selected to be MeOH, the selectivity is much lower and the temperature defines the chemoselectivity of the reaction, as we will see further in the analysis. The exact mechanism of the hydrogenation depicted in Scheme 14, cannot be revealed (path I or path II) by this first stage hydrogenation, because of the keto-enol tautomerism observed by the $^1H$ NMR spectrum. [16] The dominant form is that of structure (b).

## 6.2. Second stage hydrogenation of the C5=C6 double bond

To better understand the mechanism we need to examine the results of the second stage of the hydrogenation process. Application of MeOH as solvent, a temperature as high as 80°C and hydrogen pressure at 60bar affords the highest yield of the C3-C7 and C5-C6 bi-hydrogenation (Scheme 15). As we observe in Scheme 15 there is still an extended keto-enol tautomerism, observed once again in the $^1H$ NMR spectrum, which does not allow for an enantioselective hydrogenation. The most important observation, however, is that the 4-methoxy compound, formed with a 24% yield approximately, does not proceed to a second stage hydrogenation as the 4-hydroxy compound does.

From this we derive that after the first stage hydrogenation there is a competitive methylation of the 4-hydroxy group. The catalyst/MeOH combination provides the necessary acidity for this reaction to proceed. The most important observation, however, is that the second stage hydrogen addition does not proceed in the case of the 4-methoxy derivative. This can be attributed to steric effects that inhibit the Ru-substrate bonding. Also, there is no possibility to the formation of the 4-ketone tautomer. This indicates that the C5-C6 hydrogenation proceeds with a 1,4-addition. The Ru catalyst attaches to the C4=O carbonyl and the C6 carbon of the 1st stage hydrogenated product (a) of Scheme 14. That very same mechanism could also be the case for the 1st stage hydrogenation depicted in Scheme 14, path (II). This mechanism is very close to the one reported by Chandiran et al., presented in Scheme 2. [18] The difference in the case of the $H_4Ru_4(CO)_9[(S)\text{-BINAP}]$ complex is that the metal hydrides are already formed, on four separate nuclei, increasing the activity of the catalyst.

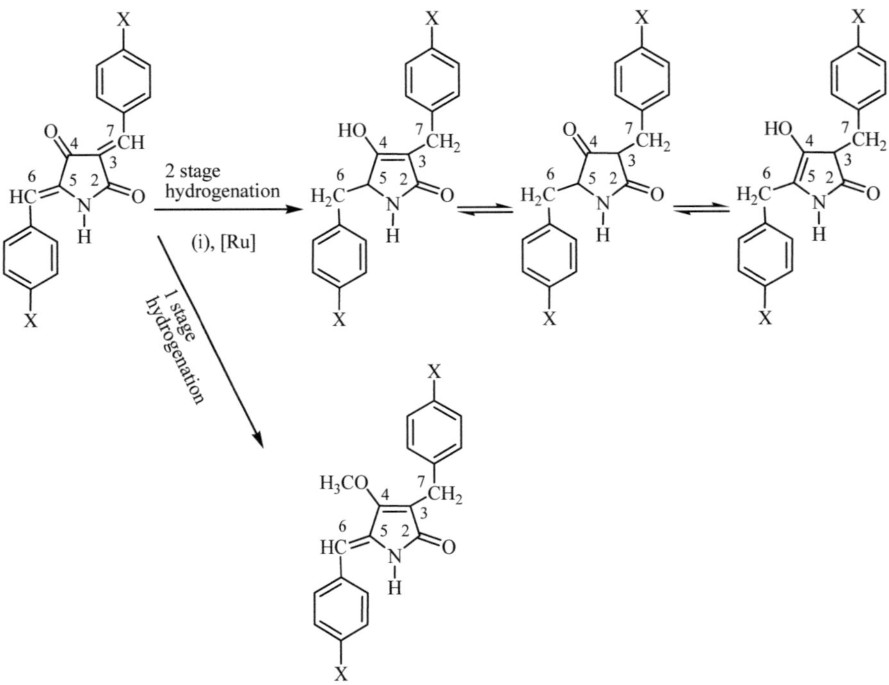


(i) $H_4Ru_4(CO)_9[(S)\text{-BINAP}]$, MeOH, $H_2$ X = H, $OCH_3$

**Scheme 15.** Second stage hydrogenation of 3,5-bisarylidene tetramic acids and keto-enol tautomerism of the bi-hydrogenated product.

Returning to the original hypothesis of the *α,β*-unsaturated ketone-blocks existing in the heterocyclic ring we can conclude the following:

- There is no chemoselective hydrogenation of the aromatic olefins, but rather a hydrogenation of the *α,β*-unsaturated ketone to a saturated ketone. The enolic form appears later, as a keto-enol tautomerism.

- The hydrogenation does not occur as a 1,2-addition but rather as a 1,4-addition involving the complexation of a Ru nucleus to the C4 carbonyl group.
- The C4=O-C3=C7 unsaturated structure is more activated than the other unsaturated groups within the molecular structure of the substrate.
- The C2=O-C3=C7 unsaturated ketone is not active under the particular reaction conditions, probably because of the electron withdrawing nitrogen atom.

The above clues illustrate a direct relationship between the electron density of the conjugated double bonds and the antagonistic activity of the separate unsaturated centers coexisting in the substrate structure.

## 6.3. Proposed mechanism for the hydrogenation of 3,5-bisarylidene tetramic acids

Clapham et al. have extensively investigated the mechanisms of the hydrogenation of polar bonds catalyzed by Ruthenium hydride complexes. [50] We will focus in the case of hydrogenation via hydrogen gas, and not transfer hydrogenation, of carbonyl compounds. The reaction types are divided in: Inner sphere hydrogenation and outer sphere hydrogenation (Figures 7 and 8). One deficiency of the catalysts that operate by inner sphere hydride transfer is that they are often not very selective for C=O bonds over C=C bonds in, for example, the reduction of *α,β*-unsaturated ketones and aldehydes. The outer sphere mechanism occurs when the C=O bind has a low hydride affinity so that electrophilic activation is required either by an external electrophile or an internal electrophile attached to an ancillary ligand. When the ancillary ligand provides a proton to the ligand, during the hydride transfer, then the mechanism is characterized as "metal-ligand bifunctional catalysis". It is discovered that complex hydrides of the type $Ru(H)_2H_2(PPh_3)_3$ react readily with ketones at low temperatures (20°C), in contrast to the CO and Cl substituted Ru complexes, such as $Ru(H)(Cl)(CO)(PPh_3)_3$ that react with ketones only at high temperatures (approximately 100°C) because of the electron withdrawing groups CO and Cl.

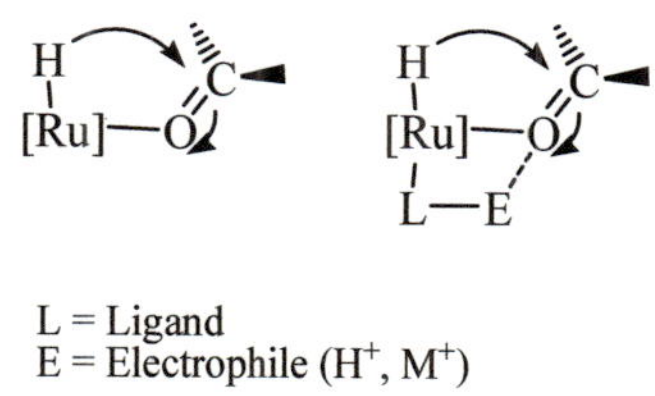


**Figure 7.** Inner sphere hydrogenation mechanism.

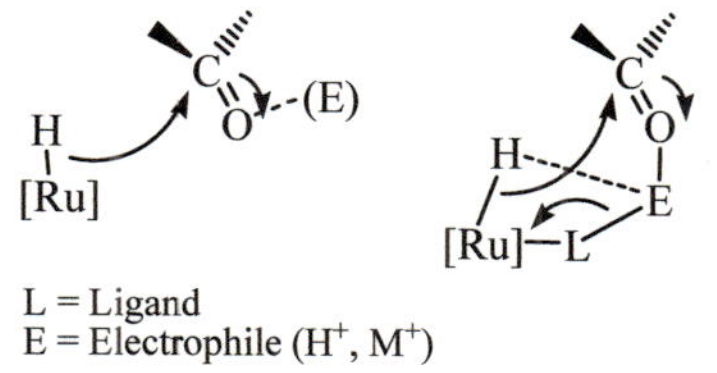


**Figure 8.** Outer sphere hydrogenation mechanism.

The generalized catalytic cycle for the inner sphere hydrogenation mechanism of C=O bonds is depicted in Scheme 16.

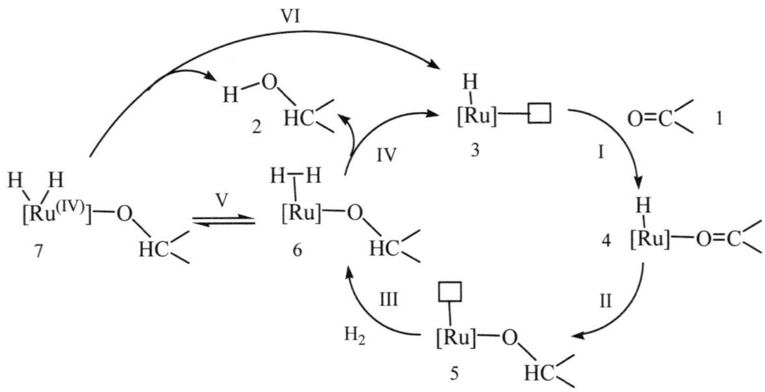

**Scheme 16.** Generalized catalytic cycle for the inner sphere hydrogenation of C=O bonds.

The cycle starts with the addition of the substrate (1) to the coordinatively unsaturated Ru(II) hydride species (3) giving the complex (4) (step I). The hydride species (3) is usually formed from a catalyst precursor at the very beginning of the catalytic reaction and is not isolated itself (see below). A hydride migration (step II) affords the new unsaturated ruthenium species (5) to which dihydrogen coordinates (step III) affording the dihydrogen species (6). A substrate insertion (step I) and a hydride migration (step II) are usually very fast so only the product (7) can be observed. Complex (6) can further react in two ways: Protonation of the coordinated substrate affords the product (2) releasing the regenerated catalyst (3) (step IV) or the coordinated dihydrogen can oxidatively add to the Ru(II) center giving a dihydride ruthenium(IV) species (7) (step V), followed by elimination of the product (2) and regeneration of the active catalyst (3). It should be noted that it is often impossible to experimentally distinguish between path IV and path V → VI. Catalytic reactions for which the inner sphere hydrogenation mechanism has been proposed have several features in common, regardless of the solvent used (organic, water or biphasic system). They all require relatively high temperatures (50–100°C, in some instances even higher) and high dihydrogen pressures (around 50atm). The catalyst-to-substrate ratio is usually small and no additives are necessary for the reaction to proceed. Step III in the Scheme 16 is generally recognized as the turn-over-limiting step.

A non-classical outer sphere mechanism, for the hydrogenation of polar multiple bonds, is shown in Scheme 17. This mechanism involves a hydride on the ruthenium catalyst and a proton on one of the ancillary ligands in a position to form a hydridic–protonic interaction (structure 8). The substrate (1) coordinates in step I by forming an outer sphere interaction between the atoms of its polar multiple bond and the proton and hydride of the complex (9). This interaction allows for the simultaneous transfer of the hydride and the proton (step II) producing the hydrogenated substrate (2) and a ruthenium complex with a vacant coordination site, (10). This 16-electron ruthenium center is usually stabilized by $\pi$-donation from the deprotonated ligand into the empty d-orbital. Hydrogen gas can then coordinate at this open site (III) producing a dihydrogen complex intermediate or transition state (11). The dihydrogen ligand heterolytically cleaves in step IV to re-form the original hydride complex 10. This exact type of mechanism is suggested by Chaplin and Dyson for the Bis-phosphine Ruthenium(II)-Arene chemoselective catalytic hydrogenation of aldehydes, in the presence of olefinic bonds. In fact, the complex [RuCl(PPh$_3$)(P($p$-tol)$_3$)(p-cymene)]PF6 affords a 82% selectivity for the hydrogenation of the C=O double bond of a 1:1 styrene/benzaldehyde mixture, while when we add 5 equivalents of NEt$_3$ in the original reaction mixture the selectivity moves to 90% towards the hydrogenation of the olefinic bond C=C. [51]

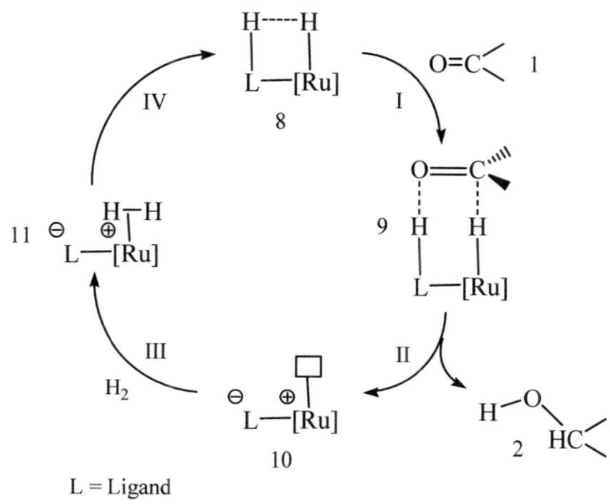


**Scheme 17.** Generalized catalytic cycle for the outer coordination sphere hydrogenation of the C=O bond catalyzed by ruthenium catalysts where the hydride addition to the substrate is assisted by an ancillary ligand.

The fact that the ruthenium complex H$_4$Ru$_4$(CO)$_9$[($S$)-BINAP] used for the hydrogenation of 3,5-bisarylidene tetramic acids consists of four nuclei allows for a 1,4-interaction of the substrate with the catalyst, following a mechanism similar to that of the outer coordination sphere (Scheme 17). However, the presence of multiple Ru nuclei may prevent the ligands of the catalyst complex from playing a role to the direct interactions with the unsaturated sites: The Ru-H- group is evidently more functionalized, than the Ru-L-H- group, and will preferably bond at the unsaturated center. It is possible that this is the preferred mechanism for both hydrogenation stages. Scheme 18 depicts the possible catalyst-substrate interactions for the two hydrogenation stages according to the previous assumption.

The presence of the 4-ketone is essential for the complexion of the substrate on the catalyst. The first stage has two possibilities of complexion and as we will discuss later in the text the activation energy of the first stage hydrogenation is much lower than that of the second stage hydrogenation. In the case of path (B) it is discovered by $^1$H NMR analysis that the 2-ketone-4-enol form is the dominant one. [16] This form, however, makes the second stage hydrogenation a more difficult to accomplish process. Since the presence of 4-ketone is required, and this form appears only in a small percentage compared to the 4-enol form, it is kinetically more difficult to achieve this second stage hydrogenation. So, longer reaction times are required and higher temperature-pressure conditions. This mechanism also explains why the 4-methoxy derivative it is not hydrogenated on a second stage. Hydrogen-transfer mechanisms are not considered in the analysis, since the solvents employed (DCM, THF, EtOH, MeOH) are unlikely to be proton donors under the reaction conditions.

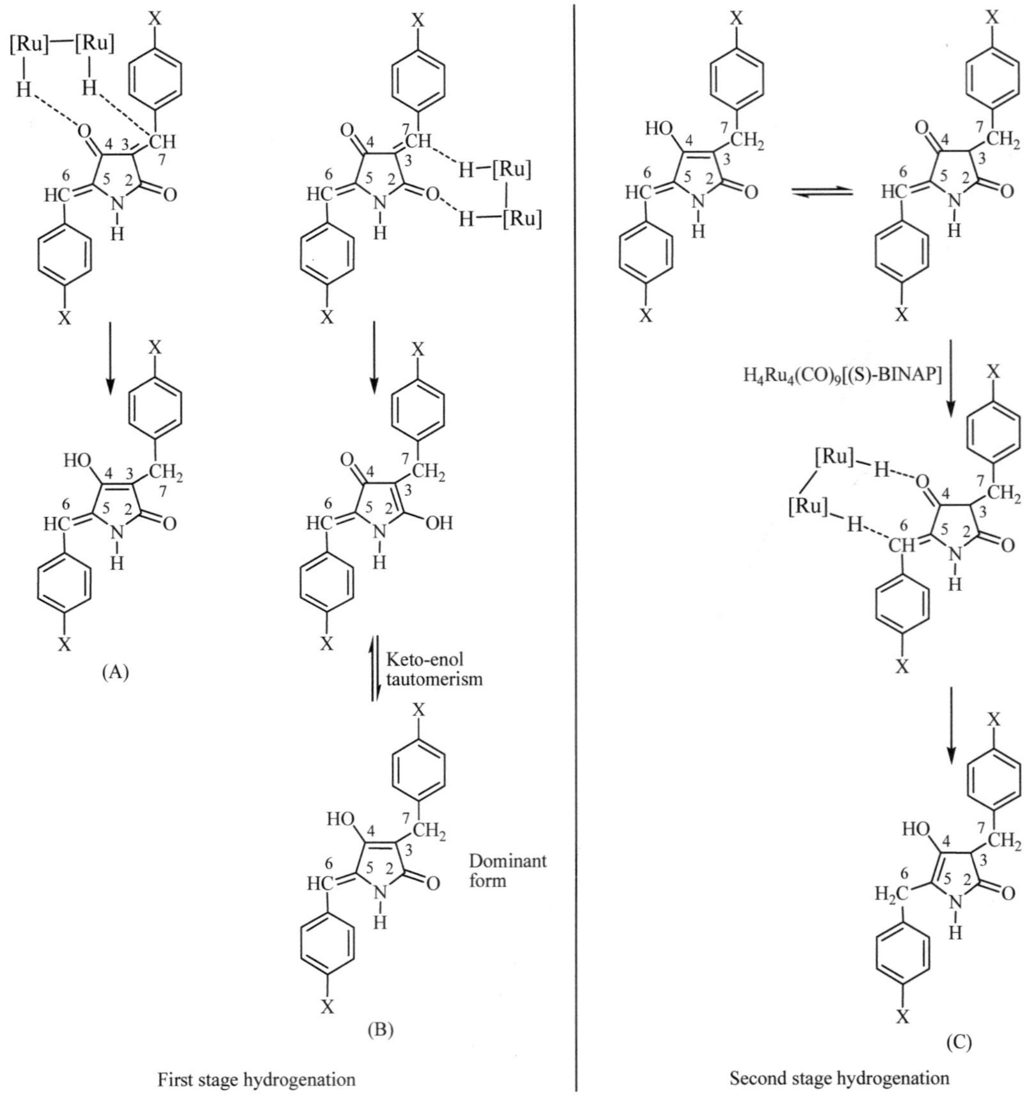


**Scheme 18.** Possible catalyst-substrate interactions for the $H_4Ru_4(CO)_9$[(*S*)-BINAP] catalyzed two-stage hydrogenation of 3,5-bisarylidene tetramic acids.

Studies on the mechanism of hydrogenation of acrylamide using a Rh(I) complex performed by Verdolino et al. describe the behavior of the conjugated acrylamide system in hydrogenation conditions. [52] It is certain that the catalytic mechanism for Rh(I) is different than that of Ru(II). Still, the study reveals that the hydrogenation of acrylamide (Figure 9) is directed to the olefinic double bond, while the carbonyl double bond remains intact. This is in accordance with the observation that the 2-ketone of the 3,5-bisarylidene tetramic acids is not activated by the hydrogenation conditions applied. This is also confirmed in the case of 3-ethoxy pyrrolidinone hydrogenation we have already mentioned. [15]

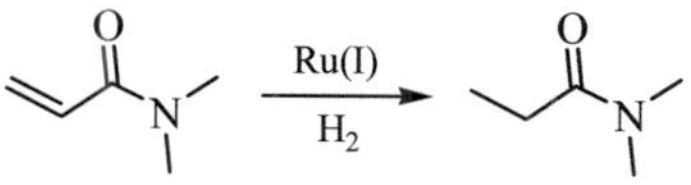


**Figure 9.** Ru(I) catalyzed hydrogenation of acrylamide.

## 6.4. Solvent, temperature and pressure effect on the Ru(II) catalyzed 3,5-bisarylidene tetramic acids' hydrogenation

The hydrogenation results for 3,5-bisarylidene tetramic acids depending on the reaction solvent applied are presented in Table 2. It is apparent that the mechanism for the hydrogenation of polar bonds is enabled in the presence of polar solvents. Employment of DCM as the sole reaction solvent does not afford any hydrogenated product. THF provides moderate results. With THF we can obtain good conversion for the first stage of the process, but results are very poor for the second stage. Ethanol gives very good results for the first stage hydrogenation, but fails to achieve second stage hydrogenation, even at elevated temperature and pressure. However, it serves as a tool for controlling the reaction outcome in terms of chemoselectivity. Ethanol is a safe chemical solvent, appropriate for green chemistry applications. For this reaction, it is the solvent of choice if the mono-hydrogenated product is the desired one. Methanol is the most polar amongst the solvents of this study. Methanol makes accessible the second stage hydrogenation, which indicates that stronger polarity conditions are required for the second stage hydrogenation. On the other hand, using methanol makes it more difficult to control the chemoselectivity of the reaction, and it will require a lower reaction temperature (60°C) to obtain a higher percentage of the mono-hydrogenated product (3 or 5). [16]

| Entry | Substrate | S/C | Concentration (M) | Solvent/DCM (v/v) | Temp (°C) | Press (bar) | Time (h) | Conversion | |
|---|---|---|---|---|---|---|---|---|---|
| | | | | | | | | 3 | 4 |
| 1 | 1 | 250 | 0.0018 | MeOH (24:1) | 100 | 60 | 20 | 4.7 | 94.2 |
| 2 | 1 | 500 | 0.0012 | EtOH (24:1) | 100 | 60 | 19 | 94.0 | 5.9 |
| 3 | 1 | 493 | 0.0034 | THF (20:1) | 100 | 60 | 20 | 94.4 | 5.5 |
| | | | | | | | | 5 | 6 |
| 4 | 2 | 120 | 0.0010 | MeOH (24:1) | 100 | 60 | 20 | 24.8 | 73.9 |
| 5 | 2 | 405 | 0.0012 | EtOH (24:1) | 80 | 60 | 20 | 95.0 | 5.0 |
| 6 | 2 | 700 | 0.0020 | THF (20:1) | 100 | 60 | 20 | 25.4 | 8.9 |

S/C = Substrate/Catalyst molar ratio.

**Table 2.** Solvent effect on the Ru(II) catalyzed 3,5-bisarylidene tetramic acids' hydrogenation.

Numbering of the substrates and products of Table 2 is with respect to Figure 10.

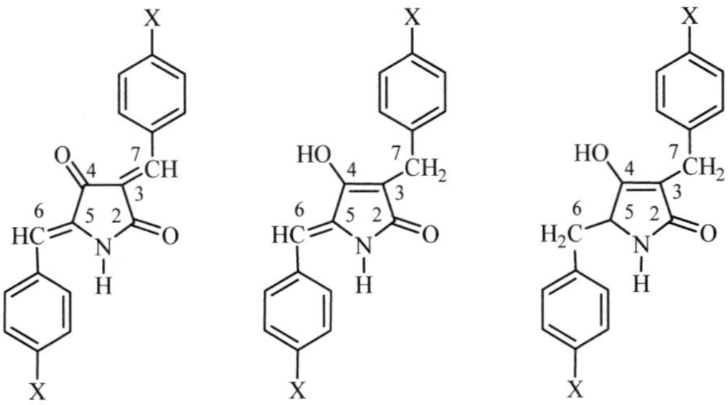

1: X = H
2: X = $OCH_3$

3: X = H
5: X = $OCH_3$

4: X = H
6: X = $OCH_3$

**Figure 10.** Substrates and products as they appear in Table 2.

A very interesting case of pronounced solvent effect is reported by Haddad et al. in the catalytic asymmetric hydrogenation of heterocyclic ketone-derived hydrazones. This study reveals an inversion in the enantioselectivity of the hydrogenation when the solvent of the reaction is changed from MeOH (85% e.e.) to DCE (-27% e.e.). Extensive study ruled out the hydrazone geometry or the hydrogenation via an endocyclic alkene as possible factors for the inversion of the configuration. This study also supports the fact that polar solvents as MeOH and EtOH afford higher selectivity values, whereas pentanol and DCM give moderate results. [53]

To return to the 3,5-bisarylidene tetramic acids hydrogenation, the applied temperature and pressure do affect the reaction outcome, but not as strongly as the solvent. [16] As we mentioned before, elevated temperature and pressure (above 80°C and 40bar respectively) is required for the ruthenium catalyst to achieve the hydrogenation of the substrate. In this case, mild hydrogenation conditions are not sufficient for a successful outcome of the process.

Mechanism decryption can be achieved, in most cases, by specially designed experiments and computational methods. Isotope kinetic studies and DFT computational methods promise to enlighten the catalytic mechanism of $H_4Ru_4(CO)_9$[(*S*)-BINAP] catalyst for the hydrogenation of olefin and carbonyl double bonds. Future implementation of these methods will provide additional data to bring us one step closer to understanding the catalytic activity of this tetra-Ru-nuclei complex.

## 7. Heterogeneous versus homogeneous catalytic hydrogenation

We have already encountered many cases of heterogeneous and homogeneous catalytic hydrogenation reactions. Heterogeneous catalysts are commonly used in the form of powders for slurry and fluidized bed reactions or as formed bodies for fixed bed hydrogenations. The addition of promoters and adjustments in particle size and porosity allow for these catalysts to be fine tuned for specific reactions. Homogeneous catalysts are also very flexible where the selection of the transition metal, ligands and reaction conditions can lead to highly selective hydrogenations. The separation problems associated with homogeneous catalysts have led to the development of heterogenized homogeneous catalysts via the fixation of the active complexes on organic or inorganic supports or via application in biphasic systems. While there has been some success in this area, there still remains a considerable amount of work to be done. [12]

At this point we can refer to a few more interesting heterogeneous catalytic reaction cases. The recent study presented by Bridier and Pérez-Ramírez demonstrates the possibility of selective catalytic hydrogenation of conjugated dienes and ene-ynes. [54] Reactions are carried out in ambient conditions. Interesting observations arise from the results of valylene gas-phase hydrogenation to isoprene. Depending on the type of catalyst, Pd or Cu-Ni, and the $H_2$/Substrate ratio it is possible to selectively derive either isoprene or mono-olefines. At low ratios oligomers are formed at a high percentage, while in high ratios mixtures of isoprene and mono-olefins are derived. Since Pd highly favors the formation of active hydrogen species, it shows rather higher performance compared to the Cu-Ni system and modified Pd catalysts. However, very high $H_2$/Substrate ratios ultimately lead to cracking. Most of the catalysts employed favor the isomerization of the hydrogenated products leading to 1,4-hydrogen addition rather than 1,2-hydrogen addition. Cu is the only catalyst for which the isomerization is not observed.

A comparative study for the hydrogenation of binary activated ketones mixtures over modified and unmodified Pt/alumina heterogeneous catalyst sheds light on the mechanism of chiral and racemic hydrogenation of ketones. [55] Binary mixtures of ethyl pyruvate, methyl benzoylformate, ketopantolactone, pyruvic aldehyde dimethyl acetal and trifluoroacetophenone are subjected to hydrogenation in either toluene or acetic acid, on platinum/alumina unmodified catalyst (racemic hydrogenation) or modified with cinchonidine (chiral hydrogenation). The experimental conditions are those of the Orito reaction. Depending on the racemic/chiral condition a reverse is observed in the hydrogenation rate of the activated ketones. This verifies the adsorption model for the hydrogenation reaction that is proposed in the Orito reaction. The intermediate complexes are determined by the chiral modifier, the substrate and the reaction conditions. This indicates that still the heterogeneous catalytic processes need further clarification and more experimental data to describe the phenomena in more details.

## 8. Conclusion

The catalytic hydrogenation of olefinic and carbonylic double bonds depends on a number of parameters that shape the outcome of the reaction. During this analysis we encountered many examples where the careful choice of parameters and design of the hydrogenation process leads to the desired product, eliminating any byproducts. Summarizing the most important parameters we can generate the following list:

- Catalyst
- Substrate interactions
- Unsaturated center type
- Reaction solvent
- Reaction conditions

The substrate interactions and the nature of the unsaturated center constitute the first most important parameter. It defines the strategy for the successful hydrogenation. Interactions involve intramolecular electron effects, as a result of the presence of an heteroatom, a functional group or conjugated double bonds. Internal hydrogen bonds and large substitution groups define the activity and approachability of an unsaturated center. Intermolecular interactions mostly refer to the active sites of the substrate that may form stable bonds with the catalyst, the solvent or another functional moiety to the formation of an intermediate. Knowledge of the structure specificities can direct the entrepreneur to the right catalyst, solvent and conditions selection. Olefinic double bonds and especially conjugated C=C bonds, have proven to be very susceptible to hydrogenation, under various conditions, with high yield and good selectivity. On the other hand, carbonylic double bonds are harder to hydrogenate, and the breakthrough by Knowles and Noyori was the trigger for the successful design of novel versatile homogeneous catalysts which can hydrogenate C=O bonds under mild conditions.

Homogeneous catalysts of Rh, Ru and Ir afford the best results in terms of yield and selectivity. Given the fact that chiral ligands have a constantly increasing number, they provide us with more possibilities and now we are practically able to design a particular ligand for a particular hydrogenation. Heterogeneous modified Ni, Cu, Pt and Pd catalysts are most popular for industrial synthetic applications, because of their lower cost and the recovery option. They present good results for the hydrogenation of olefins and conjugated olefins, but give only moderate results for C=O hydrogenation.

The reaction solvent is the most important of the reaction parameters, as it can affect the yield, the selectivity and the overall outcome of the hydrogenation. Polar bonds, like C=O, usually demand the presence of a polar solvent (alcohol). Non polar solvents can be chosen for reasons of solubility, or when polar conditions can affect the substrate and the enantioselectivity of the product. In most cases, polar solvents give better results, but in order to avoid undesired hydrogenolysis or isomerization phenomena, a less polar solvent must be chosen (e.g. 2-propanol or THF). Employment of a Ru homogeneous catalyst usually demands the presence of a polar solvent that participates in the hydrogenation mechanism, enabling the intermediate hydride formation. Choosing the right Metal-Ligand combination minimizes the need for intense temperature and pressure hydrogenation conditions. There are applications, however, when elevated temperature (over 80°C) and pressure (over 50bar) is requested for the hydrogenation to succeed. Most studies indicate that higher selectivity is related to mild reaction conditions.

3,5-Arylidene tetramic acids are catalytically hydrogenated in two consecutive stages, by the $H_4Ru_4(CO)_9$[(*S*)-BINAP] complex. This substrate consists of an extended conjugated system based on the heterocyclic lactame nucleus. The four-Ru-nuclei complex is able to form direct hydride bonds to the C=O and C=C unsaturated centers and a 1,4-hydrogenation mechanism is proposed. This process delivers high yield and chemoselectivity in polar solvents and it can be controlled by choosing the alcohol solvent (MeOH or EtOH) and the reaction temperature. The hydrogenation of heterocyclic compounds has not yet been fully charted. A complex structure like that of an unsaturated, heterocyclic compound needs to be carefully manipulated to avoid side reactions during the hydrogenation process.

Innovation in the field of pharmaceuticals and fine chemicals drives the research of catalytic complexes and organic substrates in new fascinating areas and the future is very promising for the development of chemical catalysts that will function in a protein-like manner for the highly selective hydrogenation of unsaturated centers, leaving the rest of the substrate structure intact and with a high atom economy. Modern isotope kinetic studies and DFT theoretical calculations are tools in the service of catalyst and process design for optimizing the hydrogenation results and could be cooperatively employed to afford maximum results.

## Author details

Christos S. Karaiskos, Dimitris Matiadis and Olga Igglessi-Markopoulou*
*Laboratory of Organic Chemistry, Department of Chemical Engineering,*
*National Technical University of Athens, Zografos Campus, Athens , Greece*

John Markopoulos*
*Laboratory of Inorganic Chemistry, Department of Chemistry,*
*University of Athens, Panepistimiopolis, Athens, Greece*

## 9. References

[1] Heaton B T (2005) Mechanisms in Homogeneous Catalysis: A Spectroscopic Approach. Weinheim, Germany: Wiley-VCH.

[2] Ikariya T, Murata K, Noyori R (2006) Bifunctional Transition Metal-Based Molecular Catalysts for Asymmetric Syntheses. Org. Biomol. Chem. 4: 393-406.

[3] Morris R H (2007) Handbook of Homogeneous Hydrogenation. In: de Vries J G, Elsevier C J, editors. volume 1. chapter 3. Weinheim, Germany: Wiley-VCH. pp. 45-70.

[4] Smit C, Fraaije M W, Minnaard A J (2008) Reduction of Carbon–Carbon Double Bonds Using Organocatalytically Generated Diimide. J. Org. Chem. 73: 9482-9485.

[5] Han A, Jiang X, Civiello R L, Degnan A P, Chaturvedula P V. Macor J E, Dubowchik G M (2009) Catalytic Asymmetric Syntheses of $\alpha$-Amino and $\alpha$-Hydroxyl Acid Derivatives. J. Org. Chem. 74: 3993-3996.

[6] Li X, Li L, Tang Y, Zhong L, Cun L, Zhu J, Liao J, Deng J (2010) Chemoselective Conjugate Reduction of $\alpha,\beta$-Unsaturated Ketones Catalyzed by Rhodium Amido Complexes in Aqueous Media. J. Org. Chem. 75: 2981-2988.

[7] Kosal A D, Ashfeld B L (2010) Titanocene-Catalyzed Conjugate Reduction of alpha,beta-Unsaturated Carbonyl Derivatives. Org. Lett. 12: 44-47.

[8] Llamas T, Arrayás R G, Carretero J C (2007) Catalytic Asymmetric Conjugate Reduction of β,β-Disubstituted $\alpha,\beta$-Unsaturated Sulfones. Angew. Chem. Int. Ed. 46: 3329-3332.

[9] Ngai M Y, Kong J R, Krische M J (2007) Hydrogen-Mediated C–C Bond Formation: A Broad New Concept in Catalytic C–C Coupling. J. Org. Chem. 72: 1063-1072.

[10] Genet J P (2008) Modern Reduction Methods. In: Andersson P G, Munslow I J, editors. Chapter 1. Weinheim, Germany: Wiley VCH. pp. 1-38.

---

*Corresponding Author

[11] Dobbs D A, Vanhessche K P, Brazi E, Rautenstrauch V V, Lenoir J Y, Genêt J P, Wiles J, Bergens S H (2000) Industrial Synthesis of (+)-cis-Methyl Dihydrojasmonate by Enantioselective Catalytic Hydrogenation; Identification of the Precatalyst. Angew. Chem. Int. Ed. 39: 1992-1995.

[12] Chen B, Dingerdissen U, Krauter J G E, Lansink Rotgerink H G J, Möbus K, Ostgard D J, Panster P, Riermeier T H, Seebald S, Tacke T, Trauthwein H (2005) New Developments in Hydrogenation Catalysis Particularly in Synthesis of Fine and Intermediate Chemicals. Applied Catalysis A: General. 280: 17-46.

[13] Trost B M, Frederiksen M U, Rudd M T (2005) Ruthenium-Catalyzed Reactions – A Treasure Trove of Atom-Economic Transformations. Angew. Chem. Int. Ed. 44: 6630-6666.

[14] Knowles W S (2002) Asymmetric Hydrogenations (Nobel Lecture). Angew. Chem. Int. Ed. 41: 1998-2007.

[15] Tang W, Zhang X (2003), New Chiral Phosphorous Ligands for Enantioselective Hydrogenation. Chem. Rev. 103: 3029-3069.

[16] Karaiskos C S, Matiadis D, Markopoulos J, Igglessi-Markopoulou O (2011) Ruthenium-Catalyzed Selective Hydrogenation of Arylidene Tetramic Acids. Application to the Synthesis of Novel Structurally Diverse Pyrrolidine-2,4-diones. Molecules. 16: 6116-6128.

[17] Palmer C J, Casida J E (1990) Selective Catalytic Hydrogenation of an Olefin Moiety in the Presence of a Terminal Alkyne Function. Tetrahedron Letters. vol.31. no.20: 2857-2860.

[18]Chandiran T, Vancheesan S (1992) Homogeneous Hydrogenation of Conjugated Dienes Catalysed by Naphthalenetricarbonylchromium. Journal of Molecular Catalysis. 71: 291-302.

[19] Cho I S, Alper H (1995) Selective Hydrogenation of Simple and Functionalized Conjugated Dienes Using a Binuclear Palladium Complex Catalyst Precursor. Tetrahedron Letters. vol.36. no.32: 5673-5676.

[20] Mizugaki T, Ooe M, Ebitani K, Kaneda K (1999) Catalysis of Dendrimer-Bound Pd(II) Complex. Selective Hydrogenation of Conjugated Dienes to Monoenes. Journal of Molecular Catalysis A: Chemical. 145: 329-333.

[21] Dahlén A, Hilmersson G (2003) Selective Reduction of Carbon–Carbon Double and Triple Bonds in Conjugated Olefins Mediated by SmI2/H2O/Amine in THF. Tetrahedron Letters. 44: 2661-2664.

[22] Alonso F, Osante I, Yus M (2007) Highly Selective Hydrogenation of Multiple Carbon–Carbon Bonds Promoted by Nickel(0) Nanoparticles. Tetrahedron. 63: 93-102.

[23] Tao X, Li W, Ma X, Li X, Fan W, Xie X, Ayad T, Ratovelomanana-Vidal V, Zhang Zh (2012) Ruthenium-Catalyzed Enantioselective Hydrogenation of Aryl-Pyridyl Ketones. J. Org. Chem. 77: 612-616.

[24] Spogliarich R, Farnetti E, Graziani M (1991) Effect of Charge Distribution on Selective Hydrogenation of Conjugated Enones Catalyzed by Iridium Complexes. Tetrahedron. vol.47. no.10/11: 1965-1976.

[25]Spogliarich R, Vidotto S, Farnetti E, Graziani M, Verma Gulati N (1992) Highly Selective Catalytic Hydrogenation of Cyclic Enones. Tetrahedron: Asymmetry. vol.3, no.8: 1001-1002.

[26] Van Druten G M R, Aksu L, Ponec V (1997) On the Promotion Effects in the Hydrogenation of Acetone and Propanal. Applied Catalysis A: General. 149: 181-187.

[27] Van Druten G M R, Ponec V (2000) Hydrogenation of Carbonylic Compounds Part I: Competitive Hydrogenation of Propanal and Acetone Over Noble Metal Catalysts. Applied Catalysis A: General. 191: 153-162.

[28] Ponec V (1997) On the Role of Promoters in Hydrogenations on Metals; $\alpha,\beta$-Unsaturated Aldehydes and Ketones. Applied Catalysis A: General. 149: 27-48.

[29] Kazuyuki H, Hironao S, Kosaku H (2001) Chemoselective Control of Hydrogenation Among Aromatic Carbonyl and Benzyl Alcohol Derivatives Using Pd/C(en) Catalyst. Tetrahedron. 57: 4817-4824.

[30] Chen J-X, Daeuble J F, Brestensky D M, Stryker J M (2000) Highly Chemoselective Catalytic Hydrogenation of Unsaturated Ketones and Aldehydes to Unsaturated Alcohols Using Phosphine Stabilized Copper(I) Hydride Complexes. Tetrahedron. 56: 1288-1298.

[31] Enthaler S, Erre G, Junge K, Michalik D, Spannenberg A, Marras F, Gladiali S, Beller M (2007) Enantioselective Rhodium-Catalyzed Hydrogenation of Enol Carbamates in the Presence of Monodentate Phosphines. Tetrahedron: Asymmetry. 18: 1288-1298.

[32] Ravasio N, Antenori M, Gargano M, Mastrorilli P (1996) Cu/SiO2: an Improved Catalyst for the Chemoselective Hydrogenation of $\alpha,\beta$-Unsaturated Ketones. Tetrahedron Letters. vol.37. no.20: 3529-3532.

[33] Milone C, Ingoglia R, Pistone A, Neri G, Frusteri F, Galvagno S (2004) Selective Hydrogenation of $\alpha,\beta$-Unsaturated Ketones to $\alpha,\beta$-Unsaturated Alcohols on Gold-Supported Catalysts. Journal of Catalysis. 222: 348-356.

[34] Milone C, Crisafulli C, Ingoglia R, Schipilliti L, Galvagno S (2007) A Comparative Study on the Selective Hydrogenation of $\alpha,\beta$-Unsaturated Aldehyde and Ketone to Unsaturated Alcohols on Au Supported Catalysts. Catalysis Today. 122: 341-351.

[35] Khan F.-A, Vallat A, Süss-Fink G (2012) Highly Selective C=C Bond Hydrogenation in $\alpha,\beta$-Unsaturated Ketones Catalyzed by Hectorite-Supported Ruthenium Nanoparticles. Journal of Molecular Catalysis A: Chemical. 355: 168-173.

[36] Xu Y, Alcock N W, Clarkson G J, Docherty G, Woodward G, Wills M (2004) Asymmetric Hydrogenation of Ketones Using a Ruthenium(II) Catalyst Containing BINOL-Derived Monodonor Phosphorus-Donor Ligands. Org. Lett. 6. 22: 4105-4107.

[37] Arai N, Ooka H, Azuma K, Yabuuchi T, Kurono N, Inoue Ts, Ohkuma T (2007) General Asymmetric Hydrogenation of $\alpha$-Branched Aromatic Ketones Catalyzed by TolBINAP/DMAPEN Ruthenium(II) Complex. Org. Lett. 9. 5: 939-941.

[38] Li W, Sun X, Zhou L, Hou G, Yu Sh, Zhang X (2009) Highly Efficient and Highly Enantioselective Asymmetric Hydrogenation of Ketones with TunesPhos/1,2-Diamine-Ruthenium(II) Complexes. J. Org. Chem. 74: 1397-1399.

[39] Palmer A M, Chiesa V, Holst H Ch, Le Paih J, Zanotti-Gerosa A, Nettekoven U (2008) Synthesis of Enantiopure 3,6,7,8-Tetrahydrochromeno[7,8-d]Imidazoles via Asymmetric Ketone Hydrogenation in the Presence of RuCl2[Xyl-P-Phos][DAIPEN]. Tetrahedron: Asymmetry. 19: 2102-2110.

[40] Xie J B, Xie J H, Liu X Y, Kong W L, Li Sh, Zhou Q L (2010) Highly Enantioselective Hydrogenation of r-Arylmethylene Cycloalkanones Catalyzed by Iridium Complexes of Chiral Spiro Aminophosphine Ligands. J. Am. Chem. Soc. 132: 4538-4539.

[41] Fogassy G, Tungler A, Lévai A (2003) Enantioselective Hydrogenation of Exocyclic $\alpha,\beta$-Unsaturated Ketones Part III. Hydrogenation with Pd in the Presence of Cinchonidine. Journal of Molecular Catalysis A: Chemical. 192: 189-194.

[42] He Sh, Shultz C Sc, Lai Zh, Eid R, Dobbelaar P H, Ye Zh, Nargund R P (2011) Catalytic Asymmetric Hydrogenation to Access Spiroindane Dimethyl Acetic Acid. Tetrahedron Letters. 52: 3621-3624.

[43] Solladié-Cavallo A, Roje M, Baram A, Šunjić E (2003) Partial Hydrogenation of Substituted Pyridines and Quinolines: A Crucial Role of the Reaction Conditions. Tetrahedron Letters. 44: 8501-8503.

[44] Ito M, Sakaguchi A, Kobayashi Ch, Ikariya T (2007) Chemoselective Hydrogenation of Imides Catalyzed by Cp*Ru(PN) Complexes and Its Application to the Asymmetric Synthesis of Paroxetine. J. Am. Chem. Soc. 129: 290-291.

[45] Molnár Á, Sárkány A, Varga M (2001) Hydrogenation of Carbon–Carbon Multiple Bonds: Chemo-, Regio- and Stereo-Selectivity. Journal of Molecular Catalysis A: Chemical. 173: 185-221.

[46] Ranu B C, Sarkar A (1994) Regio- and Stereoselectlve Hydrogenation of Conjugated Carbonyl Compounds via Palladium Assisted Hydrogen Transfer by Ammonium Formate. Tetrahedron Letters. vol.35. no.46: 8549-8650.

[47] Ikawa T, Sajiki H, Hirota K (2005) Highly Chemoselective Hydrogenation Method Using Novel Finely Dispersed Palladium Catalyst on Silk-Fibroin: Its Preparation and Activity. Tetrahedron. 61: 2217-2231.

[48] Sahoo S, Kumar Pr, Lefebvre F, Halligudi S B (2007) Immobilized Chiral Diamino Ru Complex as Catalyst for Chemo- and Enantioselective Hydrogenation. Journal of Molecular Catalysis A: Chemical. 273: 102-108.

[49] Fujita K, Kitachuji Ch, Furukawa Sh, Yamaguchi R (2004) Regio- and Chemoselective Transfer Hydrogenation of Quinolines Catalyzed by a Cp*Ir Complex. Tetrahedron Letters. 45: 3215-3217.

[50] Clapham S E, Hadzovic A, Morris R H (2004) Mechanisms of the $H_2$-Hydrogenation and Transfer Hydrogenation of Polar Bonds Catalyzed by Ruthenium Hydride Complexes. Coordination Chemistry Reviews. 248: 2201-2237.

[51] Chaplin A B, Dyson P J (2007) Catalytic Activity of Bis-phosphine Ruthenium(II)-Arene Compounds: Chemoselective Hydrogenation and Mechanistic Insights. Organometallics. 26: 4357-4360.

[52] Verdolino V, Forbes A, Helquist P, Norby P-O, Wiest O (2010) On the Mechanism of the Rhodium Catalyzed Acrylamide Hydrogenation. Journal of Molecular Catalysis A: Chemical. 324: 9-14.

[53] Haddad N, Qu B, Rodriguez S, van den Veen L, Reeves D C, Gonnella N C, Lee H, Grinberg N, Ma Sh, Krishnamurthy Dh, Wunberg T (2011) Catalytic Asymmetric Hydrogenation of Heterocyclic Ketone-Derived Hydrazones, Pronounced Solvent Effect on the Inversion of Configuration. Tetrahedron Letters. 52: 3718-3722.

[54] Bridier B, Pérez-Ramírez J (2011) Selectivity Patterns in Heterogeneously Catalyzed Hydrogenation of Conjugated Ene-yne and Diene Compounds. Journal of Catalysis. 284: 165-175.

[55] Balázsik K, Szőri K, Szőllősi G, Bartók M (2011) New Phenomenon in Competitive Hydrogenation of Binary Mixtures of Activated Ketones Over Unmodified and Cinchonidine-Modified Pt/alumina Catalyst. Catalysis Communications. 12: 1410-1414.

Chapter 5

# Selective Hydrogenation and Transfer Hydrogenation for Post-Functional Synthesis of Trifluoromethylphenyl Diazirine Derivatives for Photoaffinity Labeling

Makoto Hashimoto, Yuta Murai,
Geoffery D. Holman and Yasumaru Hatanaka

Additional information is available at the end of the chapter

http://dx.doi.org/10.5772/48730

## 1. Introduction

### 1.1. Photoaffinity labeling

Elucidation of protein functions on the basis of structure–activity relationships can reveal the mechanisms of homeostasis functions in life and is one of the greatest interests of scientists. In the human body, many proteins are activated and/or inactivated by ligands to maintain homeostasis. Understanding the mechanism of molecular interactions between small bioactive ligands and proteins is an important step in rational drug design and discovery.

Photoaffinity labeling, which is one of the most familiar approaches for chemical biology analysis, was initiated using diazocarbonyl derivatives in 1962 (Singh et al., 1962). Many researchers have subsequently tried to establish alternative approaches for the direct identification of target proteins for the bioactive small ligands. These approaches are based on the affinity between the ligand and the target protein *(Figure 1)*. Several reviews are published for the recent applications of photoaffinity labeling (Tomohiro et al., 2005; Hashimoto & Hatanaka, 2008).

To archive photoaffinity labeling, researchers have to prepare photoaffinity labeling ligands. The native ligands must be modified by photoreactive compounds (photophores) by organic synthesis.

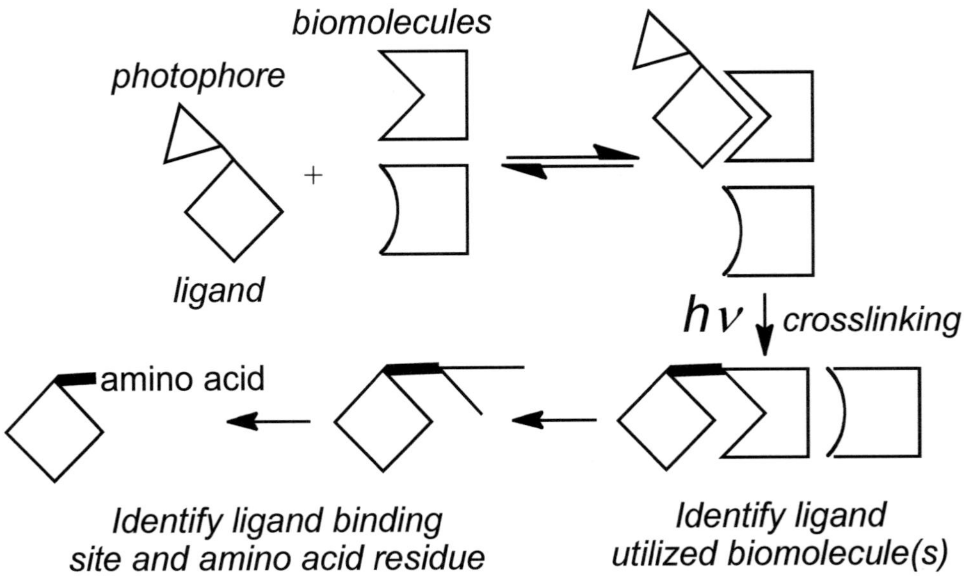


**Figure 1.** Schematic representation of photoaffinity labeling

## 1.2. Photophore synthesis and their properties

### *1.2.1. Selections of photophores*

It is important which photophores are used for effective photoaffinity labeling (Figure 2). Typically, aryl azide, benzophenone, or trifluoromethylphenyldiazirine (TPD) have been used.

Aryl azides are photoactivated below a wavelength of 300 nm, which sometimes causes damage to biomolecules. In addition, these generate nitrenes (Platz, 1995) as active species and these sometimes rearrange to ketimines as undesired side products (Karney & Borden, 1997).

Benzophenones are photoreactivated with light over 350 nm and generate reactive triplet carbonyl states (Galardy et al., 1973). These regenerate ground-state carbonyl compounds and so benzophenone ligands are reusable for other photolabeling experiments, although the photophores sometimes need long photoirradiation times for labeling.

TPD, with a three membered ring and nitrogen-nitrogen double bond, are also photoreactivated with light over 350 nm. These generate carbenes, which are more highly reactive species than other photophores, and rapidly form cross-links to biomolecules with short photoirradiation times (Smith & Knowles, 1973). It has been reported that the photolysis of diazirines can cause diazo isomerization, giving undesired intermediates in photoaffinity labeling. Diazo isomerization can be suppressed by introduction of a trifluoromethyl group into a diazirinyl three-membered ring (Brunner et al., 1980; Nassal, 1983).

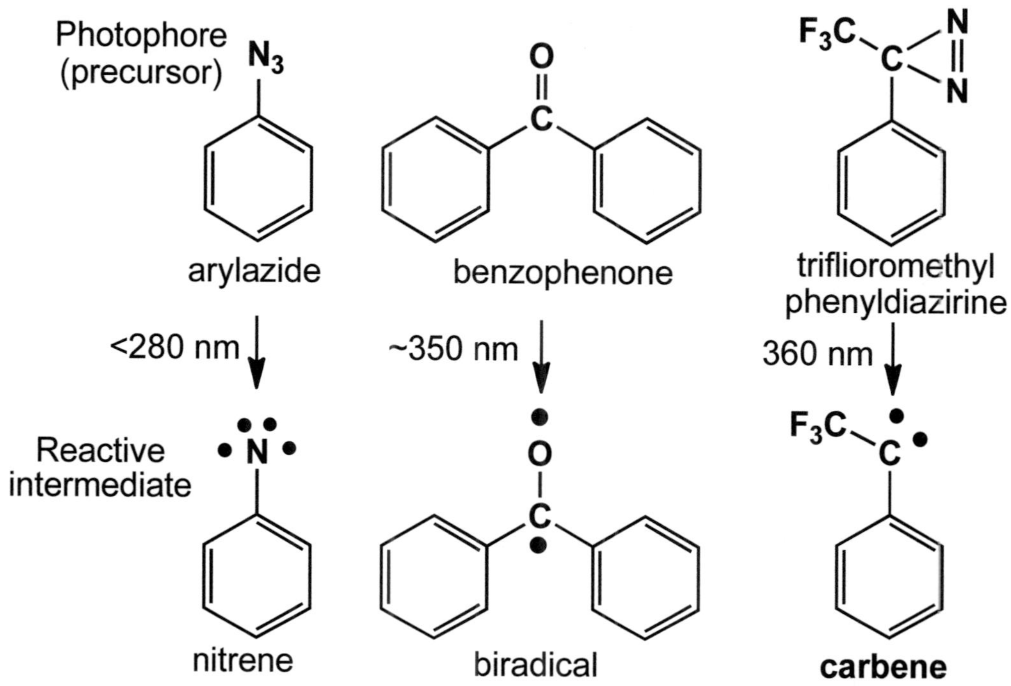


**Figure 2.** Photophores and their reactive intermediates following irradiation

Comparative irradiation studies of these three photophore types in living cells suggested that the irradiation needed for the generation of active species from azide and benzophenone caused cell death because long irradiation times are needed to incorporate the photophores into cell membrane surface biomolecules. By contrast, a carbene precursor – trifluoromethylphenyldiazirine (TPD) – did not cause cell death in the generation of active species (Hashimoto et al., 2001). Never-the-less, benzophenones (such as those attached to γ-secretase) are sometimes preferred for photoaffinity labeling experiments in vitro (Fuwa et al., 2006 & 2007).

### *1.2.2. Synthesis of trifluoromethylphenyldiazirine (TPD)*

There are more several steps involved in the the constructions of the TPD skeleton than are needed for synthesis of other photophores. Synthesis of the TPD three-membered ring required at least five steps from the corresponding aryl halide derivatives (Figure 3).

Although TPD is commercially available many are very expensive (1200 USD/0.5g for the simple TPD). In many previous synthetic routes the functional groups, which can be connected to ligands, tags and isotopes, should be pre-installed onto the benzene ring before constructions of three membered rings. The repeated construction of a diazirine moiety for each derivative is a drawback for application of the photophore for photoaffinity labeling.

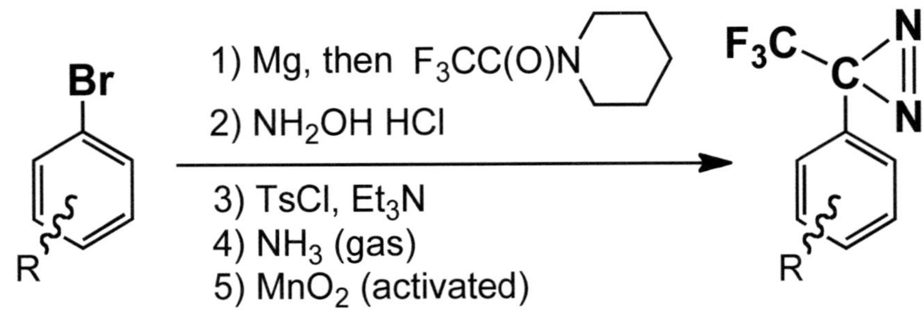


**Figure 3.** Synthesis of trifluoromethylphenyldiazirine derivatives

### *1.2.3. Post-functional synthesis of TPD derivatives*

Our breakthrough work on "post-functional" adaptation of diazirinyl compounds (Hatanaka et al., 1994, a & b) revealed that the trifluoromethyl-substituted three-membered ring was stable under many organic reaction conditions. Although the 3-(trifluoromethyl)diazirinyl moiety is categorized as an alkyl substituent, polarization means that the quaternary carbon atom is slightly positively charged, so the moiety is less activated towards electrophilic aromatic substitution than its unsubstituted counterpart (Hashimoto et al., 2006). We first selected the m-methoxy- substituted TPD (Fig. 4 R = $OCH_3$) as the mother skeleton, because: 1) the methoxy group strongly activates for electrophilic aromatic substitution, 2) the orientation of the substitution favors the o-position against the methoxy group, because the p-position is sterically hindered by the 3-(trifluoromethyl)diazirinyl moiety, and 3) demethylation of *m*-methoxy-TPD was easier than for *p*-methoxy-TPD, and realkylation of phenol derivative after demethylation was utilized for introduction of the tag.

**Figure 4.** Post-functional synthesis of trifluoromethylphenyldiazirine derivatives

It is somewhat difficult to derivatize unsubstituted TPD (Fig. 4, R=H) as this is less susceptible to aromatic substitution than *m*-methoxy TPD. It would need harsh conditions for the substitutions on aromatic ring. For example, the formylation with dichloromethyl methyl ether was performed using titanium chloride in dichloromethane for the 3-methoxy diazirine at 0 °C while the unsubstituted TPD did not afford formyl derivatives under the same condition. It is only archived when the trifluoromethanesulfonic acid, which is stronger acid than titanium chloride, was used as promoter for the reaction. These considerations directed our synthesis strategies towards derivatizations on the benzene ring after constructions of the trifluoromethyl diazirinyl ring (post-functional derivatizations). During the course of the studies, applications of hydrogenations to the TPD derivatives for post-functional synthesis are very important in associated derivatizations. However, the diazirinyl group consists of a nitrogen-nitrogen double bond in the structure and could be easily hydrogenated under the certain conditions. In this chapter, we would like to present a comprehensive summary for the hydrogenations that are compatible with or incompatible with the reaction conditions used for the trifluoromethylphenyl diazirinyl derivative. These considerations lead to effective post-functional derivatization approaches.

## 2. Selective hydrogenation methods over diazirinyl moiety for post-functional synthesis of TPD derivatives

### 2.1. Selective hydrogenation of carbon-iodine bond to carbon-hydrogen bond with $H_2$-Pd/C

It has been reported that hydrogenation of diazirinyl compounds under $H_2$-Pd/C at atmospheric pressure caused diazirinyl moiety reduction to diaziridine and further reduction of diaziridine moiety over a long time of treatment. Ambroise et al. found that hydrogenations of carbon-iodine bond in iodoarene are chemoselective.

This occurs selectively over other easily reducible functional groups using Pd/C (10 mol%) under a hydrogen atmosphere, in the presence of triethylamine and within an hour (Ambroise et al., 2000). The selective hydrogenations can be applied for TPD derivatives. The iodoarene TPD derivative (**1**) was subjected to hydrogenation under the $H_2$-Pd/C condition at atmospheric pressure (Fig.5).

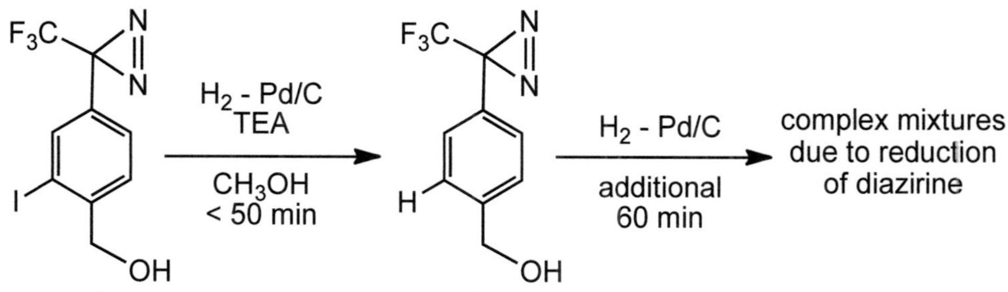


1 2

**Figure 5.** Selective hydrogenation of iodoarene TPD derivatives (**1**). Selectivity for carbon- iodine bond to carbon-hydrogen bonds (**2**) occurs on Pd/C under a hydrogen atmosphere

Detailed analysis revealed that the hydrogenation of carbon-iodine bond proceeded in parallel to the consumption of the starting material for 50 min. The hydrodeiodinated product (**2**) was subjected to further hydrogenolysis at the diazirinyl nitrogen-nitrogen double and compound **2** was completely consumed within an additional hour of hydrogenation. The chemoselective hydrogenation was applied to the synthesis of radiolabeled tritium TPD compounds from the corresponding iodoarene derivatives

(Ambroise et al., 2001) (Fig. 6). The synthesis of compounds with isotope incorporation has also been studied with other photophores including phenylazides and benzophenones (Faucher et al., 2002).

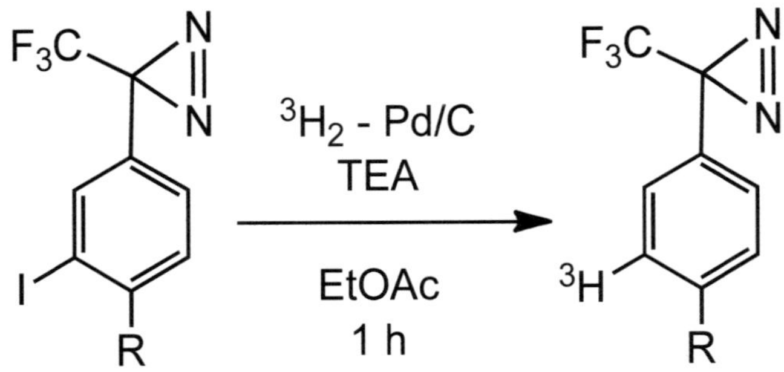


**1** R = $CH_2OH$
**3** R = COOH
**4** R = $CH_2NH_2$

**[$^3$H]-2** R = $CH_2OH$ (68%)
**[$^3$H]-5** R = COOH (64%)
**[$^3$H]-6** R = $CH_2NH_2$ (65%)

**Figure 6.** Selective tritiations of iodoarene TPD derivatives (**1, 3** and **4**) for carbon-iodine bond to carbon-tritium bond on Pd/C under tritium atmosphere. Parentheses are isolated yields.

Sammelson et al. performed selective hydrogenation for iodoarene derivative (**7**) over the chloroarene and trifluoromethyldiazirinyl group in the synthesis of photoreactive fipronil (**8**) using Pd/C under a $H_2$ or $^3H_2$ atmosphere. The resulting compound was a high-affinity probe for GABA receptor (Sammelson and Casida, 2003) (Fig. 7).

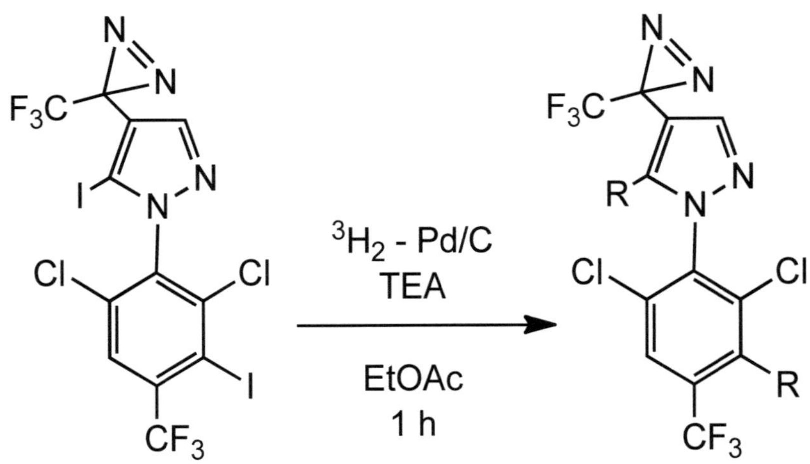


**7**

**8** R = H (45%)
**[$^3$H]-8** R = $^3$H

**Figure 7.** Selective hydrogenation or tritiations of carbon-iodine bond over carbon-chlorine bond and trifluoromethyldiazirinyl group of **7** with Pd/C under hydrogen or tritium atmosphere.

## 2.2. Selective hydrogenation of carbon-nitrogen double bonds (imines, Schiff's bases, reductive amination)

Imine (Schiff's base) TPD derivatives have been readily prepared from aldehyde (**9**) and primary amine (**10**). Catalytic hydrogenations of imines with $H_2$-Pd/C were potentially available to afford amines, but side reactions at the nitrogen-nitrogen double bond on TPD derivatives prevented use of these catalytic hydrogenations. Hydride reductions for imines are acceptable for TPD derivatives and sodium cyanoborohydride has been used for the reductive amination leading to (**11**) (Fig. 8) (Daghish et al., 2002).

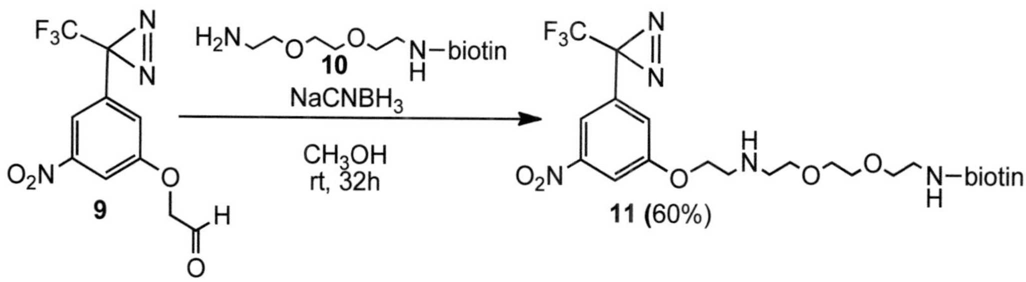


**Figure 8.** Reductive amination of TPD derivative (**9**) with biotin derivative (**10**)

Although this type of reaction is distinct from hydrogenation, we would like to briefly summarize reductions with hydride for use in TPD derivatization. $NaBH_4$ or $LiAlH_4$, which were most common hydride sources, were compatible for TPD derivative chemistry that involved reduction of carbonyl groups. However those reduction reagents that incorporated a cofactor (ie $CoCl_2$, $NiCl_2$ etc) promoted destruction of the diazirinyl ring (Hashimoto, unpublished results).

Many other hydride sources were compatible with TPD derivatizations.

Hydrazones derivatives of TPD (**13**) have been prepared with moderate yield from the corresponding TPD acetophenone (**12**) using hydrazine hydrate (1.5 eq). In early stages the acetophenone moieties were more reactive for the nucleophilic substitution with hydrazine hydrate than the reaction involving reduction of the diazirinyl group to diaziridines.

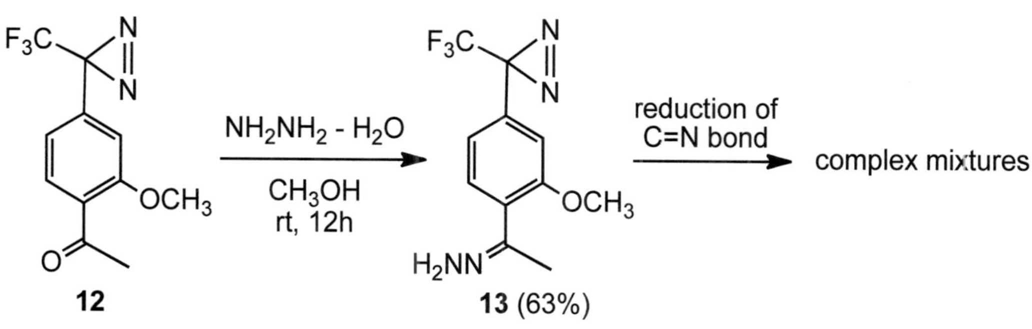


**Figure 9.** Selective TPD hydrazone formation (**13**) from acetophenone derivative (**12**)

The selective reduction of the carbon-nitrogen double bond in hydrazone to form carbon-nitrogen single bond was not archived under various conditions (alcoholic KOH with reflux, t-BuOK-DMSO at room temperature, or t-BuOK-toluene with reflux). Instead the side reaction involving loss of the diazirinyl group occurred (Hashimoto, unpublished results). (Fig. 9)

## 2.3. Selective hydrogenation of carbon-oxygen double bonds (carbonyl and carboxyl) for the TPD derivatives.

Many methods for the reduction of carbon-oxygen double bonds have been reported. The carbonyl groups, which can be introduced by Friedel-Crafts acylation, are one of the most important synthetic methods for the post-functional synthesis. Friedel-Crafts reactions of TPD derivatives are not attainable because the trifluoromethyldiazirinyl moiety has slight electron withdrawing properties (due to polarity of the quaternary carbon, which is connected directly to benzene ring). Furthermore the diazirinyl moiety was not stable over 25 °C in the presence of Lewis acids, which are the conditions generally used for catalysis in Friedel-Crafts reaction (Moss et al., 2001). TPD derivatives (**14** and **15**) can react at room temperature with the reactive acyl donor acetyl chloride when using aluminum chloride to introduce acetyl moiety (**12** and **16**) (Hashimoto et al., 2003, 2004) (Fig. 10) .

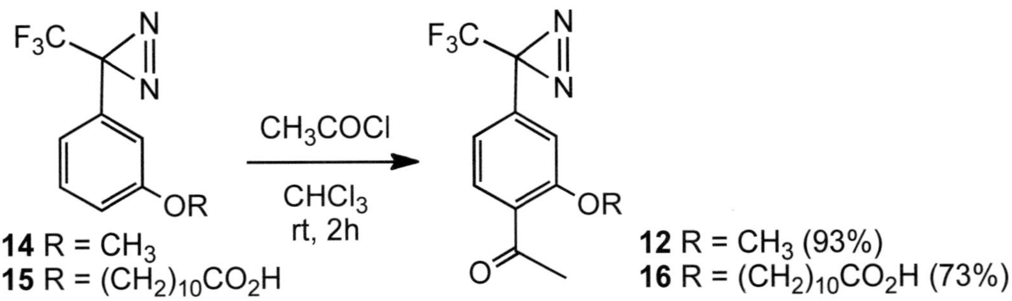


**Figure 10.** Friedel-Crafts acetylation of TPD derivatives (**14** and **15**) with acetyl chloride and aluminum chloride.

On the other hand incorporation of less active acyl donors such as dichloromethyl methyl ether has to use stronger the Lewis acid, $TiCl_4$. These conditions allow reaction with compound **14** to proceed (Hashimoto et al. 1997). Dichloromethane was used as solvent in early synthesis of this type but dichloromethyl methyl ether can also be used as solvent. This has enabled improvement in the yield of compound **17**. (Fig. 11)

Hydrogenation of the Friedel-Crafts acylated products has been studied. Clemmensen reduction, Wolff-Kishner reduction and catalytic hydrogenation with Pd/C cannot be applied to synthesis of TPD derivative as these conditions lead to breakage of the diazirinyl moiety.

During the course of these trial screening reactions, it was found that transfer hydrogenation with triethylsilane in trifluoroacetic acid could be applied to TPD derivatives (**12**). The conversions from benzyl carbonyl to methylene are very smooth and afforded the product (**18**) in very high yield without breaking the diazirinyl ring. (Fig. 12) (Hashimoto et al. 2003 & 2004)

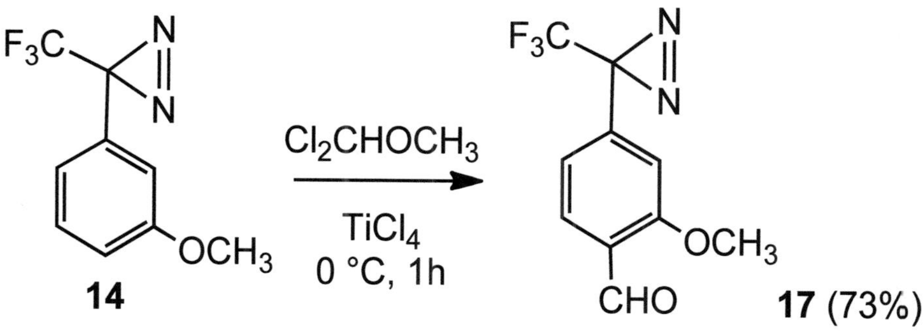


**Figure 11.** Synthesis of benzaldehyde TPD derivative (**17**) with Friedel-Crafts alkylation, followed by hydrolysis from m-methoxy TPD (**14**)

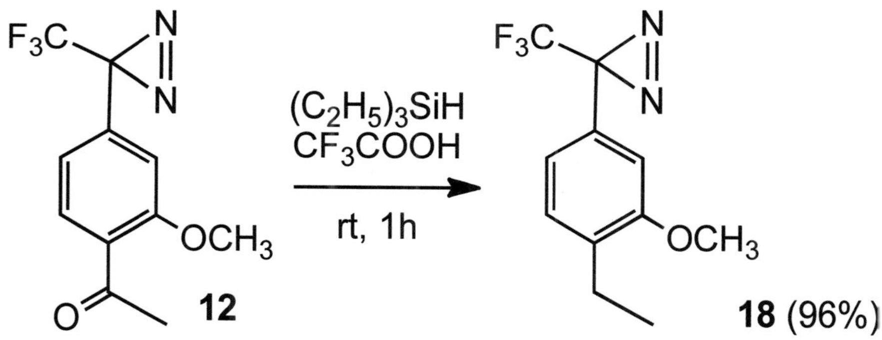


**Figure 12.** Transfer hydrogenation of TPD acetophenone derivative (**12**) with triethylsilane and trifluoroacetic acid

A sequence of reactions involving Friedel-Crafts acylation followed by reduction of the benzyl carbonyl to methylene enables us to stereocontrol synthesis of trifluoromethyldiazirinyl homo- and bishomo- phenylalanine derivatives. Synthesis of homo-phenylalanine has been reported using various methodologies including enzymatic methods (Zhao et al., 2002), Suzuki-coupling (Barfoot, et al., 2005), diastereoselective Michel addition (Yamada et al., 1998) and catalytic asymmetric hydrogenation (Xie, et al., 2000). These methods require the preparation of special reagents or precursors for the asymmetric synthesis of both enantiomers, especially aromatic compounds. Therefore one has to spend time and effort on establishment of TPD derivatizations without decomposition of diazirine.

Amino acids are one of the most popular precursors and easily available compounds for stereo controlled synthesis using the asymmetric center. Friedel-Crafts reactions between aromatics and a side chain of aspartic acid (Asp) or glutamic acid (Glu) are some of the key reactions for asymmetric synthesis for both homo- or bishomo- phenylalanine enantiomers' skeletons. (Reifenrath, et al. 1976; Nordlander et al., 1985; Melillo et al., 1987; Griesbeck & Heckroth, 1997; Xu et al., 2000; Lin et al., 2001)

It has been reported that synthesis of homophenylalanine using a Friedel-Crafts reaction of Asp anhydride (*N*-unprotected or *N*-protected) with $AlCl_3$ requires use of large excesses of aromatics and reflux in organic solvent for long durations (Xie, et al., 2000). These synthesis conditions cannot apply the equivalent condition of amino acid and TPD derivatives. Furthermore, the diazirinyl ring did not tolerate heating in the presence of Lewis acids. After Friedel-Crafts acylation, the constructed benzyl carbonyl group was hydrogenated to methylene under $H_2$-Pd/C, which is not suitable for TPD derivatives. These difficulties were overcome Friedel-Crafts acylation of TPD derivative (**14**) and side chain derivatives of Asp (**19**) or Glu (**20**) using trifluoromethanesulfonic acid followed by ionic hydrogenation of benzylcarbonyl group to methylene with triethylsilane - trifluoroacetic acid. After constructions of the homo- (**23**) or bishomo- (**24**) phenylalanine skeletons, removal of the protective groups afforded TPD containing homo- (**25**) or bishomo- (**26**) phenylalanine while maintaining the stereochemistry of starting Asp or Glu (Murai et al., 2009; Murashige et al. 2009) (Fig. 13).

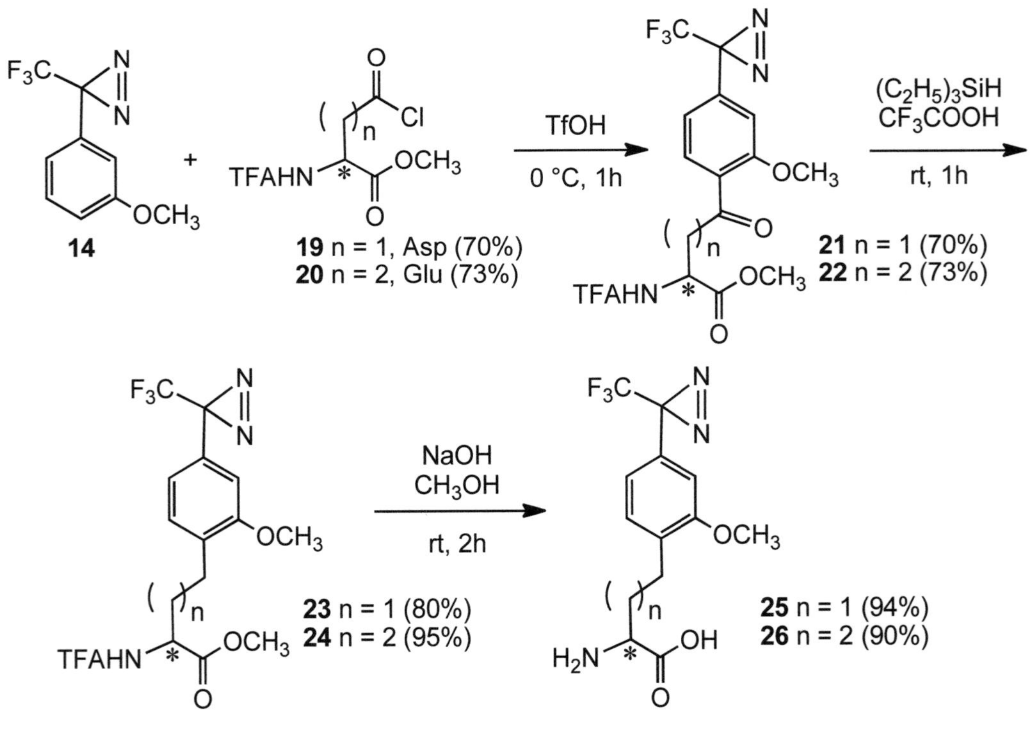


**Figure 13.** Stereo controlled synthesis of homo- (**25**) and bishomo- (**26**) phenylalanine TPD derivatives from m-methoxy TPD (**14**) and optically pure Asp (**19**) or Glu (**20**) derivatives

### 2.3.1. *Selective hydrogenation of carbon-oxygen double bonds with stable isotope labeling*

Established methods for the post-functional synthesis (described in the previous section) have facilitated the preparation of stable isotope labeled TPD. Stable isotopes act as a tag for the exogenous ligand derivatives on mass spectrometry. The methodologies will be very useful for the field of photoaffinity labeling to detect the labeled components. Friedel-Crafts acylation with 1-$^{13}$C acetyl chloride, which is a relatively inexpensive reagent compared with other $^{13}$C labeled compounds, afforded $^{13}$C labeled acetophenone derivative ([$^{13}$C]-**12**) in moderate yield. Hydrogenations by deuterium atom of the acetophenone has been applied using various conditions. Deuterium was effectively introduced to the methylene moiety by deuterium labeled triethylsilane ($Et_3SiD$) and unlabeled trifluoroacetic acid ($CF_3COOH$) to afford [1-$^{13}$C-1, 1-$D_2$]-**18**). It is not necessary to use deuterium labeled trifluoroacetic acid for the deuteration. (Hashimoto & Hatanaka, 2004)

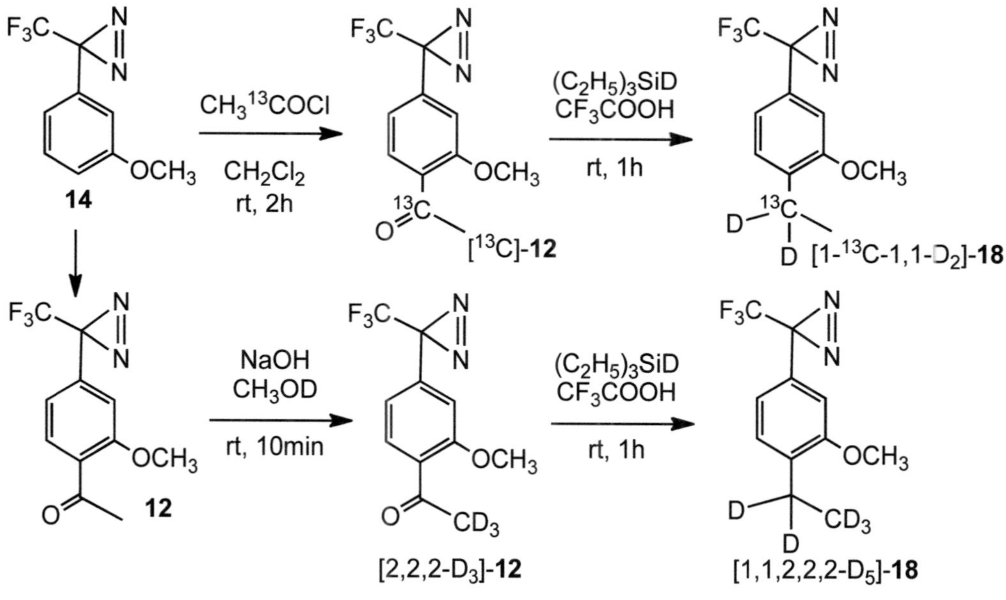


**Figure 14.** Synthesis of stable isotope labeled TPD derivatives with transfer hydrogenation

The α-position of the carbonyl groups was susceptible to very fast hydrogen-deuterium exchange using sodium hydroxide (NaOH) and methanol-OD ($CH_3OD$) at room temperature. There are no serious decrements of deuterium incorporation with various work up to synthesis [2,2,2-$D_3$]-**12**. After that, ionic hydrogenation with $Et_3SiD$ and trifluoroacetic acid afforded 5 deuterium incorporated TPD derivatives ([1,1,2,2,2-$D_5$]-**18**). (Fig. 14). These synthetic methodologies have also been applied to synthesis of deuterium incorporated photoreactive fatty acid derivatives. (Murai et al. 2010)

## 2.4. Selective hydrogenation of carbon-carbon double bonds for the TPD derivatives

The synthetic strategies for the Wittig reaction, followed by hydrogenation are amongst the major methods for carbon elongation derivatizations. These synthetic methods have not been compatible with synthesis of the TPD derivatives. This is because conditions for establishment for the selective hydrogenation (reduction) for the carbon-carbon double bond over that for nitrogen- nitrogen double bond on TPD are not easily achieved.

We found Wilkinson's catalyst, chlorotris(triphenylphosphine)rhodium(I) in methanol has specificity for the target reaction. The alkene containing TPD derivatives (**27-29**), which are synthesized from Wittig reaction for TPD aldehyde (**17**) and stable ylides, were subjected to hydrogenation with $H_2$-Wilkinson's catalyst at atmospheric pressure. It was observed that 25mol% of Wilkinson's catalyst required for complete hydrogenation. The α, β-unsaturated ester (**27**), nitrile (**28**) and aldehyde (**29**) were also hydrogenated under these conditions. The aldehyde carbonyl group conversion to primary alcohol (**33**) was only partially complete. (Fig. 15)

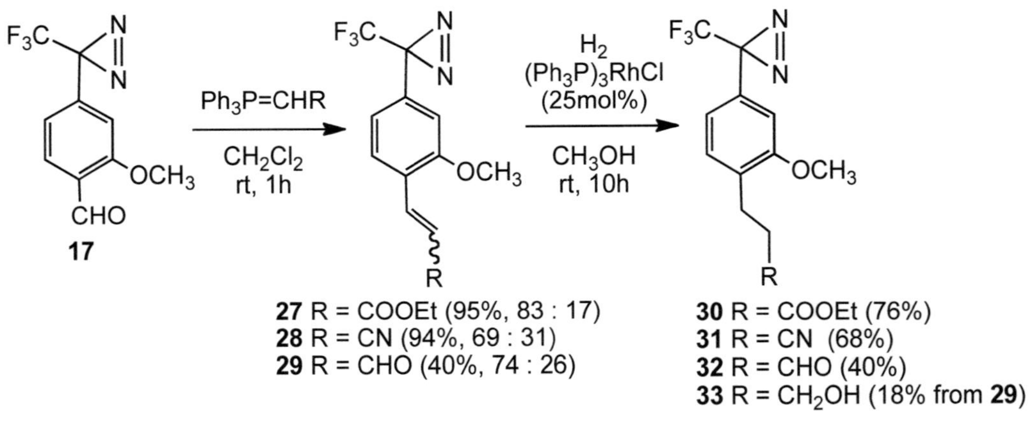


**Figure 15.** Synthesis of α, β-unsaturated carbonyl TPD derivatives and their hydrogenation with Wilkinson's catalyst

The hydrogenation of **27** and **28** with deuterium gas allowed effective incorporation of the deuterium atom intro these compounds (Hashimoto et al., 2007).

## 3. Conclusions

Hydrogenations are very important for post-functional synthesis of TPD compounds.

It is very important for synthesis of TPD compounds that a range of hydrogenation methods are investigated. Selective hydrogenations in the presence of nitrogen-nitrogen double bonds in TPD have been established. Very strict conditions are necessary as the important nitrogen-nitrogen double bond can easily be lost. The establishments of a range of hydrogenation methods, together with the limitations of these methods that are described in this review, will facilitate further progress in the post-functional preparations of TPD. These chemical considerations could generate further widespread use of these biochemically ideal photoaffinity labels.

## Author details

Makoto Hashimoto and Yuta Murai
*Graduate School of Agriculture, Hokkaido University, Kita 9, Nishi 9, Kita-ku, Sapporo, Japan*

Geoffery D. Holman
*Department of Biology and Biochemistry, University of Bath; Claverton Down, Bath BA2 7AY, U.K.*

Yasumaru Hatanaka
*Graduate School of Medicine and Pharmaceutical Sciences, University of Toyama, Sugitani, Toyama, Japan*

## Acknowledgement

This research was partially supported by a Ministry of Education, Science, Sports and Culture Grant-in-Aid for Scientific Research in a Priority Area, 18032007 and for Scientific Research (C), 19510210, 21510219 (M.H.).

## 4. References

Ambroise, Y.; Mioskowski, C.; Djéga-Mariadassou, G. & Rousseau, B. (2000). Consequences of affinity in heterogeneous catalytic reactions: highly chemoselective hydrogenolysis of iodoarenes. *J. Org. Chem.* 65, 7183-7186.

Barfoot, C. W.; Harvey, J. E.; Kenworthy, M. N.; Kilburn, J. P.; Ahmed, M. & Taylor, R. J. K. (2005). Highly functionalised organolithium and organoboron reagents for the preparation of enantiomerically pure α-amino acids. *Tetrahedron*, 61, 3403-3417.

Brunner, J.; Senn, H. & Richards, F. M. (1980). 3-Trifluoromethyl-3-phenyldiazirine. A new carbene generating group for photolabeling reagents. *J. Biol. Chem.* 255, 3313–3318

Daghish, M.; Hennig, L.; Findeisen, M.; Giesa, S., Schumer, F.; Hennig, H.; Beck-Sickinger, A. G. & Welzel, P. (2002) Tetrafunctional photoaffinity labels based on Nakanishi's *m*-nitroalkoxy-substituted phenyltrifluoromethyldiazirine. *Angew. Chem. Int. Ed.* 41, 2293-2297

Faucher, N.; Ambroise, Y.; Cintrat, J. -C.; Doris, E.; Pillon, F. & Rousseau, B. (2002). Highly chemoselective hydrogenolysis of iodoarenes. *J. Org. Chem.* 67, 932-934.

Fuwa, H.; Hiromoto, K.; Takahashi, Y.; Yokoshima, S.; Kan, T.; Fukuyama, T.; Iwatsubo,T.; Tomita, T. & Natsugari, H. (2006). Synthesis of biotinylated photoaffinity probes based on arylsulfonamide γ-secretase inhibitors, *Bioorg. Med. Chem. Lett.*16, 4184–4189.

Fuwa, H.; Takahashi, Y.; Konno, Y.; Watanabe, N.; Miyashita, H.; Sasaki, M.; Natsugari,H.; Kan, T.; Fukuyama, T.; Tomita, T. & Iwatsubo, T. (2007). Divergent synthesis of multifunctional molecular probes to elucidate the enzyme specificity of dipeptidic γ-secretase inhibitors. *ACS Chem. Biol.* 2, 408–418.

Galardy, R. E.; Craig, L. C. & Printz, M. P. (1973). Benzophenone triplet: a new photochemical probe of biological ligand-receptor interactions. *Nat. New Biol.* 242, 127–128.

Griesbeck, A. G. & Heckroth, H. (1997). A simple approach to ω-amino acids by acylationof arenes with *N*-acyl aspartic anhydrides. *Synlett* 11, 1243-1244.

Hashimoto, M.; Kanaoka, Y. & Hatanaka, Y. (1997). A versatile approach for functionalization of 3-aryl-3-trifluoromethyldiazirine photophore. *Heterocycles*, 46, 119-123.

Hashimoto, M.; Yang, J. & Holman, G. D. (2001). Cell-surface recognition of biotinylated membrane proteins requires very long spacer arms : An example from glucose-transporter probes. *CHEMBIOCHEM* 2, 52-59.

Hashimoto, M.; Hatanaka, Y. & Nabeta, K. (2003). Effective synthesis of a carbon-linked diazirinyl fatty acid derivative via reduction of the carbonyl group to methylene with triethylsilane and trifluoroacetic acid. *Heterocycles* 59, 395-398.

Hashimoto, M. & Hatanaka, Y. (2004). Simple synthesis of deuterium and $^{13}C$ labeled trifluoromethyl phenyldiazirine derivatives as stable isotope tags for mass spectrometry. *Chem. Pharm. Bull.* 52, 1385—1386.

Hashimoto, M.; Kato, Y. & Hatanaka, Y. (2006) Simple method for the introduction of iodo-label on (3-trifluoromethyl) phenyldiazirine for photoaffinity labeling. *Tetrahedron Lett.* 47, 3391–3394.

Hashimoto, M.; Kato, Y. & Hatanaka, Y. (2007). Selective hydrogenation of alkene in (3-trifluoromethyl) phenyldiazirine photophor with Wilkinson's catalyst for photoaffinity labeling. *Chem. Pharm. Bull.* 55, 1540-1543.

Hashimoto, M. & Hatanaka, Y. (2008). Recent progress in diazirine-based photoaffinity labeling. *Eur. J. Org. Chem.* 2513–2523

Hatanaka, Y.; Hashimoto, M.; Kurihara, H.; Nakayama, H. & Kanaoka, Y. (1994a). A novel family of aromatic diazirines for photoaffinity labeling. *J. Org. Chem.* 59, 383–387.

Hatanaka, Y.; Hashimoto, M.; Nakayama, H. & Kanaoka, Y. (1994b). Syntheses of nitro-substituted aryl diazirines. An entry to chromogenic carbene precursors for photoaffinity labeling. *Chem. Pharm. Bull.* 42, 826–831.

Karney, W. L. & Borden, W. T. (1997). Why does o-fluorine substitution raise the barrier to ring expansion of phenylnitrene? *J. Am. Chem. Soc.* 119, 3347–3350.

Lin, W.; He, Z.; Zhang, H.; Zhang, X.; Mi, A. & Jiang, Y. (2001). Amino acid anhydride hydrochlorides as acylating agents in Friedel-Crafts reaction: a practical synthesis of L-homophenylalanine. *Synthesis* 7, 1007-1009.

Melillo, D. G.; Larsen, R. D.; Mathre, D. J.; Shukis, W. F.; Wood, A. W. & Colleluori, J. R. (1987). Practical enantioselective synthesis of a homotyrosine derivative and (*R,R*)-4-propyl- 9-hydroxynaphthoxazine, a potent dopamine agonist. *J. Org. Chem.* 52, 5143-5150.

Moss, R. A.; Fedé, J. –M. & Yan S. (2001). $SbF_5$-mediated reactions of oxafluorodiazirines. *Org. Lett.* 3, 2305–2308.

Murai, Y.; Hatanaka, Y.; Kanaoka, Y. & Hashimoto, M. (2009). Effective synthesis of optically active 3-phenyl-3-(3-trifluoromethyl) diazirinyl bishomophenylalanine derivatives, *Heterocycles* 79, 359-364.

Murai, Y.; Takahashi, M.; Muto, Y.; Hatanaka, Y. & Hashimoto, M. (2010). Simple deuterium introduction at $\alpha$-position of carbonyl in diazirinyl derivatives for photoaffinity labeling. *Heterocycles*, 82, 909 - 915.

Murashige, R.; Y. Murai, Y.; Hatanaka, Y. & Hashimoto, M. (2009). Effective synthesis of optically active trifluoromethyldiazirinyl homophenylalanine and aroylalanine derivatives with Friedel-Crafts reactions in triflic acid. *Biosci. Biotechnol. Biochem.* 73, 1377-1380.

Nassal, M. (1983). 4-(1-Azi-2,2,2-trifluoroethyl)benzoic acid, a highly photolabile carbene generating label readily fixable to biochemical agents. *Liebigs Ann. Chem.* 1510– 1523.

Nordlander, J. E.; Payne, M. J.; Njoroge, F. G.; Vishwanath, V. M.; Han, G. R.; Laikos, G. D. & Balk, M. A. (1985) A short enantiospecific synthesis of 2-amino-6,7-dihydroxy- 1,2,3,4-tetrahydronaphthalene (ADTN). *J. Org. Chem.* 1985, 50, 3619-3622.

Platz, M. S. (1995). Comparison of phenylcarbene and phenylnitrene. *Acc. Chem. Res.* 28, 487–492.

Reifenrath, W. G.; Bertelli, D. J.; Micklus M. J. & Fries, D. S. (1976). Stereochemistry of friedel-crafts addition of phthalylaspartic anhydride to benzene. *Tetrahedron Lett.* 17, 1959-1962.

Sammelson, R. E. & Casida, J, E. (2003). Synthesis of a tritium-labeled, fipronil-based, highly potent, photoaffinity probe for the GABA receptor. *J. Org. Chem.* 68, 8075-8079.

Singh, A.; Thornton, E. R. & Westheimer, F. H. (1962). The photolysis of diazoacetyl-chymotrypsin. *J. Biol. Chem.* 237, PC3006-PC3008.

Smith, R. A. G. & Knowles, J. R. (1973). Aryldiazirines. potential reagents for photolabeling of biological receptor sites. *J. Am. Chem. Soc.* 95, 5072–5073.

Tomohiro, T.; Hashimoto, M. & Hatanaka, Y. (2005). Cross-linking chemistry and biology: development of multifunctional photoaffinity probes. Chemical Record, 5, 385–395

Yamada, M.; Nagashima, N.; Hasegawa, J. & Takahashi, S. (1998). A highly efficient asymmetric synthesis of methoxyhomophenylalanine using michael addition of phenethylamine. *Tetrahedron Lett.* 39, 9019-9022.

Xie, Y.; Lou, R.; Li, Z.; Mi, A. & Jiang, Y. (2000). DPAMPP in catalytic asymmetric reactions: enantioselective synthesis of L-homophenylalanine. *Tetrahedron: Asymm.* 11, 1487-1494.

Xu, Q.; Wang, G.; Wang, X.; Wu, T.; Pan, X.; Chan, A. S. C. & Yang, T. (2000). The synthesis of L-(+)-homophenylalanine hydrochloride. *Tetrahedron: Asymm.* 11, 2309-2314.

Zhao, H.; Luo, R. G.; Wei, D. & Malhotra, S. V. (2002). Concise synthesis and enzymatic resolution of L-(+)-homophenylalanine hydrochloride, *Enantiomer*, 7, 1-3.

Chapter 6

# Terminal and Non Terminal Alkynes Partial Hydrogenation Catalyzed by Some $d^8$ Transition Metal Complexes in Homogeneous and Heterogeneous Systems

Domingo Liprandi, Edgardo Cagnola, Cecilia Lederhos,
Juan Badano and Mónica Quiroga

Additional information is available at the end of the chapter

http://dx.doi.org/10.5772/47742

## 1. Introduction

The synthesis and manufacture of food additives, flavours and fragrances, as well as pharmaceutical, agrochemical and petrochemical substances, examples of fine and industrial chemicals, are closely related to selective alkyne hydrogenation [1,2].

Regarding alkyne partial hydrogenation, the main goal is to avoid hydrogenation to single bond and in the case of non-terminal alkynes is to give priority to the highest possible conversion and selectivity to the *(Z)*-alkene [3-5].

These kind of reactions are carried out by means of a catalytic process where control over conversion and selectivity can be exerted in different ways, e.g.: by varying a) the active species or b) the support, and/or by adding c) a promoter / a poison / a modifier, and finally, and not less important, by modifying the reaction temperature. Examples of factor b) are: mesoporous [6] and siliceous [7] materials, a pumice [8], carbons [9], and hydrotalcite [3]. Cases of factor c) are the typical Lindlar catalyst (palladium heterogenized on calcium carbonate poisoned by lead acetate or lead oxide, Pd-$CaCO_3$-Pb) [10] and the presence of quinoline and triphenylphosphine [11,12]. Research on factors a) and c) include bi-elemental systems such as Ni-B, Pd-Cu, etc. [13-19]. An example of the effect of the reaction temperature is a paper by Choi and Yoon, who found that the selectivity to *(Z)*-alkene increases when the temperature decreases using a Ni catalyst [20].

Besides, transition metal complexes are a group of substances widely used as catalysts that could be considered as a new active species or as a metal conditioned by its ligands, a kind of "poison" or a modifier for the metal atom [21].

These coordination compounds, used as catalysts, have gained increasing importance for such reactions [22-28] because they allow getting higher activities and selectivities, even under mild conditions of temperature and pressure [29-33].

The $d^8$ metals, e.g. Rh(I), Ir(I), Pd(II), Ni(II) and Pt(II), form complexes for which the square planar geometry is specially favoured. They are important in catalysis as the central atom can increase its coordination number by accepting ligands in the apical sites [34] or by interacting with the support. These complexes have also the ability to dissociate molecular dihydrogen, and stabilize a variety of reaction intermediates through coordination as ligands in relatively stable but reactive complexes. This is made possible by promoting rearrangements within their coordination spheres [35].

On the other hand, regarding the physical condition, the desired product can be obtained in homogeneous or heterogeneous systems. The latter, in the context of transition metal complexes used as catalysts, have some practical-economical benefits as follows: a) the easy and cheap way in which the catalyst is removed from the remaining solution after ending the hydrogenation reaction; b) the main product does not need further purification due to a possible contamination with a heavy metal compound when no complex leaching is detected; and lastly, c) there is no need for a costly temperature control.

The purpose of this chapter is to illustrate, based on results already published, the previous ideas using: a) 1-heptyne and 3-hexyne as the substrates to be partially hydrogenated, examples of terminal and non-terminal alkynes respectively, b) $[PdCl_2(NH_2(CH_2)_{12}CH_3)_2]$ and $[RhCl(NH_2(CH_2)_{12}CH_3)_3]$ as the catalytic species with coordination spheres having tridecylamine as an electron-donating $\sigma$ ligand and chloride as an electron-withdrawing $\sigma/\pi$ ligand and Pd(II) and Rh(I) as the central atoms respectively and c) $\gamma$-$Al_2O_3$ Ketjen CK 300 and RX3, a commercial carbonaceous material from NORIT, as supports for the heterogeneous catalytic tests.

Last but not least, the complex catalytic performances are compared, at the same operational conditions, against those obtained with the Lindlar catalyst which is accepted as a standard one.

## 2. Experimental

### 2.1. Complexes preparation and purification

$[PdCl_2(TDA)_2]$ and $[RhCl(TDA)_3]$ (TDA = $NH_2(CH_2)_{12}CH_3$) were prepared in a glass equipment with agitation and reflux in a purified argon atmosphere using tridecylamine (TDA) and $PdCl_2$ or $RhCl_3$ according to the case. $[PdCl_2(TDA)_2]$ (yellow-orange) was obtained at 338 K after 4.0 h with a molar ratio $TDA/PdCl_2 = 2$, while $[RhCl(TDA)_3]$ (yellow) was got at 348 K after 4.5 h with a molar ratio $TDA/RhCl_3 = 6$ using toluene and carbon tetrachloride as solvents respectively. The purification was made by column chromatography with silica gel as the stationary phase using chloroform for $[PdCl_2(TDA)_2]$ and chloroform/methanol (5/1 vol/vol) for $[RhCl(TDA)_3]$ as the corresponding eluting solvents. All the aliquots were tested to determine the presence of free TDA by thin layer chromatography. After drying the TDA-free solution in a rotary evaporator each complex, in solid state, was obtained.

## 2.2. Blank test

In each preceding complex preparation, a blank experiment, using only the corresponding salt and solvent, was run verifying that there was no product obtained from them at all.

## 2.3. Complexes immobilization

Anchoring of the complexes was carried out on γ-alumina (Ketjen CK 300), previously calcinated in air at 773 K for 3 h, or on RX3 (NORIT), a pelletized commercial carbon, by means of the incipient wetness technique. The solvents used for impregnation were as follows: a) chloroform for $[PdCl_2(TDA)_2]$ and b) chloroform-methanol 5/1 (vol/vol) for $[RhCl(TDA)_3]$ using a suitable concentration to obtain 0.3 wt % M (M = Pd or Rh). Then, solvents were let evaporate in a desiccator at 298 K, until constant mass was verified.

In order to check for a possible leaching of the immobilized complexes, each fresh-supported system was subjected to a 100-hour run in the corresponding reaction solvent at 353 K. After the tests, none Pd or Rh metal was detected in the remaining solution by Atomic Absorption spectroscopy, thus revealing a strong complex adherence to each support. In this respect, the constancy of M/Al or M/C atomic ratios obtained by XPS before and after the mentioned tests (see Table 2), ratifies that there was no leaching at all and that the complexes species remained anchored on both supports.

## 2.4. Complexes characterization

### 2.4.1. *Elemental composition*

The presence and weight percent of metal (Pd or Rh), chlorine and nitrogen elements were evaluated for each pure complex on a C- and H-free base, according to standard methods [36-39] to determine the stoichiometric ratios of the main atoms and to give a minimum formula for each complex.

### 2.4.2. *X-Ray photoelectron spectroscopy (XPS)*

XPS spectra were carried out to evaluate: a) the electronic state of atoms, b) the atomic ratios, for the pure complexes and for the supported complexes before and after the reaction and (c) the atomic ratios M/Al or M/C (where M = Pd or Rh) for the supported complexes before and after reactions. This was done to get each pure complex minimum formula and some insight in the way the complexes were immobilized on the supports; verifying at the same time that the coordination compounds were maintained after anchoring (fresh catalysts) and after the catalytic evaluations (run catalysts) with the final purpose of demonstrating that the complexes are the real catalytic active species.

A Shimadzu ESCA 750 Electron Spectrometer coupled to a Shimadzu ESCAPAC 760 Data System was used. The C 1s line was taken as an internal standard at 285.0 eV so as to correct possible deviations caused by electric charge on the samples. The superficial electronic states of the atoms were studied according to the position of the following peak maxima: Rh $3d_{5/2}$ and Pd $3d_{5/2}$, N 1s for the TDA molecule and Cl $2p_{3/2}$ for the complexes in any condition. In order to ensure that there was no modification on the electronic state of the species, the sample introduction was made according to the operational procedure reported elsewhere [40]. Exposing the samples to the atmosphere for different periods of time confirmed that there were no electronic modifications. Determination of the atomic ratios x/Metal (x = N, Cl) and Metal/Z (Z = Al or C, depending on the support) were made by comparing the areas under the peaks after background subtraction and corrections due to differences in escape depths [41] and in photoionization cross sections [42].

### 2.4.3. *Infrared spectroscopy (FTIR and IR)*

Pure complexes and TDA IR spectra were taken, in the 4100-900 $cm^{-1}$ range, to determine the presence of tridecylamine as a ligand in the complexes coordination spheres. The analysis was carried out using the TDA characteristic normal wavenumbers [43-45]. Besides, pure $[PdCl_2(TDA)_2]$ IR spectra was also taken and analyzed, below 600 $cm^{-1}$, with the purpose to elucidate if the obtained complex was the cis or trans isomer, and consequently to be able to assign a correct local site symmetry for this species.

The high wavenumber spectra were taken in a Shimadzu FTIR 8101/8101M single beam spectrometer; the equipment has a Michelson type optical interferometer. Two chambers are available to improve the quality of the spectra. The first one has a pyroelectric detector made of a high sensitivity LiTaO element, and the other has an MCT detector and the possibility to create a controlled $N_2$ (or dry air) atmosphere. On the other hand due to the low detector sensitivity below 500 $cm^{-1}$, a Perkin-Elmer 580 B double beam spectrometer was also used.

All the samples were dried at 353 K and were examined either in potassium bromide or cesium iodide disks in a concentration ranging from 0.5 to 1 wt% to ensure non-saturated spectra [41].

### 2.4.4. *Supports characterization*

The porosity of the supports was characterized by physical adsorption of $N_2$ (77 K) and $CO_2$ (273 K). Gas adsorption is useful to calculate specific surface area and pore volume. The use of both adsorptives ($N_2$ and $CO_2$) allows estimating the pore volume distribution of pores up to about 7.5 nm diameter [46]. By applying the Dubinin-Raduskevich equation to the $CO_2$ adsorption isotherm at 273 K, the volume of micropores with a diameter less than 0.7 nm ($V_{micro}$) can be obtained. On the other hand, the volume of supermicropores ($V_{supermicro}$), diameter ranging from 0.7 to 2 nm, is obtained by subtraction of $V_{micro}$ from the volume calculated by applying the Dubinin-Raduskevich method to the $N_2$ adsorption isotherm at 77 K [46]. The volume of mesopores with diameter between 2 and 7.5 nm was calculated from the $N_2$ adsorption isotherm at 77 K. In this respect, the volume of gas adsorbed between 0.2 and 0.7 relative pressure corresponds to the mesopore range of porosity. The wider porosity, macropores ($V_{macro}$) and part of the mesopores (with diameter from 7.5 to 50 nm), was determined by mercury porosimetry, using a Carlo Erba 2000 porosimeter. This equipment reaches a maximum pressure of 196 MPa, which allows estimating the volume of pores with a diameter longer than 7.5 nm. The addition of the mesopore volumes determined from $N_2$ adsorption isotherm and by mercury porosimetry gives the total mesopore volume ($V_{meso}$) [46]. Finally, with the BET equation applied to the $N_2$ adsorption isotherm at 77 K it is possible to evaluate the specific surface area.

This information can be used to know the support profitable sites where the complexes can be located and the concentration of the substrates would turn out augmented by a physicochemical adsorption process favouring the partial hydrogenation.

### *2.4.5. Catalytic runs*

The catalytic runs were made using 100 mL of a 2 % v/v alkyne in toluene solution, in a PTFE-coated batch stainless steel stirred tank reactor, operated at 600 rpm during 120 min. The weight of the supported complex catalysts was 0.075 g in all cases. In the catalytic evaluation of the unsupported complexes, a suitable mass of these ones was used to provide the same amount of metal (Pd or Rh) as in the corresponding supported catalysts. The same criterion was used for the Lindlar catalyst. Every catalytic test was carried out in triplicate at P = 150 kPa with a relative experimental error of about 3 %.

Detection of possible diffusional limitations during the catalytic runs was taken into account according to the procedures described in the literature [47-48]. External diffusional limitations were examined by varying the stirring velocity in the range of 180-1400 rpm. Conversion and selectivity constancy verified above 500 rpm allows to say that this type of limitation was absent at the selected rotatory speed. On the other hand, intraparticle mass transfer limitations were considered by crushing the heterogenised complex catalyst up to ¼ the original size and using the obtained samples to carry out the partial hydrogenation reactions. Conversion and selectivity values, equal to those obtained with the uncrushed heterogenised catalyst, permitted to state that this type of limitation was also absent in the physical operational conditions of our work. Last but not least, the catalyst cylinders were properly treated and weighted after the end of reactions. The difference in the mass of catalyst cylinders (before and after the test reactions) was not appreciable within the experimental error of the analytical balance method, meaning that there was no mass loss from the cylinders. Thus, it can be considered that the attrition effect was absent or was negligible enough to play a role in determining an additional mass transfer limitation. The analysis of reactants and products was made by gas chromatography, using a FID and a CP WAX 52 CB capillary column.

The following substrates and conditions were used to perform the catalytic tests:

- 1-heptyne (a terminal alkyne) partial hydrogenation was used to evaluate the catalytic performance of both complexes in homogeneous condition and anchored on $\gamma$-$Al_2O_3$ at 303 K with a 1-heptyne/Pd or Rh molar ratio equal to 7.3 $10^3$ and 7.0 $10^3$, respectively.

- 3-hexyne (a non-terminal alkyne) partial hydrogenation was used to evaluate the catalytic performance of $[RhCl(TDA)_3]$ in homogeneous condition and anchored on RX3 at 275, 290 and 303 K with a 3-hexyne/Rh molar ratio equal to 8.1 $10^3$.

*2.4.6. Atomic absorption spectroscopy*

The possible presence of M (Rh or Pd), provoked by a solvent leaching effect in each solution after catalytic evaluation of the heterogeneous systems, was analyzed by means of the Atomic Absorption technique.

## 3. Results and discussion

### 3.1. Pd or Rh complex minimum formula

Table 1 shows M (Pd or Rh), N and Cl elemental composition on a C- and H- free base, as well as Cl/M and N/M molar ratios obtained for the pure complexes.

| Complex | Elemental Composition (weight %) | | | Atomic ratios | |
|---|---|---|---|---|---|
| | M | Cl | N | Cl/M | N/M |
| $[PdCl_2(TDA)_2]$ | 52.0 | 34.3 | 13.7 | 1.99 | 2.00 |
| $[RhCl(TDA)_3]$ | 56.9 | 19.6 | 23.5 | 1.00 | 3.04 |

**Table 1.** M (Pd or Rh), N and Cl (on a C- and H-free base) elemental composition and Cl/M and N/M molar ratios for the pure complexes.[49]

The **Pd:Cl:N and Rh:Cl:N** molar ratios for the pure complexes calculated from the weight percent values (Table 1) and the molar masses of these elements, can be expressed as ca. **1:2:2 and 1:1:3** respectively.

Additionally, Table 2 presents the M $3d_{5/2}$, N 1s and Cl $2p_{3/2}$ peaks binding energies (BE), and the atomic ratios N/M, Cl/M for the pure and for the fresh and run supported complexes, obtained by XPS.

Table 2 also includes the superficial molar ratio M/Al or M/C for the heterogenized complexes. XPS binding energies for the pure substances are in accordance with the literature values [50,51]; showing that Pd or Rh, N and Cl are present in the corresponding products obtained after the synthesis and purification stages. Besides, the electronic states of these atoms may be considered as follows: a) **n+ for Pd or Rh, with n = 2 and n close to 1 respectively**; this is based on data in Table 2 and the literature values 338.3 eV for $[PdCl_2(NH_3)_2]$ and ranging from 307.3 to 308.5 eV for Rh(I) [50,51]; b) **-3 for N but as an ammonium-like nitrogen** as the 1s binding energies in Table 2 fall within the values found in the literature (400.9 to 402 eV [50,51]) corresponding to a $NH_4^+$ species; **this information suggests a bonding character for the N lone pair towards an electrophilic centre, in this case the Pd or Rh atom**; and c) **-1 for Cl as in a chloride compound** because the $2p_{3/2}$ binding energies in Table 2 are included within the literature values (197.9 to 198.5 eV [50,51]). At the same time, from the mentioned table, the atomic ratios for each complex indicate that these elements appear in the proportion **Pd:Cl:N equal to 1:2:2 and Rh:Cl:N equal to 1:1:3. These numbers are equal to those obtained via Elemental Composition results.**

| Complex | Condition | Binding energies (eV) | | | Atomic ratios (at/at) | | |
|---|---|---|---|---|---|---|---|
| | | M $3d_{5/2}$ | N 1s | Cl $2p_{3/2}$ | N/M | Cl/M | M/Al or C |
| $[PdCl_2(TDA)_2]$ | pure [49] | 338.2 | 401.9 | 198.3 | 2.00 | 1.99 | - |
| | γ-$Al_2O_3$ fresh [49] | 338.3 | 401.7 | 198.2 | 2.01 | 2.00 | 0.09 |
| | γ-$Al_2O_3$ run[a] [49] | 338.2 | 402.0 | 198.1 | 1.99 | 1.99 | 0.09 |
| | RX3 fresh [53] | 338.4 | 401.9 | 198.2 | 2.02 | 2.00 | 0.10 |
| | RX3 run[b] | 338.3 | 402.0 | 198.3 | 2.01 | 1.99 | 0.10 |
| $[RhCl(TDA)_3]$ | pure [49] | 307.1 | 402.1 | 198.1 | 3.00 | 1.01 | - |
| | γ-$Al_2O_3$ fresh [49] | 307.2 | 402.2 | 198.3 | 2.99 | 1.02 | 0.05 |
| | γ-$Al_2O_3$ run[a] [49] | 307.1 | 402.2 | 198.2 | 2.99 | 0.99 | 0.05 |
| | RX3 fresh [52] | 307.1 | 401.9 | 198.0 | 3.00 | 1.02 | 0.06 |
| | RX3 run[b] | 307.0 | 402.0 | 198.0 | 2.99 | 1.00 | 0.06 |

[a] 1-heptyne partial hydrogenation
[b] 3-hexyne partial hydrogenation

**Table 2.** XPS binding energies and XPS atomic ratios for pure, fresh and run heterogenized Pd or Rh complexes.

On the other hand, Figure 1 shows the pure TDA, $[PdCl_2(TDA)_2]$ and $[RhCl(TDA)_3]$ FTIR spectra, while Figure 2 depicts the pure $[PdCl_2(TDA)_2]$ IR spectrum in the range below 600 $cm^{-1}$. As observed at high wavenumbers in Figure 1, the following characteristic peaks of a primary aliphatic amine [44], are present: (A) $NH_2$ "stretching" (3600-3100 $cm^{-1}$), CH "stretching" (3000-2800 $cm^{-1}$), (B) $NH_2$ "bending" (1700-1600 $cm^{-1}$), CH "bending" (1500-1300 $cm^{-1}$) and (C) CN "stretching" (1200-1000 $cm^{-1}$). In particular, labels A, B and C, related to the nitrogen atom, are taken as a reference because they are sensitive to its environment. Besides Figure 1 shows that $[PdCl_2(TDA)_2]$ and $[RhCl(TDA)_3]$ FTIR peaks globally agree with those corresponding to pure TDA. **Anyhow, differences are found in the labelled wavenumbers indicated above: A, B and C as they show a slight shift to lower frequencies with respect to the pure ligand, meaning an interaction between the nitrogen lone pair and the Pd or Rh atom. This argument is reinforced by the fact that when a primary amine is bonded, the $NH_2$ stretching peak is considerably different in shape and intensity from the original $NH_2$ band [45], as seen in the shown spectra. This information confirms that the TDA molecule is one of the ligands in the complexes coordination spheres.**

The IR spectrum presented in Figure 2 shows the peaks corresponding to the Pd-ligand vibrations for the pure complex. **$[PdCl_2(TDA)_2]$ can be considered as the trans-isomer** because of the presence of single peaks, which obey the principle of mutual exclusion, typical of centre-symmetric species [54].

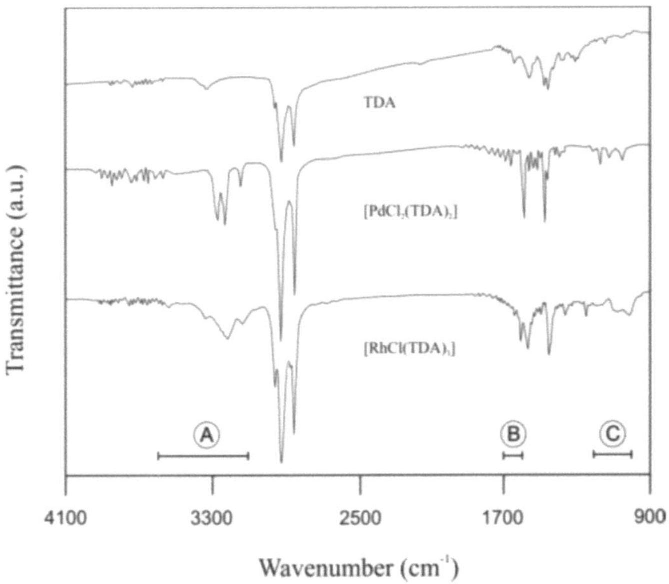


**Figure 1.** FTIR spectra corresponding to pure TDA, $[PdCl_2(TDA)_2]$ and $[RhCl(TDA)_3]$.[49]

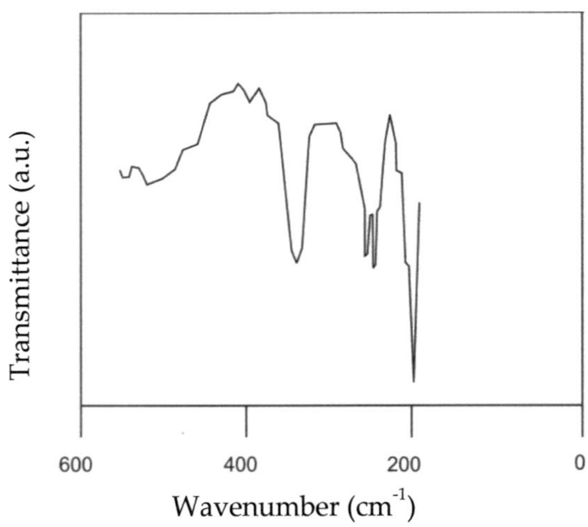


**Figure 2.** IR spectrum below 600 $cm^{-1}$ for pure $[PdCl_2(TDA)_2]$. [54]

*At this point, the complex minimum formula, on the basis of the preceding Elemental Composition, XPS and FTIR arguments, can be expressed by* $[PdCl_2(TDA)_2]$ *or* $[RhCl(TDA)_3]$.

## 3.2. Palladium and rhodium local site symmetries, HOMO-LUMO electron configurations and complexes dimensions

Molecular orbitals with symmetries corresponding to the irreducible representations of the molecular point group automatically satisfy the Fock equation. For complex species the terminal atom symmetry orbital (TASO)/molecular orbitals (MO) and the metal atomic orbitals are taken into account to explain metal-ligand bonding according to their symmetry properties. In this respect, the (n-1)d and ns metal atomic orbitals are those that match best the energy of the TASO/MOs. Based on this, complex antibonding MOs have considerably more metal character while complex bonding MOs have more ligand character; with the former lying higher in energy. On the other hand, tetra-coordinated palladium(II) and rhodium(I) ($d^8$ species) complexes have a square-planar geometry [55]. On this background and knowing that $[PdCl_2(TDA)_2]$ is the trans isomer two facts can be considered:

i. HOMO-LUMO Electron Configurations

Knowing that $[PdCl_2(TDA)_2]$ and $[RhCl(TDA)_3]$ have a $D_{2h}$ and $C_{2v}$ local site symmetries respectively and taking the main rotation axis along the z cartesian axis, the Angular Overlap Model (AOM) [56] can be used to predict the HOMO-LUMO frontier orbitals in an increasing order of energy for each case:

- $[PdCl_2(TDA)_2]$
  Non-bonding ($d_{xy}$), antibonding double degenerate $2e^*_\pi$ (($d_{xz}$, $d_{yz}$)*), $e^*_\sigma$ (($d_{z2}$)*) and $3e^*_\sigma$ (($d_{x2-y2}$)*) [49]. Assigning the eight electrons to this scheme, it turns out that ($d_{z2}$)* (z direction) and ($d_{x2-y2}$)* (x and y directions) are the HOMO and LUMO frontier orbitals, respectively.
- $[RhCl(TDA)_3]$
  Non-bonding ($d_{xy}$), antibonding double-degenerate $e^*_\pi$ (($d_{xz}$, $d_{yz}$)*), 7/4 $e^*_\sigma$ (($d_{z2}$)*) and 9/4 $e^*_\sigma$ (($d_{x2-y2}$)*). Assigning the eight electrons to this scheme, it turns out that ($d_{z2}$)* (z direction) and ($d_{x2-y2}$)* (x and y directions) are the HOMO and LUMO frontier orbitals, respectively.

The Rh(I) complex HOMO-LUMO frontier orbitals lie higher in energy than those corresponding to the Pd(II) complex because of the lower oxidation number of the central atom. The HOMO is useful to produce the cleavage of the H–H bonding, generating hydrogen atoms and the LUMO is available to receive electron density from the substrate molecule, weakening the C–C triple bond; both concepts are key factors during the catalytic cycle leading to the hydrogenation of the substrate.

ii. Complexes Dimensions

The approximate molecular size of the metal complexes can be estimated in order to study structural aspects related to their location in the supports porosity. This can be done taking into account the square planar geometry, typical covalent radii, a 109.5° C–C–C angle and basic trigonometry.

- $[PdCl_2(TDA)_2]$
  The lengths TDA–Pd–TDA and Cl–Pd–Cl are ca. 4 and 0.7 nm, respectively.
- $[RhCl(TDA)_3]$
  The lengths TDA–Rh–TDA and Cl–Rh–TDA are ca. 4 and 2.3 nm, respectively.

### 3.3. Supported Pd and Rh complexes structures

Table 3 presents the BET surface area and the pore volume distribution for both supports.

| Sample | $S_{BET}$ ($m^2$ $g^{-1}$) | $V_{micro}$ (mL $g^{-1}$) [<0.7 nm] | $V_{supermicro}$ (mL $g^{-1}$) [0.7-2 nm] | $V_{meso}$ (mL $g^{-1}$) [2-7.5 nm] | $V_{macro}$ (mL $g^{-1}$) [7.5-50 nm] |
|---|---|---|---|---|---|
| γ-$Al_2O_3$ | 180 | 0.048 | 0.030 | 0.487 | 0.094 |
| RX3 | 1411 | 0.356 | 0.333 | 0.098 | 0.430 |

**Table 3.** BET surface area and supports pore volume distributions.[53]

According to this information it can be remarked that RX3 possesses a $S_{BET}$ that is 7.84 times greater than the γ-$Al_2O_3$ surface area. Besides, γ-$Al_2O_3$ can be considered basically a mesoporous support, while RX3 has a pore volume distribution in the range of micro, supermicro and macro pores.

On the other hand, XPS results of binding energies and atomic ratios for the anchored complexes, Table 1, show that: a) Cl/M and N/M atomic ratios are equal to those obtained for the pure complexes and in total accordance with the elemental composition results for the pure complexes; and b) there is a constancy of the M $3d_{5/2}$, N 1s and Cl $2p_{3/2}$ XPS BEs with respect to the corresponding values for the pure complexes, meaning that their electronic states remain unchanged. Item a) indicates that the supported complexes may be considered as tetra-coordinated, maintaining its chemical identity after anchoring, and item b) suggests that the metal (Pd or Rh) is not in contact with the support surface. In this way, as the inductive influence of the hydrocarbon chain on the nitrogen atom is exerted up to the second/third carbon atom, the immobilization of the complexes takes place via a physicochemical interaction between the last part of the TDA molecule and the alumina or carbon basal planes, i.e. an anchoring showing a kind of "table" arrangement as seen in Figure 3.

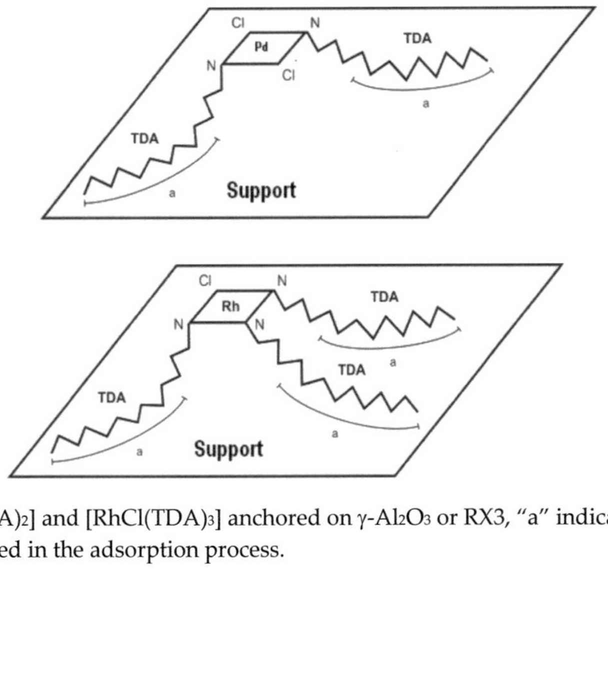


**Figure 3.** [$PdCl_2(TDA)_2$] and [$RhCl(TDA)_3$] anchored on γ-$Al_2O_3$ or RX3, "a" indicates the TDA carbon chain portion involved in the adsorption process.

Based on the ideas of the previous paragraph it can be stated that the anchored complexes maintain the local site symmetries around the central atoms, that is $D_{2h}$ for $[PdCl_2(TDA)_2]$ and $C_{2v}$ for $[RhCl(TDA)_3]$, with the same HOMO-LUMO electron configurations.

Besides, as the longest dimension of the coordination compounds is the same for both complexes, and according to the $Al_2O_3$ or RX3 distribution pore sizes (Table 3), it can be concluded that the species can be located only in the meso and macropores (2–7.5 and 7.5–50 nm respectively) for both supports, thus occupying ca. 88 % and 43 % of the support total pore volume, respectively.

## 3.4. Catalytic tests

### *3.4.1. 1-heptyne partial hydrogenation (terminal alkyne)*

1-heptene and n-heptane were the only products detected by GC during the catalytic runs using: (1) commercial Lindlar catalyst, (2) $[PdCl_2(TDA)_2]$ or $[RhCl(TDA)_3]$ (homogeneous condition), and (3) $[PdCl_2(TDA)_2]$ or $[RhCl(TDA)_3]$ supported on $\gamma$-$Al_2O_3$ (heterogeneous condition) at 303 K, 150 kPa and 2 % v/v 1-heptyne/toluene solution. In Figures 4 and 5, the conversion ($X_e$) and the selectivity ($S_e$) to 1-heptene versus 1-heptyne total conversion ($X_T$) are plotted for all of the catalytic systems.

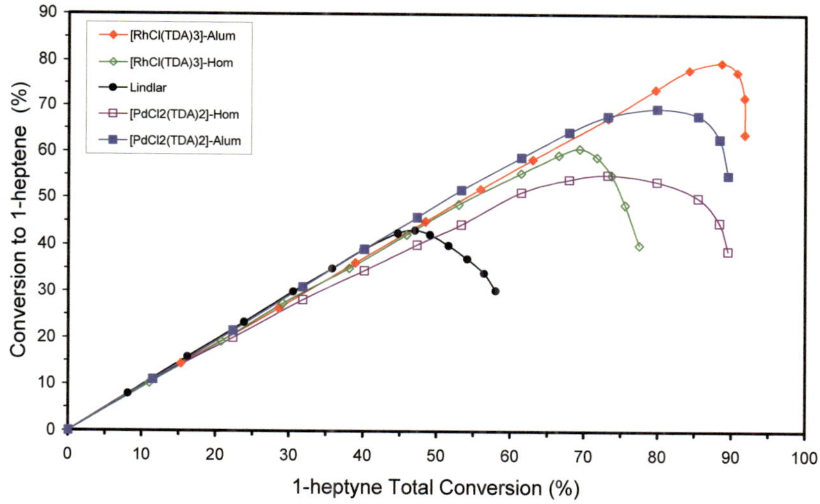


**Figure 4.** Conversion to 1-heptene vs. 1-heptyne total conversion for: Lindlar catalyst, $[PdCl_2(TDA)_2]$, $[PdCl_2(TDA)_2]$/ $\gamma$-$Al_2O_3$, $[RhCl(TDA)_3]$, $[RhCl(TDA)_3]$/ $\gamma$-$Al_2O_3$.[49] 1-heptyne/Pd = 7.3 $10^3$ and 1-heptyne/Rh = 7.0 $10^3$.

It can be seen, from Figure 4, that the catalytic systems show an initial part with an almost 45° linear slope. From that part onwards, all the systems show a similar profile shape, with increasing conversion to 1-heptene up to a maximum value, after which this variable falls.

The selectivity plots displayed in Figure 5 show a plateau-shaped behaviour in a very important range of 1-heptyne total conversion followed then by a decreasing tendency.

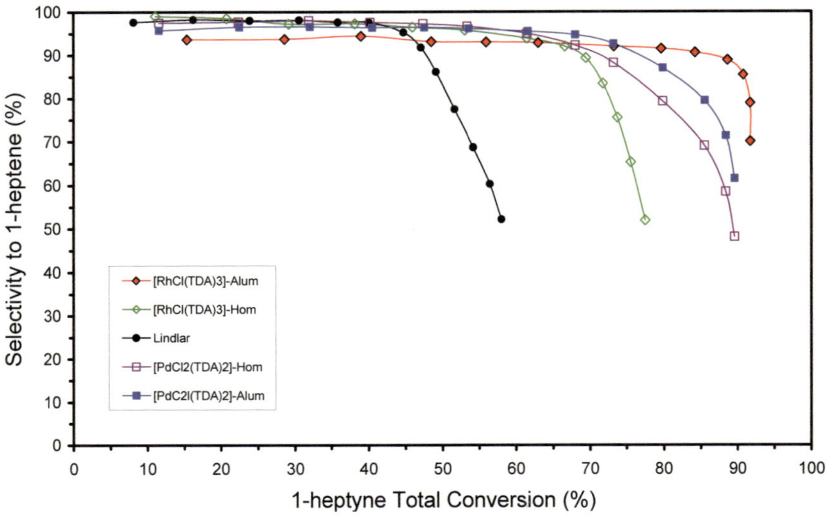


**Figure 5.** Selectivity to 1-heptene vs. 1-heptyne total conversion for: Lindlar catalyst, $[PdCl_2(TDA)_2]$, $[PdCl_2(TDA)_2]/\gamma\text{-}Al_2O_3$, $[RhCl(TDA)_3]$, $[RhCl(TDA)_3]/\gamma\text{-}Al_2O_3$.[49] 1-heptyne/Pd = 7.3 $10^3$ and 1-heptyne/Rh = 7.0 $10^3$.

On the other hand, high selectivities to 1-heptene ($S_e$), which are obtained up to 50 min of reaction time, and their corresponding conversions to 1-heptene ($X_e$) and 1-heptyne total conversions ($X_T$), are presented in Table 4.

| Catalytic System | Condition | $S_e$ (%) | $X_e$ (%) | $X_T$ (%) |
|---|---|---|---|---|
| $[RhCl(TDA)_3]$ | Heterogeneous | ≅ 92-93 | 73.6 | 80.0 |
| | Homogeneous | ≥ 92 | 59.4 | 66.5 |
| $[PdCl_2(TDA)_2]$ | Heterogeneous | ≥ 95 | 64.3 | 68.0 |
| | Homogeneous | ≥ 92 | 54.0 | 67.9 |
| Lindlar | Heterogeneous | ≥ 92 | 43.2 | 47.1 |

**Table 4.** Selectivity and conversion to 1-heptene and 1-heptyne total conversion values for the catalytic systems up to 50 min of reaction. 1-heptyne/Rh = 7.0 $10^3$ and 1-heptyne/Pd = 7.3 $10^3$.

With this information, trends and selection of the best catalytic systems are drawn taking into account two factors: a) high selectivity values to 1-heptene and b) the range of 1-heptyne total conversion in which the high selectivity values are maintained.

A general trend, based on the better catalytic performances (Table 4), can be established for the involved systems:

**$[RhCl(TDA)_3]/Al_2O_3$ > $[PdCl_2(TDA)_2]/Al_2O_3$ > $[RhCl(TDA)_3]$ > $[PdCl_2(TDA)_2]$ >> Lindlar**

*From this trend, it can be stated that $[PdCl_2(TDA)_2]$ and $[RhCl(TDA)_3]$ in heterogeneous or homogeneous conditions are better options than the Lindlar catalyst for the 1-heptyne partial hydrogenation under mild operational conditions.*

Besides, depending on the central element or the physical condition two new trends can be written:

**Central Element: Rh systems are better than Pd systems**

**Physical Condition: Heterogeneous systems are better than Homogeneous systems**

*From these tendencies, it can be stated that the best combination is the Rh(I) complex heterogeneous system, which is in the first place in the general trend and the worst option is the Lindlar catalyst.*

### 3.4.2. 3-hexyne Partial Hydrogenation (Non-Terminal Alkyne)

*(Z)*-3-hexene, *(E)*-3-hexene and n-hexane were the only products detected by GC during the reaction tests using the catalytic systems: (1) commercial Lindlar catalyst (2) [RhCl(TDA)$_3$] (homogeneous condition) , and (3) [RhCl(TDA)$_3$]/RX3 (heterogeneous condition) at 275, 290 and 303 K, 150 kPa and 2 % v/v 3-hexyne/toluene solution. In Figure 6(a) the conversions to *(Z)*-3-hexene and *(E)*-3-hexene are shown as a function of the 3-hexyne total conversion for the Lindlar catalyst and for Rh(I) homogeneous and heterogeneous complex for the three temperatures, while Figure 6(b), for the sake of clarity, is presented at the optimum temperature 303 K. It can be noted the predominant formation of the *(Z)*-alkene stereo-isomer, the desired product. In this respect, it can be seen in Figure 6, that all of the catalytic systems show again an initial part with an almost linear slope, which takes a value of 45° for the [RhCl(TDA)$_3$]/RX3 catalyst. After that initial part, all of the systems have a similar shape with an increasing 3-hexyne total conversion, showing a maximum value of conversion to *(Z)*-3-hexene. There was also a relatively low amount of the side products: *(E)*-3-hexene formed either as initial product or via $Z \rightarrow E$ isomerization, and n-hexane (not plotted in Figure 6 because of the low values obtained and for the sake of clarity) produced either by hydrogenation of the alkyne or the alkene isomers [7,57]. Last but not least, [RhCl(TDA)$_3$]/RX3 showed the lowest conversion values to the *(E)* isomer and to the alkane.

In Figure 7, a detail from Figure 6, it can be observed that, for a given catalytic system, the variation of conversion to *(Z)*-3-hexene vs. 3-hexyne total conversion follows an increasing tendency as the temperature is raised. However, it can be noted that the performance of Rh(I) complex heterogeneous system is slightly sensitive to temperature changes while the homogeneous system and the Lindlar catalyst are considerably sensitive to temperature changes.

For a given temperature, the [RhCl(TDA)$_3$]/RX3 system shows the highest conversions to *(Z)*-3-hexene at the highest 3-hexyne total conversions (maximum value: $X_{(Z)}$ = 95.0% at $X_T$ = 99.8%), followed by [RhCl(TDA)$_3$] and then by the Lindlar catalyst.

In Figure 8 the selectivity to *(Z)*-3-hexene vs. the 3-hexyne total conversion values are presented. The selectivity plots show an initial plateau-shaped behaviour followed by a marked decreasing tendency for the increasing 3-hexyne total conversion. The [RhCl(TDA)$_3$]/RX3 system allows to obtain a practically constant value of a very high selectivity (not lower than 98.5%) up to a very high 3-hexyne total conversion (ca. 85%); after that, the selectivity decays to a value ca. 82 %. Meanwhile, in the case of [RhCl(TDA)3] and the Lindlar catalyst, high values of selectivities (ca. 89.4 and ca. 94.2 respectively) were obtained for lower 3-hexyne total conversions up to ca. 44%; then both systems show a monotonously decreasing profile shape, which is more pronounced in the case of the Lindlar catalyst.

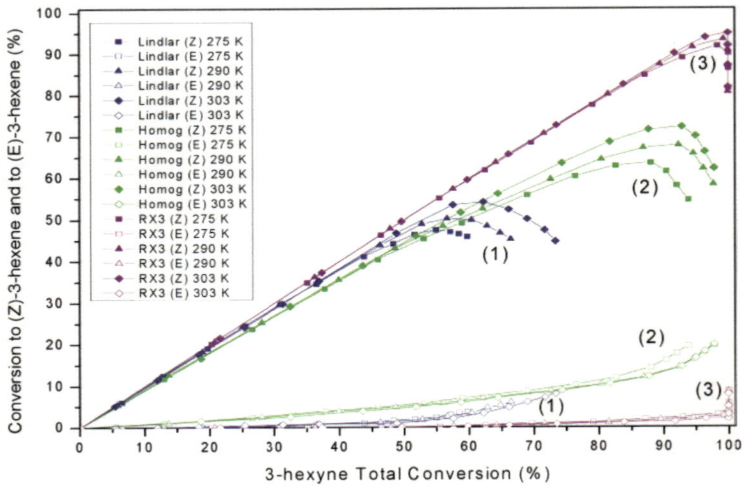


(a)

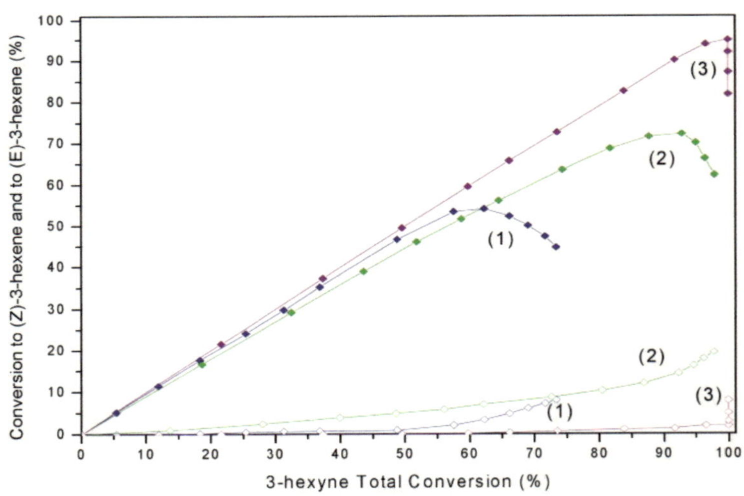


(b)

**Figure 6.** (a) Conversion to *(Z)*-3-hexene and to *(E)*-3-hexene vs. 3-hexyne total conversion for: (1) Lindlar catalyst, (2) [RhCl(TDA)3], (3) [RhCl(TDA)3]/RX3; filled square/open square 275 K, filled triangle/open triangle 290 K, filled diamond/open diamond 303 K. Open symbols: *(E)*-3-hexene, solid symbols *(Z)*-3-hexene.[54] 3-hexyne/Rh = 8.1 $10^3$.
(b) Conversion to *(Z)*-3-hexene and to *(E)*-3-hexene vs. 3-hexyne total conversion for: (1) Lindlar catalyst, (2) [RhCl(TDA)3], (3) [RhCl(TDA)3]/RX3; at the optimum temperature 303 K. Open symbols: *(E)*-3-hexene, solid symbols *(Z)*-3-hexene.[54] 3-hexyne/Rh = 8.1 $10^3$.

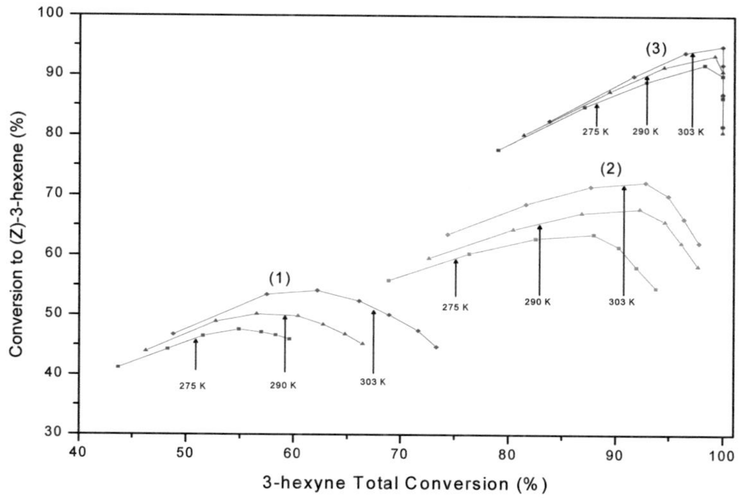


**Figure 7.** Conversion to *(Z)*-3-hexene vs. 3-hexyne total conversion for: (1) Lindlar catalyst, (2) [RhCl(TDA)3], (3) [RhCl(TDA)3]/RX3 (a detail of Figure 6 in the zone where the three systems present the most remarkable differences).[54] 3-hexyne/Rh = 8.1 10[3].

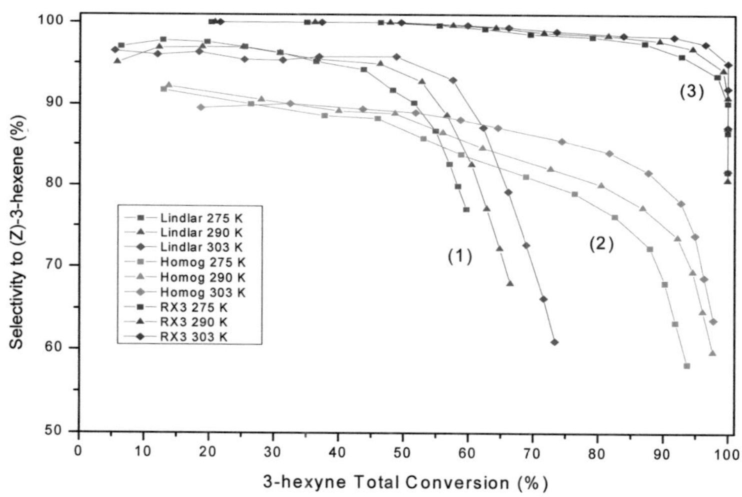


**Figure 8.** Selectivity to *(Z)*-3-hexene vs. 3-hexyne total conversion for: (1) Lindlar, (2) [RhCl(TDA)3], (3) [RhCl(TDA)3]/RX3; (●) 275 K, (▲) 290 K, (♦) 303 K.[54] 3-hexyne/Rh = 8.1 10[3].

To reinforce the conclusions given in the previous paragraphs, some relevant conversion and selectivity values for diverse reaction conditions are summarized in Table 5.

Considering Figures 6, 7 and 8 and Table 5 the different catalytic systems can be ordered in a descendent $X_{(Z)}$ production as follows: [RhCl(TDA)3]/RX3 > [RhCl(TDA)3] >> Lindlar catalyst. Finally, for each system the higher the temperature the higher the selectivity and the higher the conversion to the *(Z)*- isomer, although the [RhCl(TDA)3]/RX3 system behaves as the less sensitive catalyst to that variable.

| Reaction time (min) | Catalyst | T (K) | $X_T$ (%) | $X_{(Z)}$ (%) | $S_{(Z)}$ (%) | $S_{(E)}$ (%) | $S_n$ (%) |
|---|---|---|---|---|---|---|---|
| 50 | Lindlar | 275 | 48 | 44 | 92 | 3 | 5 |
| | | 290 | 53 | 49 | 93 | 3 | 4 |
| | | 303 | 57 | 53 | 93 | 4 | 3 |
| | [RhCl(TDA)3] | 275 | 76 | 61 | 79 | 13 | 8 |
| | | 290 | 81 | 64 | 80 | 13 | 7 |
| | | 303 | 82 | 69 | 84 | 13 | 3 |
| | [RhCl(TDA)3]/RX3 | 275 | 87 | 85 | 98 | 2 | 0.3 |
| | | 290 | 89 | 87 | 98 | 2 | 0.3 |
| | | 303 | 92 | 90 | 98 | 1.5 | 0.5 |
| 120 | Lindlar | 275 | 60 | 46 | 77 | 6 | 17 |
| | | 290 | 66 | 45 | 68 | 9 | 23 |
| | | 303 | 73 | 44 | 61 | 11 | 28 |
| | [RhCl(TDA)3] | 275 | 94 | 55 | 58 | 21 | 21 |
| | | 290 | 98 | 58 | 59 | 20 | 21 |
| | | 303 | 98 | 62 | 64 | 20 | 16 |
| | [RhCl(TDA)3]/RX3 | 275 | 99.9 | 82 | 82 | 9 | 9 |
| | | 290 | 99.9 | 81 | 81 | 8 | 11 |
| | | 303 | 99.9 | 82 | 82 | 8 | 10 |

**Table 5.** 3-hexyne total conversion ($X_T$), conversions to *(Z)*-3-hexene ($X_{(Z)}$) and selectivities to *(Z)*-3-hexene ($S_{(Z)}$), *(E)*-3-hexene ($S_{(E)}$) and n-hexane ($S_n$) for the following catalysts: Lindlar, [RhCl(TDA)3] complex unsupported and anchored on RX3. 3-hexyne/Rh = 8.1 $10^3$.

*It can be remarked that for the 3-hexyne (non-terminal alkyne) partial hydrogenation, [RhCl(TDA)3] is a much better option than the Lindlar catalyst to obtain the desired product.*

## 3.5. The optimum catalytic system for both test reactions

### *3.5.1. Chemical considerations*

Based on the information obtained from the previous sections, it can be concluded that [RhCl(TDA)3], supported either on γ-$Al_2O_3$ or on RX3 in a "table" arrangement structure (Section 3.3), is the best option to carry out the 1-heptyne or 3-hexyne partial hydrogenation to obtain high conversion values. As it was said in Section 3.2, this complex has a $d^8$ central element with a combination of ligands L'/L = 3 (L' = TDA, L = Cl), with a square planar geometry associated to a $C_{2v}$ local site symmetry. This type of coordination compound, at some point of the reaction mechanism, has to release one of the coordination sphere ligands; in this respect, the trans-effect series indicates that the labile ligand is TDA opposite to the Cl ligand. Some features related to the central atom, the TDA ligand, the complex coordination number, the site symmetry and the supported condition, could explain this optimum performance.

According to the Atomic Orbital Model (AOM) mentioned in Section 3.2, the $(d_{z2})^*$ and $(d_{x2-y2})^*$ are the HOMO/LUMO frontiers orbitals, respectively. The former, along the z axis with a high electron density, is ready to overlap with the hydrogen σ antibonding orbital, thus favouring H-H bond breaking and the latter, an empty orbital extended on the xy plane, is available to receive electron density from the substrate molecule triple bond, thus weakening its π bonds. These two factors, that are important in the hydrogenation catalytic cycle, will be particularly favoured in this case as the complex antibonding orbitals have considerably more metal than ligand character with a relatively high energy because of the low oxidation state of the central element.

On the other hand, the complex bonding molecular orbitals have a predominant character from the TASO/MOs of the ligands in the coordination sphere constituted by tridecylamine (an electron donating σ species) and chloride (an electron-withdrawing σ/π species) in a 3/1 ratio what means a net electronic enrichment on the Rh atom, a fact that will also contribute to the H-H bond rupture. Besides, the TDA ligand presents another remarkable feature related to the Dispersion forces which are relevant when the TDA/solvent interaction is analyzed. This is important as soon as the TDA molecule is released, due to the trans-effect, and it is simultaneously stabilized by the solvent via a solvatation process, a fact that is greatly favoured because of the long hydrocarbon chain of the TDA ligand.

Besides, the presence of a support, either $\gamma$-$Al_2O_3$ or RX3, can favour the catalytic process by activating some extra substrate molecules (1-heptyne or 3-hexyne) because of their interactions with the support surface via Acid/Base Lewis or Dipersion forces for $\gamma$-$Al_2O_3$ or RX3 respectively. In any case, this situation turns out to be an important factor as it makes the alkyne (1-heptyne or 3-hexyne) concentration around the supported complex higher than in the bulk solution.

Finally, the high selectivity values could be explained considering the interaction between the substrate triple or double bond (alkyne or alkene) with the complex species LUMO frontier orbital, and, additonally, for the heterogeneous system, with the support chemical sites. For both factors the stronger the interaction, the more favourable the hydrogenation process, but in each case a triple bond will give place to the strongest interaction. Thus, the selectivity to 1-heptene or *(Z)*-3-hexene will be kept in a very high value inasmuch as the alkyne concentration is high enough to consider the alkene hydrogenation negligible, after that the situation is reversed.

### *3.5.2. Practical considerations*

On the other hand, regarding that $[RhCl(NH_2(CH_2)_{12}CH_3)_3]$ anchored on $\gamma$-$Al_2O_3$ or RX3 is the optimum catalyst, some practical-economical benefits can be mentioned: a) the easy and cheap way in which the catalyst is removed from the remaining solution after ending the hydrogenation reaction; b) the main product does not need further purification due to a possible contamination with a heavy metal compound as no complex leaching was detected; and lastly, c) there is no need for a costly temperature control due to the mild operational conditions.

## 4. Conclusions

Experimental results demonstrate that $[PdCl_2(NH_2(CH_2)_{12}CH_3)_2]$ and $[RhCl(NH_2(CH_2)_{12}CH_3)_3]$, $d^8$ transition metal complexes, can be used as catalysts to partially hydrogenate 1-heptyne and 3-hexyne (terminal and non-terminal alkynes respectively) in homogeneous and heterogeneous systems ($\gamma$-$Al_2O_3$ and RX3 as supports) at mild operational conditions (P = 150 kPa and T up to 303K) with very good catalytic performances even better than the Lindlar catalyst. Analyses based on Elemental Composition, XPS, IR, Atomic Absorption Spectroscopy show that the active catalytic species is the complex itself in each case, with a minimum formula as written above.

Tetracoordinated electron-rich transition elements, as well as the presence and the relative quantity of a good electron donating ligand such as $NH_2(CH_2)_{12}CH_3$ with a long-chain hydrocarbon substituent and the heterogeneous condition, contribute to obtain a catalytic system with a high activity and selectivity performance. According to this and supported by experimental results, the optimum catalyst turns out to be $[RhCl(NH_2(CH_2)_{12}CH_3)_3]$ supported either on $\gamma$-$Al_2O_3$ or RX3. This behaviour can be understood in terms of Coordination Sphere parameters, Complex Dimensions, Local Site Symmetry, HOMO/LUMO frontier orbitals and some Support features

## Author details

Domingo Liprandi[1,*], Edgardo Cagnola[1], Cecilia Lederhos[2],
Juan Badano[2] and Mónica Quiroga[1,2]
[1]*Inorganic Chemistry, Departament of Chemistry, Faculty of Chemical Engineering, National University of Litoral (UNL), Santa Fe, Argentina,*
[2]*Institute of Catalysis and Petrochemistry Research, INCAPE (CONICET- UNL), Santa Fe, Argentina*

## Acknowledgement

UNL and CONICET financial supports are greatly acknowledged.

## 5. References

[1] Chen B, Dingerdissen U, Krauter JGE, Lansink Rotgerink HGJ, Móbus K, Ostgard DJ, Panste P, Riermeir TH, Seebald S, Tacke T, Trauthwein H (2005) New developments in hydrogenation catalysis particularly in synthesis of fine and intermediate chemicals. App.Catal. A: Gen. 280:17-46.

[2] Elsevier CJ, Kluwer AM (2007) Homogeneous hydrogenation of alkynes and dienes. In: de Vries JG, Elsevier CJ (eds) Handbook of homogeneous hydrogenation. vol 1, ch 14. Wiley-VCH, Darmstadt, p 375

---

* Corresponding Author

[3] Mastalir A, Király Z (2003) Pd nanoparticles in hydrotalcite: mild and highly selective catalysts for alkyne semihydrogenation. J. Catal. 220: 372-381.

[4] Marín-Astorga N, Pecchi G, Fierro JLG, Reyes P (2003) Alkynes hydrogenation over Pd-supported catalysts. Catal. Lett. 91 (1–2):115-121.

[5] Semagina N, Kiwi-Minsker L (2009) Palladium Nanohexagons and Nanospheres in Selective Alkyne Hydrogenation. Catal. Lett. 127 (3–4):334-338.

[6] Choudary BM, Kantam ML, Reddy NM, Rao KK, Haritha Y, Bhaskar V, Figueras F, Tuel A (1999) Hydrogenation of acetylenics by Pd-exchanged mesoporous materials. App. Catal. A; Gen. 181:139-144.

[7] Li F, Yi X, Fang W (2009) Effect of Organic Nickel Precursor on the Reduction Performance and Hydrogenation Activity of $Ni/Al_2O_3$ Catalysts. Catal. Lett. 130:335-340.

[8] Gruttadauria M, Noto R, Deganello G, Liotta LF (1999) Efficient semihydrogenation of the C-C triple bond using palladium on pumice as catalyst. Tetrahedron Lett. 40:2857-2858.

[9] Lennon D, Marshall R, Webb G, Jackson SD (2000) The effects of hydrogen concentration on propyne hydrogenation over a carbon supported palladium catalyst studied under continuous flow conditions. Stud. Surf. Sci. Catal. 130: 245-250.

[10] Lindlar H, Dubuis R, Jones FN, McKusick FC (1973) Palladium catalyst for partial reduction of acetylenes. Org. Synth. Coll. 5: 880.

[11] Yu J, Spencer JB (1998) Discovery that quinoline and triphenylphosphine alter the electronic properties of hyfrogenation catalysts. Chem. Comm. 1103-1104

[12] Nijhuis TA, van Koten G, Kapteijn F, Moulijn JA (2003) Separation of kinetics and mass-transport effects for a fast reaction: the selective hydrogenation of functionalized alkynes. Catal. Today 79–80: 315-321.

[13] Huang W, Li A, Lobo RF, Chen JG (2009) Effects of Zeolite Structures, Exchanged Cations, and Bimetallic Formulations on the Selective Hydrogenation of Acetylene Over Zeolite-Supported Catalysts. Catal. Lett. 130: 380-385.

[14] Badano JM, Quiroga M, Betti C, Vera C, Canavese S, Coloma-Pascual F (2010) Resistence To Sulfur And Oxygenated Compounds Of Supported Pd, Pt, Rh, Ru Catalysts. Catal. Lett. 137: 35-44.

[15] Nijhuis TA, van Koten G, Moulijn JA (2003) Optimized palladium catalyst systems for selective liquid-phase hydrogenation of functionalized alkynes. App. Catal. A: Gen. 238: 259-271.

[16] Chen JG, Qi S-T, Humbert MP, Menning CA, Zhu Y-X (2010) Rational design of low-temperature hydrogenation catalysts: Theoretical predictions and experimental verification. Acta Phys. Chim. Sin. 26 (4): 869-876.

[17] Shiju RN, Guliants VV (2009) Recent developments in catalysis using nanostructured materials. Appl. Catal. A: Gen. 356: 1-17.

[18] Volpe MA, Rodríguez P, Gigola CE (1999) Preparation of $Pd/Pb/\alpha\text{-}Al_2O_3$ catalysts for selective hydrogenation using $PbBu_4$: the role of metal-support boundary atoms and the formation of a stable surface complex. Catal. Lett. 61: 27-32

[19] Zhang W, Li L, Du Y, Wang X, Yang P (2009) Gold/Platinum Bimetallic Core/Shell Nanoparticles Stabilized by a Fréchet-Type Dendrimer: Preparation and Catalytic Hydrogenations of Phenylaldehydes and Nitrobenzenes. Catal. Lett. 127 (3–4): 429-436.

[20] Choi J, Yoon NM (1996) An excellent nickel boride catalyst for the cis-selective semihydrogenation of acetylenes. Tetrahedron Lett. 37 (7): 1057-1060.

[21] Crespo-Quesada M, Dykeman RR, Laurenczy G, Dyson PJ, Kiwi-Minsker L (2011) Supported nitrogen-modified Pd nanoparticles for the selective hydrogenation of 1-hexyne. J. Catal. 279: 66-74.

[22] Costa M, Pelagatti P, Pelizzi C, Rogolino D (2002) Catalytic activity of palladium(II) complexes with tridentate nitrogen ligands in the hydrogenation of alkenes and alkynes. J. Molec. Catal. A: Chem. 178: 21-26.

[23] de Wolf E, Spek AL, Kuipers BWM, Philipse AP, Meeldijk JD, Bomans PHH, Frederik P.M., Deelman B.J., van Koten G. (2002) "Fluorous derivatives of [Ru(COD)(dppe)]$BX_4$ (X=F, Ph): synthesis, physical studies and application in catalytic hydrogenation of 1-alkenes and 4-alkynes. Tetrahedron 58: 3911-3922.

[24] Edvrard D, Groison K, Mugnier Y, Harvey PD (2004) The $Pd_4(dppm)_4(H)_2^{2+}$ cluster: a precatalyst for the homogeneous hydrogenation of alkynes. Inorg. Chem. 43: 790-796.

[25] Frediani P, Giannelli C, Salvini A, Ianelli S (2003) Ruthenium complexes with 1,1'-biisoquinoline as ligands. Synthesis and hydrogenation activity. J. Organomet. Chem. 667: 197-208.

[26] Kerr JM, Suckling CJ (1988) Selective hydrogenation by novel palladium(II) complex. Tetrahedron Lett. 29 (43): 5545-5548.

[27] Park JW, Chung YM, Suh YW, Rhee HK (2004) Partial hydrogenation of 1,3-cyclooctadiene catalyzed by palladium-complex catalysts inmobilized on silica. Catal. Tod. 93-95: 445-450.

[28] Santra PK, Sagar P (2003) Dihydrogen reduction of nitroaromatics, alkenes, alkynes using Pd(II) complexes both in normal and high pressure conditions. J. Molec. Catal. A: Chem. 197 (1-2): 37-50.

[29] L'Argentière PC, Cagnola EA, Cañón MG, Liprandi DA, Marconetti DV (1998), A nickel tetra-coordinated complex as catalyst in heterogeneous hydrogenation. J. Chem. Technol .Biotechnol. 71: 285-290.

[30] Quiroga ME, Cagnola EA, Liprandi DA, L'Argentière PC (1999) Supported Wilkinson's complex used as a high active hydrogenation catalyst. J. Mol. Catal. A: Chem. 149: 147-152.

[31] Liprandi DA, Quiroga ME, Cagnola EA, L'Argentière PC (2002) A new more sulfur-resistannt rhodium complex as an alternative to the traditional Wilkinson's catalyst. Ind. Eng. Chem. Res. 41: 4906-4910

[32] Cagnola EA, Quiroga ME, Liprandi DA, L'Argentière PC (2004) Immobilized Rh, Ru, Pd and Ni complexes as catalysts in the hydrogenation of cyclohexene. Appl. Catal.. A: Gen. 274: 205-212.

[33] Hamilton CA, Jackson SD, Kelly GJ, Spence R, Bruin D (2002) Competitive Reactions in Alkyne Hydrogenation. App. Catal. A: Gen. 237: 201-209.

[34] Bailar J C, (ed) (1975) Comprehensive Inorganic Chemistry. Vol 3, p. 1234.

[35] Halpern J (1968) in: Homogeneous Catalysis. American Chemical Society: Chap. 1.

[36] Kolthoff IM, Sandell EB, Meehan EJ, Bruckenstein S (1969) Quantitative Chemical Analysis, fourth ed., Interscience Publishers, New York.

[37] Anderson SN, Basolo F (1963) Inorg. Synth. 7: 214-220.

[38] Vogel AI (1951) A Text Book of Quantitative Inorganic Analysis, Longmans, Green and Co, London.

[39] Livingstone S, in: Bailar JC Jr, Emeléus H., Nyholm R, Trotman-Dickenson AF (Eds.) (1973) The Chemistry of Ruthenium, Rhodium, Palladium, Osmium, Iridium and Platinum, Comprehensive Inorganic Chemistry, Pergamon Press, Oxford.

[40] Mallat T, Petrov J, Szabó S, Sztatisz J (1985) Palladium–cobalt catalyst: phase structure and activity in liquid phase hydrogenations. Reac. Kinet. Catal. Lett. 29: 353-361.

[41] Borade R, Sayari A, Adnot A, Kaliaguine S (1990) J. Phys. Chem. 94: 5989.

[42] Scofield JH (1976) J. Electron. Spectrosc. Relat. Phenom. 8: 129.

[43] Nakamoto K (1986) Infrared and Raman Spectra of Inorganic and Coordination Compounds, 4th Ed., Wiley, New York, parts I and III.

[44] Silverstein RM, Clayton Basler G, Morril TC, Spectrometric Identification of Organic Compounds, 5th Ed. Wiley, New York, 1991, chapter III.

[45] Pouchert CJ (1981) The Aldrich Library of Infrared Spectra Ed. (III): 1562 D.

[46] Rodríguez-Reinoso F, Linares-Solano A (1988) in: Chemistry and Physics of Carbon, Vol. 21, Walker PL Jr. (Ed) ,Marcel Dekker, New York p1.

[47] Holland FA, Chapman FS (1976) Liquid Mixing and Processing in Stirred Tanks, Reinhold, New York, Chap 5.

[48] Le Page JF (1978) Catalyse de Contact, Editions Technip, Paris, Chap 2.

[49] Quiroga M, Liprandi D, Cagnola E, L'Argentière P (2007) 1-heptyne semihydrogenation catalyzed by palladium or rhodium complexes. Influence of: metal atom, ligands and the homo/heterogeneous condition. Appl. Catal. A: Gen. 326:121–129.

[50] Wagner CD, Riggs WM, Davis RD, Moulder JF (1978). In Handbook of X-ray Photoelectron Spectroscopy. Muilenberg. G.E., Ed. Perkin-Elmer: Eden Preirie, MN.

[51] NIST X-ray Photoelectron Spectroscopy Database NIST Standard Reference Database 20, Version 3.5 (Web Version), National Institute of Standards and Technology, USA, 2007.

[52] Liprandi DA, Cagnola EA, Paredes JF, Badano JM, Quiroga M E (2012) A High (Z)/(E) Ratio Obtained During the 3-Hexyne Hydrogenation with a Catalyst Based on a Rh(I) Complex Anchored on a Carbonaceous Support. Catal. Lett. 142: 231–237.

[53] L'Argentiere P, Quiroga M, Liprandi D, Cagnola E., Román Martínez MC, Díaz Auñón JA, Salinas Martínez de Lecea C. (2003) Activated carbon heterogenized $[PdCl_2(NH_2(CH_2)_{12}CH_3)_2]$ for the selective hydrogenation of 1-heptyne. Catal. Lett. 87: 97 – 101.

[54] Liprandi DA, Cagnola E A, Quiroga M E, L'Argentière PC (2009) Influence of the Reaction Temperature on the 3-Hexyne Semi-Hydrogenation Catalyzed by a Palladium(II) Complex. Catal. Lett. 128: 423–433

[55] Cotton FA, Wilkinson G (1988) Advanced Inorganic Chemistry, fifth ed, John Wiley and Sons, New York, pp 901-902.

[56] Purcell KF, Kotz JC (1977) Inorganic Chemistry, Holt-Saunders International Editions: Philadelphia, pp 543-549.

[57] Papp A, Molnár A, Mastalir A (2005) Catalytic investigation of Pd particles supported on MCM-41 for the selective hydrogenations of terminal and internal alkynes. Appl. Catal. A: Gen. 289: 256-266.

Chapter 7

# Kinetic Study of the Partial Hydrogenation of 1-Heptyne over Ni and Pd Supported on Alumina

M. Juliana Maccarrone, Gerardo C. Torres, Cecilia Lederhos, Carolina Betti, Juan M. Badano, Mónica Quiroga and Juan Yori

Additional information is available at the end of the chapter

http://dx.doi.org/10.5772/48699

## 1. Introduction

Selective hydrogenation reactions are industrially used for the partial hydrogenation of unsaturated organic compounds in order to form more stable products or intermediate materials for different processes. The production of final organic products of high added value or intermediate compounds for the synthesis of fine chemicals is both of industrial and academic importance [1]. Alkenes are much appreciated products used in the food industry (flavours), the pharmaceutical industry (sedatives, anesthetises, vitamins) and in the perfumes industry (fragrances). They are also used for the production of biologically active compounds [2], resins, polymers and lubricants, etc.

The partial hydrogenation of acetylenic compounds using homogeneous or heterogeneous metallic catalysts provides a very viable and economically feasible way for the obtaining of these olefinic compounds. Selective catalysts and optimum operational conditions are necessary in order to avoid the complete hydrogenation of the unsaturated bond. Certain transition metals anchored on different solids have demonstrated to be very active and selective catalysts for this type of reaction. They also have the advantage that they can be operated under milder reaction conditions. It is well documented that palladium is a highly active catalyst for hydrogenation [3]. In this sense, Lindlar catalyst ($Pd/CaCO_3$, 5 wt % of Pd modified with $Pb(OAc)_2$) has been used since 1952 as an excellent commercial catalyst for this type of reactions [4]. The argued reasons for the differences in reactivity of Pd indicate that when the metal is electron deficient it becomes less active because alkynes are more weakly adsorbed [5].

During decades a lot of research has been carried out modifying this type of catalysts in order to increase the activity and selectivity: different supports as alumina, coal, silica [5-7] have been tried while modified palladium [8], or nanoparticles of Pd have also been investigated [9-11]. In a recent work of our group very high activities and selectivities of W monometallic and W-Pd bimetallic catalysts were found during the partial hydrogenation of a long chain terminal alkyne [12]. In this contribution it was reported that important variations in the activity and selectivity were produced when the employed pretreatment conditions were changed. The published reports do not give a clear explanation of the effect of each metal.

Providing an economic, active and selective metallic catalyst for the production of alkenes via selective hydrogenation is a very important challenge. Besides the bibliography on the kinetics of the reaction of heavy alkynes is scarce [13-15]. In this sense, the kinetic study of the liquid phase hydrogenation reaction of 1-heptyne using monometallic nickel and palladium catalysts anchored on γ-alumina made in the current work had three principal objectives: i) to gain knowledge about the reaction system, ii) to obtain kinetic expressions which enable the design of partial hydrogenation reactors, and iii) to give an explanation of the catalytic effect of different metals. The kinetic modelling was used in this work to shed light on these issues.

## 2. Experimental

### 2.1. Catalyst preparation

The catalysts were prepared by the incipient wetness technique using Ketjen CK300 γ-$Al_2O_3$ as support (cylinders of 1.5 mm, calcined at 823K during 4 h, 180 $m^2$ $g^{-1}$ BET surface area). To study the influence of metals, acidic solutions with pH=1 were prepared using $NiCl_2$ (Merk Cat. No. 806722), and $Pd(NO_3)_2$ (Fluka, Cat. No. 76070) to prepare the monometallic catalysts. The concentration of the solutions was adjusted in order to obtain metal loading of 4 and 0.4 wt % of Ni and Pd, respectively. The impregnated solids were dried during 24 h at 373 K and then they were calcined in an air flow at 823 K during 3 h. Prior to reaction, the catalysts were reduced in hydrogen flow (105 mL $min^{-1}$) during 1 h at 673 K for Ni/$Al_2O_3$ and at 573 K for Pd/$Al_2O_3$.

### 2.2. Characterization

#### *2.2.1. Chemical analysis*

The metal loadings of the catalysts were determined by digesting the samples and then analyzing the liquors by inductively coupled plasma analysis (ICP) in an OPTIMA 2100 DV Perkin Elmer equipment.

#### *2.2.2. X-Ray Photoelectron Spectroscopy (XPS)*

XPS determinations were made in Multitech UniSpecs XR-50 equipment with a dual Mg/Al X-ray source. A SPECS PHOIBOS 150 hemisphere analyzer was used in the Fixed Analyzer Transmission mode (FAT). The samples were treated *ex situ* with a hydrogen flow during 1 h at the corresponding reduction temperature and *in situ* in the load camera of the instrument with a $H_2/Ar$ (5% v/v) flow during 10 min. The spectra were obtained using pass energy of 30 eV and a Mg anode operated at 200 W. During the tests the pressure was less than 2 $10^{-12}$ MPa.

### 2.2.3. X-Ray Diffraction (XRD)

X-Ray Diffraction (XRD) measurements of the samples were obtained using a Shimadzu XD-D1 instrument with CuK$\alpha$ radiation ($\lambda$ = 1.5405A°) in the $15 < 2\theta < 85°$ at a scan speed of 1° $min^{-1}$. Samples were powdered and reduced under a hydrogen flow, then they were immediately put into the chamber of the equipment and the data acquisition was started.

### 2.2.4. Temperature Programmed Reduction (TPR)

The tests were performed in an Ohkura 2002 S apparatus equipped with a thermal conductivity detector. A cold water trap was placed before the thermal detector to condense water. Before the TPR tests the samples were dried *in situ* at 673 K for 30 min in an oxygen flow (AGA purity 99.99%). After that the samples were cooled down up to 298 K in an argon flow. Then the temperature was increased up to 1223 K at 10 K $min^{-1}$ in a $H_2/Ar$ (5% v/v) gas flow.

### 2.2.5. Hydrogen chemisorption

Hydrogen chemisorption was performed by means of the dynamic pulse method using a Micromeritics Auto Chem II apparatus equipped with a thermal conductivity detector. 0.2 g of the samples were reduced 1 h *in situ* at the above mentioned temperatures using a $H_2/Ar$ (5% v/v) flow. The samples were degassed *in situ* for 2 h under an argon flow (AGA, 99.99%) at a temperature equal to that of the corresponding reduction step and then cooled down to room temperature. In the case of the palladium catalysts they were cooled down to 373 K to make the formation of palladium hydride negligible [16]. After that the chemisorption of hydrogen was performed until total saturation of the samples.

## 2.3. Catalytic tests

The partial hydrogenation of 1-heptyne was carried out in a stainless steel stirred batch reactor equipped with a magnetically coupled stirrer with two blades in counter-rotation that was operated at 800 rpm. The inner wall of the reactor was completely coated with PTFE lining in order to prevent the catalytic action of steel reported by other authors [17]. The tests were performed using the following conditions: partial pressure of $H_2$ equal to 1.4, 1.9, 2.4 bar; reaction temperature: 293, 303, 323 K; initial 1-heptyne concentration of 0.1019, 0.1528, 0.2038 mol $L^{-1}$; particle size: fractions of 60 or 120 mesh and 1500 μm. 1-heptyne (Fluka, Cat. No. 51950, >98%), the reactant, was dissolved in toluene (Merck, Cat. No. TX0735-44, >99%). 75 mL of the stock solution and catalyst samples of 0.3 g of $Ni/Al_2O_3$ and 0.03 g of $Pd/Al_2O_3$ previously reduced in hydrogen for 1 h at 673 and 573K, respectively, were used in the different catalytic tests.

The evolutions of reactant and products concentrations with reaction time were followed by Gas Chromatography using a flame ionization detector (FID) and a capillary column 30 m J&W INNOWax 19091N-213.

## 3. Results and discussion

### 3.1. Catalysts characterization

The metal loadings of the catalysts determined by ICP were 3.5 and 0.35 wt % of Ni and Pd, respectively.

For the monometallic Ni/$Al_2O_3$ catalyst no hydrogen consumption was detected during the chemisorption analysis, in total accordance with previously reported results [18]. For the Pd/$Al_2O_3$ catalyst a chemisorption value of 18 μmol $H_2$ $g_{cat}^{-1}$ was measured.

Figures 1 and 2 show the XPS spectra of Ni $2p_{3/2}$ and Pd 3d for Ni/$Al_2O_3$ and Pd/$Al_2O_3$ catalysts, respectively. It can be seen that the monometallic nickel catalyst presents two

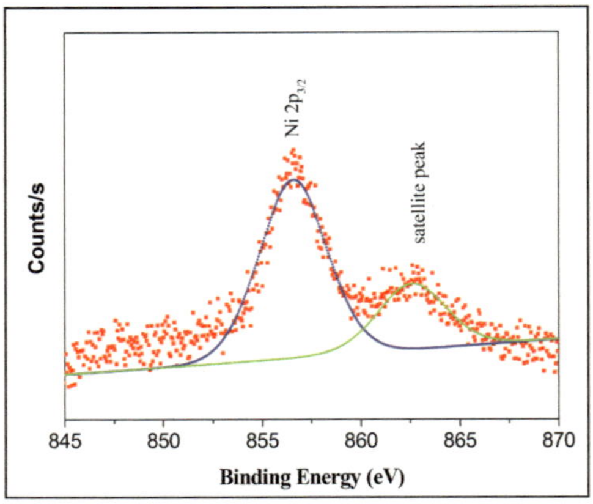


**Figure 1.** XPS Ni $2p_{3/2}$ spectra of Ni/$Al_2O_3$.

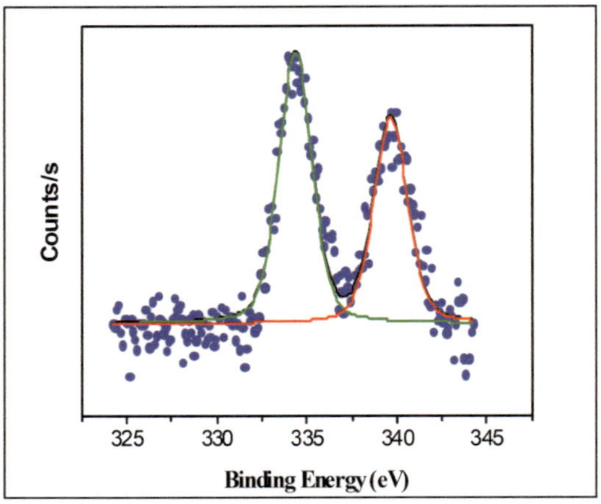


**Figure 2.** XPS Pd 3d spectra of Pd/$Al_2O_3$

peaks, with maximums at 855.4 eV and 862.6 eV. The former can be associated to $Ni^{2+}$ species [19-22]; the other peak corresponds to the shake up of Ni(II) [25,28,29]. According to the literature, the maximum BE of the Pd $3d_{5/2}$ peak of the monometallic palladium catalyst at 334.9 eV, with its $3d_{3/2}$ doublet peak at 340.1 eV, corresponds to $Pd^0$ [18-19,22-23].

The XRD difractograms of the catalysts only present the γ-alumina peaks at 2θ = 37.7, 46.0 and 67.0º [24-26]. For this reason the difractograms are not shown. Because of the low amount of Pd or Ni on the $Pd/Al_2O_3$ or $Ni/Al_2O_3$ catalysts, the crystalline phases of palladium or nickel were undetectable. Several authors stated that high charges of nickel (>>15 wt %) are necessary to observe the peaks at 2θ = 43.3°; 63.0°; 75.5° and 79.5° of NiO [27,28].

Figure 3 shows the TPR traces of the studied catalysts. The TPR trace of the Pd monometallic catalyst had a main hydrogen consumption peak at 287 K, which can be attributed to the reduction of PdO species and to the formation of palladium hydrides [29]. This catalyst also has an inverted reduction peak at 339 K that could be related to the decomposition of the β-PdH phase [20,21,24]. As these species interact weakly with the support the palladium hydrides are completely eliminated during reduction. Figure 3 also shows the profile of reduction of $Ni/Al_2O_3$ catalyst, the principal reduction peak begins at 700 K, finishes over 900 K and presents a maximum at 802 K which corresponds to the reduction of NiO ($Ni^{2+}$ species) interacting with the alumina [30]. The second peak of minor intensity is observed at higher temperatures (>1000 K), and corresponds to the reduction of nickel aluminates [31-32]. At the reduction temperature used during the preparation of the catalysts, the obtained TPR spectra suggest the presence of species $Ni^{2+}$ and $Pd^0$ on $Ni/Al_2O_3$ and $Pd/Al_2O_3$ catalysts, respectively. These results are in total agreement with the XPS results.

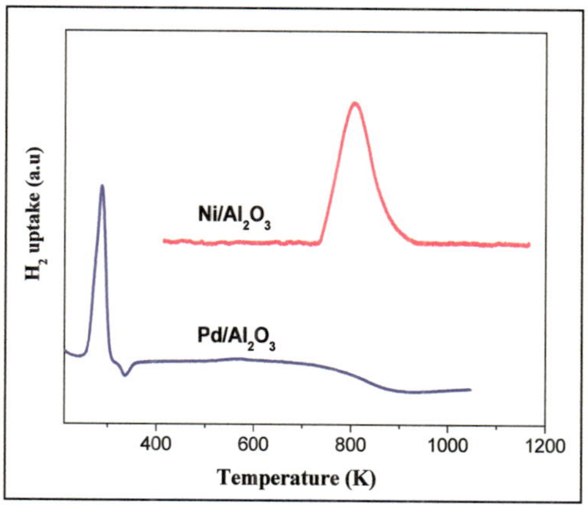


**Figure 3.** TPR traces of the $Ni/Al_2O_3$ and $Pd/Al_2O_3$ catalysts.

It must be noted that the characterization techniques suggest that, after the pretreatment conditions employed during the preparation of the catalysts, only one type of active site is present on each catalytic system.

## 3.2. Catalytic results

### *3.2.1. Partial hydrogenation of 1-heptyne*

Before considering kinetic expressions or comparing catalyst performances it is necessary to check whether the selected reaction system proceeds in kinetic regime. The possibility of external and internal diffusional limitations during the catalytic tests was thus experimentally assessed.

#### *3.2.1.1. Experimental verification of the absence of external and internal mass transfer limitations*

In order to eliminate external diffusional limitations, experiences were carried out using different stirring speeds in the range of 180-1400 rpm. It was found that at stirring rates higher than 500 rpm, 1-heptyne conversion values remained constant, indicating that external gas-liquid limitations were absent. A stirring rate of 800 rpm was therefore chosen for all the kinetic tests. On the other hand and in order to ensure that the catalytic results were not influenced by external and intraparticle mass transfer limitations, the catalyst pellets were milled to samples of different particle size: a fraction bigger than 100 mesh (<150 μm), a fraction of 60-100 mesh (250-150 μm) and pellets of 1500 μm (not milled). The obtained values of 1-heptyne conversion were the same for the two milled fractions indicating the absence of internal diffusional limitations. Then particles with sizes smaller than 250 μm were used in all tests.

#### *3.2.1.2. Catalytic activity results*

The catalytic activity results for the partial hydrogenation of 1-heptyne are shown in Figures 4 and 5, where it is represented the variation of 1-heptyne ($C_A$) and 1-heptene ($C_B$) concentration as a function of the reaction time for the $Ni/Al_2O_3$ and $Pd/Al_2O_3$ catalysts.

It can be clearly seen that Pd is more active than Ni, even when using one tenth of the catalyst mass of the Ni catalyst. Reasons for the differences in reactivity can be found in the literature. Most authors report that when the metal is more electron deficient it becomes less active because alkynes are more weakly adsorbed [5,33]. Nothing however is commented about the role of hydrogen. The published reports do not give a clear explanation of the effect of each metal and for this reason kinetic modelling was used in this work to shed light on these issues.

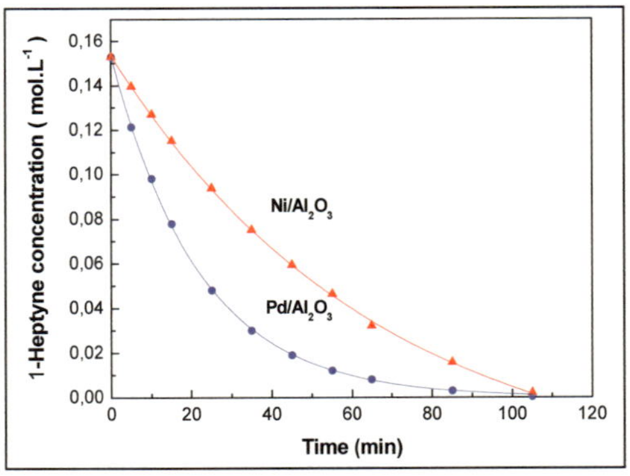


**Figure 4.** $C_A$ vs. reaction time. $P_{H2}$ =1.4 bar, $C^0_A$= 0.1528 mol $L^{-1}$, T= 323 K, 800 rpm, $W_{cat}$= 0.3 $g_{Ni/Al2O3}$ or 0.03 g $_{Pd/Al2O3}$, S/Ni = 64 and S/Pd = 29385.

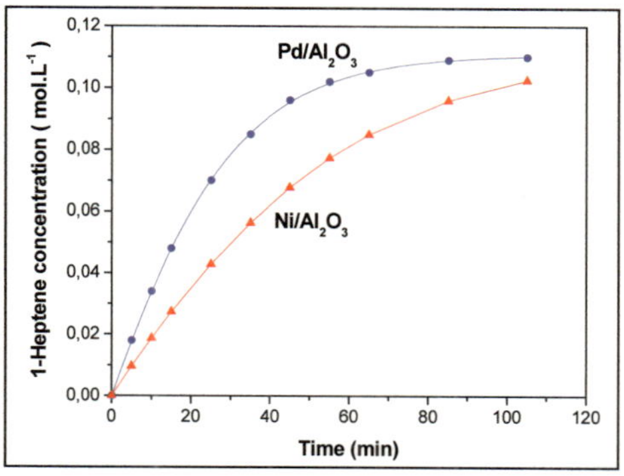


**Figure 5.** $C_B$ vs. reaction time. $P_{H2}$=1.4 bar, $C_{0A}$= 0.1528 mol $L^{-1}$, T= 323 K, 800 rpm, $W_{cat}$= 0.3 $g_{Ni/Al2O3}$ or 0.03 g $_{Pd/Al2O3}$, S/Ni = 64 and S/Pd = 29385.

# 4. Kinetic modeling

## 4.1. Reaction network

A series-parallel reaction network was proposed for partial hydrogenation of 1-heptyne [14], as indicated in Figure 6.a. This is composed of three hydrogenation reactions that can be a priori considered reversible. The equilibrium constant for each of the previous reactions were calculated using Joback's group contribution method [34]. The values at 323 K were calculated as $K_1$=3.35 $10^{21}$, $K_2$=1.87 $10^{35}$ and $K_3$=5.58 $10^{13}$. These values indicate that the individual reactions in Figure 6.a can be considered as irreversible. The experimentally obtained values of total conversion of 1-heptyne confirmed this prediction.

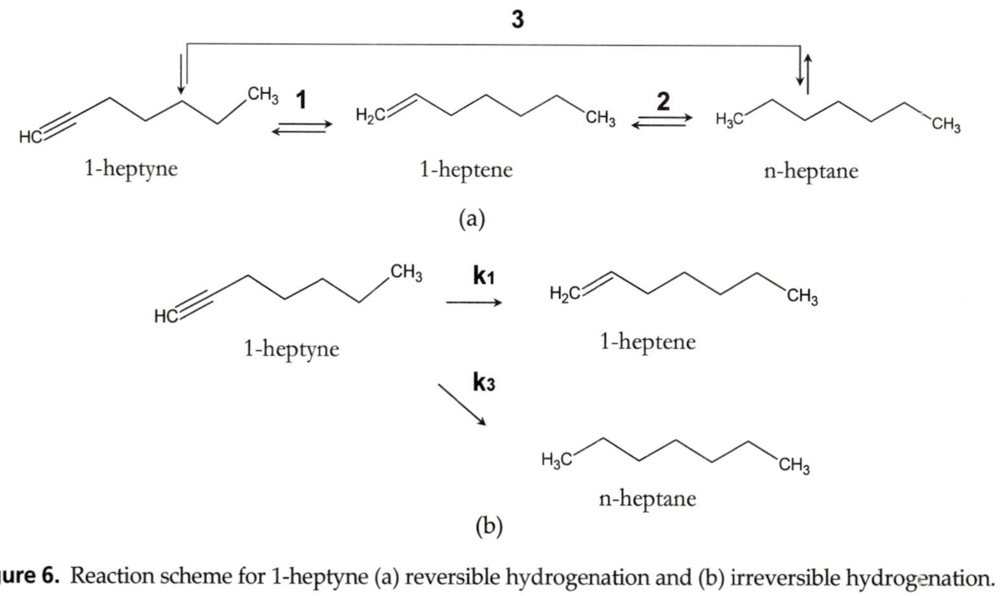


**Figure 6.** Reaction scheme for 1-heptyne (a) reversible hydrogenation and (b) irreversible hydrogenation.

Experimentally, it is observed that while 1-heptyne is present in the reaction medium, 1-heptene concentration always increases, showing higher concentration than n-heptane.

After all 1-heptyne was consumed, 1-heptene concentration begins to decrease very slowly and n-heptane concentration equally increases. These profiles are consistent with two reaction schemes: i) two parallel irreversible reactions (steps 1 and 3), and ii) series-parallel irreversible reactions, with a $k_1/k_2$ value higher than 100. The latter considerations, allows us to disregard step 2. Therefore a simplified network of parallel reactions for 1-heptyne hydrogenation can be assumed in Figure 6.b.

## 4.2. Langmuir-Hinshelwood-Hougen-Watson (LHHW) models

Models of heterogeneous reactions were outlined using the Langmuir-Hinshelwood-Hougen-Watson formalism (LHHW models). Taking into account the previously presented characterization results of $Ni/Al_2O_3$ and $Pd/Al_2O_3$, in all the models only one type of active sites was considered to be present. Six different models with their respective basic hypotheses are presented in Table 1. The elementary steps with $H_2$ dissociative or non-dissociative adsorption reaction mechanism are presented in Table 2.

| Mo-del | Hypothesis of the model | Simplified Rate | Parameters |
|---|---|---|---|
| I | Controlling step: adsorption of $H_2$. Dissociative adsorption of $H_2$ [42]. Competitive adsorption of 1-heptyne and $H_2$. Total coverage of active sites. | $r=\frac{P_3}{[C_A+P_1.C_B+P_2.C_C]^2}$ | $P_1=\frac{K_B}{K_A}$ $P_2=\frac{K_C}{K_A}$ <br> $P_3=\frac{k_{H_2}.C_{H_2}.C_S^2}{K_A^2}$ |
| II | Controlling step: adsorption of 1-Heptyne. Dissociative adsorption of $H_2$ [42]. Competitive adsorption of 1-heptyne and $H_2$. Total coverage of active sites. | $r=\frac{P_6.C_A}{[1+P_4.C_B+P_5.C_C]}$ | $P_4=\frac{K_B}{\sqrt{K_{H_2}.C_{H_2}}}$ <br> $P_5=\frac{K_C}{\sqrt{K_{H_2}.C_{H_2}}}$ .. |
| III | Controlling step: surface chemical reaction. Dissociative adsorption of $H_2$ [42]. Competitive adsorption of 1-heptyne and $H_2$. Total coverage of active sites. | $r_1=\frac{P_{10}.C_A}{[1+P_7.C_A+P_4.C_B+P_5.C_C]^3}$ <br> $r_3=\frac{P_{11}.C_A}{[1+P_7.C_A+P_4.C_B+P_5.C_C]^3}$ | $P_6=\frac{k_A.C_S}{\sqrt{K_{H_2}.C_{H_2}}}$ <br> $P_7=\frac{(1+K.K_{H_2}.C_{H_2}).K_A}{\sqrt{K_{H_2}.C_{H_2}}}$ <br> $P_{10}=\frac{k_1.K_A.C_S^3}{\sqrt{K_{H_2}.C_{H_2}}}$ <br> $P_{11}=\frac{k_3.K_A.K.C_S^3}{\sqrt{K_{H_2}.C_{H_2}}}$ |
| IV | Controlling step: adsorption of $H_2$. Non dissociative adsorption of $H_2$ [43]. Competitive adsorption of 1-heptyne and $H_2$. The active sites are not completely covered. | $r=\frac{P_{12}}{[1+K_A.C_A+K_B.C_B+K_C.C_C]}$ | $P_{12}=k'_{H_2}.C_{H_2}.C_S$ |

| Mo-del | Hypothesis of the model | Simplified Rate | Parameters |
|---|---|---|---|
| V | Controlling step: adsorption of 1-Heptyne. Non dissociative adsorption of $H_2$ [43]. Competitive adsorption of 1-heptyne and $H_2$. The active sites are not completely covered. | $r = \frac{P_{14}.C_A}{\left[1+P_{13}+K_B.C_B+K_C.C_C\right]}$ | $P_{13} = K^*_{H_2}.C_{H_2}$ <br> $P_{14} = k_A.C_S$ |
| VI | Controlling step: surface chemical reaction. Non dissociative adsorption of $H_2$ [43]. Competitive adsorption of 1-heptyne and $H_2$. The active sites are not completely covered. | $r_1 = \frac{P_{16}.C_A}{\left[1+P_{13}+P_{15}.C_A+K_B.C_B+K_C.C_C\right]^2}$ <br> $r_3 = \frac{P_{17}.C_A}{\left[1+P_{13}+P_{15}.C_A+K_B.C_B+K_C.C_C\right]^2}$ | $P_{15} = \left(1+K^*.K^*_{H_2}.C_{H_2}\right).K_A$ <br> $P_{16} = k_1.K_A.K^*_{H_2}.C_{H_2}.C_S^2$ <br> $P_{17} = k_3.K_A.K^*.K^{*2}_{H_2}.C^2_{H_2}.C_S^2$ |

**Table 1.** LHHW kinetic models.

| $H_2$ dissociative adsorption. | | $H_2$ non dissociative adsorption. | |
|---|---|---|---|
| $H_2+2S \Leftrightarrow 2\,HS \quad K_{H_2} = \frac{C^2_{HS}}{C_{H_2}.C_S^2}$ | (1) | $H_2+S \Leftrightarrow H_2S \quad K^*_{H_2} = \frac{C_{H_2S}}{C_{H_2}.C_S}$ | (8) |
| $A + S \Leftrightarrow AS \quad K_A = \frac{C_{AS}}{C_A.C_S}$ | (2) | $A + S \Leftrightarrow AS \quad K_A = \frac{C_{AS}}{C_A.C_S}$ | (9) |
| $AS + 2HS \rightarrow BS + 2S \quad K_1 = \infty$ | (3) | $AS + H_2S \rightarrow BS + S \quad K_1 = \infty$ | (10) |
| $AS + 2HS \Leftrightarrow AH_2S + 2S \quad K = \frac{C_{AH_2S}.C_S^2}{C_{AS}.C^2_{HS}}$ | (4) | $AS + H_2S \Leftrightarrow AH_2S + S \quad K^* = \frac{C_{AH_2S}.C_S}{C_{AS}.C_{H_2S}}$ | (11) |
| $AH_2S + 2\,HS \rightarrow CS + 2S \quad K_3 = \infty$ | (5) | $AH_2S + H_2S \rightarrow CS + S \quad K_3 = \infty$ | (12) |
| $BS \Leftrightarrow B + S \quad \frac{1}{K_B} = \frac{C_B.C_S}{C_{BS}}$ | (6) | $BS \Leftrightarrow B + S \quad \frac{1}{K_B} = \frac{C_B.C_S}{C_{BS}}$ | (13) |
| $CS \Leftrightarrow C + S \quad \frac{1}{K_C} = \frac{C_C.C_S}{C_{CS}}$ | (7) | $CS \Leftrightarrow C + S \quad \frac{1}{K_C} = \frac{C_C.C_S}{C_{CS}}$ | (14) |

**Table 2.** Elementary steps with $H_2$ dissociative or non-dissociative adsorption reaction mechanism.

### 4.3. Mass balances

The following mass balances for components in the liquid phase were considered for the reaction scheme of Figure 6.b, for 1-heptyne (A), 1-heptene (B) and n-heptane (C):

$$dC_A / dt = -r_1 - r_3 \tag{15}$$

$$dC_B / dt = r_1 \tag{16}$$

$$dC_C / dt = r_3 \tag{17}$$

initial conditions were: t = 0 min, $C^0{}_A = 0.1528$ mol $L^{-1}$, $C^0{}_B = C^0{}_C = 0$ mol $L^{-1}$.

## 4.4. Numerical resolution and statistics

The system of differential equations (15)-(17) was solved numerically using the Runge-Kutta-Merson algorithm. The model parameter estimation was performed by a non-linear regression, using a Levenberg-Marquardt algorithm which minimized the objective function:

$$SCD = \sum_{j}^{n}\left(C_{i,j} - C_{i,j}^{CALC}\right)^2 \tag{18}$$

where $C_{i,j}$ and $C_{i,j}^{CALC}$ are the experimental and the predicted concentration values, respectively, "$i$" is the chemical compound and "$j$" is the reaction time.

The model adequacy and the discrimination between models were determined using the model selection criterion (MSC), according to the following equation:

$$MSC = ln\left(\frac{\sum_{j}^{n}\left(C_{i,j} - \overline{C}_i\right)^2}{\sum_{j}^{n}\left(C_{i,j} - C_{i,j}^{CALC}\right)^2}\right) - \left(\frac{2p}{n}\right) \tag{19}$$

where $\overline{C}_i$ is the average relative concentration; p is the amount of parameters fitted and n is the number of experimental data. In order to compare different models, the selected one is that leading to the highest MSC value.

The Standard Deviation (S) was calculated with the following equation:

$$S = \sqrt{s^2} = \sqrt{\frac{\sum_{j}^{n}\left(C_{i,j} - C_{i,j}^{CALC}\right)^2}{n-p}} \tag{20}$$

## 4.5. Model discrimination

The first main requisite for appropriateness of a model should be that of physical significance. A priori this means that the model parameters adopt feasible real values. A second requisite is that of adequate statistical confidence, i.e. the parameters should lie in one as small as possible confidence interval.

The practical criteria for the selection of the kinetic models were:

1. The estimated values of the parameters must be positive and different from zero.
2. The upper and lower extremes of the confidence interval (95%) must be positive.

3. The amplitude of the confidence interval must be lower than the value of the estimated parameter.

The final model is selected from the set of models complying the above 1 to 3 conditions, as the model with the lowest SCD, the summation of squares of the deviations. Another condition is that the standard deviation is smaller than the value of the parameter. If differences are not big, then the model selection criterion (MSC) should be used. Appropriate models should have a MSC value greater than 4.

## 4.6. Kinetic models for the reaction

Preliminary tests were performed in order to check the influence of the different variables on the reaction rate. These results will be used later in the model selection stage. The variables screened were the partial pressure of hydrogen, the initial concentration of 1-heptyne and the reaction temperature. In order to analyze the influence of each variable a pseudo homogeneous reaction model was proposed in which the reaction rate was assumed to follow a potential law. The initial reaction rate should thus be written as:

$$r_A^0 = k.(C_A^0)^{\alpha}.(P_{H2})^{\beta} \tag{21}$$

An Arrhenius dependent was supposed for the specific constant of reaction:

$$k = A.e^{\frac{-E_A}{R.T}} \tag{22}$$

### *4.6.1. Influence of the partial pressure of hydrogen*

Several tests were performed at varying hydrogen partial pressures (1.4, 1.9 and 2.4 bar) keeping all other variables constant ($C^0_A$= 0.1528 mol $L^{-1}$, T= 303 K, 800 rpm, $W_{cat}$= 0.3 g $Ni/Al_2O_3$ or 0.03 $Pd/Al_2O_3$). The partial pressure of hydrogen was calculated as the difference between the total pressure and the partial pressure of the solvent, since the partial pressures of 1-heptyne and the products was negligible. The pressure of the solvent at 303 K was determined using Antoine's equation (0.048 bar).

Experimental values of conversion of 1-heptyne as a function of time are plotted in Figures 7.a and 7.b. Both Figures show that as $P_{H2}$ rises, 1-heptyne total conversion increases for $Ni/Al_2O_3$ catalyst, but decrease for $Pd/Al_2O_3$. The reaction order with respect to hydrogen (β) was calculated from the linearized form of equation (21).

$$ln(r_A^0) = ln\left[k.\left(C_A^0\right)^{\alpha}\right] + \beta.ln(P_{H2}) \tag{23}$$

The initial reaction rates of 1-heptyne were calculated by polynomial differentiation of the traces of Figures 7.a and 7.b and extrapolation to zero reaction time. From the plot of $ln(r^0_A)$ vs. $ln(P_{H2})$, shown in Figures 8.a and 8.b, reaction order in hydrogen of 1.3 and -2.6 were calculated for $Ni/Al_2O_3$ and for $Pd/Al_2O_3$, respectively. The results indicate that: a) for

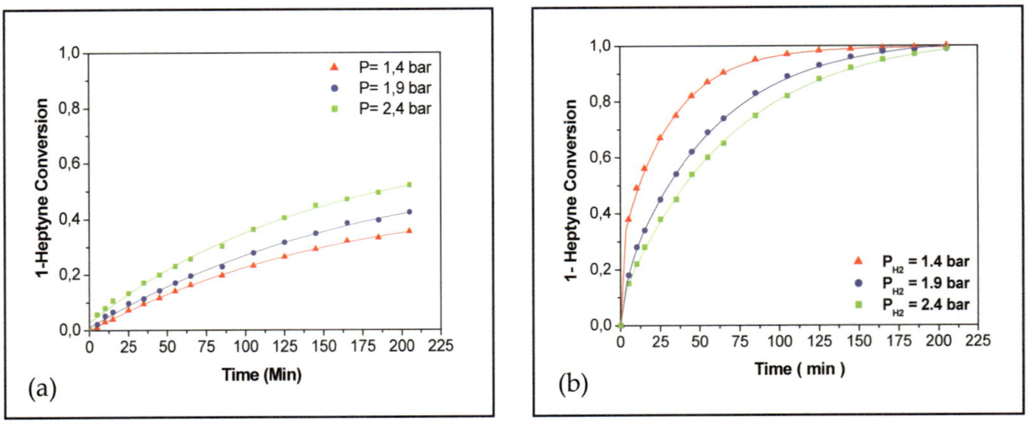


**Figure 7.** Effect of $P_{H2}$ on the catalytic activity for (a) $Ni/Al_2O_3$ and (b) $Pd/Al_2O_3$. $C^0_A$= 0.1528 mol $L^{-1}$, T=303 K, 800 rpm, $W_{cat}$= 0.3 $g_{Ni/Al2O3}$ or 0.03 g $_{Pd/Al2O3}$, S/Ni = 64 and S/Pd =29385.

$Ni/Al_2O_3$ catalyst, high hydrogen partial pressures are beneficial for the reaction kinetics, probably both adsorption and surface reaction elementary steps could be enhanced; and b) an increase in the partial pressure of hydrogen negatively affects the reaction rate for $Pd/Al_2O_3$.

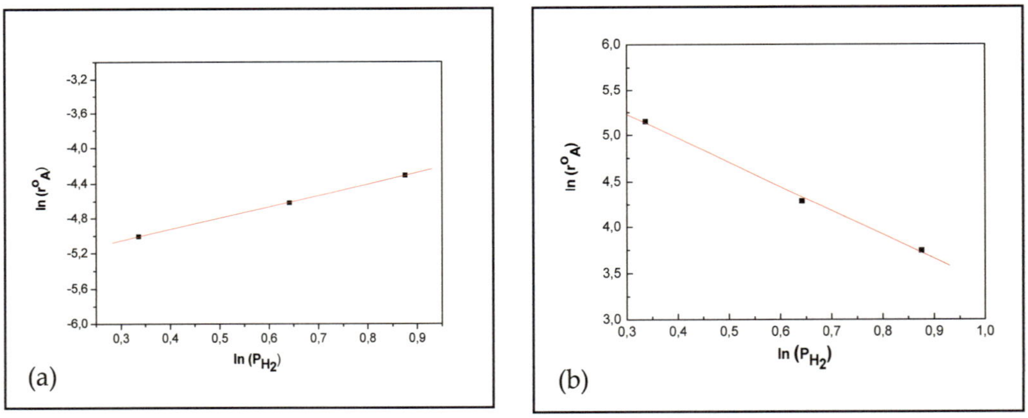


**Figure 8.** Initial reaction rate as a function of hydrogen pressure for (a) $Ni/Al_2O_3$ and (b) $Pd/Al_2O_3$. $C^0_A$=0.1528 mol $L^{-1}$, T=303 K, 800 rpm, $W_{cat}$= 0.3 $_{gNi/Al2O3}$ or 0.03 g $_{Pd/Al2O3}$, S/Ni = 64 and S/Pd =29385.

### *4.6.2. Influence of the initial concentration of 1-heptyne*

Catalytic tests were performed varying the initial concentration of 1-heptyne: 0.1019, 0.1528 and 0.2038 mol $L^{-1}$, keeping all the rest of the variables constant ($P_{H2}$=1.4 bar, T= 303 K, 800 rpm, $W_{cat}$ = 0.3 g $Ni/Al_2O_3$ or 0.03 $Pd/Al_2O_3$).

The obtained values of conversion of 1-heptyne as a function of time are plotted in Figures 9.a and 9.b. It can be seen that for both catalysts the catalytic activity is decreased when the initial concentration of 1-heptyne is increased.

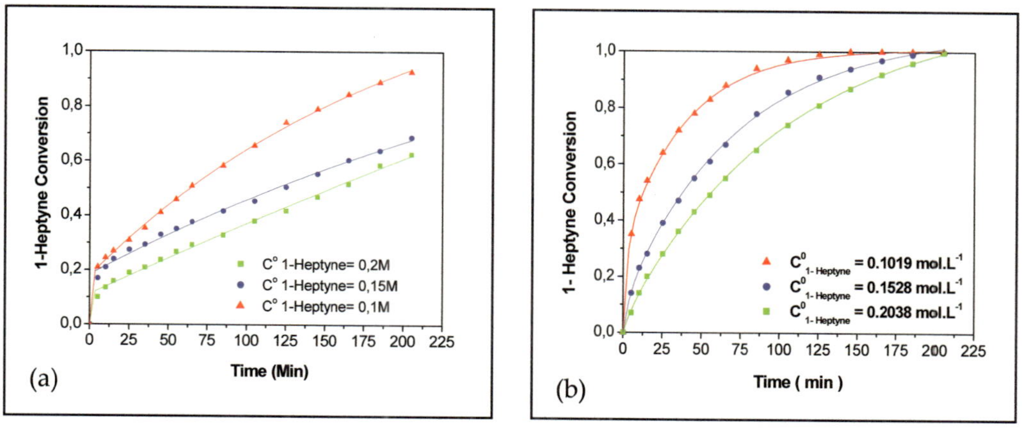


**Figure 9.** Effect of $C^0_A$ on the catalytic activity for (a) $Ni/Al_2O_3$ and (b) $Pd/Al_2O_3$. $P_{H2}$=1.4 bar, T=303 K, 800 rpm, $W_{cat}$= 0.3 $g_{Ni/Al2O3}$ or 0.03 g $_{Pd/Al2O3}$, S/Ni = 42, 64 and 85, S/Pd =19596, 29335 and 39192.

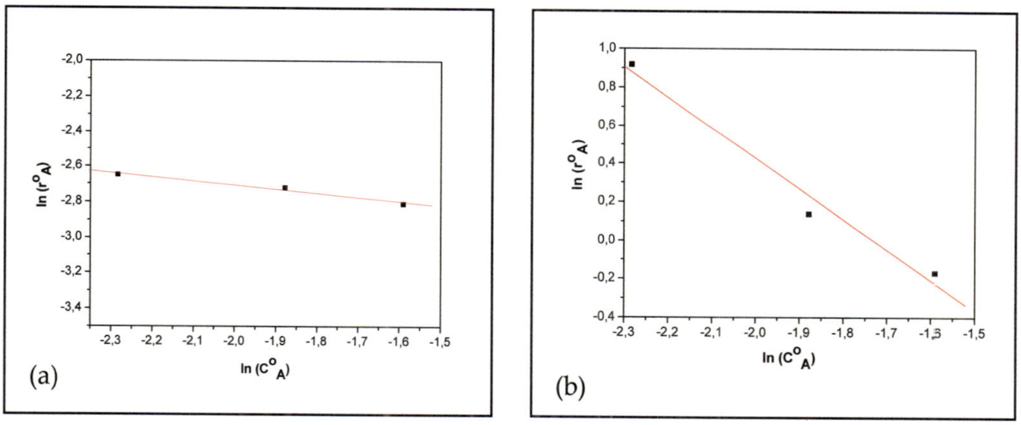


**Figure 10.** Initial reaction rate of 1-heptyne concentration for (a) $Ni/Al_2O_3$ and (b) $Pd/Al_2O_3$. $P_{H2}$=1.4 bar, T=303 K, 800 rpm, $W_{cat}$= 0.3 $g_{Ni/Al2O3}$ or 0.03 g $_{Pd/Al2O3}$, S/Ni = 42, 64 and 85, S/Pd =19596, 29335 and 39192.

The value of the reaction order in 1-heptyne ($\alpha$) can be calculated along the lines described in the previous section:

$$ln(r_A^0) = ln\left[k.(P_{H2})^{\beta}\right] + \alpha.ln(C_A^0) \tag{24}$$

The graph of $\ln(r^0_A)$ vs. $\ln(C^0_A)$, Figures 10.a and 10.b, yields value of order of reaction of 1-heptyne equal to -0.22 and -1.5 for $Ni/Al_2O_3$ and $Pd/Al_2O_3$, respectively. The results indicate that an increase in the initial concentration of 1-heptyne is detrimental to the reaction rates.

### *4.6.3. Influence of the reaction temperature*

Catalytic tests were performed varying the reaction temperature: 293, 303 and 323 K, and keeping the rest of the variables constant ($P_{H2}$=1.4 bar, $C^0_A$ = 0.1528 mol $L^{-1}$, 800 rpm, $W_{cat}$ = 0.3 g $Ni/Al_2O_3$ or 0.03 $Pd/Al_2O_3$).

The experimental values of conversion of 1-heptyne as a function of time at different temperature values are plotted in Figures 11.a and 11.b. As expected the activity of the catalyst is increased while the reaction temperature is raised up.

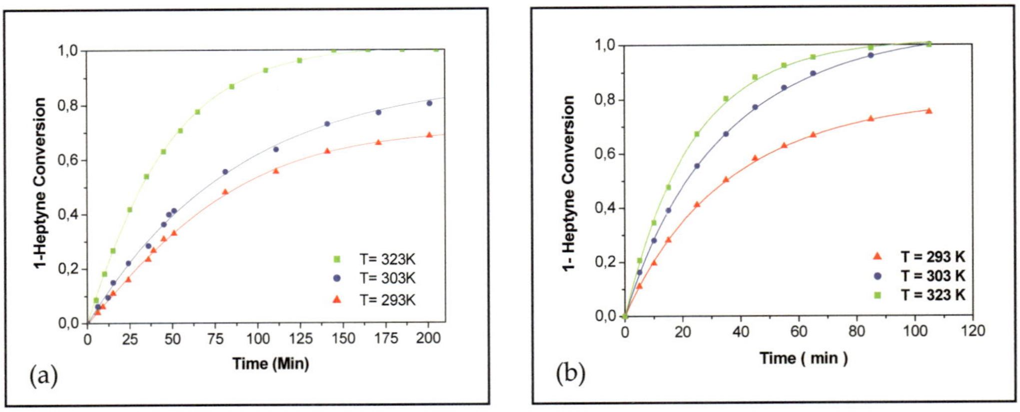


**Figure 11.** Effect of T on the catalytic activity for (a) $Ni/Al_2O_3$ and (b) $Pd/Al_2O_3$. $P_{H2}$=1.4 bar, 800 rpm, $W_{cat}$= 0.3 $g_{Ni/Al2O3}$ or 0.03 g $_{Pd/Al2O3}$, S/Ni = 64 and S/Pd = 29385.

When equation (21) is linearized a value of "apparent" activation energy ($E_A$) can be got, as indicated in Eq. (25):

$$ln(r_A^0) = ln\left[A.(P_{H2})^{\beta}.(C_A^0)^{\alpha}\right] - \frac{E_A}{R.T} \tag{25}$$

The initial reaction rates of 1-heptyne were calculated as in the previous sections. The value of the apparent activation energy were obtained from the plots presented in Figure 12 of $\ln(r^0_A)$ as a function of 1/T. The calculated values were 24 and 18 KJ $mol^{-1}$ for $Ni/Al_2O_3$ and $Pd/Al_2O_3$, respectively. These values have not a real physical meaning and are only apparent.

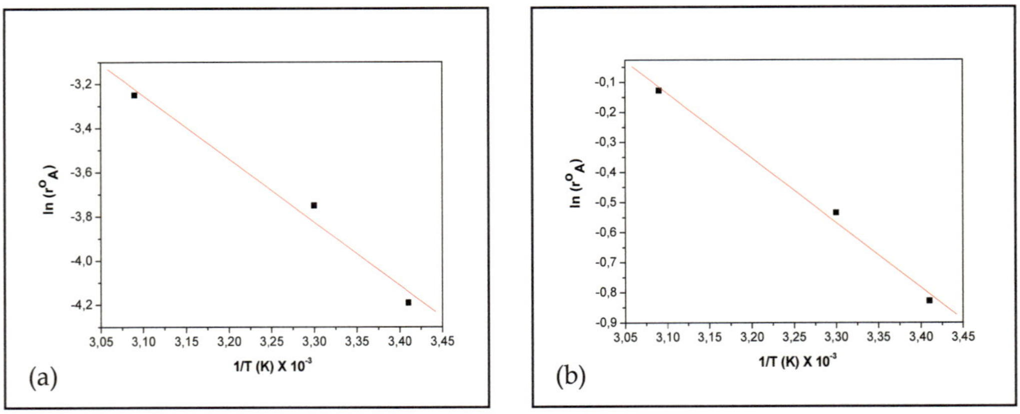


**Figure 12.** Temperature dependence of the reaction rate for (a) $Ni/Al_2O_3$ and (b) $Pd/Al_2O_3$. $P_{H2}$=1.4 bar, T=303 K, 800 rpm, $W_{cat}$= 0.3 $g_{Ni/Al2O3}$ or 0.03 g $_{Pd/Al2O3}$, S/Ni = 64 and S/Pd = 29385.

### 4.6.4. *Model discrimination for Ni/$Al_2O_3$ catalyst*

Models II, III, V and VI of Table 1 were discarded because they could not explain the negative and positive orders in 1-heptyne and hydrogen obtained experimentally. The parameters estimated for the models I and IV are indicated in Table 3. A statistical analysis was performed to discriminate between the different models, by means of the selection criteria described in Section 4.5. The results of this analysis are detailed in Table 3. It can be concluded that the best fit is achieved with model I- B. In this model the value of $P_2$ is equal to zero, indicating that n-heptane is not adsorbed. The model IV-D also shows a good fit of the experimental data, but a pseudo homogeneous kinetic expression is obtained with different reaction orders than those previously calculated.

| Model | Option | Estimated parameter (*) | SCD | MSC | Parameter | Parameter sign | Discrimination | Viability |
|---|---|---|---|---|---|---|---|---|
| I | A | $P_1 = 4.4823237 \pm 168.14985673$ | $2.1.10^{-3}$ | 3.7 | $P_1$ | (+) | IC < 0, CL > VE | Not viable |
| | | $P_2 = 3.44577501 \pm 335.92334499$ | | | $P_2$ | (+) | IC < 0, CL > VE | |
| | | $P_3 = 0.0001211222587 \pm 0.00003376836$ | | | $P_3$ | (+) | IC < 0, CL > VE | |
| | B | $P_2 = 0$ | $2.10^{-3}$ | 4.5 | $P_2=0$ | | | Viable |
| | | $P_1 = 5.93108434 \pm 0.97784253$ | | | $P_1$ | (+) | IC > 0, CL < VE | |
| | | $P_3 = 0.000113720472 \pm 0.000026730661$ | | | $P_3$ | (+) | IC > 0, CL < VE | |
| IV | A | $K_A = 8.34221367 \pm 355.9610964$ | $2.47.10^{-4}$ | 5.17 | $K_A$ | (+) | IC < 0, CL > VE | Not viable |
| | | $K_B = -1097.26187 \pm 7313.33134$ | | | $K_B$ | (-) | IC < 0, CL > VE | |
| | | $K_C = 2384.39316 \pm 12837.26474$ | | | $K_C$ | (+) | IC < 0, CL > VE | |
| | | $P_{12} = 0.00654481117 \pm 0.15591798$ | | | $P_{12}$ | (+) | IC < 0, CL > VE | |
| | B | $K_A = -6.36443252 \pm 3.54535551$ | $2.47.10^{-4}$ | 5.23 | $K_A$ | (-) | IC < 0, CL > VE | Not viable |
| | | $K_B = 0$ | | | $K_B = 0$ | | | |
| | | $K_C = -17.1237873 \pm 50.0933591$ | | | $K_C$ | (-) | IC < 0, CL > VE | |
| | | $P_{12} = 7.82119826.10^{-5} \pm 1.570645.10^{-3}$ | | | $P_{12}$ | (+) | IC < 0, CL > VE | |
| | C | $K_A = -5.14858364 \pm 0.29479977$ | $2.47.10^{-4}$ | 5.21 | $K_A$ | (-) | IC < 0, CL > VE | Not viable |
| | | $K_B = 0$ | | | $K_B = 0$ | | | |
| | | $K_C = 0$ | | | $K_C = 0$ | | | |
| | | $P_{12} = 0.000613057924 \pm 3.87.10^{-5}$ | | | $P_{12}$ | (+) | IC < 0, CL > VE | |
| | D | $K_A = 0$ | | | $K_A = 0$ | | | Viable |
| | | $K_B = 0$ | | | $K_B = 0$ | | | |
| | | $K_C = 0$ | | | $K_C = 0$ | | | |
| | | $P_{12} \neq 0$ | | | $P_{12} \neq 0$ | | | |
| | | A pseudohomogeneous model is obtained. | | | | | | |

Reaction conditions: $P_{H2}$=1.4 bar, T=323 K, $w_{cat}$=0.3 g, stirring rate=800 rpm, $C^0_A$=0.1528 mol.L$^{-1}$, S/Ni = 64.
(*) 95% confidence interval. IC: confidence interval; CL: confidence level; VE: estimated parameter value

**Table 3.** Estimated parameters and model discrimination for Ni/$Al_2O_3$.

Figure 13 contains experimental values of the concentration of 1-heptyne, 1-heptene and n-heptane along with theoretical values (solid line) estimated with model I-B, as a function of time. A good fit between the two sets of values can be seen. The same regression with model I-B was done with experimental data obtained at other reaction temperatures in the 293-323 K range. In all cases and as a consequence of the fit, parameters different from zero were obtained for a confidence interval of 95% and with values of the MSC parameter greater than 4.0. The thermodynamic consistency of the $P_1$ and $P_3$ parameters was graphically evaluated by plotting $\ln P_1$ and $\ln P_3$ as a function of 1/T. In both cases a straight line was obtained (Figure 14) indicating that the constants have an Arrhenius dependence on temperature.

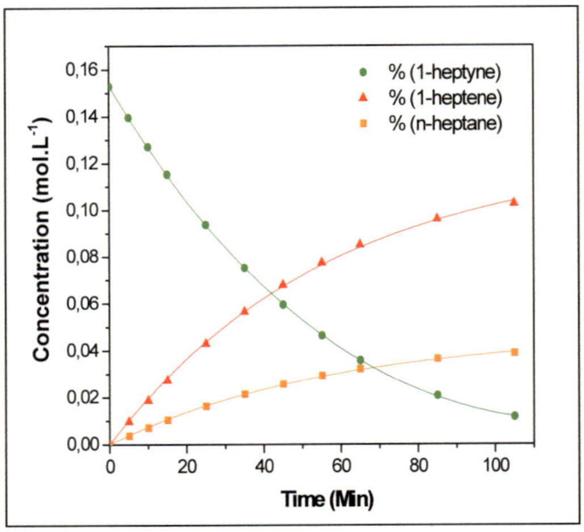


**Figure 13.** Fit of the experimental data of $C_i$ as a function of time with model I-B for $Ni/Al_2O_3$ catalyst. $P_{H2}$=1.4 bar, $C^0_A$=0.1528 mol $L^{-1}$, 323 K, $W_{cat}$=0.3 g, S/Ni = 64.

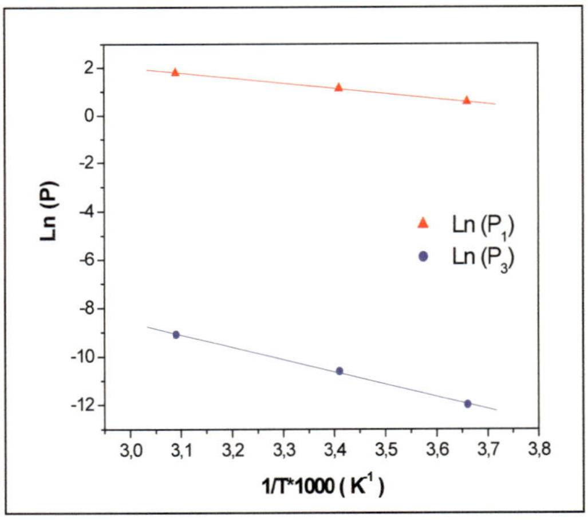


**Figure 14.** Temperature dependence of parameters $P_1$ and $P_3$ for $Ni/Al_2O_3$ catalyst

The slopes of the straight lines correspond to the values of the enthalpies of adsorption of 1-heptyne and 1-heptene and the value of the energy of activation for the dissociative adsorption of hydrogen on the active sites. Considering the definition of $P_1$ and $P_3$, the following equations can be obtained:

$$P_1 = \frac{K_B}{K_A} = \left(\frac{A_B}{A_A}\right).Exp\left[\frac{|\Delta H_B| - |\Delta H_A|}{R.T}\right] \quad (26)$$

$$P_3 = \frac{C_{H2}.k_{H2}.C_S^2}{K_A^2} = \left(\frac{C_{H2}.A_{H2}.C_S^2}{A_A^2}\right).Exp\left[\frac{-(E_{H2} + 2.|\Delta H_A|)}{R.T}\right] \quad (27)$$

In equations (26) and (27) the enthalpies of adsorption of 1-heptyne and 1-heptene were expressed in absolute values. The obtained values from the slopes of the lines in Figure 14 were:

$$|\Delta H_B| - |\Delta H_A| = -17.91 \text{ KJ mol}^{-1} \quad (28)$$

$$E_{H2} + 2.|\Delta H_A| = 58 \text{ KJ mol}^{-1} \quad (29)$$

Considering that $|\Delta H_B| \approx 0$, in accord with the experimental results, a value of the enthalpy of adsorption for 1-heptyne can be obtained from equation (28): $\Delta H_A$ = -17.91 KJ mol$^{-1}$. Introducing this value in equation (29) a value of $E_{H2}$ of 22.2 KJ mol$^{-1}$ can be estimated.

From the results it could be concluded that:

1. Model I-B that supposes dissociative adsorption of hydrogen as the rate-controlling step of reaction and a single type of active sites with total coverage, is the one that best fits experimental data with statistical and thermodynamic consistency.
2. The model does not allow to directly obtain the enthalpies of adsorption of 1-heptyne and 1-heptene and the activation energy for the adsorption of hydrogen.
3. From equation (28) it can be inferred that the enthalpy of adsorption of 1-heptyne is greater than that of 1-heptene, in agreement with the information available in the literature on the partial hydrogenation of alkynes [34].
4. This model also supposes that the alkane is not adsorbed. If we additionally suppose that the enthalpy of adsorption of 1-heptene is negligible, in accordance with experimental results, a value of the enthalpy of adsorption of 1-heptyne and activation energy for the $H_2$ adsorption can be obtained from eqs. (28) and (29).

Then, it can be concluded that the dissociative adsorption of $H_2$ is the rate limiting step for the $Ni/Al_2O_3$ catalyst, and that the active sites are preferentially occupied by 1-heptyne.

### *4.6.5. Model discrimination for the $Pd/Al_2O_3$ catalyst*

Models I, II, IV and V of Table 1 were discarded because they could not explain the negative orders in 1-heptyne and hydrogen obtained experimentally. The parameters estimated for the models III and VI are indicated in Table 4. A statistical analysis was performed to discriminate between the different models, by means of the selection criteria described in Section 4.5. The results of this analysis are detailed in Table 4; these results indicate that model III-C gives the best fit of the experimental data. In this model the value of the parameters $P_4$ and $P_5$ are equal to zero. Therefore the only species being adsorbed on the $Pd/Al_2O_3$ catalyst are 1-heptyne and hydrogen.

| Model | Option | Estimated parameter (*) | SCD | MSC | Parameter | Parameter sign | Discrimination | Viability |
|---|---|---|---|---|---|---|---|---|
| III | A | $P_4$ = -53.0866055±3399.8626255 | $2.2.10^{-4}$ | 5.206 | $P_4$ | (-) | IC < 0, CL > VE | Not viable |
| | | $P_5$ = 221.348971±7134.846189 | | | $P_5$ | (+) | IC < 0, CL > VE | |
| | | $P_7$ = 28.7488806±370.8944874 | | | $P_7$ | (+) | IC < 0, CL > VE | |
| | | $P_{10}$ = 4.68720094±147.04798306 | | | $P_{10}$ | (+) | IC < 0, CL > VE | |
| | | $P_{11}$ = 1.92159358±60.33982722 | | | $P_{11}$ | (+) | IC < 0, CL > VE | |
| | B | $P_4$ = 0 | $2.2.10^{-4}$ | 5.266 | | | | Not viable |
| | | $P_5$ = 73.3086809±919.4831331 | | | $P_5$ | (+) | IC < 0, CL > VE | |
| | | $P_7$ = 23.0235065±284.1617705 | | | $P_7$ | (+) | IC < 0, CL > VE | |
| | | $P_{10}$ = 2.75603228±79.34470502 | | | $P_{10}$ | (+) | IC < 0, CL > VE | |
| | | $P_{11}$ = 1.12978185±32.52404635 | | | $P_{11}$ | (+) | IC < 0, CL > VE | |
| | C | $P_4$ = 0 $P_5$ = 0 | $9.1.10^{-5}$ | 6.23 | | | | Viable |
| | | $P_7$ = 0.554228500±0.330992359 | | | $P_7$ | (+) | IC > 0, CL < VE | |
| | | $P_{10}$ = 0.0368778035±0.0032648682 | | | $P_{10}$ | (+) | IC > 0, CL < VE | |
| | | $P_{11}$ = 0.0150940473±0.0014453604 | | | $P_{11}$ | (+) | IC > 0, CL < VE | |
| VI | A | $K_B$ = -16.3714023±661.1021023 | $2.2.10^{-4}$ | 5.145 | $K_B$ | (-) | IC < 0, CL > VE | Not viable |
| | | $K_C$ = 48.4132846±2509.3986254 | | | $K_C$ | (+) | IC < 0, CL > VE | |
| | | $P_{13}$ = 6.35365882±135.97712018 | | | $P_{13}$ | (+) | IC < 0, CL > VE | |
| | | $P_{15}$ = 7.14818320±324.0314528 | | | $P_{15}$ | (+) | IC < 0, CL > VE | |
| | | $P_{16}$ = 2.13263324±47.39641486 | | | $P_{16}$ | (+) | IC < 0, CL > VE | |
| | | $P_{17}$ = 0.874143654±19.422821246 | | | $P_{17}$ | (+) | IC < 0, CL > VE | |
| | B | $K_B$ = 0 | $2.2.10^{-4}$ | 5.206 | | | | Not viable |
| | | $K_C$ = 53.7394284±6350.0751416 | | | $K_C$ | (+) | IC < 0, CL > VE | |
| | | $P_{13}$ = 3.53653444±247.34456656 | | | $P_{13}$ | (+) | IC < 0, CL > VE | |
| | | $P_{15}$ = 19.8231433±1876.7295967 | | | $P_{15}$ | (+) | IC < 0, CL > VE | |
| | | $P_{16}$ = 1.71129670±50.9810815 | | | $P_{16}$ | (+) | IC < 0, CL > VE | |
| | | $P_{17}$ = 0.702039726±20.910298274 | | | $P_{17}$ | (+) | IC < 0, CL > VE | |
| | C | $K_B$ = 0 $K_C$ = 0 | $2.2.10^{-4}$ | 5.26 | | | | Not viable |
| | | $P_{13}$ = 2.93941739±18.79975681 | | | $P_{13}$ | (+) | IC < 0, CL > VE | |
| | | $P_{15}$ = 1.86782613±10.08702077 | | | $P_{15}$ | (+) | IC < 0, CL > VE | |
| | | $P_{16}$ = 0.541342090±5.18514542 | | | $P_{16}$ | (+) | IC < 0, CL > VE | |
| | | $P_{17}$ = 0.221642448±2.121511182 | | | $P_{17}$ | (+) | IC < 0, CL > VE | |
| | D | $K_B$ = 0 $K_C$ = 0 $P_{13}$ = 0 | $2.2.10^{-4}$ | 5.327 | | | | Not viable |
| | | $P_{15}$ = 0.610445795±0.825496865 | | | $P_{15}$ | (+) | IC < 0, CL > VE | |
| | | $P_{16}$ = 0.0357150918±0.0052744265 | | | $P_{16}$ | (+) | IC > 0, CL < VE | |
| | | $P_{17}$ = 0.0146412833±0.0023224598 | | | $P_{17}$ | (+) | IC > 0, CL < VE | |
| | E | $K_B$ = 0 $K_C$ = 0 $P_{13}$ = 0 $P_{15}$ = 0 | $2.4.10^{-4}$ | 5.303 | | | | Viable |
| | | $P_{16}$ = 0.0320548839±0.0012299287 | | | $P_{16}$ | (+) | IC > 0, CL < VE | |
| | | $P_{17}$ = 0.0130965076±0.0007651406 | | | $P_{17}$ | (+) | IC > 0, CL < VE | |

Reaction conditions: $P_{H2}$=1.4 bar, T=323 K, $w_{cat}$=0.03 g, stirring rate=800 rpm, $C^0_A$=0.1528 mol.L$^{-1}$, S/Pd = 29385.
(*) 95% confidence interval. IC: confidence interval; CL: confidence level; VE: estimated parameter value.

**Table 4.** Estimated model parameters and model discrimination for $Pd/Al_2O_3$

The heterogeneous model VI-E also produces a good fit of the experimental data but its parameters $K_B$, $K_C$, $P_{13}$ and $P_{15}$ are equal to zero, thus transforming into a pseudo homogeneous reaction rate expression in which the reaction orders in 1-heptyne and hydrogen are positive. This is not in agreement with the observed results, so this model was discarded.

Figure 15 contains values of concentration of 1-heptyne, 1-heptene and n-heptane as a function of time, experimentally (symbols) or theoretically obtained (solid line) with the III-C model. There is an excellent fit.

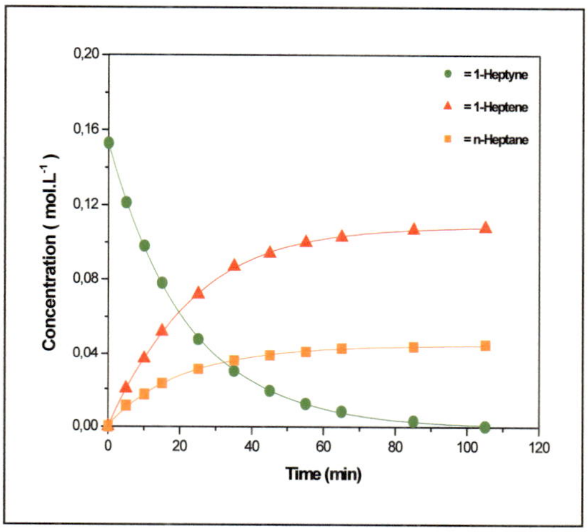


**Figure 15.** Fit of the experimental data of $C_i$ as a function of time with model III-C for $Pd/Al_2O_3$ catalyst. $P_{H2}$=1.4 bar, T=323 K, 800 rpm, $W_{cat}$= 0.03 g, S/Pd = 29385.

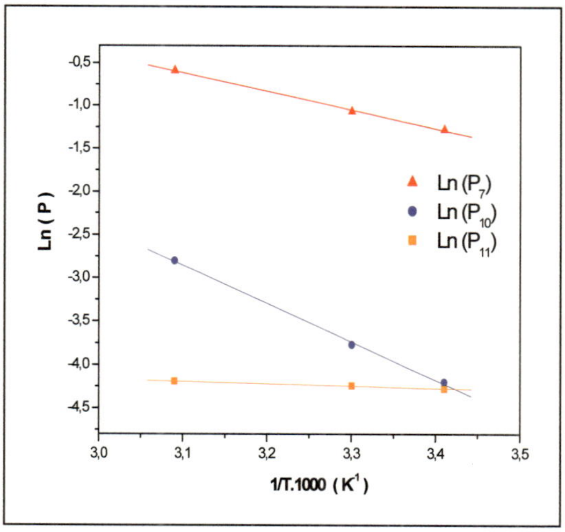


**Figure 16.** Temperature dependence of parameters $P_7$, $P_{10}$ and $P_{11}$ for $Pd/Al_2O_3$ catalyst

The same fit with model III-C was repeated with experimental data at other reaction temperatures in the 293-323 K range. In all cases and as a consequence of the fit, parameters were obtained with values different from zero in a confidence interval of 95% and with values of the MSC parameter greater than 4.0. The thermodynamic consistence of parameters $P_7$, $P_{10}$ and $P_{11}$ was evaluated by plotting $\ln P_7$, $\ln P_{10}$ and $\ln P_{11}$ as a function of 1/T. In all cases straight lines were obtained confirming the hypotheses of Arrhenius dependence with respect to temperature (Figure 16).

The slopes of the straight lines obtained correspond to the enthalpies of adsorption of 1-heptyne and $H_2$ and the energies of activation of the surface reactions of hydrogenation. Taking into account the definition of the parameters $P_7$, $P_{10}$ and $P_{11}$ (Table 1) the following equations can be obtained:

$$P_7 = \frac{(1 + K.K_{H2}.C_{H2}).K_A}{\sqrt{K_{H2}.C_{H2}}} \tag{30}$$

Supposing that $1 >> K.K_{H2}.C_{H2}$ (this assumption is made for the purpose to estimate some values of the involved constants).

The previous equation is reduced to:

$$P_7 = \frac{K_A}{\sqrt{K_{H2}.C_{H2}}} = \frac{A_A}{\sqrt{A_{H2}.C_{H2}}}.\mathrm{Exp}\left[\frac{-\left(\frac{1}{2}.|\Delta H_{H2}| - |\Delta H_A|\right)}{R.T}\right] \tag{31}$$

$$P_{10} = \frac{k_1.K_A.C_S^3}{\sqrt{K_{H2}.C_{H2}}} = \frac{A_1.A_A.C_S^3}{\sqrt{A_{H2}.C_{H2}}}.\mathrm{Exp}\left[\frac{-\left(E_1 - |\Delta H_A| + \frac{1}{2}.|\Delta H_{H2}|\right)}{R.T}\right] \tag{32}$$

$$P_{11} = \frac{k_3.K_A.K.C_S^3}{\sqrt{K_{H2}.C_{H2}}} = \frac{A_3.A_A.A.C_S^3}{\sqrt{A_{H2}.C_{H2}}}.\mathrm{Exp}\left[\frac{-\left(E_3 - |\Delta H_A| + \Delta H + \frac{1}{2}.|\Delta H_{H2}|\right)}{R.T}\right] \tag{33}$$

In equations (31), (32) and (33) the enthalpies of adsorption of 1-heptyne and $H_2$ are expressed in absolute values. The values of the respective slopes were calculated as follows:

$$\frac{|\Delta H_{H2}|}{2} - |\Delta H_A| = 18.02 \text{ KJ mol}^{-1} \tag{34}$$

$$E_1 + \frac{|\Delta H_{H2}|}{2} - |\Delta H_A| = 36.6 \text{ KJ mol}^{-1} \tag{35}$$

$$E_3 + \Delta H + \frac{|\Delta H_{H2}|}{2} - |\Delta H_A| = 2.18 \text{ KJ mol}^{-1} \tag{36}$$

The enthalpy of $H_2$ adsorption on Pd has been reported in Ref. [35] (-75.31 KJ mol$^{-1}$), then from equation (34): $|\Delta H_A| = 19.64$ KJ mol$^{-1}$. Replacing eq. (34) into eq. (35): $E_1 = 18.58$ KJ mol$^{-1}$. From equations (34) and (36):

$$E_3 + \Delta H = -15.84 \text{ KJ mol}^{-1} \tag{37}$$

From the results it can be concluded that:

5. Model III-C that poses the surface chemical reaction as the limiting step is the one that best fits the experimental data. The model also poses the dissociative adsorption of $H_2$ and the competition with 1-heptyne for the active sites. The model presents statistical and thermodynamic consistency.
6. The model enables obtaining directly the values of the enthalpies of adsorption of 1-heptyne and the activation energy for the hydrogenation of 1-heptyne to 1-heptene ($E_1$). Neither the calculation of the activation energy for the hydrogenation of 1-heptyne to n-heptane ($E_3$).
7. The model indicates that 1-heptyne and $H_2$ are the only species adsorbed on the active sites. The enthalpy of adsorption of 1-heptyne over Pd (-19.64 KJ mol$^{-1}$) is not much different from the value reported by Semagina et al [36] for the hydrogenation of 1-hexyne over Pd nanoparticles.
8. Equation (34) shows that the enthalpy of adsorption of hydrogen is higher than that of 1-heptyne over the $Pd/Al_2O_3$ catalyst. This suggests that there are not thermodynamic limitations for the adsorption of $H_2$. This was confirmed by the tests of hydrogen chemisorptions as Pd is able to chemisorb an important amount of $H_2$ at room temperature, suggesting that there is not a kinetic impediment as that observed for Ni. Consequently, the dissociative adsorption of hydrogen is fast and then the controlling step is the surface chemical reaction.
9. The value obtained for the activation energy of the hydrogenation reaction of 1-heptyne to 1-heptene ($E_1$) turned out to be quite low (18.58 KJ mol$^{-1}$). This coincides with the fact that the reaction can proceed at low temperatures.
10. The kinetic modelling of the reactions over the Ni and Pd catalysts gives an explanation of the different reactivity of the catalysts. Pd is more active than Ni for partial hydrogenation reactions because there is a kinetic limitation for the adsorption of hydrogen on Ni. Hydrogen is more strongly chemisorbed on Pd, so there is a great coverage of the surface by $H_2$, therefore making the surface reaction step as the rate-controlling one.

## 5. Conclusions

$Pd/Al_2O_3$ was more active and selective than $Ni/Al_2O_3$ for the partial hydrogenation of 1-heptyne to 1-heptene.

In order to analyze the influence of the different variables (hydrogen partial pressure, initial concentration of 1-heptyne and reaction temperature) on the reaction rate a pseudo homogeneous model was proposed based on power law kinetics. Reaction orders for hydrogen and 1-heptyne were obtained as well as the apparent activation energy. For the $Ni/Al_2O_3$ catalyst, reaction orders of 1.3 in hydrogen and -0.22 in 1-heptyne, and apparent activation energy of 24 KJ $mol^{-1}$ were obtained. For the $Pd/Al_2O_3$ catalyst, reaction orders of -2.6 and -1.5 in hydrogen and 1-heptyne, respectively, and apparent activation energy of 18 KJ $mol^{-1}$ were obtained.

In order to elucidate the role of Ni and Pd on the reaction rate, the kinetic data were fitted with six different heterogeneous LHHW models. The results obtained indicate that for the $Ni/Al_2O_3$ catalyst the controlling step is the dissociative adsorption of hydrogen over the metal active sites and the reaction rate can be expressed by:

$$r = \frac{P_3}{\left[C_A + P_1.C_B\right]^2}$$

If it is assumed that the adsorption enthalpy of 1-heptene can be considered negligible, a value of -17.91 KJ $mol^{-1}$ is obtained for the adsorption enthalpy of 1-heptyne. In the same way, the value of activation energy for the hydrogen adsorption is 22.2 KJ $mol^{-1}$.

For the $Pd/Al_2O_3$ catalyst the controlling steps are the surface hydrogenation reactions (1-heptyne to 1-heptene and 1-heptyne to n-heptane). The corresponding reaction rates are:

$$r_1 = \frac{P_{10}.C_A}{\left[1 + P_7.C_A\right]^3} \qquad r_3 = \frac{P_{11}.C_A}{\left[1 + P_7.C_A\right]^3}$$

Using a value for the adsorption enthalpy of hydrogen on Pd reported in literature, a value of -19.64 KJ $mol^{-1}$ is obtained for the adsorption enthalpy of 1-heptyne ($\Delta H_A$) over Pd. Besides, the value obtained for the activation energy for the hydrogenation reaction of 1-heptyne to 1-heptene ($E_1$) was 18.58 KJ $mol^{-1}$. This coincides with the fact that the reaction can proceed at low temperatures.

The different activity levels of the $Pd/Al_2O_3$ and $Ni/Al_2O_3$ catalysts are due to a kinetic limitation for the adsorption of hydrogen on Ni. In the case of Pd this limitation does not exist.

## Author details

M. Juliana Maccarrone, Cecilia Lederhos and Carolina Betti
*Institute of Catalysis and Petrochemistry Research, INCAPE (CONICET- UNL), Santa Fe, Argentina*

Gerardo C. Torres
*Departament of Chemical Engineering Reactions, Faculty of Chemical Engineerin, Santa Fe, Argentina*

Juan M. Badano and Mónica Quiroga*
*Institute of Catalysis and Petrochemistry Research, INCAPE (CONICET- UNL), Santa Fe, Argentina*
*Inorganic Chemistry, Departament of Chemistry, Faculty of Chemical Engineering National University of Litoral (UNL), Santa Fe, Argentina*

Juan Yori
*Institute of Catalysis and Petrochemistry Research, INCAPE (CONICET- UNL), Santa Fe, Argentina*
*Departament of Chemical Engineering Reactions, Faculty of Chemical Engineering, Santa Fe, Argentina*

## Acknowledgement

The experimental assistance of C. A. Mázzaro and the financial support provided by UNL and CONICET of Argentina are greatly acknowledged.

## Notation

$\alpha$: order of reaction with respect to 1-heptyne.
$\beta$: order of reaction with respect to hydrogen.
A: Arrhenius preexponential factor.
$A_A$: preexponential factor of the heat of adsorption of 1-heptyne.
$A_B$: preexponential factor of the heat of adsorption of 1-heptene.
$A_{H2}$: preexponential factor of the heat of adsorption of hydrogen.
$A_1$: preexponential factor of the specific reaction rate constant, $k_1$.
$A_3$: preexponential factor of the specific reaction rate constant, $k_3$.
$C_A^0$: initial concentration of 1-heptyne (mol $L^{-1}$).
$C_A$: concentration of 1-heptyne (mol $L^{-1}$).
$C_B$: concentration of 1-heptene (mol $L^{-1}$).
$C_C$: concentration of n-heptane (mol $L^{-1}$).
$C_{H2}$: concentration of hydrogen in the liquid phase (mol $L^{-1}$).
$C_S$: concentration of free sites.
$E_A$: apparent activation energy (KJ $mol^{-1}$).
$E_{H2}$: activation energy for the hydrogen adsorption (KJ $mol^{-1}$).
$E_1$: activation energy of the reaction of hydrogenation of 1-heptyne to 1-heptene (KJ $mol^{-1}$).
$E_3$: activation energy of the reaction of hydrogenation of 1-heptyne to n-heptane (KJ $mol^{-1}$).
$\Delta H$: reaction heat (KJ $mol^{-1}$)
$|\Delta H_{H2}|$: absolute value of the enthalpy of hydrogen adsorption (KJ $mol^{-1}$).
$|\Delta H_A|$: absolute value of the enthalpy of adsorption of 1-heptyne (KJ $mol^{-1}$).
$|\Delta H_B|$: absolute value of the enthalpy of adsorption of 1-heptene (KJ $mol^{-1}$).
k: specific reaction rate constant.

* Corresponding Author

K: equilibrium constant (equation (6)).
$K_A$: adsorption constant for 1-heptyne.
$K_B$: adsorption constant for 1-heptene
$K_C$: adsorption constant for n-heptane
$K_{H2}$: adsorption constant for hydrogen
$k_A$: direct rate specific constant for 1-heptyne adsorption.
$k_{H2}$: direct rate specific constant for $H_2$ dissociative adsorption.
$k'_{H2}$: direct rate specific constant for $H_2$ non disociative adsorption.
$k_1$: direct rate specific constant for 1-heptyne to 1-heptene surface hydrogenation reaction.
$k_3$: direct rate specific constant for 1-heptyne to n-heptane surface hydrogenation reaction.
$P_{H2}$: partial pressure of hydrogen (bar).
R: ideal gas universal constant (0.082 $L.atm.mol^{-1}.K^{-1}$).
$r_1$: rate of step 1 (mol $L^{-1}.h^{-1}$).
$r_3$: rate of step 3 (mol $L^{-1}.h^{-1}$).
$r^0_A$: 1-heptyne initial hydrogenation rate (mol $L^{-1}.h^{-1}$).
T: absolute temperature (K)
t: time.

## 6. References


[1] L'Argentière PC, Cagnola EA, Quiroga ME, Liprandi DA (2002) A palladium tetra-coordinated complex as catalyst in the selective hydrogenation of 1-heptyne. Appl. Catal. A: Gen. 226: 253-263.

[2] Lennon D, Marshall R, Webb G, Jackson S D (2000) The effect of hydrogen concentration on propyne hydrogenation over a carbon supported palladium catalyst studied under continuous flow conditions. Stud. Surf. Sci. Catal. 130: 245-250.

[3] Nishimura S. (2001) Handbook of Heterogeneous Catalytic Hydrogenation for Organic Synthesis. John Wiley & Sons, Inc. Canadá: ISBN 0-47139698-2, pp. 148-169.

[4] Lindlar H, Dubuis R (1966) Palladium catalyst for partial reduction of acetylenes. Org Synth 46: 89–92.

[5] Lederhos CR., L'Argentière PC, Fígoli NS (2005) 1-Heptyne Selective Hydrogenation over Pd Supported Catalyst. Ind. Eng. Chem. Re.s Devel 44: 1752-1756.

[6] Bennett JA, Creamer NJ, Deplanche K, Acaskie LE, Shannon IJ, Wood J (2010) A comparison of $Pd/Al_2O_3$ and bio-Pd in the hydrogenation of 2-pentyne. J. Chem. Eng. Sc. 65: 282-290.

[7] Sárkány A, Beck A, Horváth A, Révay Zs, Guczi L (2003) Acetylene hydrogenation on sol-derived $Pd/SiO_2$. Appl. Catal. A: Gen. 253: 283–292.

[8] Anderson JA, Mellor JL, Wells RPK (2009) Pd catalysed hexyne hydrogenation modified by Bi and by Pb. J. Catal. 261: 208-216.

[9] Evangelisti C, Panziera N, D'Alessio A, Bertinetti L, Botavina M, Vitulli G (2010) New monodispersed palladium nanoparticles stabilized poly-(N-vinyl-2-pyrrolidone): Preparation, structural study and catalytic properties. J. Catal. 272: 246-252.

[10] Mastalir A, Király Z (2003) Pd nanoparticles in hydrotalcite: mild and highly selective catalysts for alkyne semihydrogenation. J. Catal. 220: 372–381.
[11] Lederhos CR, Badano JM, Quiroga ME, Coloma-Pascual F, L'Argentière PC. (2010) Influence of Ni addition to a low-loaded palladium catalyst on the selective hydrogenation of 1-heptyne. Quim. Nova 33(4): 816-820.
[12] Lederhos CR, Maccarrone MJ, Badano JM, Coloma-Pascual F, Yori JC, Quiroga ME (2011) Hept-1-yne partial hydrogenation reaction over supported Pd and W catalysts. Appl. Catal. A: Gen. 396: 170-176.
[13] Alves JA, Bressa SP, Martínez OM, Barreto GF (2007) Kinetic study of the liquid-phase hydrogenation of 1-butyne over a commercial palladium/alumina catalyst. Chem. Eng. J. 125 (3): 131-138.
[14] Crespo-Quesada M, Dykeman RR, Laurenczy G, Dyson PJ, Kiwi-Minsker L (2011) Supported nitrogen-modified Pd nanoparticles for the selective hydrogenation of 1-hexyne. J. Catal. 279 (1): 66-74.
[15] Maccarrone MJ, Lederhos C, Badano J, Quiroga ME, Yori J C (2011) Liquid-phase selective hydrogenation of 1-heptyne over Ni/$Al_2O_3$. Effect of the reaction temperatura. Avances en Ciencias e Ingeniería 2 (2): 59-68.
[16] Boudart M, Hwang HS (1975) Solubility of hydrogen in small particles of palladium J. Catal. 39: 44-52.
[17] Hu S, Chen Y (1998) Partial hydrogenation of benzene: A review. J. Chin. Chem. Eng. 29: 387-396.
[18] L'Argentière P C, Cañón MM, Fígoli NS, Ferrón (1993) AES and XPS Studies of the Influence of Ni Addition on Pd/$Al_2O_3$ Catalytic Activity and Sulfur Resistance. J. Appl. Surf. Sci. 68: 41-47.
[19] NIST X-ray Photoelectron Spectroscopy Database NIST Standard Reference Database 20. Version 3.5 (Web Version). National Institute of Standards and Technology. USA. (2007)
[20] Telkar MM, Nadgeri JM, Rode CV, Chaudhari RV (2005) Role of a co-metal in bimetallic Ni–Pt catalyst for hydrogenation of m-dinitrobenzene to m-phenylenediamine. Appl. Catal. A: Gen. 295: 23–30.
[21] Juan-Juan J, Roman-Martinez MC, Illan-Gomez MJ (2004) Catalytic activity and characterization of Ni/$Al_2O_3$ and NiK/$Al_2O_3$ catalysts for $CO_2$ methane reforming. Appl. Catal. A: Gen. 264: 169-174.
[22] Web Page: www.lasurface.com/accueil/index.php
[23] Wagner CD, Riggs WM, Davis RD, Moulder JF (1978). In Handbook of X-ray Photoelectron Spectroscopy. Muilenberg. G.E., Ed. Perkin-Elmer: Eden Preirie, MN.
[24] Dantas Ramos AL, da Silva Alves P, Aranda DAG, Schmal M (2004) Characterization of carbon supported palladium catalysts: inference of electronic and particle size effects using reaction probes. App. Catal. A: Gen. 277 (1-2): 71-81.
[25] Benitez VM, Querini CA, Fígoli NS, Comelli RA (1999) Skeletal isomerization of 1-butene on $WO_x/\gamma$-$Al_2O_3$. Appl. Catal. A: Gen. 178: 205-218.
[26] Huang S, Zhang C, He H (2008) Complete oxidation of o-xylene over Pd/$Al_2O_3$ catalyst at low temperature. Catal Today 139: 15-23.

[27] Salagre P, Fierro JLG, Medina F, Sueiras JE (1996) Characterization of nickel species on several $\gamma$ -alumina supported nickel samples. J. Molec. Catal. A: Chem. 106 125-134.

[28] Heracleous E, Lee AF, Wilson K, Lemonidou AA (2005) Investigation of Ni-based alumina-supported catalysts for the oxidative dehydrogenation of ethane to ethylene: structural characterization and reactivity studies. J. Catal. 231: 159-171.

[29] Ferrer V, Moronta A, Sánchez J, Solano R, Bernal S, Finol D (2005) Effect of the reduction temperature on the catalytic activity of pd-supported catalysts. Catal. Today 107: 487-492.

[30] Hou Z, Yokota O, Tanaka T, Yashima T (2003) Characterization of Ca-promoted Ni/$\gamma$-$Al_2O_3$ catalyst for $CH_4$ reforming with $CO_2$. Appl. Catal. A: Gen 253: 381-387.

[31] Hoffer BW, van Langeveld AD, Janssens JP, Bonne RLC, Lok CM, Moulijn JA (2000) Stability of highly dispersed Ni/$Al_2O_3$ catalysts: Effects of pretreatment. J. Catal. 192: 432-440.

[32] Li F, Yi X, Fang W (2009) Effect of Organic Nickel Precursor on the Reduction Performance and Hydrogenation Activity of Ni/$Al_2O_3$ Catalysts. Catal. Let.t 130: 335–340.

[33] Mallat T, Baiker A (2000) Selectivity enhancement in heterogeneous catalysis induced by reaction modifiers. App. Catal. A: Gen. 200: 3-22.

[34] Joback KG. Unified Approach to Physical Property Estimation Using Multivariate Statistical Techniques, M.S. Thesis, MIT, Cambridge, MA, (1984).

[35] Sierra Jimenez F, "Algunos aspectos modernos del fenómeno de la adsorción". 22 (1945) Universidad de Murcia, Servicio de Publicaciones.

[36] Semagina N, Renken A, Kiwi-Minsker L (2007) Monodispersed Pd-nanoparticles on carbon fiber fabrics as structured catalyst for selective hydrogenation. Chem. Eng. Sc. 62 (18-20): 5344-5348.

Section 2

# Hydrogenation Reactions in Environmental Chemistry and Renewable Energy

Chapter 8

# Hydroconversion of Triglycerides into Green Liquid Fuels

Rogelio Sotelo-Boyás, Fernando Trejo-Zárraga
and Felipe de Jesús Hernández-Loyo

Additional information is available at the end of the chapter

http://dx.doi.org/10.5772/48710

## 1. Introduction

Due to the depletion of crude reserves and the increasing demand for clean hydrocarbon fuels the production of renewable materials-based fuels has emerged to solve at least partially this problem in the past decade and it is expected to continue [1,2]. Green fuels can be classified as naphtha, jet fuel, and diesel. In the case of green diesel, its increasing demand could reach 900 million tons by 2020 [3].

The common ways to produce diesel-type fuel from biomass are a) by transesterification of triglycerides to obtain biodiesel, and b) by hydroprocessing to synthesize green diesel. Biodiesel is a mixture of fatty acid methyl esters (FAME) while green diesel is a mixture of hydrocarbons, mainly heptadecane and octadecane. Both fuels can be used as additives to petro-diesel. FAME can be used to enhance the lubricity of petro-diesel while green diesel can boost the cetane number. Biodiesel can be obtained with an alkaline liquid catalyst at 60 °C and atmospheric pressure, while green diesel requires a bifunctional solid catalyst (acid/metal), temperatures around 300 °C and ca. 5 MPa of hydrogen in a continuous flow process.

The technology for producing green fuels from triglycerides has been used in petroleum refineries for about 60 years, for instance the same catalyst, reactor type and separation equipment used in the hydrotreating of vacuum gas oil can effectively be used for hydrotreatment of fats and vegetable oils. Thus, during the last decade it has been an increasing interest in research and development for their optimal production, trying to find the best catalyst and the most favorable operating conditions. Many countries and major petroleum companies are nowadays considering the use of vegetable oils and fats as raw materials for production of hydrogenated green fuels. The main drawbacks seem to be the availability of raw material and the large consumption of hydrogen during the process, which increases the production cost. However there are positive aspects of this technology that overcome the technical difficulties. In what follows these aspects are comprehensively reviewed, starting from the raw material and closing with the current commercial processes.

## 2. Triglycerides

Triglycerides make up the structure of all vegetable oils and fats found in nature. They are primarily composed of long chains of fatty acid esters as shown in Figure 1. The side chains of triglycerides are either saturated, monounsaturated or polyunsaturated. They can be classified by the length and saturation degree of their side chains. The acid portion of the ester linkage (fatty acids) usually contains an even number of carbon atoms in a linear chain of 12 to 24 carbon atoms with up to three unsaturated bonds usually in the position 9, 12 and 15 with cis orientation, and prevalently in the 9 and 12 position such as in a linoleic oil [4].

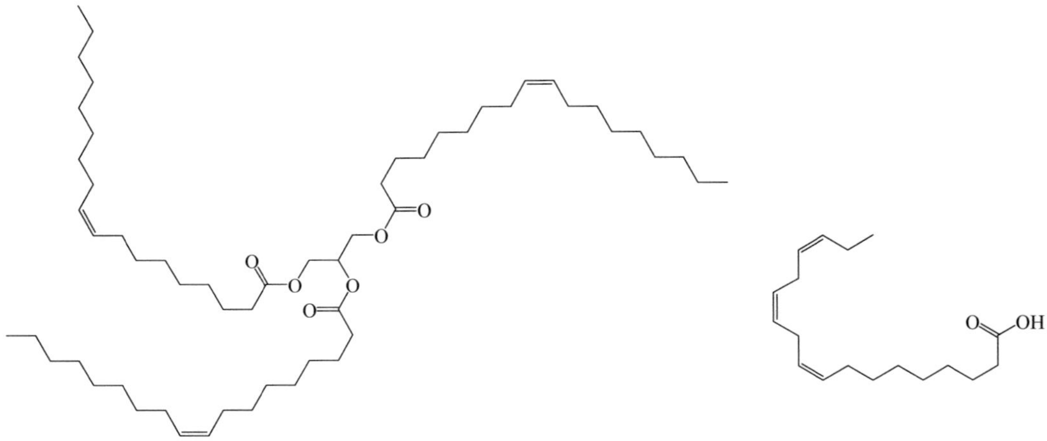

*Triolein* *Linolenic acid*

**Figure 1.** Basic structure of a triglyceride and a fatty acid commonly present in fats and vegetable oils

The properties of triglycerides depend on the fatty acid composition and on the relative location of fatty acids on the glycerol. Both fats and oils are composed of triglycerides. The difference between them is simply one of melting point: fats are solid at room temperature (20 °C) whereas oils are liquids. Thus, natural fats and oils are designated as the triglyceride type in terms of saturated and unsaturated acids and isomeric forms. Saturated oils have better oxidation stability and higher melting points than unsaturated ones. A higher degree of unsaturation also means a higher reactivity [5].

The vegetable oil composition is commonly described by its content of fatty acids as it is shown in Table 1 for palm, canola, jatropha, soybean and sunflower oils. This composition is usually determined by derivatitation of the esters obtained by total esterification of the oil with a strong acid, e.g. boron trifluoride, as it is shown elsewhere [10]. The molecular weight is given for the fatty acid and for the triglyceride that contains three side chains of the corresponding acid. The average molecular weight of the vegetable oils listed in this table

| Source | Structure | Molecular weight (MW) | | Typical composition, wt.% | | | | |
|---|---|---|---|---|---|---|---|---|
| | | Fatty acid | Triglyceride | Jatropha | Palm | Canola | Soybean | Sunflower |
| Capric | C10:0 | 172.3 | 554.8 | 0.0 | 0.0 | 0.6 | 0.0 | 0.0 |
| Lauric | C12:0 | 200.3 | 639.0 | 0.0 | 0.0 | 0.0 | 0.0 | 0.0 |
| Myristic | C14:0 | 228.4 | 723.2 | 0.0 | 2.5 | 0.1 | 0.0 | 0.0 |
| Palmitic | C16:0 | 256.4 | 807.3 | 15.9 | 40.8 | 5.1 | 11.5 | 6.5 |
| Palmitoleic | C16:1 | 254.4 | 801.3 | 0.9 | 0.0 | 0.0 | 0.0 | 0.2 |
| Stearic | C18:0 | 284.5 | 891.5 | 6.9 | 3.6 | 2.1 | 4.0 | 5.8 |
| Oleic | C18:1 | 282.5 | 885.4 | 41.1 | 45.2 | 57.9 | 24.5 | 27.0 |
| Linoleic | C18:2 | 280.4 | 879.4 | 34.7 | 7.9 | 24.7 | 53.0 | 60.0 |
| Linolenic | C18:3 | 278.4 | 873.3 | 0.3 | 0.0 | 7.9 | 7.0 | 0.2 |
| Arachidic | C20:0 | 312.5 | 975.6 | 0.0 | 0.0 | 0.2 | 0.0 | 0.3 |
| Eicosenoic | C20:1 | 310.5 | 969.6 | 0.2 | 0.0 | 1.0 | 0.0 | 0.0 |
| Behenic | C22:0 | 340.6 | 1059.8 | 0.0 | 0.0 | 0.2 | 0.0 | 0.0 |
| Erucic | C22:1 | 338.6 | 1053.8 | 0.0 | 0.0 | 0.2 | 0.0 | 0.0 |
| | | | Estimated MW: | 869.7 | 847.0 | 876.9 | 871.9 | 876.7 |

**Table 1.** Typical composition of various vegetable oils.
Nomenclature: Cn:m describes a fatty acid with n carbon atoms and m double bonds. Adapted from refs. [6-9].

was estimated by considering that the concentration of a given triglyceride in the oil is the same as the concentration of its corresponding fatty acids, i.e. the three side chains of the triglyceride come from this fatty acid. Thus the estimated average molecular weight for jatropha oil is ca. 870. Another way to determine the molecular weight of a vegetable oil includes the acid and saponification values [11]. The properties of vegetable oils and fats can be modified by hydrogenation.

Hydrogenation of edible vegetable oils has been around for about 100 years in the fat and oils industry, the hydrogenation takes place in a multi-phase catalytic reactor (e.g. slurry type) on a supported Ni catalysts in which the side chains of the triglycerides present in the oil are saturated and converted into a solid or semisolid product [12]. Thus the oil gets a better oxidation stability, and if it is partially hydrogenated it can be used as margarine. This chapter does not focus on food applications, but on the use of vegetable oils for production of liquid transportation fuels by means of a hydrogenation highly enough to crack the triglycerides into hydrocarbons, i.e. hydrocracking.

## 3. Fundamentals of hydroconversion processes

Hydroprocessing is an important class of catalytic processes in a refinery scheme that comprises a set of reactions in which hydrogen is passed through a bifunctional catalyst (metal/acid). Hydroprocessing is used to convert a variety of petroleum distillates into clean transportation fuels and heating oil. The reactions that occur in hydroprocessing can be classified in two groups: a) hydrocracking, and b) hydrotreating.

Hydrocracking involves destructive hydrogenation and is characterized by the conversion of the higher molecular weight components in a feedstock to lighter products. Isomerization and cracking of C-C bonds in bigger molecules occur at some extent to produce hydrocarbons within the boiling range of gasoline and diesel. Such treatment requires high temperature and the use of high hydrogen pressures to minimize the condensative chain polymerization reactions that lead to coke formation [5]. From a catalytic viewpoint, hydrocracking is carried out on acid supports, i.e., amorphous supports (alumino-silicates), silicoaluminophosphates (SAPO), and crystalline supports (zeolites) [13]. Figure 2 shows schematically an example of hydrocracking of a model molecule present in vacuum gas oil.

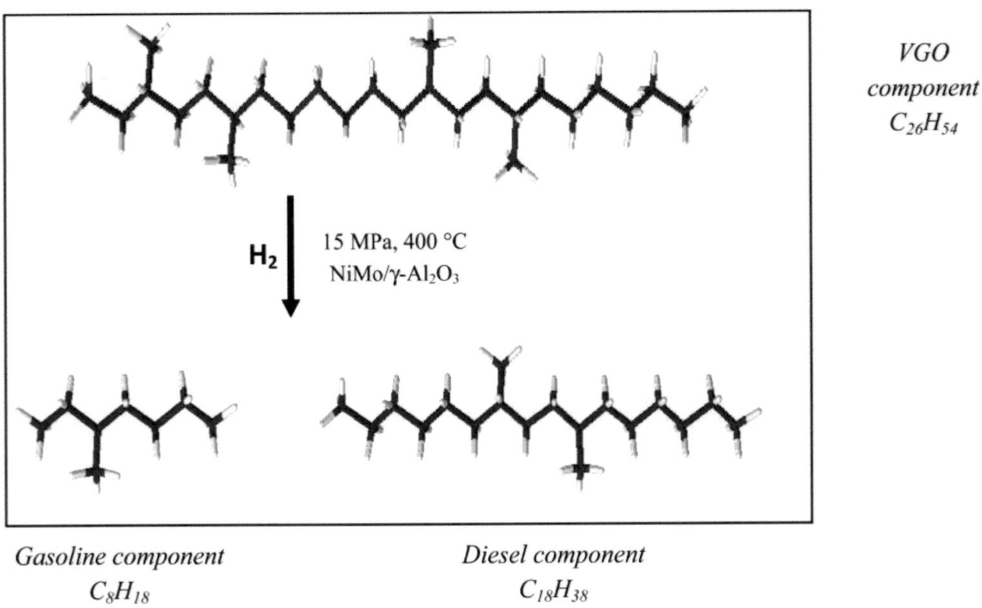


**Figure 2.** Hydrocracking of gas oil into gasoline and diesel on a bifunctional catalyst

Hydrotreating or hydrofining involves non-destructive hydrogenation and is used to improve the quality of petroleum distillates without significant alteration of the boiling range. Nitrogen, sulfur, and oxygen compounds undergo hydrogenolysis to remove ammonia, hydrogen sulfide, and water, respectively. Mild temperature and hydrogen pressures are employed so that only the more unstable compounds that might lead to the formation of gums, or insoluble materials, are converted to more stable compounds [5]. Hydrotreating takes place on the metal active sites of a catalyst, e.g., NiMo or CoMo in sulfide state supported on $\gamma$–$Al_2O_3$ [13]. Other catalysts different to metal sulfides have been used in hydrotreating, such as supported noble metal catalysts [14] and Ni-Mo/$\gamma$-$Al_2O_3$ catalyst in which its acidity was improved with addition of fluorine [15]. The NiMo/$\gamma$-$Al_2O_3$

is actually one of the most commonly used catalysts in the hydroprocessing of middle and heavy distillates at petroleum refineries. This catalyst has a high hydrogenation activity and mild acidity which are also appropriate for the hydroconversion of triglycerides into diesel hydrocarbons [16].

## 4. Hydroprocessing of triglycerides

As an alternative renewable raw material, the triglycerides present in fats and vegetables oils can be industrially hydroconverted at the petroleum refineries.

Five of the most common approaches for upgrading hydrotreaters for clean-fuels production are (1) upgrading feedstock and integrating processes, (2) implementing a higher-activity catalyst, (3) replacing reactor internals for increased efficiency, (4) adding reactor capacity, and (5) increasing $H_2$ partial pressure. Refiners also have the option to implement advanced process control, and simulations to optimize operation and commercial catalyst developments have been accelerating [17].

### 4.1. Green fuels from triglycerides: Technological advantages

The hydroprocessing of triglycerides involves the hydrogenation of the double bonds of the side chains and the removal of oxygen on the metal sites of the catalysts. The hydrotreating of most vegetable oils leads to the production of $C_{15}$-$C_{18}$ hydrocarbons, i.e. a liquid mixture within the boiling point range of diesel which is commonly called "green diesel", "renewable diesel" or "bio-hydrogenated diesel", and therefore with the same chemical nature than petroleum-derived diesel [18].

In comparison with the process to produce fatty acid methyl esters (FAME or biodiesel), the hydroprocessing of vegetable oils for production of green diesel has the following advantages[19]:

- The product is compatible with existing engines.
- Flexibility with the feedstock, e.g. the content of free fatty acids in the vegetable oil does not matter.
- Higher cetane number.
- Higher energy density.
- Higher oxidation stability (zero $O_2$ content).
- It does not increase the emissions of $NO_x$.
- It does not require water.
- There are not byproducts that require additional treatment (e.g. glycerol).
- The distribution of the renewable diesel does not cause additional pollution since it can be transported through the same pipelines that are currently used for distribution of petrodiesel.
- Better performance in cold weather.

To obtain lighter hydrocarbons such as those within the boiling point range of jet fuel or gasoline, catalysts with stronger acid sites (e.g. zeolites) can be used. The zeolite supported catalysts usually contain higher acidity and therefore a more severe cracking activity than those supported on alumina. The acid sites of the catalyst increase also the isomerization degree of the molecules, thus boosting the properties of the green liquid fuels, such as a lower pour point and a higher octane number.

## 4.2. Raw materials used as source of triglycerides

Different sources of triglycerides have been hydrotreated using bifunctional catalysts to produce green fuels such as palm, sunflower, jatropha, rapeseed, and soybean oils among others. Waste cooking oils have also been used to produce green fuels. Some of these vegetable oils and its composition are shown in Table 1. Most of the vegetable oils found in nature contain linear paraffinic side chains with 16 and 18 carbon atoms, predominantly monounsaturated chains (C18:1). The hydroprocessing of long-chain triglycerides highly unsaturated requires higher hydrogen consumption than that required for less unsaturated triglycerides to reach the same level of hidroconversion into liquid fuels. This fact is also dependent on the catalyst used and reaction conditions at which triglycerides are processed.

Hydroconversion of triglycerides can be carried out in two ways. The first one consists in hydroprocessing of triglycerides only and the second one in co-feeding to the process triglycerides and (petroleum-based) vacuum gas oil.

### *4.2.1. Hydroprocessing of triglycerides only*

During hydroprocessing of triglycerides, the type of catalyst is one of the most important factors to determine the yield and composition of liquid products, such as green naphtha ($C_5$-$C_{10}$), green jet fuel ($C_{11}$- $C_{13}$), and green diesel ($C_{14}$- $C_{20}$), and even green liquid petroleum gas (LPG). A severe hydrocracking catalyst would lead to a high production of green naphtha whereas a mild-hydrocracking catalyst is prone to produce mainly green diesel. The reaction temperature plays an important role for the yield and quality of hydroprocessed oils as well. It has been observed that diesel selectivity decreases with increasing reaction temperature while naphtha selectivity increases as result of the thermal cracking of diesel hydrocarbons. Therefore, high temperatures and strong acid catalysts are preferred if naphtha is the desired product. Conversely, moderate temperatures and catalysts with mild acid sites are needed if middle distillates are the desired products [20]. The yield of green gasoline can also be increased by using a two-step process, i.e. hydrotreating followed by hydrocracking. In the first one, oxygen is removed from biomass as water. The deoxygenated product is then separated by distillation, and the heavier fractions are further brought to the second step to convert them into (lighter) molecules within the boiling range of naphtha, as reported elsewhere [21]. Figure 3 shows a schematic two-step process in which green naphtha, jet fuel, and diesel are obtained. The first reactor is packed with a hydrotreating catalyst to remove oxygen. A low extent hydroisomerization also occurs in the first reactor. Then, the oxygen-free product is processed in a second reactor packed with a selective hydrocracking catalyst where both cracking of larger molecules and hydroisomerization take place. Distillation of the product yields three major liquid streams, i.e. naphtha, jet fuel, and diesel.

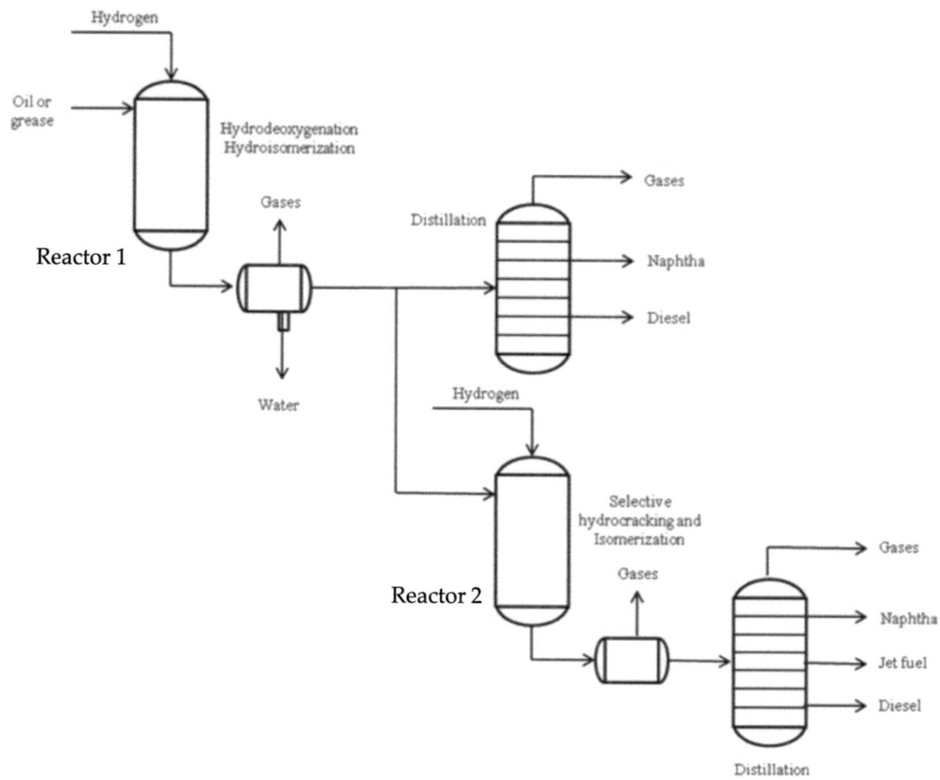


**Figure 3.** Schematic representation of a two-step process for obtaining green naphtha, jet fuel, and diesel coupling hydrodeoxygenation and selective hydrocracking.

In comparison with the use of different catalysts to produce green jet fuel or diesel, the use of catalysts for production of green naphtha through simultaneous cracking and hydrogenation has not been abundantly reported in literature. It has been stated that green gasoline is obtained when using NiMo/zeolite (klinoptilolite type) at 300-320 °C in a batch reactor after 1-2 h of reaction time. The strong acid sites in the zeolite favored the cracking of the large molecules in the vegetable oil into naphtha; however, middle distillates ($C_{13}$-$C_{19}$) were also obtained at some extent [22]. Another acidic support able to produce green naphtha is HZSM-5 zeolite. It has been reported that a synthesized zeolite having a $SiO_2/Al_2O_3$ ratio of 30 readily produces a high amount of gasoline from palm oil due to the large number of available Brønsted acid sites that increase the selectivity toward cracking reactions [23].

The green diesel obtained directly by hydrodeoxygenation is mainly composed by $C_{17}$ and $C_{18}$ n-paraffins that have a high cetane number but poor cold flow properties as these hydrocarbons have a high melting (freezing) point in between 20 and 28 °C. Therefore, its quality needs to be boosted by bringing the hydrodeoxygenated product through a second process in which selective hydroisomerization and cracking take place [14,15,24]. In this process, not only quality-improved diesel is obtained but also jet fuel, as the acid sites of the catalyst promote the conversion of larger alkyl chains into smaller chains.

Hydroisomerization is a key parameter to obtain methyl-branched paraffins mainly. This goal is achieved by using a shape-selective catalyst such as zeolites or other acidic supports. Normal paraffins in the boiling point range of diesel have generally a higher cetane number than that of their branched isomers. Conversely, iso-paraffins have lower freezing points than n-paraffins. Thus there is a trade-off in the quality of a fuel rich in paraffins, i.e. either with good combustion properties or with good cold flow properties.

Several studies have reported green diesel having a cetane numbers higher than petrodiesel and biodiesel, with values from 65 to 104 [24-27]. Hancsók et al. [24,26] and Simacek et al.[27] have studied the influence of the reaction temperature on the cetane number. As the reaction temperature was increased the cetane number of the liquid product was observed to decrease. This can be related with an increasing thermal cracking and therefore a reduced concentration of large paraffins in the diesel. Though green diesel is evidently a premium quality diesel and could be used as an effective additive enhancer, improved cold flow properties are also necessary for both jet and diesel fuels to be used at very low temperatures (e.g. below 0 °C).

Hydroisomerization is desirable for green diesel to have a lower freezing point though with lower cetane number. The hydroisomerization of a green diesel for instance containing mainly n-$C_{17}$ and n-$C_{18}$ paraffins and with a cetane number close to 100 and a freezing point close to 20 °C may produce a green diesel with a cetane number close to 70 and a freezing point lower than -5 °C, as it is shown by the variation of these two properties as function of carbon number in Figure 4. It is also observed than 5-methyl iso-paraffins have better cold flow properties that 2-methyl iso-paraffins.

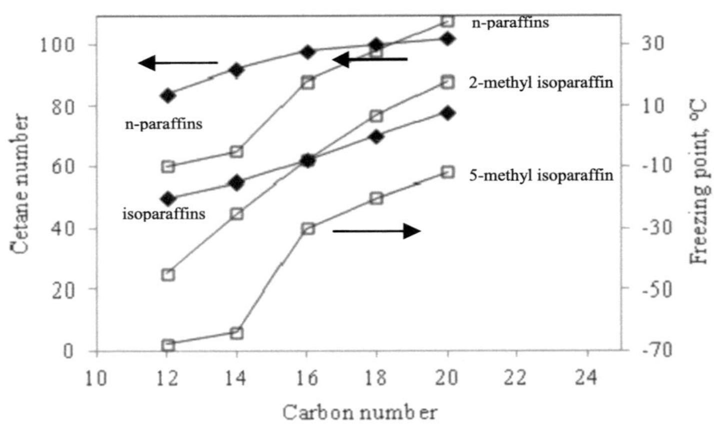


**Figure 4.** Cetane number and freezing points as function of carbon number in linear and methyl-branched paraffins. (Adapted from ref. [28]).

The interest for the production of both green jet fuel and diesel has been constantly increasing during the last decade, and gained such an importance nowadays that two-step processes are industrially available. Neste Oil in Europe and UOP Honeywell in America have built plants based on these types of processes to convert distinct vegetable oils into green jet fuel and green diesel. Obviously, these green fuels need to accomplish with quality standards to be used along with current fossil fuels in conventional engines. Table 2 presents some of the most important properties of jet fuel and diesel.

| Green jet fuel specifications | | | |
|---|---|---|---|
| Property | Value | Test method | |
| | | ASTM | IP |
| Total acidity, mg KOH/g | 0.10 (Max.) | D3242 | 354 |
| Distillation temperature, °C | Report | D2887, D86 | 123 |
| Flash point, °C | 38 (Min.) | D56, D3828 | 34 |
| Density @15°C, kg/m$^3$ | 775-840 | D1298, D4052 | 365 |
| Freezing point, °C | -40 Jet A (Max.)<br>-47 Jet A-1 (Max.) | D2386, D5972, D7153, D7154 | 16 |
| Viscosity @ -20°C, cSt | 8.0 (Max.) | D445 | 71 |
| Neat heat of combustion, MJ/kg | 42.8 (Min.) | D3338, D4529, D4809 | 12, 355, 381 |
| **Diesel fuel specifications** | | | |
| Property | Value | Test method | |
| | | ASTM | EN |
| Flash point, °C | 38 (Min.) | D93 | 2719 |
| Kinematic viscosity, mm$^2$/s @40°C | 1.9 (Min.)<br>4.1 (Max.) | D445 | 3104 |
| Ash, wt% | 0.01 (Max.) | D482 | 6245 |
| Cetane number | 40 (Min.) | D613 | 5165 |
| Cloud point, °C | Report | D2500 | ISO 3015 |

**Table 2.** Some properties of green jet fuel and diesel specified in test methods (Adapted from test methods ASTM D-975-11b, ASTM D-7566-11a).

Once green diesel has an improved performance in cold weather it may be used directly in vehicles or it can be mixed with petro-diesel. In the case of jet fuel, it has been mixed at 50% with petroleum-derived jet fuel or other percentages in one turbine and used in some documented commercial flights as shown in Table 3. The green jet fuel produced by UOP/ENI technology has been proved to meet all the main properties of petroleum-derived aviation fuels such as flash point, cold temperature performance, and good air oxidation stability [29].

The studies regarding to one-step process in which a catalyst is able to enhance both hydroisomerization and cracking are scarce and only a few reports have appeared in the literature. Microporous silicoaluminophosphates (SAPO) have been reported to possess acid sites of sufficient strength to be highly selective toward hydroisomerization. Methyl iso-paraffins are the predominant products obtained with Pt/SAPO-11 and Pt/SAPO-31 catalysts [30] and Pd/SAPO-31 [2]. In this case, it was possible to obtain hydrodeoxygenated and hydroisomerized liquid products over only one type of catalyst, which it would be more desirable and economical at industrial scale. $NiMo/Al_2O_3-SiO_2$ catalyst has also been used to hydrogenate unsaturated carbons of jatropha oil followed by deoxygenation to produce propane and $C_{15}$-$C_{18}$ n-paraffins, which were further hydroisomerized and cracked to generate $C_{15}$-$C_{18}$ iso-paraffins. All products were synthesized in a one-step process [7].

| Airline company | Jet model | Date | Oil source | Percentage of blend in one motor |
|---|---|---|---|---|
| Virgin Atlantic | B747 | Feb. 23, 2008 | Coconut, babassu | 20% |
| Air New Zealand | B747 | Dec. 30, 2008 | Jatropha | 50% |
| Continental Airlines | B747 | Jan. 9, 2009 | Jatropha | 50% |
| Japan Airlines | B747 | Jan. 30, 2009 | Camelina | 50% |
| Aeromexico | B777 | Aug. 2, 2011 | Jatropha | 30% |

**Table 3.** Different demonstration commercial flights employing mixtures of green jet fuel mixed with aviation turbine fuel from petroleum (Adapted from several web pages of different airlines companies).

### *4.2.2. Hydroprocessing of triglycerides plus petroleum gas oil simultaneously*

Vegetable oils can be mixed with petroleum fractions such as straight run gas oil and/or vacuum gas oil to be hydroprocessed on conventional hydrotreaters used in refineries to reduce operation costs by using the existing installations. In this case, two main reactions would occur, e.g. hydrodeoxygenation of triglycerides and hydrodesulfurization of gas oil. Both reactions are carried out on the same hydrotreating catalyst. It has been observed that $NiMo/Al_2O_3$ catalysts promote hydrodeoxygenation having high selectivity toward diesel range products compared with $NiW/SiO_2-Al_2O_3$; however, the NiW sulfides reduced the hydrogen consumption and increased the production of jet fuel. In a mixture of soybean oil and gas oil, it was observed that hydrodeoxygenation of soybean and hydrodesulfurization of gas oil were not competing reactions. In spite of similarities in active sites for both reactions their number is high enough to avoid inhibition among them [31]. In other studies it was reported that simultaneous hydrodeoxygenation of gas oil with high sulfur content with vegetable oil had a synergistic effect up to 15% of oil in the blend. With higher contents of vegetable oil in the mixture, desulfurization of the petroleum stream was reduced [32]. The mixture of rapeseed oil and light gas oil after being hydroprocessed yielded high conversions of sulfur and oxygen to obtain a diesel fuel with improved properties [33]. The addition of sunflower oil to gas oil to be simultaneously processed showed an enhancement of cetane number by 1-7 units when hydroprocessing took place at 420 °C and 18 MPa by which the fuel obtained can be considered as an excellent bio-component for the diesel fuel blending [27]. Figure 5 shows the co-feeding of vegetable oil of 5 to 20 wt% and vacuum gas oil (VGO) in a reactor operating at typical hydroprocessing conditions in a refinery. The cetane number is expected to be higher than that of the diesel obtained by only processing VGO, for about 10 to 20 units.

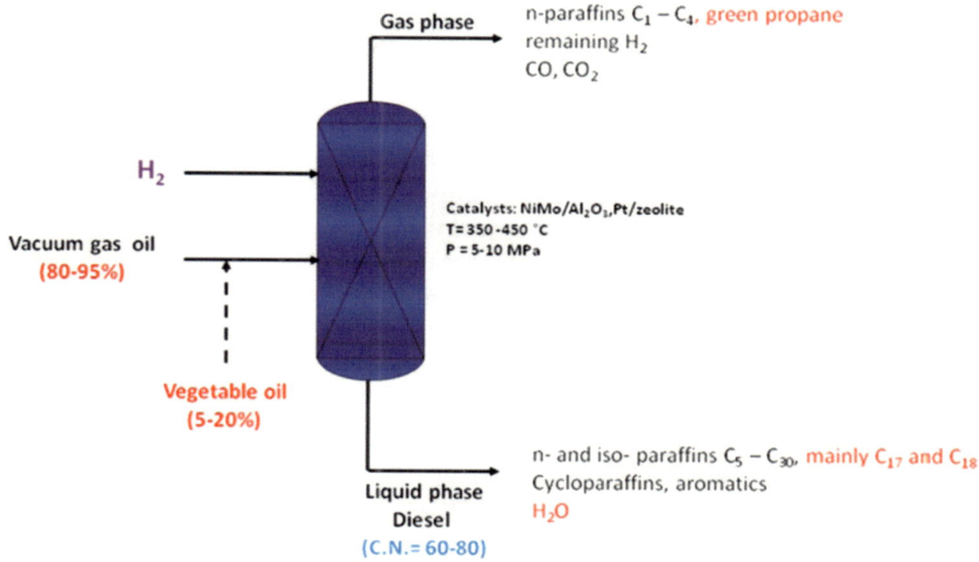


**Figure 5.** Co-feeding of vacuum gas oil and vegetable oil in a hydrotreater.

## 5. Chemistry of the hydrotreating of vegetable oils

The understanding of the chemistry of the hydroprocessing of triglycerides is essential to formulate a kinetic model that could lead to the design of the reactor and its subsequent simulation and optimization. The triglycerides are converted into hydrocarbons, mainly to n-paraffins at the temperatures between 300 and 450 °C and hydrogen pressures above 3 MPa leaving CO, $CO_2$ and water as by-products. The mechanism of the reaction is complex and consists of a series of consecutive steps (Viz Fig. 6). Oxygen removal from triglycerides occurs through different reactions such as hydrodeoxygenation, decarboxylation, and decarbonylation and influences on the distribution of hydrocarbon products [33,34].

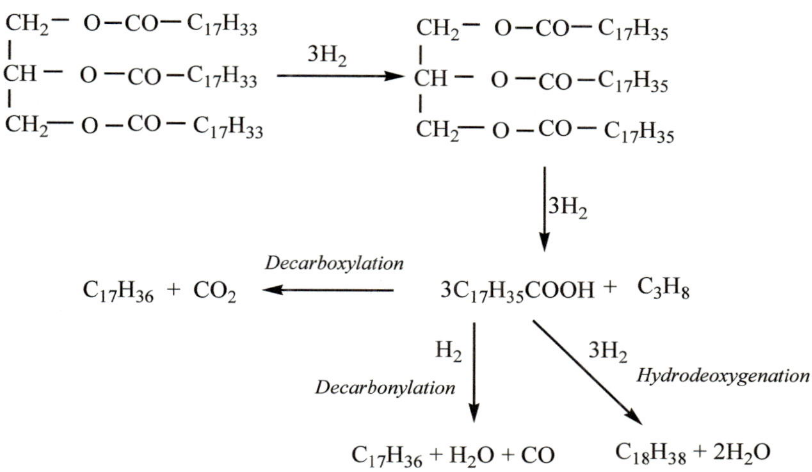


**Figure 6.** Molecular reactions occurring in the hydroconversion of a triglyceride (e.g. triolein)

The production of 2 moles of n-heptadecane, 1 mole of n-octadecane and 1 mole of propane requires 10 moles of hydrogen per each mole of triolein in the feedstock (Viz Figure 6). More unsaturated triglycerides will require a larger amount of hydrogen to produce heptadecane and octadecane. Hydrocarbons having one carbon atom less than the parent fatty acid chain, i.e., pentadecane and heptadecane, are the products of the decarboxylation and decarbonylation (with CO and $CO_2$ as gaseous by-products), whereas the liquid hydrocarbons with the same carbon number as the original fatty acid, i.e., hexadecane and octadecane, are the products of hydrodeoxygenation reaction (with $H_2O$ as secondary product). In most cases, the extent of both pathways is elucidated from the liquid hydrocarbon distributions and knowing the value of the $C_{17}/C_{18}$ ratio is consequently a common way to determine the dominant path of the reaction [16,31,35]. The product distribution is also influenced by the reaction pressure since at a high hydrogen pressure the hydrodeoxygenation will be the preferred pathway. On the other hand, at a lower hydrogen pressure the decarboxylation reaction will be enhanced. Thus, the $CO_2/CO$ ratio in the product distribution could also be used to determine the selectivity for decarboxylation to decarbonylation reactions.

Under reaction conditions it is likely that limitations by diffusion take place, i.e., by hydrogen mass transfer through the stagnant liquid film formed by reactants and/or products on the catalyst surface [36]. Figure 7 shows a probable mechanism in which triglycerides are converted into linear paraffins. In this case, the oil is considered to be composed by triolein, tripalmitin, and trilinolein having distinct unsaturations. As first step, free fatty acids are formed by scission of propane from the glycerol backbone of the triglyceride molecules in presence of hydrogen. Three moles of oleic, palmitic, and linoleic acid are formed, respectively. In the second step, hydrogenation takes place to saturate the oleic and linoleic acids. The side chain of palmitic acid is already completely saturated. Then, the three common reported reactions to eliminate oxygen may occur. Decarbonylation and decarboxylation form hydrocarbons having one carbon atom less than the parent free fatty acid (FFA) whereas hydrodeoxygenation removes the oxygen atom keeping the same carbon atoms as in the original FFA. In this way, the linear paraffins are comprised in the range of $C_{15}$-$C_{18}$.

Thermodynamic data for production of linear $C_{17}$ hydrocarbons from stearic acid at 300 °C indicated that Gibbs free energy and standard enthalpy of reaction for decarboxylation were -83.5 kJ/mol and 9.2 kJ/mol, respectively. Whereas these values for decarbonylation were -7.0 kJ/mol and 179.1 kJ/mol, respectively, and for hydrogenation of -86.1 kJ/mol and -115.0 kJ/mol, respectively. Apart from deoxygenation reactions, it is likely that other reactions, i.e., hydroisomerization, dehydrogenation, and cyclization occur simultaneously. However, the extent of these reactions depends on the catalyst type and hydroprocessing reaction conditions [37]. Formation of alcohols and esters has also been reported to take place during hydroprocessing of triglycerides [38]. The use of sulfided catalysts during hydrodeoxygenation of triglycerides under typical hydrotreating conditions (P=50-100 MPa, T=320-420 °C and $NiMo/Al_2O_3$ or $CoMo/Al_2O_3$ as catalyst) typically produces carboxylic acids as intermediates [39]. However, during deoxygenation, some degree of cracking of the fatty acid chains is carried out in order to form more hydrocarbons from triglycerides in which C-C bond scission is involved to produce lower hydrocarbons. The deoxygenation of tristearin on carbon-supported Ni, Pd, and Pt catalysts produced n-pentadecane through β- and γ- scissions [40].

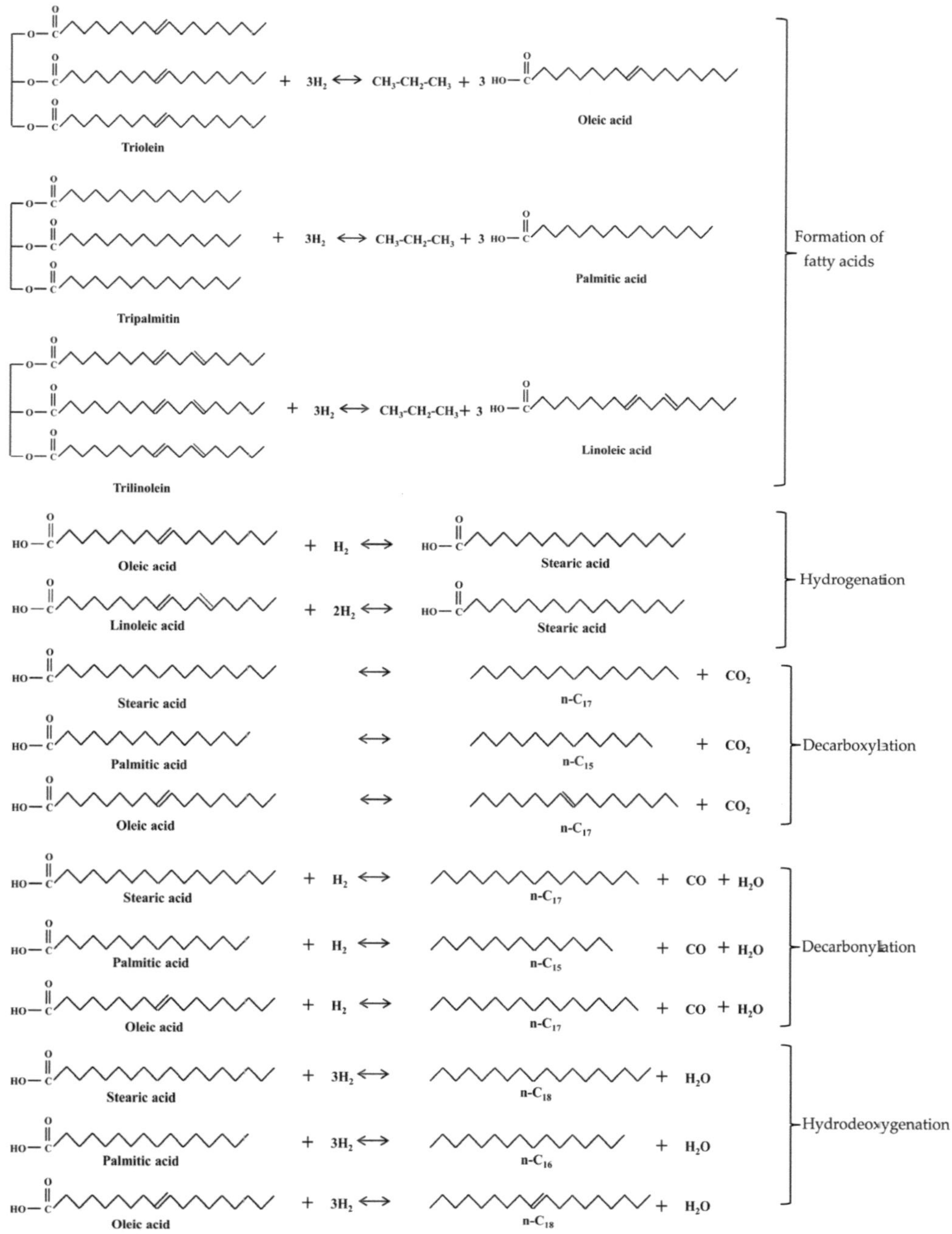


**Figure 7.** Reaction pathways during hydroprocessing of three triglycerides (Adapted from Ref. [38]).

The products from hydrotreating of fats and vegetable oils will rather be linear paraffins. As explained in Section 4.2.1, the hydroisomerization of these n-paraffins is needed to improve the cold properties of green fuels. For hydroisomerization to take place a bifunctional catalyst is needed, thus the saturation of the intermediate iso-olefins is carried out on the metal sites. It is generally acknowledged that isomerization of paraffins takes place by a mechanism involving carbenium ions as intermediate with protonated cyclopropane (PCP) isomerization as the rate determining step. When secondary or tertiary carbocations are formed, cracking of the paraffin by β-scission would also appear. Thus, a catalyst with a suitable high acid activity would both isomerize and crack these carbenium ions on the acid sites yielding after hydrogenation on the metal sites, lighter hydrocarbons such as propane and gasoline range hydrocarbons.

## 6. Catalysts used for the hydrotreating of vegetable oils

There are three methods to crack long chain of hydrocarbons to short chain of hydrocarbons. The first method is known as thermal cracking and occurs with the aid of heat to produce a lighter product. The second process is the catalytic cracking that is conducted in the presence of an acid catalyst without the use of hydrogen and needs less thermal energy than thermal cracking. The third method is known as catalytic hydrocracking and occurs in the presence of a bifunctional catalyst and a high hydrogen pressure. The catalytic hydrocracking process consumes less thermal energy and the presence of hydrogen minimizes coke formation and therefore reduces catalyst deactivation by pore blockage [41].

In spite of having so far optimized the different factors that improve the conversion of triglycerides into green fuels, the properties of the catalyst for upgrading vegetable oils still need to be enhanced. While the current commercial catalysts used in obtaining renewable fuels have been optimized to process petroleum feedstocks, the new catalysts to process vegetable oils may be synthesized taking into account the following considerations: a) high activity toward deoxygenation, b) minimization of coke formation, c) water resistance, d) capability to regenerate in single processes, e) high tolerance to chemical poisons, f) scalability in any commercial process [18].

The first study on hydrocracking of vegetable oils was presented by Nunez [42] in 1984 in his Doctoral thesis and in which he reported the use of rhodium and ruthenium supported catalysts for the hydrocracking of soybean oil in a batch reactor. Two years later, Nunez et al. [43] reported that the hydrocracking of soybean oil was observed to begin at about 400 °C. At this temperature, they observed the decarbonylation/decarboxylation of fatty acids to take place with a marked hydrogenolysis on a bifunctional catalyst, yielding mainly normal paraffins and an overall conversion of 83 wt% (including the gas fraction CO, $CO_2$, $C_1$-$C_4$) under favorable conditions of pressure and temperature. Since then and especially in the last 6 years, many types of catalysts have been used to crack the long chain of hydrocarbons of vegetable oils to produce short chain of hydrocarbons, such as paraffins, olefins and aromatics by catalytic cracking. Several of these studies on hydrocracking of vegetable oil (Viz Table 4) have involved the use of zeolites and conventional hydrotreating catalysts used in petroleum refining as well. The strong acidity of zeolites promotes the hydroisomerization and cracking of triglycerides up to gasoline range hydrocarbons.

Charusiri et al. [44] have studied the catalytic conversion of waste cooking vegetable oil in the temperature range of 380 to 430 ºC and hydrogen pressure about 1-2 MPa with reaction time from 45 to 90 min, using catalysts of zeolite HZSM-5, sulfated zirconia and a hybrid catalyst HZSM-5-sulfated zirconia. Their results showed that the catalyst with the higher conversion to gasoline fuel was the hybrid catalyst with a yield of 26.57 wt% at 430 °C. In the process, three parallel reactions were observed to occur: hydrogenation, hydrodeoxygenation and decarboxylation of carboxylic acids as it is shown in Figure 7. By the hydrodeoxygenation path, n-paraffins with an even number of carbon atom, i.e. n-$C_{16}$ and n-$C_{18}$, corresponding to the side chains of the fatty acids originally present in the triglycerides, were formed along with water and propane. In a similar experiment carried out in a 70 $cm^3$ batch micro-reactor over a temperature range of 400 to 430 °C and reaction time from 30 to 90 min over sulfated zirconia, Charusiri et al. [45] found that longer reaction times than 90 minutes favored the production of light gases and aromatics. They worked at a range of initial hydrogen pressures between 10 and 30 bar and reported a maximum conversion to gasoline at the lower limit of 10 bar. Due to the strong acid sites of the zirconia catalyst the conversion of vegetable oil into gasoline was possible at a relatively low hydrogen pressure. High temperature favored the production of liquid hydrocarbons consisting mainly of gasoline. They concluded that initial hydrogen pressure was responsible for the catalytic cracking step, and hydrogenation and hydrocracking were possible for the cracking and rearrangement to yield light hydrocarbon molecules.

A study to produce biogasoline from palm oil through simultaneous catalytic cracking and hydrogenation reactions was conducted for Nasikin et al. [22], using a batch reactor at atmospheric pressure in presence of hydrogen. The reaction was studied at 300 and 320 °C, while the reaction times were 1, 1.5 and 2 hours for each temperature. A catalyst with cracking and hydrogenation activity, i.e. NiMo/zeolite (klinoptilolite type) was used in a feed/catalyst ratio of 75 wt/wt. Biogasoline containing $C_8$ to $C_{10}$ was produced, with a volumetric yield of 11.93%. Green diesel was also obtained with a volumetric yield of 13.1%.

In another study [46], the hydroprocessing of two mixtures 80:20 and 60:40 wt% of gas oil:sunflower oil, respectively, with sulfided NiO (3%)–$MoO_3$ (12%) as precursors supported on $\gamma$-$Al_2O_3$ incorporating β-zeolite (BEA) in concentrations of 0, 15 or 30 wt% was carried out at 320-350 °C; 3–6 MPa, and weight hourly space velocities (WHSV) of 1–4 $h^{-1}$. The catalyst containing 30 wt% BEA achieved nearly 100% conversion into diesel hydrocarbons, compared to 95.5% conversion obtained by using the Ni–Mo/alumina catalyst without β-zeolite.

Hancsók et al. [24] used Pt/HZSM-22/$\gamma$-$Al_2O_3$ catalysts for the isomerization of pre-hydrogenated sunflower oil for production of diesel over catalysts containing 0.25–1.1% platinum on HZSM-22 at temperatures of 280 to 370 °C, total pressures of 3.5 to 8 MPa and liquid hourly space velocities (LHSV) of 1.0 to 4.0 $h^{-1}$. Under favorable conditions, they obtained a yield of liquid products higher than 90 %, with a cetane number in the range 81–

84, and with a *cold filter plugging point* (CFPP) in between -18 and -14 °C. The ratio of iso- to n-paraffins was in the range 3.7:1–4.7:1. This catalyst seems to be very efficient to produce an excellent green diesel fuel with not only a high cetane number, but also a low freezing point.

Some studies conducted by Simacek et al. [47], showed that hydroprocessing of rapeseed oil can be converted into diesel fuel. The study was carried out at various temperatures (260–340 °C) under a pressure of 7 MPa in a laboratory continuous flow reactor. Three Ni–Mo/alumina catalysts with different concentrations of NiO and $MoO_3$, respectively, i.e. A (3.8 and 17.3 wt%), B (2.6 and 15.7 wt%) and C (2.6 and 8.8 wt%), were used. At reaction temperatures higher than 310 °C, an organic liquid product was obtained containing only hydrocarbons of the same nature as those found in diesel fuel.

Recently, Liu et al. [7] produced bio-hydrogenated diesel by the hydrotreatment of vegetable oils over Ni-Mo based catalysts in a high pressure fixed bed flow reactor system at 350 °C under 4 MPa of hydrogen. *Jatropha curcas L.* oil was converted into paraffins by one step hydrotreatment process. Ni-Mo/$SiO_2$ catalyst favored the production of n-$C_{18}H_{38}$, n-$C_{17}H_{36}$, n-$C_{16}H_{34}$ and n-$C_{15}H_{32}$. These long normal hydrocarbons have a high melting point and thus the liquid hydrocarbon product has poor cold properties (pour point higher than 20 °C). The use of Ni-Mo supported on zeolites, i.e. Ni-Mo/H-Y or Ni-Mo/H-ZSM5, resulted in the production of a large amount of gasoline hydrocarbons.

Sotelo-Boyás et al. [16] studied the hydrocracking of rapeseed oil on three different types of bifunctional catalysts: Pt/H-Y, Pt/H-ZSM-5, and sulfided NiMo/$\gamma$-$Al_2O_3$. Experiments were carried out in a batch reactor over a temperature range of 300 to 400 °C and initial hydrogen pressures from 5 to 11 MPa. The reaction time was limited to 3 h to prevent a high degree of cracking. The Pt-zeolite catalysts had a strong catalytic activity for both cracking and hydrogenation reactions, and therefore a higher severity was required to reach a relatively high oil conversion into liquid hydrocarbons. Among the three catalysts, hydrocracking on Ni-Mo/$\gamma$-$Al_2O_3$ gave the highest yield of liquid hydrocarbons in the boiling range of the diesel fraction, i.e., green diesel, containing mainly n-paraffins from $C_{15}$ to $C_{18}$, and therefore with a high cetane number but poor cold flow properties. While for both zeolitic catalysts, hydrotreating of rapeseed oil produced more iso- than n-paraffins in the boiling range of $C_5$ to $C_{22}$, which included significant amounts of green gasoline.

On the other hand, Liu et al.[48], worked on the production of bio-hydrogenated diesel by hydrotreatment of high-acid-value waste cooking oil over a ruthenium catalyst supported on Al-polyoxocation-pillared montmorillonite at 350 °C and 2 MPa. Free fatty acids and the triglycerides in the waste cooking oil were simultaneously deoxygenated. The predominant hydrocarbon products (98.9 wt %) were n-$C_{18}H_{38}$, n-$C_{17}H_{36}$, n-$C_{16}H_{34}$ and n-$C_{15}H_{32}$.

Silicoaluminophosphates materials (SAPO), have also been tested [2] in the hydroconversion of sunflower oil on a bifunctional Pd/SAPO-31 catalyst as a perspective technological way for single-stage production of hydrocarbons in the diesel fuel range. Transformation of sunflower oil was performed at temperatures of 310 to 360 °C and WHSV of 0.9 to 1.6 $h^{-1}$, under a pressure of 2.0 MPa in a laboratory flow reactor. At temperatures in between 320

and 350 °C, the liquid phase product contained only hydrocarbons, and the main components were identified as $C_{17}$ and $C_{18}$ n- and iso-paraffins. The Pd/SAPO-31 catalyst demonstrated a high initial activity for the hydroconversion of the feed and good isomerization activity, but its deactivation occurred after several hours of operation.

Murata et al. [49] obtained renewable green diesel-type alkanes by hydrotreating jatropha oil at standard hydrotreating conditions (i.e. 270-300 °C, 2 MPa) with Pt/H-ZSM-5 and rhenium-modified Pt/H-ZSM-5 catalysts. The non-modified Pt/H-ZSM-5 was not effective to produce a high amount of $C_{15}$-$C_{18}$ paraffins at high oil/catalyst ratios. Rhenium-modified Pt/H-ZSM-5 catalysts were found to be much more effective for hydrotreating jatropha oil even at a high oil/cat ratio of 10, and 80% conversion and 70% selectivity to $C_{18}$ were achieved. The deoxygenation activity of the catalyst was therefore improved with the addition of rhenium.

The above results show that vegetable oils using hydrocracking catalyst can be converted into liquid fuels and research continues to find the best catalyst that promotes a high selectivity to liquid fuels with high cetane number and good cold flow properties.

As summary, different oil sources, reactor type, reaction conditions, catalysts, and main products obtained during the hydroprocessing of vegetable oils are shown in Table 4.

| Only oil as feedstock | | | | | | |
|---|---|---|---|---|---|---|
| **Oil source** | **Reactor type** | **Reaction conditions** | **Catalyst** | **Main products** | **Performance** | **Ref.** |
| Jatropha | Fixed bed | T=350°C<br>P=4 MPa<br>LHSV=7.6 $h^{-1}$<br>$H_2$/oil ratio=800 $Nm^3/m^3$ | $NiMo/Al_2O_3$-$SiO_2$ | $C_{15}$-$C_{18}$ n-paraffins<br>LPG | Conversion: 100 %<br>Yield: 83.5 wt% | [7] |
| | | | $NiMo/SiO_2$<br>$NiMo/\gamma$-$Al_2O_3$<br>NiMo/H-Y<br>NiMo/H-ZSM-5 | $C_{11}$-$C_{20}$<br>$C_{11}$-$C_{20}$<br>$C_{11}$-$C_{20}$<br>$C_5$-$C_{10}$ | | |
| | Batch | T=270°C<br>P=6.5 MPa<br>t=12 h<br>Catalyst/oil wt ratio=1 | Pt/H-ZSM-5 | $C_{10}$-$C_{20}$ n-paraffins | Conversion:83.8%<br>Yield: 67.7 wt % | [49] |
| | | | Pt/USY | $C_{10}$-$C_{20}$ n-paraffins | Conversion: 100%<br>Yield: 90 wt% | |
| Soybean | Batch | T=400°C<br>P=9.2MPa<br>t=1 h<br>Catalyst/oil wt ratio=0.044, 0.088 | $NiMo/\gamma$-$Al_2O_3$ | $C_{15}$-$C_{18}$ n-paraffins | Conversion:92.9%<br>Yield $C_{15}$-$C_{18}$: 64.45 wt % | [50] |
| | | | $Pd/\gamma$-$Al_2O_3$ | $C_{15}$-$C_{17}$ n-paraffins | Conversion:91.9%<br>Yield $C_{15}$-$C_{17}$ : 79.22 wt% | |

| Only oil as feedstock | | | | | | |
|---|---|---|---|---|---|---|
| **Oil source** | **Reactor type** | **Reaction conditions** | **Catalyst** | **Main products** | **Performance** | **Ref.** |
| | | | CoMo/ γ-$Al_2O_3$ | $C_{15}$-$C_{17}$ n-paraffins | Conversion:78.9% Yield $C_{15}$-$C_{17}$: 33.67 wt% | |
| | | | Ni/$Al_2O_3$-$SiO_2$ | $C_{15}$-$C_{17}$ n-paraffins | Conversion:60.8% Yield $C_{15}$-$C_{17}$: 39.24 wt% | |
| | | | Pt/ γ-$Al_2O_3$ | $C_{15}$-$C_{17}$ n-paraffins | Conversion:50.8% Yield $C_{15}$-$C_{17}$: 37.71 wt % | |
| | | | Ru/ γ-$Al_2O_3$ | $C_{15}$-$C_{17}$ n-paraffins | Conversion:39.7% Yield: 32.00 wt% | |
| | Batch | T=350°C P=0.7 MPa $N_2$ t=4 h Stirring rate=1000 rpm | Ni/$Al_2O_3$ | ≥ $C_{18}$ | Conversion: 68% Yield ≥ $C_{18}$: 51.20 wt% | [40] |
| | | | NiAl/LDH | $C_8$-$C_{17}$ | Conversion :74% Yield $C_8$-$C_{17}$: 52.90 wt% | |
| | | | MgAl/LDH | $C_8$-$C_{17}$ | Conversion: 72% Yield $C_8$-$C_{17}$:47.80 wt% | |
| Rapeseed | Fixed bed | T=340°C P=4.0 MPa LHSV=1 $h^{-1}$ $H_2$/oil ratio=500-1000 $Nm^3/m^3$ | NiMo/γ-$Al_2O_3$ | $C_{15}$-$C_{18}$ n-paraffins | Conversion:93% Yield $C_{15}$-$C_{18}$ n-paraffins: 54.52 wt% | [51] |
| | | | NiW/$TiO_2$ NiMo/$TiO_2$ NiW/$ZrO_2$ NiW/NaY | | | |
| | Batch | T=300-400°C P=5-11 MPa t=3 h | NiMo/$Al_2O_3$ | $C_7$-$C_{18}$ n-paraffins | Yield: 70-80 % wt% | [16] |
| | | | Pt/H-Y Pt/H-ZSM-5 | $C_5$-$C_{22}$ n- and i-paraffins | Yield: 20-40 % | |
| Sunflower | Fixed bed | T=360-420°C P=18 MPa Oil flow=49 g/h $H_2$ flow=0.049 $Nm^3/h$ | Sulfided catalyst (not specified) | $C_{15}$-$C_{20}$ n- and iso-paraffins | Yield: 64.7 wt% (360 °C) | [27] |

| Only oil as feedstock | | | | | | |
|---|---|---|---|---|---|---|
| **Oil source** | **Reactor type** | **Reaction conditions** | **Catalyst** | **Main products** | **Performance** | **Ref.** |
| | Fixed bed | T=360-380°C<br>P=6-8 MPa<br>LHSV=1.0-1.2 $h^{-1}$<br>$H_2$/oil ratio=450 $Nm^3/m^3$ | $CoMo/Al_2O_3$<br><br>$NiMo/Al_2O_3$<br><br>$NiW/Al_2O_3$ | Gases<br>$C_5$-$C_{10}$ n-paraffins<br>$C_{11}$-$C_{19}$ n- and iso-paraffins | Conversion: 94-99.8<br>Yield: 63.1-71.5 wt%<br><br>Conversion: 81.8-97.4%<br>Yield: 42-51.9 wt%<br><br>Conversion: 86.7-95.6 %<br>Yield: 9.4%-49.3% | [28] |
| | Fixed bed | T=350-370°C<br>P=2-4 MPa<br>LHSV=1.0 $h^{-1}$<br>$H_2$/oil ratio=500 $Nm^3/m^3$ | $NiMo/Al_2O_3$-F | $C_{15}$-$C_{18}$ n-paraffins | Yield: 73.2-75.6 wt% | [15] |
| | Fixed bed | T=380°C<br>P=4-6 MPa<br>LHSV=1.0 $h^{-1}$<br>$H_2$/oil ratio=500-600 $Nm^3/m^3$ | $CoMo/Al_2O_3$ | $C_{14}$-$C_{19}$ n-paraffins | Conversion: 100%<br>Yield: 73.7-73.9% | [52] |
| | Fixed bed | T=310-360°C<br>P=2 MPa<br>LHSV=0.9-1.2 $h^{-1}$<br>$H_2$/oil ratio=1000 $Nm^3/m^3$ | Pd/SAPO-31 | $C_{16}$-$C_{18}$ n- and iso-paraffins | Yield: 89.3-73.4 wt% | [2] |
| Safflower | Batch | T=340°C<br>P=0.98 MPa $H_2$-$N_2$<br>$H_2$-$N_2$ ratio: 1:4<br>Cat./oil wt ratio= 0.0025<br>t=3 h | Pt/SBA-15<br>NiMoC/SBA-15<br>CoMoC/SBA-15<br>Triflic acid/SBA-15 | Gases<br>$C_5$-$C_9$ n-paraffins<br>$C_{10}$-$C_{14}$ n-paraffins<br>$C_{15}$-$C_{23}$ n-paraffins | Conversion: 25%<br><br>Conversion: 99% | [53] |
| Palm | Fixed bed (pilot plant) | T=350°C<br>P=4-9 MPa<br>LHSV=2 $h^{-1}$<br>TOS=0-14 days | $NiMo/Al_2O_3$ | $C_{16}$-$C_{18}$ n-paraffins | Molar yield: 100% | [38] |

| Oil source | Reactor type | Reaction conditions | Catalyst | Main products | Performance | Ref. |
|---|---|---|---|---|---|---|
| **Only oil as feedstock** | | | | | | |
| | Batch | T=300-320°C<br>Atm. pressure<br>t=1-2 h<br>Catalyst/oil wt ratio=0.0133 | NiMo/zeolite (klinoptilolite type) | $C_8$-$C_{19}$ n-paraffins | Vol. yield: 11.93 % | [22] |
| | Fixed bed | T=450°C<br>WHSV=2.5 $h^{-1}$ | HZSM-5<br>MCM-41 | Naphtha, kerosene, and diesel-range paraffins | | [54] |
| Waste cooking oil | Fixed bed | T=350°C<br>P=2 MPa<br>LHSV=15.2 $h^{-1}$<br>$H_2$/oil ratio=400 $Nm^3/m^3$ | Ru/Al13-montmorillonite | $C_{15}$-$C_{18}$ n-paraffins | Conversion: 100%<br>Yield: 82.1-84% | [48] |
| | Batch | T=380-430°C<br>P=1-2 MPa<br>Catalyst/oil wt ratio=0.0028<br>t=0.75-1.5 h | HZSM-5<br>Sulfated zirconia | Naphtha, kerosene, and diesel-range paraffins | Yield:79.17 wt% | [44] |
| **Mixture of gas oil + oil as feedstock** | | | | | | |
| Rapeseed | Fixed bed | T=300°C<br>P=4.5 MPa<br>LHSV=1.5 $h^{-1}$<br>$H_2$/oil ratio=250 $Nm^3/m^3$ | $NiMo/Al_2O_3$ | $C_{15}$-$C_{18}$ n-paraffins | Conversion: 97% | [33] |
| | Fixed bed | T=400-420°C<br>P=18 MPa<br>WHSV=1 $h^{-1}$<br>$H_2$/oil ratio=1000 $Nm^3/m^3$ | $NiMo/Al_2O_3$ | $C_{15}$-$C_{20}$ n-paraffins | At 420 °C<br>Yield: 55.4 wt % | [55] |
| Palm | Fixed bed | T=310-350°C<br>P=3.3 MPa<br>WHSV=0.7-1.4 $h^{-1}$<br>$H_2$/oil ratio=1500 $Nm^3/m^3$ | $CoMo/Al_2O_3$ | Diesel-range paraffins | Conversion 100% | [56] |
| | | T = 300-320 ºC,<br>P = 1 atm | NiMo/zeolite klinoptilolite type | Diesel-range paraffins | Yield: 11.93 wt% | [22] |
| Sunflower | Fixed bed | T=360-380°C<br>P=8 MPa<br>LHSV=1 $h^{-1}$<br>$H_2$/oil ratio=600 $Nm^3/m^3$ | $NiMo/Al_2O_3$ | $C_{11}$-$C_{22}$ n-paraffins | Yield: 82-90 wt% | [32] |

| Mixture of gas oil + oil as feedstock | | | | | | |
|---|---|---|---|---|---|---|
| **Oil source** | **Reactor type** | **Reaction conditions** | **Catalyst** | **Main products** | **Performance** | **Ref.** |
| | Fixed bed | T=350°C<br>P=13.8 MPa<br>LHSV=1.5 $h^{-1}$<br>$H_2$/oil ratio=1060 $Nm^3/m^3$ | Hydrotreating catalyst (not specified) | Diesel-range paraffins | Conversion: 85 %<br>Yield: 42 wt% | [57] |
| | Fixed bed | T=350°C<br>P=6.9 MPa<br>LHSV=1.5 $h^{-1}$<br>$H_2$/oil ratio=1068 $Nm^3/m^3$ | Hydrocracking commercial catalysts (not specified) | Naphtha, kerosene, and diesel-range paraffins | Conversion catalyst B: 64 %<br>Conversion catalyst C: 37.5 % | [20] |
| | Fixed bed | T=300-450°C<br>P=5 MPa<br>LHSV=4.97 $h^{-1}$<br>$H_2$/oil ratio=1600 $Nm^3/m^3$ | $NiMo/Al_2O_3$ | $C_{15}$-$C_{18}$ n-paraffins | Yield: 54 – 75wt% | [35] |
| | Fixed bed | T=320-350°C<br>P=3-6 MPa<br>WHSV=1-4 $h^{-1}$<br>$H_2$/oil ratio=1068 $Nm^3/m^3$ | $NiMo/Al_2O_3$-β-zeolite | $C_{17}$-$C_{18}$ n-paraffins | Conversion: >90% | [46] |
| Soybean | Fixed bed | T=340-380°C<br>P=5 MPa<br>LHSV=2.4 $h^{-1}$<br>$H_2$/oil ratio = 1500 $Nm^3/m^3$ | $NiW/Al_2O_3$-$SiO_2$<br>$NiMo/Al_2O_3$ | $C_{15}$-$C_{20}$ n-paraffins | Yield: 85-95 wt% | [31] |
| Waste cooking oil | Batch | T=325°C<br>P=2 MPa<br>Stirring=900 rpm<br>t=1, 2, 5, 20 h | Pt/ γ-$Al_2O_3$<br><br>Ni/ γ-$Al_2O_3$<br><br>Pt/γ-$Al_2O_3$ | $C_{15}$-$C_{18}$ n-paraffins | Conversion: 100%<br>Yierld:60 molar%<br><br>Conversion: 76.8%<br><br>Conversion: 100% | [58] |
| | Fixed bed | T=330-398°C<br>P=8.3 MPa<br>LHSV=1 $h^{-1}$<br>$H_2$/oil ratio=710 $Nm^3/m^3$ | Hydrotreating catalyst (not specified) | $C_8$-$C_{29}$ n- and iso-paraffins | At 350°C<br>Conversion: 72.62%<br>Yield: 71 wt%<br><br>At 390 °C<br>Conversion 81.88% | [59] |

**Table 4.** Summary of type of oil, reactor, reaction conditions, catalyst, and main products obtained during hydroprocessing of pure oil and oil plus gas oil mixture (Adapted from several references).

## 7. Current commercial processes

During the last 10 years, several petroleum related companies have had an increasing interest in producing renewable green liquid fuels from the hydroprocessing of various lipid feedstocks, and just recently the developed technology has started to be commercialized.

UOP Honeywell Co. with its 90 years of refining technology experience is offering an alternative process to produce green fuels from various biofeedstocks. The UOP/ENI Ecofining process [29] has been designed to convert non-edible second generation natural oils to green diesel, which can be used in any percentage in the existing fuel tanks. Since green diesel has the same chemical properties of petro-diesel, it can be used in today's tanks, pipelines, trucks, pumps and automobiles without major infrastructure changes. UOP have agreements in place with China, India and the United Arab Emirates for biofuel development projects that will utilize Ecofining process along with green jet fuel technology to introduce new energy solutions and develop new biofuel economies [29]. The UOP/Eni Ecofining process is based on the hydrotreating of triglycerides along with free fatty acids, which results in deoxygenation via hydrodeoxygenation and decarboxylation [60,61]. After the hydrotreating step, an isomerization process is performed to produce an isoparaffin rich diesel fuel to obtain a fuel with good cold flow properties. In a life cycle analysis reported in a 2009 [25], the renewable diesel produced by the Ecofining process has been reported to be economically and environmentally competitive with biodiesel production.

Haldor Topsøe, a Danish catalyst company, has also developed a new hydrotreating technology for production of green diesel and jet fuel from raw tall. In contrast to other feedstocks used for renewable diesel production, tall is a non-edible material, and thus the process does not raise the problems of global food shortage. The basic engineering for applying the process in Preem Gothenburg Sweden Refinery has been completed by Topsøe[62].

The Neste Oil Co. has claimed to have developed a technology capable of producing high-quality diesel fuel from vegetable oils and animal fats as well. The NExBTL process produces renewable diesel by hydrotreating vegetable oils or waste fats, resulting in a 40-80% lifecycle reduction in $CO_2$ depending on the feedstock. Neste Oil recently opened a plant in Singapore using NExBTL technology that is intended to produce more than 800,000 tons per year of renewable diesel from feedstocks such as palm oil and waste animal fat [63]. Neste Oil's own tests, together with those carried out by engine and automotive manufacturers, have shown that NExBTL renewable synthetic diesel-based fuel performs very well in both car and truck engines [64], as it is shown by the following properties:

- A high cetane value in between 84 and 99 means that NExBTL can be used to upgrade off-spec diesel fuel and make it suitable for on-road use; and also to create a superior diesel product.
- A low cloud point value (as low as 30 °C below zero) allows NExBTL to be used year round, even in the colder northern states and Canada, without any concern.
- NExBTL is a stable hydrocarbon that can be stored for extended periods.

Another company is Tyson Foods, Inc., which is taking a strategic step in its quest to be a leader in renewable energy. Tyson and Syntroleum Corporation, a based synthetic fuels technology company, have developed a technology to produce synthetic fuels targeting the renewable diesel, jet, and military fuel markets. Dynamic Fuels, which is a joint venture between Syntroleum and Tyson Foods, recently opened a plant in Geismar, LA USA for the production of renewable diesel through hydrotreating from non-food grade animal fats. The animal fat includes beef tallow, pork lard and chicken fat. Syntroleum reported that the carbon footprint of the renewable diesel produced at the plant is 75% less than that of petroleum diesel. The plant was designed to produce up to 75 million gallons of renewable diesel fuel per year [65].

Valero Energy Corporation announced that a new plant at St. Charles Refinery in Norco LA., will be utilized to convert used cooking oil and animal fat into renewable diesel via hydrogenation and isomerization processes [66]. The fat and oil feedstocks will be supplied by Darling International, who is Valero's partner in the venture [67].

ConocoPhillips has begun commercial production of renewable diesel fuel at the company's Whitegate Refinery in Cork, Ireland. The production process, developed by ConocoPhillips, hydrogenates vegetable oils to produce a renewable diesel fuel component that meets European Union standards. The refinery is producing 1,000 barrels per day of renewable diesel fuel for sale into the Irish market. The fuel is produced with vegetable oil as feedstock using existing equipment at the refinery, and is being blended and transported with petro-diesel. In December 2006, ConocoPhillips conducted a successful renewable diesel production test at Whitegate. Soybean oil was intended to be the primary feedstock, although the plant can use other vegetable oils and animal fats. It has been reported that this renewable diesel burns cleaner than conventional diesel, it has a lower sulfur content and emits less nitrous oxide [68,69]

Toyota Motor Corporation (TMC), Hino Motors, the Tokyo Metropolitan Government and Nippon Oil Corporation (NOC) started a joint project aimed at commercializing what they are calling bio hydrofined diesel (BHD), a second-generation renewable diesel fuel produced by hydrogenating a vegetable oil feedstock. Nippon Oil and Toyota have worked jointly on the development of BHD technology since 2005. At the 16$^{th}$ Annual Catalysts in Petroleum Refining & Petrochemicals symposium in Saudi Arabia, Dr. Akira Koyama of Nippon Oil presented NOC's findings on the reactivity, distillate yields, evaluation of the fuel (now called BHD) and its applicability as an automotive fuel. The use of refinery-based hydrogenation processes to produce a synthetic, second-generation renewable diesel is driven by several issues, including some technical considerations over the properties and effects of first-generation fatty acid methyl ester biodiesel (storage, oxidation, possible effect on fuel handling systems). In its studies, Nippon Oil explored reaction temperatures ranging from 240 °C to 360 °C, with reaction pressures of 6 MPa and 10 MPa, and used a common hydrodesulfurization catalyst. The resulting fuel is claimed to be aromatics- and sulfur-free, with a cetane number of 101 [70].

| Technology | Process | Feedstock | Product |
| --- | --- | --- | --- |
| UOP/Eni (Ecofining process) | Hydrotreating and Isomerization | Triglycerides and/or free fatty acids | Green diesel and jet fuel |
| Haldor Topsøe | Hydrotreating | Raw tall oil | Green diesel and jet fuel |
| The Neste Oil. (NExBTL process) | Hydrotreating | Palm oil and waste animal fat | Green diesel |
| Tyson Foods Inc. and Syntroleum Corporation | Hydrotreating | Animal fat includes beef tallow, pork lard, chicken fat and greases | Green diesel and jet fuel |
| Valero Energy Corporation | Hydrogenation and Isomerization | Used cooking oil and animal fat | Green diesel |
| ConocoPhillips | Hydrotreating | Vegetable oil like soybean oil, grape seed oil and other vegetable oils and animal fats | Green diesel |
| Toyota Motor Corporation, Hino Motors, Nippon Oil Corporation. | Hydrotreating | Vegetable oils | Green diesel |

**Table 5.** Main commercial processes to produce green fuels.

## 8. Conclusions

Hydroprocessing of vegetable oils and fats in green transportation fuels is a prominent technology that will probably continue to develop in the present decade, as different countries or petroleum companies realize that it is both an economical and environmentally efficient way to produce energy and decrease the accelerated consumption of fossil fuels. Green fuels are currently not a substitute of fossil fuels, but they can be used effectively as additives to boost the properties of their corresponding fossil fuels. The use of green jet fuel is projected to increase as the European Union introduces strict environmental regulations in the use of transportation fuels.

Technologically there are some factors that need to take into consideration when dealing with different raw materials. The catalyst for instance plays an important role in the reaction process, and it is a key point to optimize the process. In this context, the use of simulation and in general computer-aided tools can be used for optimization of the hydroconversion process of vegetable oils. This however needs the use of kinetics models, which at the same time requires a deep understanding of the chemistry involved in hydrocracking, hydroisomerization and hydrodeoxygenation. This is clearly not an easy task, and in spite some authors [45,71,72] have derived kinetic models for oxygenated molecules, they do not represent a comprehensive kinetics occurring in the hydrocracking of vegetable oils. A deeper study on the hydrocracking chemistry and the subsequent development of a suitable kinetic model seems the way to follow for future work in this outstanding technology.

## Author details

Rogelio Sotelo-Boyás
*Instituto Politécnico Nacional, ESIQIE. UPALM Col. Zacatenco, México D.F., México*

Fernando Trejo-Zárraga
*Instituto Politécnico Nacional, CICATA-Legaria. Col. Irrigación, México D.F., México*

Felipe de Jesús Hernández-Loyo
*Universidad del Istmo, Department of Petroleum Engineering, Tehuantepec, Oaxaca, México*

## Acknowledgement

Erika de la Rosa, Alonso Licona and Rubi Morales from ESIQIE-IPN are acknowledged for their kind support in the updating of data for Table 4.

## 9. References

[1] Huber GW, Iborra S, Corma A. Synthesis of Transportation Fuels from Biomass: Chemistry, Catalysts, and Engineering. Chem. Rev. 2006;106 (9):4044-98.

[2] Kikhtyanin OV, Rubanov AE, Ayupov AB, Echevsky GV. Hydroconversion of sunflower oil on Pd/SAPO-31 catalyst. Fuel 2010;89 (10):3085-92.

[3] Benazzi AE, Cameron C. Boutique diesel in the on-road Market. Hydrocarbon Processing 2006;85:DD17-DD8.

[4] Kent J, editor. Kent and Riegel's Handbook of Industrial Chemistry and Biotechnology. New York, USA: Springer - Verlag, 2007.

[5] AIi MF, El AIi BM, Speight JG. Handbook of Industrial Chemistry: McGraw-Hill, 2005.

[6] Kubátova A, Luo Y, Šťávová J, Sadrameli SM, Aulich T, Kozliak E, Seames W. New Path in the Thermal Cracking of Triacylglycerols (Canola and Soybean Oil). Fuel 2011;90:2598-608.

[7] Liu Y, Sotelo-Boyás R, Murata K, Minowa T, Sakanishi K. Hydrotreatment of Vegetable Oil to Produce Bio-hydrogenated Diesel and Liquefied Petroleum Gas Fuel over Catalysts Containing Sulfided Ni-Mo and Solid Acids. Energy & Fuels 2011;25:4675-85.

[8] Snare M, Kubicková I, Maki-Arvela P, D. C, Eranen K, Murzin DY. Catalytic deoxygenation of unsaturated renewable feedstocks for production of diesel fuel hydrocarbons. Fuel 2008;87 (6):933-45.

[9] Haas MJ. Improving the Economics of Biodiesel Production through the Use of Low Value Lipids as Feedstocks: Vegetable Oil Soapstock. Fuel process. technol. 2005;86:1087-96.

[10] Lee DS, Noh BS, Bae SY, Kim K. Characterization of fatty acids composition in vegetable oils by gas chromatography and chemometrics. Analytica Chimica Acta 1998 (358):163-75.

[11] Hawash S, El Diwani G, Abdel EK. Optimization of Biodiesel Production from Jatropha Oil By Heterogeneous Base Catalysed Transesterification. Int. J. Eng. Sci.Tech. 2011;3 (6):5242-51.

[12] Balakos MW, Hernandez EE. Catalyst characteristics and performance in edible oil hydrogenation. Catalysis Today 1997;35 (4):415-25.

[13] Ancheyta J, Trejo F, Rana MS. Asphaltenes: Chemical Transformation during Hydroprocessing of Heavy Oils. New York: CRC Press, 2009.

[14] Lestari S, Mki-Arvela P, Bernas H, Simakova O, Sjholm R, Beltramini J, Lu GQM, Myllyoja J, Irina Simakova I, Murzin DY. Catalytic Deoxygenation of Stearic Acid in a Continuous Reactor over a Mesoporous Carbon-Supported Pd Catalyst. Energy & Fuels 2009;23 (8):3842-5.

[15] Kovacs S, Kasza T, A. T, Horbath IW, Hancsók J. Fuel production by hydrotreating of triglycerides on NiMo/$Al_2O_3$/F catalyst. Chem. Eng. J. 2011;In press: doi:10.1016/j.cej.2011.05.110.

[16] Sotelo-Boyás R, Liu Y, Minowa T. Renewable Diesel Production from the Hydrotreating of Rapeseed Oil with Pt/Zeolite and NiMo/$Al_2O_3$ Catalysts. Ind. Eng. Chem. Res. 2011;50 (5):2791-9.

[17] http://www.hydrocarbonpublishing.com/ReportP/ht.php. Future Roles of FCC and Hydroprocessing Units in Modern Refineries. Hydrocarbon Publishing Company, 2010.

[18] Choudhary TV, Phillips CB. Renewable fuels via catalytic hydrodeoxygenation. Applied Catalysis A: General 2011;397 (1-2):1-12.

[19] Mikkonen S. Second-generation renewable diesel offers advantages. Hydrocarbon Processing 2008;87 (2):63-6.

[20] Bezergianni S, Kalogianni A, Vasalos IA. Hydrocracking of vacuum gas oil-vegetable oil mixtures for biofuels production. Bioresource Technology 2009;100 (12):3036-42.

[21] Baker EG, Elliot DC. Method for Upgrading Oils Containing Hydrocarbon Compounds to Highly Aromatic Gasoline. US Patent 5,180,868, 1993.

[22] Nasikin M, Susanto BH, Hirsaman A, Wijanarko A. Biogasoline from Palm Oil by Simultaneous Cracking and Hydrogenation Reaction over NiMo/zeolite. Catalyst. World Appl. Sci. J. 2009;5:74-9.

[23] Farizul HK, Amin NAS, Suhardy D, Saiful AS, Mohd NS. Catalytic Conversion of RBD Palm Oil to Gasoline: The Effect of Silica-Alumina Ratio in HZSM-5. 1st international conference on natural resources engineering and technology. Putrajaya, Malaysia, 2006. pp. 262-73.

[24] Hancsók J, Krár M, Magyar S, Boda L, Holló A, Kalló D. Investigation of the production of high cetane number bio gas oil from pre-hydrogenated vegetable oils over Pt/HZSM-22/ $Al_2O_3$. Microporous and Mesoporous Materials 2007;101 (1-2):148-52.

[25] Kalnes TN, Koers KP, Marker T, Shonnard DR. A Technoeconomic and Environmental Life Cycle Comparison of Green Diesel to Biodiesel and Syndiesel. Environmental Progress & Sustainable Energy 2009;28 (1):111-20.

[26] Krar M, Kovacs S, Boda L, Leveles L, A. T, Horbath IW, Hancsók J. Fuel purpose hydrotreating of vegetable oil on NiMo/g-Al2O3 catalyst. Hungarian Journal of Ind. Chem. Veszprem 2009;37 (2):107-11.

[27] Simacek P, Kubicka D, Kubicková I, Homola F, Pospisil M, Chudoba J. Premium quality renewable diesel fuel by hydroprocessing of sunflower oil. Fuel 2011;90 (7):2473-9.

[28] Hancsók J, Kasza T, Kovacs S, Solymosi P, Holló A. Production of Bioparaffins by the Catalytic Hydrogenation of Natural Triglycerides. J. Cleaner Prod. 2012;In press. DOI: 10.1016/j.jclepro.2012.01.036.

[29] UOP. Green Diesel. http://www.uop.com/processing-solutions/biofuels/green-diesel/. Last accessed: February 5th, 2012.

[30] Kikhtyanin OV, Toktarev AV, Echevsky GV. Hydrotreatment of Diesel Feedstock over Pt-SAPO-31 Catalyst: from Lab to Pilot Scale. In: Gedeon A, Massiani P, Babonneau F, editors. Zeolites and Related Materials: Trends, Targets and Challenges. The Netherlands: Elsevier, 2008. pp. 1227-30.

[31] Tiwari R, Rana BS, Kumar R, Verma D, Kumar R, Joshi RK, Garg MO, Sinha AK. Hydrotreating and Hydrocracking Catalysts for Processing of Waste Soya-Oil and Refinery-Oil Mixtures. Catal. commun. 2011;12:559-62.

[32] Tóth C, Baladincz P, Kovács S, Hancsók J. Producing Clean Diesel Fuel by Co-hydrogenation of Vegetable Oil with Gas Oil. Clean technol. environ. policy 2011;13:581-5.

[33] Donnis B, Gottschalck R, Blom EP, Knudsen KG. Hydroprocessing of Bio-Oils and Oxygenates to Hydrocarbons. Understanding the Reaction Routes. Topics in Catalysis 2009;52:229-40.

[34] Huber GW, Corma A. Synergies between Bio- and Oil Refineries for the Production of Fuels from Biomass. Angew. Chem. Int. Ed. 2007;46:7184-201.

[35] Huber GW, O'Connor P, Corma A. Processing biomass in conventional oil refineries: production of high quality diesel by hydrotreating vegetable oils in heavy vaccum oil mixtures. Applied Catalysis A: General 2007 (329):120-9.

[36] Smejkal Q, Smejkalová L, Kubicka D. Thermodynamic balance in reaction system of total vegetable oil hydrogenation. Chemical Engineering Journal 2009;146 (1):155-60.

[37] Snare M, Kubicková I, Maki-Arvela P, Eranen K, Murzin DY. Heterogeneous Catalytic Deoxygenation of Stearic Acid for Production of Biodiesel. Ind. Eng. Chem. Res. 2006;45:5708-15.

[38] Guzman A, Torres JE, Prada LP, Nunez ML. Hydroprocessing of crude palm oil at pilot plant scale. Catalysis Today 2010;156 (1-2):38-43.

[39] Vonghia E, Boocock DGB, Konar SK, Leung A. Pathways for the Deoxygenation of Triglycerides to Aliphatic Hydrocarbons over Activated Alumina. Energy & Fuels 1995;9:1090-6.

[40] Morgan T, Santillan-Jimenez E, Harman-Ware AE, Ji Y, Grubb D, Crocker M. Catalytic Deoxygenation of Triglycerides to Hydrocarbons over Supported Nickel Catalysts. Chem. Eng. J. 2012;In press. doi: 10.1016/j.cej.2012.02.27.

[41] Satterfield CN. Heterogeneous Catalysis in Industrial Practice. New York:: McGraw-Hill, 1991.

[42] Nunes PP. Hydrocraquage de l'huile de soja sur des catalyseurs au rhodium et au ruthenium supportes. Paris: Université Pierre et Marie Curie, 1984.

[43] Nunes PP, Brodzki D, Bugli G, Djega-Mariadassou G. Soybean Oil Hydrocracking under Pressure: Process and General Aspect of the Transformation. Revue de L'Institute Francais du Pétrole 1986;41 (3):421-31.

[44] Charusiri W, Yongchareon W, Vitidsant T. Conversion of used vegetable oils to liquid fuels and chemicals over HZSM-5, sulfated zirconia and hybrid catalysts. Korean J. Chem. Eng. 2006;23 (3):349-55.

[45] Charusiri W, Vitidsant T. Kinetic Study of Used Vegetable Oil to Liquid Fuels over Sulfated Zirconia. Energy & Fuels 2005;19:1738-89.

[46] Sankaranarayanan TM, Banu M, Pandurangan A, Sivasanker S. Hydroprocessing of Sunflower Oil-Gas oil Blends Over Sulfided Ni-Mo-Al-Zeolite Beta Composites,. Bioresource Technology 2011;102:10717-23.

[47] Simacek P, Kubicka D, Sebor G, Pospisil M. Hydroprocessed rapeseed oil as a source of hydrocarbon-based biodiesel. Fuel 2009;88 (3):456-60.

[48] Liu Y, Sotelo-Boyás R, Murata K, Minowa T, Sakanishi K. Production of Bio-hydrogenated Diesel by Hydrotreatment of High-Acid-Value Waste Cooking Oil over Ruthenium Catalyst Supported on Al-polyoxocation-Pillared Montmorillonite. Catalysts 2012;2 (1):171-90.

[49] Murata K, Liu Y, Inaba M, Takahara I. Production of Synthetic Diesel by Hydrotreatment of Jatropha Oils Using Pt-Re/H-ZSM-5 Catalyst. Energy Fuels 2010;24 (4):2404-9.

[50] Veriansyah B, Han JY, Kim SK, Hong SA, Kim YJ, Lim JS, Shu YW, Oh SG, Kim J. Production of Renewable Diesel by Hydroprocessing of Soybean Oil: Effect of Catalysts. Fuel 2011;94: 578–85.

[51] Mikulec J, Cvengroš J, Joríková L, Banic M, Kleinová A. Second Generation Diesel Fuel from Renewable Sources. J. Cleaner Prod. 2010;18:917-26.

[52] Krár M, Kovacs S, Kalló D, Hancsók J. Fuel Purpose Hydrotreating of Sunflower Oil on CoMo/ $Al_2O_3$. Catalyst. Bioresour. Technol. 2010;101:9287-93.

[53] Barrón AE, Melo-Banda JA, Dominguez JM, Hernández E, Silva R, Reyes AI, Meraz MA. Catalytic Hydrocracking of Vegetable Oil for Agrofuels Production using Ni-Mo, Ni-W, Pt and TFA Catalysts Supported on SBA-15. Catalysis Today 2011;166:102-10.

[54] Sang OY, Twaiq F, Zakaria R, Mohamed AR, Bhatia S. Biofuel Production from Catalytic Cracking of Palm Oil. Energy Sources 2003;25:859-69.

[55] Simacek P, Kubicka D. Hydrocracking of petroleum vacuum distillate containing rapeseed oil: Evaluation of diesel fuel. Fuel 2010;In Press.

[56] Templis C, Vonortas A, Sebos I, Papayannakos N. Vegetable oil effect on gasoil HDS in their catalytic co-hydroprocessing. Applied Catalysis B: Environmental 2011;104 (3-4):324-9.

[57] Lappas AA, Bezergianni S, Vasalos IA. Production of biofuels via co-processing in conventional refining processes. Catalysis Today 2009;145 (1-2):55-62.

[58] Madsen AT, Ahmed EH, C.H. C, Fehrmann R, Riisager A. Hydrodeoxygenation of waste fat for diesel production: Study on model feed with Pt/alumina catalyst. Fuel 2011;In Press, doi: 10.1016/j.fuel.2011.06.005.

[59] Bezergianni S, Dimitriadis A, Kalogianni A, Pilavachi PA. Hydrotreating of Waste Cooking Oil for Biodiesel Production. Part I: Effect of Temperature on Product Yields and Heteroatom Removal. Bioresource Technology 2010;101:6651-6.

[60] Kalnes TN, Marker T, Shonnard DR. Green diesel: A second generation biofuel. International Journal of Chemical Reactor Engineering 2007;5 (A48).

[61] Petri JA, Marker TL. Production of diesel fuel from biorenewable feedstocks. US, 2009.

[62] TOPSOE Renewable Fuels. (http://www.topsoe.com/business_areas/refining/Renewable_fuels.aspx). Last accessed April 4th, 2012.

[63] United Press International. Singapore opens renewable diesel plant. http://www.upi.com/Science_News/Resource-Wars/2011/03/09/Singapore-opens-renewable-dieselplant/UPI-74561299678336/. Last accessed February 23rd, 2012.

[64] Neste Oil. NExBTL Diesel http://www.nesteoil.com/default.asp?path=1,41,11991,12243,12335 .Last accessed April 4th, 2012.

[65] Environmental Leader. Tyson Foods, Syntroleum Partner to Turn Grease into Fuel. http://www.environmentalleader.com/2010/11/09/tyson-foods-syntroleum-partner-to-turn-grease-into-fuel/ Last accessed April, 2012.

[66] New Orleans Net. New renewable diesel plant headed to Norco with federal backing. http://www.nola.com/politics/index.ssf/2011/01/new_biodiesel_plant_headed_to.html. Last accessed April 2nd, 2012.

[67] Diamond green diesel plant. http://www.biodieselmagazine.com/articles/7826/diamond-green-diesel-plant-secures-financing . Last accessed April 4th, 2012.

[68] Conocophillips renewable diesel. http://www.conocophillips.com/EN/about/company_reports/spirit_mag/Pages/whitegate_story.aspx. Last accessed April 4th, 2012.

[69] Conocophillips renewable diesel. http://www.biodieselmagazine.com/articles/1481/conocophillips-begins-production-of-renewable-diesel. Last accessed April 4th, 2012.

[70] Second Generation Renewable Diesel. http://www.greencarcongress.com/2007/02/tests_of_hydrog.html. Last accessed April 4th, 2012.

[71] Kubicka D, Salmi T, Tiitta M, Murzin DY. Ring-opening of decalin – Kinetic modeling. Fuel 2009;88 (2):366-73.

[72] Snare M, Kubicková I, Maki-Arvela P, Eranen K, Warna J, Murzin DY. Production of diesel fuel from renewable feeds: Kinetics of ethyl stearate decarboxylation. Chemical Engineering Journal 2007;134 (1-3):29-34.

Chapter 9

# Transition Metal Sulfide Catalysts for Petroleum Upgrading – Hydrodesulfurization Reactions

A. Infantes-Molina, A. Romero-Pérez, D. Eliche-Quesada,
J. Mérida-Robles, A. Jiménez-López and E. Rodríguez- Castellón

Additional information is available at the end of the chapter

http://dx.doi.org/10.5772/45629

## 1. Introduction

Environmental catalysis researchers worldwide have focused much attention on the development of catalytic systems capable of reducing the sulfur amount present in petroleum feedstocks until levels globally established by the recently enacted environmental protection laws. In this regard, the maximum sulfur content present in diesel fuel to obtain an Ultra Low Sulfur Diesel (ULSD) is of 10 ppm in the European Union from the beginning of 2009 with the entry into force of the Euro V fuel standard directive. Meanwhile this limit is slightly higher in the United States, 15 ppm, regulated by the Environmental Protection Agency (EPA) (Hsu & Robinson, 2006). Thus, the development of highly active and selective HDS catalysts, capable of processing these feeds, is one of the most important problems that the petroleum industry has to face nowadays.

Transition metal sulfides (TMS) have been traditionally used as active phases in hydrotreating catalysts since they are known to be efficient systems for catalyzing hydrotreating reactions. Concretely cobalt or nickel promoted molybdenum–tungsten sulfides are well established as the active species for commercial hydrodesulfuration (HDS) catalysts and mainly porous-alumina as a support. Amelioration has been achieved by modifying the properties of these sulfide systems, although the nature of the active phase has hardly been modified during many decades (Song & Ma, 2003). One direction for current research focuses on the use of new types of supports. Studies on nickel sulfided catalysts have concluded that supports, such as $Al_2O_3$, strongly interact with $Ni^{2+}$ ions avoiding their sulfidation (Gil et al., 1994). $Ni^{2+}$-alumina interactions may be weakened by using carriers such as alumina-pillared compounds, where the aluminium oxide is diluted within a layered inorganic matrix inducing a permanent porosity. These materials have been used as catalysts supports in hydrotreating reaction, showing interesting results (Kloprogge et al., 1993; Occelli & Rennard, 1988). On the other hand, mesoporous silica sieves have become a real alternative to alumina due to their hexagonal array of uniform mesopores and a very high surface area, presenting potential catalytic application for reactions involving bulky molecules, including hydrodesulfurization of petroleum fractions (Corma et al., 1995; Song & Reddy, 1999). In the same way, HMS type materials have been widely studied in this type of reactions (Nava et al., 2011; Zepeda et al., 2005). The intercalation of heteroatoms such as Al, Ti, Ga or Zr into the silica framework not only improves the material stability but also generates new acid, basic or redox functions that extend their application in new fields of catalysis. Thus, zirconium doped mesoporous silica with high surface area, mild acidity and high stability (Rodríguez-Castellón et al., 2003) have shown interesting properties as a support for catalytic fuel processing in reactions such as the hydrogenation, hydrogenolysis and hydrocracking of tetralin (Eliche-Quesada et al., 2003a, 2003b, 2004, 2005). The use of SBA-15 as a support for hydrotreating catalysts has presented several advantages with regard to HMS and MCM-41 mesoporous solids, since SBA-15 material has thicker pore walls and better hydrothermal stability, which are very important properties in hydrotreating processes due to the severe reaction conditions employed (Vradman et al., 2003). Recently, Gómez Cazalilla et al. (Gómez-Cazalilla et al., 2007) have proposed a cheap sol-gel synthesis route for SBA-15 and aluminium doped SBA-15, with sodium silicate as the silica source. The resulting materials have shown interesting properties as catalyst supports in hydrotretating reactions (Gómez-Cazallila et al., 2009a, 2009,b).

Other direction for current research focuses on the use of new active phases for developing high-performance HDS catalysts. The pioneering work of Pecoraro and Chianelli (Pecoraro & Chianelli, 1981) reported the great catalytic activity of bulk transition metals sulfides (TMS). Such metals were plotted into a curve called "volcano plot" where the HDS activity per mole of metal versus the M-S bond strength was plotted, the $RuS_2$ phase being the most active (Toulhoat et al., 1999). Nonetheless when the $RuS_2$ phase is supported, the results found in literature are diverse. On one hand a lower activity was observed due to its reduction into metallic ruthenium under the reducing conditions employed in the catalytic test (De los Reyes, et al., 1990) and if it is supported on alumina, sulfiding temperatures higher than 773 K are required to form the $RuS_2$ phase with pyrite-like structure, which is the true active phase for hydrotreating reactions. Nonetheless, it has been reported that $Ru/\gamma\text{-}Al_2O_3$ catalyst sulfided in 100% $H_2S$ at 673 K possessed about 7-fold higher thiophene conversion rates than $CoMo/\gamma\text{-}Al_2O_3$ when the surface of the active area is considered (Kuo et al., 1988).

Quartararo et al. (Quartararo et al., 2000) perfectly describe that there are many factors during the synthesis of ruthenium sulfide catalysts that must be taken into account for controlling their physicochemical properties, and as a consequence for achieving a good performance with this kind of catalysts. It is recommended no calcination after the incorporation of ruthenium chloride, while the sulfiding mixture should be $H_2S/N_2$ to achieve a high degree of sulfurization and avoid the reduction of the $RuS_2$ phase formed (De los Reyes, 2007). Furthermore, the sulfiding temperature influences the catalytic behaviour (De los Reyes et al., 1991) as well as the crystallographic orientations that induce the preference toward HDS and hydrogenation (HYD) reactions. One of the main goals to reach is the stabilization of such a phase under the reaction conditions. Ishihara et al. (Ishihara et al., 1992) were the first to report the addition of alkali metals to $RuS_2$ catalysts supported on $Al_2O_3$. The addition of NaOH did not improve the HDS reaction because of the poisoning of some sites. Nonetheless, a cesium-promoted Ru catalyst with Ru/Cs molar ratio of 1:2 exhibited HDS activities comparable to that of conventional Co-Mo catalyst (Ishihara et al., 2003). The insertion of atoms like cesium seems to enhance the number of labile sulfur atoms, aids to stabilize the $RuS_2$ active phase as it strengths Ru-S bond of ruthenium sulfide, promotes the C-S bond scission of dibenzothiophene (DBT) and therefore the catalytic activity increases (Ishihara et al. 2004). However, if a Cs excess is present, the formation of $H_2S$ and regeneration of coordinatively unsaturated sites are prevented, which results in a decrease in the catalytic activity.

With these premises, catalysts for HDS reaction based on molybdenum, tungsten and ruthenium sulfide are described. The role of promoters and material supports on the catalytic activity are reviewed. In this regard, the support effect on HDS activity on molybdenum and tungsten sulfided catalysts promoted with nickel and cobalt are evaluated by using fluorinated alumina $\alpha$-zirconium phosphate materials, zirconium doped mesoporous silica (Zr-MCM) and a commercial $\gamma$-$Al_2O_3$. Moreover the HDS activity of alternative phases such as $RuS_2$ is also described considering not only the role of the support (MCM-41, Zr-MCM-41, $\gamma$-$Al_2O_3$, SBA-15, Zr-SBA-15 and Al-SBA-15) employed but also the addition of a stabilizing agent such as Cs and the cesium precursor salt employed. The catalysts were characterized by X-ray diffraction (XRD), $N_2$ adsorption–desorption isotherms at 77 K, $NH_3$-temperature-programmed desorption ($NH_3$-TPD), X-ray photoelectron spectroscopy (XPS), $H_2$-temperature-programmed reduction ($H_2$-TPRS), transmission electron microscopy (TEM) and DRIFT spectra of adsorbed NO.

## 2. Tungsten and Molybdenum sulfide catalysts

### 2.1. HDS of Thiophene

Nickel, molybdenum and nickel-molybdenum sulfided catalysts supported on alumina-pillared $\alpha$-zirconium phosphate $Zr[Al_{3.39}O_{1.12}(OH)_{1.60}F_{4.90}]\ H_{0.57}(PO_4)_2$ (Mérida et al., 1996), with different loadings of Ni and Mo are described. The catalysts were tested in the thiophene HDS reaction at 673 K, using an automatic microcatalytic flow reactor under atmospheric pressure. A hydrogen flow of 50 $cm^3\ min^{-1}$ containing 4.0 mol% thiophene was fed to the reactor. Monometallic nickel catalysts were prepared following the incipient wetness method with ethyl alcohol solutions of nickel(II) nitrate ($Ni(NO_3)_2$) and nickel metallic loadings of 4, 8 and 12 wt%, denoted as 4wt%Ni, 8wt%Ni and 12wt%Ni, respectively. A catalyst only containing molybdenum (13wt%Mo), was also synthesized with aqueous solution of ammonium molybdate [$(NH_4)_6Mo_7O_{24}\cdot 4H_2O$]. Finally, another set of catalysts containing both nickel and molibdenum were prepared by successive impregnations with loadings 2.1-9 wt% and 3-13 wt% Ni-Mo. After the impregnation procedure, the catalysts were air dried at 333 K and calcined at 673 K for 5 h. In order to observe the influence of calcination temperature, 12wt%Ni catalyst was calcined at 623 K and expressed as 12wt%Ni (623 K). The precursors were sulfided at 673 K for 1 h under a flow of 60 $cm^3$ $min^{-1}$ $H_2S/H_2$ (10/90%).

The fluorinated alumina pillared α-zirconium phosphate support ($Al_2O_3$, 29.3 wt%; $S_{BET}$=184 $m^2$ $g^{-1}$) displays a mixed porosity, essentially in the range of mesoporous but with a micropore contribution of ≈ 0.1 $cm^3$ $g^{-1}$, and contain acid sites, mainly of Lewis type, which are active in the dehydration of isopropyl alcohol.

Evidence for formation of metallic sulfides on the support surface was provided by XRD and XPS analyses. XRD analysis revealed the formation of NiS for monometallic nickel sulfide catalysts with a loading higher than 8 wt%, showing very weak diffraction lines, suggesting that this phase should be extremely dispersed and strongly interacting with the support. XRD patterns of Mo and NiMo catalysts only show diffraction peaks corresponding to the $MoS_2$ phase, which are hardly visible, indicating that $Ni^{2+}$ would be inserted into the structure of $MoS_2$ forming a solid solution and for this reason it is not detected.

From XPS measurements, the Ni $2p_{3/2}$ and Mo $3d_{5/2}$ BE values as well as Ni/Zr, Mo/Zr, S/Mo and S/$Ni_{sulf}$ surface atomic ratios for sulfided catalysts are obtained and the corresponding values are included in Table 1. Since the BE of Zr $3d_{5/2}$ and P $2p$ core level electrons were practically constant with values of 183.3 eV and 134.1 eV, respectively, and the P/Zr atomic ratio was maintained close to the theoretical value, P/Zr = 2, it may be inferred that the impregnation-calcination-sulfidation processes did not significantly alter the host framework. The nickel species present on the catalyst surface are nickel(II) sulfide, nickel(II) oxide and $NiAl_2O_4$, as a result of the Ni $2p_{3/2}$ core level spectra decomposition, where three contributions are observed: the first one at 851.3-851.9 eV-NiS; the second one at 854.1 eV-$Ni^{2+}$ in octahedral sites of the supported NiO structure; and a third one at 855.0-855.7 eV-$NiAl_2O_4$ (Okamoto et al., 1977). This means that a fraction of Ni in the oxide form can be sulfided but other fraction remains unsulfided because of its strong interaction with the support that impedes its sulfidation at 673 K. From XPS data, the percentages of $Ni^{2+}$ sulfided are calculated (Table 1). In all cases the percentages of sulfided nickel were lower than 35%, slightly increasing with the metallic content. The S/$Ni_{sulfided}$ surface atomic ratios (Table 1) are close to 1, these values match well with NiS in agreement with XRD data. Regarding the Ni/Zr, it increases with the metallic loading attributed to the higher amount of nickel. The catalyst 12%Ni calcined at lower temperature points to a higher dispersion as from the higher Ni/Zr ratio.

On the other hand, the Ni $2p_{3/2}$ spectra of sulfided NiMo-AlZrP catalysts only show a signal at *ca.* 852.7 eV due to NiS phase indicating that $Ni^{2+}$ is completely sulfided and therefore the environment of this ion in NiMo calcined precursors is different from that of the Ni based materials (851.3-851.9 eV ). A square pyramidal arrangement of $Ni^{2+}$ ions within the $MoS_2$ framework has been proposed (Topsoe et al., 1987) and confirmed by EXAFS spectroscopy (Louwers & Prins, 1992). With regard to Mo $3d$ core level spectra of Mo and NiMo sulfided catalysts, besides the S $2s$ contribution at 226.5 eV (Arteaga et al., 1986), the spectra exhibit the typical Mo $3d_{5/2}$ component at 229.1 ± 0.1 eV characteristic of $MoS_2$ species (Arteaga et al., 1986). As no significant differences in the BE values are observed, the nature of Mo species may be essentially the same in the catalysts based on Mo and NiMo. The surface Mo/Zr, and Ni/Zr atomic ratios increase with the Mo and Ni contents (Table 1), and are also higher than on monometallic ones indicating a higher dispersion of Mo and Ni on the surface of Ni-Mo catalysts. The surface S/Mo atomic ratios for NiMo catalysts were very close to 2 which are consistent with the formation of $MoS_2$ on the catalysts surface and are higher than that observed on 13%Mo catalyst, indicating higher sulfur content in bimetallic catalysts due to the sulfur bonded to nickel.

| **Catalyst Labelling** | **Binding energy (eV)** | | **Surface atomic ratios** | | | | |
|---|---|---|---|---|---|---|---|
| | Ni $2p_{3/2}$ | Mo $3d_{5/2}$ | Ni/Zr | Mo/Zr | S/Mo | S/$Ni_{sul}$ | %$Ni_{sul}$ |
| 4%Ni | 855.7-851.8 | | 0.50 | | | 1.06 | 22.6 |
| 8%Ni | 855.6-851.5 | | 0.85 | | | 1.05 | 30.6 |
| 12%Ni | 855.3-851.3 | | 1.01 | | | 1.47 | 34.3 |
| 12%Ni (623K) | 855.4-852.4 | | 1.49 | | | 1.30 | 33.6 |
| 2.1-9%Ni-Mo | 852.7 | 229.1-232.3 | 0.67 | 2.98 | 1.94 | | 100.0 |
| 3-13%Ni-Mo | 852.8 | 229.1-232.3 | 1.16 | 3.82 | 1.98 | | 100.0 |
| 13%Mo | | 229.2-232.3 | | 3.01 | 1.68 | | |

**Table 1.** Binding Energy (BE) and surface atomic ratios as determined by XPS analysis of sulfided catalysts

The catalytic performance of these systems (Ni, Mo and NiMo-AlZrP catalysts) has been evaluated in the thiophene HDS reaction. Thus, from conversion values, the pseudo-first order constant ($k_{HDS}$) was calculated according to the equation:

$$k_{HDS} = -(F/W)\ln(1-x) \tag{1}$$

where, $F$ is the feed rate of thiophene (mol $min^{-1}$), $W$ is the catalyst weight (g) and $x$ is the fractional conversion.

Table 2 compiles $k_{HDS}$ values at the beginning and the end of the reaction (*TOS=6 h*).

As can be seen, Ni-based catalysts show promising $k_{HDS}$ values (between $6 \cdot 10^{-5}$ to $8 \cdot 10^{-5}$ mol $g^{-1}$ $min^{-1}$) without deactivation observed. The catalyst with the highest Ni loading and prepared at lower calcination temperature (623 K) shows the maximum activity with a $k_{HDS}$ value close to $10^{-4}$ mol $g^{-1}$ $min^{-1}$. This activity could be related to both its higher nickel sulfidation and dispersion, as observed from XPS. The sulfided 13%Mo catalyst initially displays a high thiophene HDS activity but undergoes a strong deactivation during the first 0.5 h showing then a similar activity than sulfided Ni catalysts. This behaviour is explained by the presence of coordination vacancies on the edges of $MoS_2$ crystallites. $H_2S$ molecules may occupy these vacancies and act as Brønsted acid sites which provoke the formation of coke and consequently a drastic decrease of the catalytic activity by blocking the pores and active sites (Yang & Satterfield, 1983). Finally, sulfided NiMo catalysts show a much higher HDS activity than sulfided Ni and Mo ones, which is attributed to both the promoting effect of $Ni^{2+}$ions in the mixed NiMo sulfided materials and the existence of sulfur vacancies (uncoordinated sites) on molybdenum. In this sense, $H_2S$ is not formed directly from the sulfur compound, but from the sulfur on the catalyst surface, being labile that bonded to both Ni and Mo. Thus, $H_2S$ released forms a new vacancy on the catalyst, as reported by Ruette and Ludena (Ruette & Ludena, 1981). These catalysts show high activity for the hydrodesulfurization of thiophene, mainly sulfided NiMo-AlZrP catalysts, which exhibit comparable or even higher activity than Ni, Mo and NiMo sulfided catalysts supported on carbons (Eijsbouts et al., 1994), zeolites (Welters et al., 1994) and alumina pillared smectites (Kloprogge et al., 1993) and tested in similar conditions.

With regards to the selectivity, it is reported in the literature (Silva-Rodrigo et al., 2004) that the reaction of thiophene with $H_2$ over supported HDS catalysts follows two main pathways: (1) direct thiophene hydrogenation leading to tetrahydrothiophene (THT), with further C–S bonding hydrogenolysis to form butane; (2) direct C-S scission to form 1,3-butadiene which is lately hydrogenated to form butene. 1-butene and *cis*- and *trans*-2-butene are intermediary products (Scheme 1). These butane isomers can lately be hydrogenated to form *n*-butane.

| Catalyst | Selectivity (%) | | | | $k_{HDS}$ x·$10^5$ (mol $g^{-1}$·$min^{-1}$) | | $k_{HYD}$·x $10^5$ (mol $g^{-1}$·$min^{-1}$) | $k_{HYD}/k_{HDS}$ |
|---|---|---|---|---|---|---|---|---|
| | n-b | 1-b | 2-t-b | 2-c-b | Initial | *TOS* 6h | | |
| 4%Ni | - | 27.5 | 35.0 | 37.5 | 7.50 | 6.22 | - | - |
| 8%Ni | - | 28.0 | 36.0 | 36.0 | 9.24 | 8.16 | - | - |
| 12%Ni | - | 29.0 | 35.0 | 36.0 | 7.75 | 7.15 | - | - |
| 12%Ni (623K) | - | - | - | - | 11.39 | 8.42 | - | - |
| 2.1-9%Ni-Mo | 17.5 | 25.0 | 27.0 | 30.5 | 48.53 | 16.01 | 4.68 | 0.30 |
| 3-13%Ni-Mo | 12.0 | 18.0 | 35.0 | 35.0 | 60.37 | 21.23 | 4.34 | 0.20 |
| 13%Mo | 12.0 | 21.0 | 33.0 | 34.0 | 22.50 | 6.07 | 7.80 | 1.28 |

n-b: n-butane; 1-b: 1-butene; 2-t-b: 2-t-butene, 2-c-b: 2-c-butene

**Table 2.** Pseudo-first-order rate constants for HDS of thiophene and selectivity values to different reaction products after 6 hours on stream

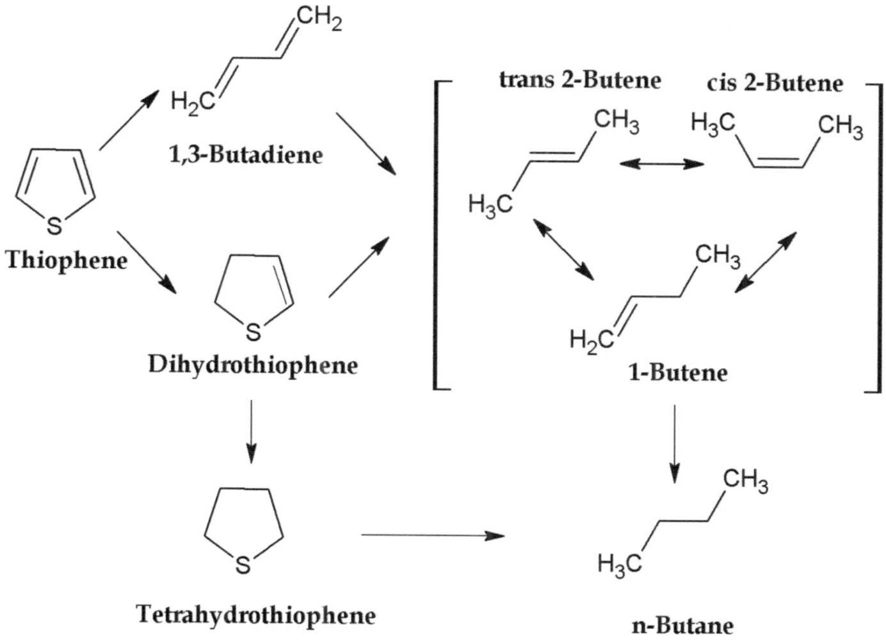


**Scheme 1.** Reaction scheme of HDS of Thiophene over Ni and NiMo sulfided catalysts

The reaction products identified by gas chromatograph where *n*-butane, 1-butene and *cis* and *trans*-butene (Table 2). It is noticeable different selectivities for these catalysts, thus monometallic Ni ones do not produce the hydrogenation product, butane. The hydrogenation ability of the catalyst can be interpreted by the presence of active hydrogen generated on molybdenum sites. Hydrogen activation takes place through intercalation in $MoS_2$ by proton permeation in the van der Waals gap and by stabilizing the electron charge in the 2*p*-band of sulfur. From $k_{HYD}/k_{HDS}$ ratios, summarized in Table 2, 13%Mo catalyst possesses the highest value confirming that it is molybdenum the responsible for the hydrogenation capability of these catalysts.

Characterization results indicated that the total sulfidation of $Ni^{2+}$ ions only occurred in NiMo catalysts, where a higher dispersion of the active species was also observed. These factors explain the high activity observed in thiophene HDS reaction, assigned to the promoter effect of Ni, which in turns diminishes the hydrogenation capability of the catalyst. NiMo catalyst with the highest metallic loading, 3-13%Ni-Mo, showed the maximum activity, $k_{HDS}$=21.2 $10^{-5}$ mol $g^{-1}$ $min^{-1}$. All the catalysts presented a similar behaviour, i.e., after an initial deactivation period, the catalyst maintained its activity for a long time attributed to the presence of Lewis acid sites in the support that avoids the formation of coke. Moreover, after reaction, no sulfur loss was detected by XPS analysis (data not shown), with BE values of Ni and Mo being in the range of the sulfided forms, pointing to the high stability of these catalysts under the conditions employed. These results show the interesting properties of alumina pillared α-zirconium phosphate to be used as hydrotreating support.

## 2.2. HDS of Dibenzothiophene (DBT)

The properties of Ni-Mo(W) and Co-Mo(W) catalysts supported on zirconium doped MCM-41 (Zr-MCM) are described and their activity in the HDS reaction of DBT compared with a catalyst supported on commercial γ-$Al_2O_3$. The HDS of DBT was carried out at 3.0 MPa of total pressure, $H_2$ flow rate of 100 $cm^3$ $min^{-1}$ and weight hourly space velocities (WHSV) of 32 $h^{-1}$. Thus Ni-W, Ni-Mo, Co-W and Co-Mo catalysts (W = 20 wt%; Mo= 11 wt%; Ni and Co = 5 wt%) were prepared by the incipient wetness method using mixed solutions of ammonium metatungstate (Aldrich) and nickel(II) citrate or cobalt(II) nitrate (Aldrich) in the case of Ni(Co)-W catalysts, or a mixed solution of ammonium heptamolybdate (Aldrich) and nickel(II) citrate or cobalt(II) nitrate for Ni(Co)-Mo catalysts. All materials, after impregnation with the metallic salts, were dried and calcined at 823 K for 4 h. These calcined precursors were then sulfided at 673 K with a $N_2/H_2S$ (90/10%) flow of 60 $cm^3$ $min^{-1}$ for 2 h prior to the catalytic test. The catalysts are labelled as Ni(Co)($x$)-Mo(W)($y$), where $x$ denotes nickel or cobalt content (wt%); and $y$, molybdenum or tungsten content (wt%).

The support chosen in this study is zirconium doped mesoporous silica (Zr-MCM) that possesses a hexagonal array of mesoporous pores (30 Å), very high surface area ($S_{BET}$= 608 $m^2$ $g^{-1}$) and an induced acidity (mild strength) due to the incorporation of zirconium into the mesoporous structure that also provides higher stability (Rodríguez-Castellón et al., 2003). After impregnation-calcination-sulfidation, the mesoporous structure is not altered as observed from XRD and $N_2$-adsorption-desorption isotherms.

Characterization results of sulfided catalyst by XRD indicate that tungsten promoted catalysts (Ni(Co)5-W20) showed not well defined diffraction peaks, pointing to a low crystalline $WS_2$ phase; while molybdenum based catalysts (Ni(Co)5-Mo11) presented well defined diffraction peaks corresponding to the $MoS_2$ phase. In general, nickel promoted catalysts present a better dispersion of Mo or W species as the lower intensity of $MoS_2$ and $WS_2$ diffraction peaks on these catalysts indicates. In no case, diffraction lines of nickel or cobalt sulfide are observed, suggesting that these phases are very dispersed or inserted into $WS_2$ or $MoS_2$ structure, forming a well dispersed Ni(Co)-W(Mo)-S solid solution, as it was observed before. The textural parameters of the support and the different sulfided catalysts (Table 3) reflect an important reduction of the specific surface area and pore volume after the incorporation of the active phase. This decrease could be attributed not only to the presence of particles of Ni(Co)-W(Mo)-S partially blocking the mesopores, but also to the increase in the density of the materials after the incorporation of a Ni(Co)-W(Mo)-S species.

The chemical species present on the surface of sulfided samples and their relative proportions were evaluated by XPS. The corresponding spectral parameters are included in Table 3. All sulfided catalysts present a maximum at 161.9 eV in the S 2$p$ energy region, which is characteristic of sulfide ($S^{2-}$) species. The W 4$f$ core level spectra indicate the presence of mainly tungsten sulfide $WS_2$ (32.4 eV) and non-sulfided W(VI) species or partially sulfided O-W-S species (36.2 eV) (Benítez et al.,1996). Mo 3$d$ core level spectra exhibit the typical Mo 3$d_{5/2}$ component characteristic of $MoS_2$ species at 228.5 eV and a weak contribution at higher B.E. (229.8 eV) due to partially sulfided O-Mo-S (Pawelec et al., 2003).

For Ni promoted catalysts, the Ni $2p_{3/2}$ photoemission line presents contributions at ca. 853.6 eV (nickel sulfide) and 856.5 eV ($NiWO_4$). The high BE for $Ni^{2+}$ forming a sulfide phase also suggests that $Ni^{2+}$ ions are embedded in the structure of $WS_2$ or $MoS_2$, probably forming the Ni-W(Mo)-S phase. Similarly, the Co promoted catalysts showed evidence for cobalt mainly in a sulfide phase (778.5 eV) (Alstrup et al., 1982) and a minor one at 781.3 eV, due to tetrahedral Co in an oxidic environment (Infantes-Molina et al., 2005).

| Sample | $S_{BET}$ ($m^2g^{-1}$)[a] | $V_p$ ($cm^3g^{-1}$)[a] | $d_p$(av) (Å)[a] | Binding Energy (eV)[b] | | | |
|---|---|---|---|---|---|---|---|
| Zr-MCM | 608 | 0.49 | 29.9 | Ni $2p_{3/2}$ | Co $2p_{3/2}$ | W $4f_{7/2}$ | Mo $3d_{5/2}$ |
| Ni5W20 | 332 | 0.25 | 24.1 | 853.6<br>856.4 | | 32.4<br>36.2 | |
| Co5W20 | 276 | 0.23 | 27.2 | - | 778.5<br>781.3 | 32.4<br>36.3 | |
| Ni5Mo11 | 271 | 0.21 | 25.9 | 853.6<br>856.4 | - | | 228.5<br>229.7 |
| Co5Mo11 | 241 | 0.21 | 29.2 | | 778.4<br>781.3 | | 228.4<br>229.9 |

**Table 3.** Textural properties[a] from $N_2$ adsorption-desorption isotherms at 77 K and spectral parameters[b] obtained by XPS analysis of sulfided catalysts

The dispersion of active phases on the sulfided catalysts was estimated from the surface atomic ratios. Table 4 compiles the Ni(Co)/(Si+Zr), Mo(W)/(Si+Zr) and S/Mo(W) atomic ratios. Ni/(Si+Zr) and Co/(Si+Zr) ratios are higher for molybdenum catalysts, showing a higher dispersion of the promoters on these catalysts (Ni(Co)-Mo). With regards to Mo(W)/(Si+Zr) atomic ratios, these are higher for nickel promoted catalysts, indicating a better superficial dispersion of these phases under the presence of nickel. The S/W atomic ratios for Ni(Co)-W catalysts are very close to 2, which is consistent with the formation of $WS_2$ on the catalyst surface, and it is also higher for nickel promoted catalyst (Ni5-W20). Meanwhile S/Mo ratios are much more high for both catalysts, being greater for Ni5-Mo11 one. These data suggest the higher degree of sulfidation for molybdenum based catalysts, as deduced from the TPRS data (*vide infra*) and also when nickel is present that also facilitates the dispersion of W and Mo.

| Surface atomic ratios | | | | | | |
|---|---|---|---|---|---|---|
| Sample | Ni/X | Co/X | W/X | Mo/X | S/W | S/Mo |
| Ni5-W20 | 0.115 | - | 0.110 | - | 2.53 | - |
| Co5-W20 | - | 0.089 | 0.096 | - | 1.85 | |
| Ni5-Mo11 | 0.196 | - | - | 0.090 | - | 5.43 |
| Co5-Mo11 | - | 0.119 | - | 0.037 | - | 4.04 |

**Table 4.** Surface atomic ratio of sulfided catalysts obtained by XPS analysis. X = Si + Zr

The nature of sulfur species as well as their stability is determined by $H_2$-TPRS measurements. From these experiments it is concluded that nickel promoted catalysts present a higher amount of $H_2S$ released at ca. 543 K. This removal of $H_2S$ comes from nickel or cobalt sulfide located at the edges of $WS_2$ or $MoS_2$ slabs forming the Ni(Co)-W(Mo)-S active phase (Magnus & Moulijn, 1994) and ascribed to the formation of coordinatively unsaturated sites (CUS) on the edge planes of the Ni(Co)-W(Mo)-S phase, being the active sites in HDS reaction. Therefore, nickel catalysts present a higher amount of active sites (CUS sites) on these catalysts. This also may explain the higher HDS activity observed for Ni5-W20 and Ni5-Mo11 catalysts (*vide infra*). Moreover, the curves are more intense for Mo catalysts, indicating a higher sulfurization degree when molybdenum is present, as obtained from XPS (*vide supra*).

The activity of these catalysts was evaluated in the DBT HDS model reaction. For DBT HDS, it has been proposed (Bataille et al., 2000) that the reaction proceeds through the hydrogenolysis pathway, the direct desulfurization route (DDS), leading to the production of biphenyl (BP); or by a second hydrogenation reaction pathway (HYD), in which one of the aromatic rings of dibenzothiophene is firstly prehydrogenated, forming tetrahydro (THDBT)- and hexahydro-dibenzothiophene (HHDBT), which is later desulfurized to form cyclohexylbenzene (CHB) (Scheme 2).

From conversion values and by applying equation (1), pseudo first order constants ($k_{HDS}$) were determined for HDS of DBT. The corresponding values are plotted in Figure 1.A, where it is observed that the activity increases with the temperature. At high reaction temperatures the activity follows the order: Ni5-Mo11> Ni5-W20> Co5-Mo11> Co5-W20, being always Ni-based catalysts much more active than their counterparts with cobalt. This fact could be related to the higher sulfidation degree and dispersion of nickel containing catalysts as observed in XPS and XRD, and the greater presence of CUS sites as observed from TPRS studies.

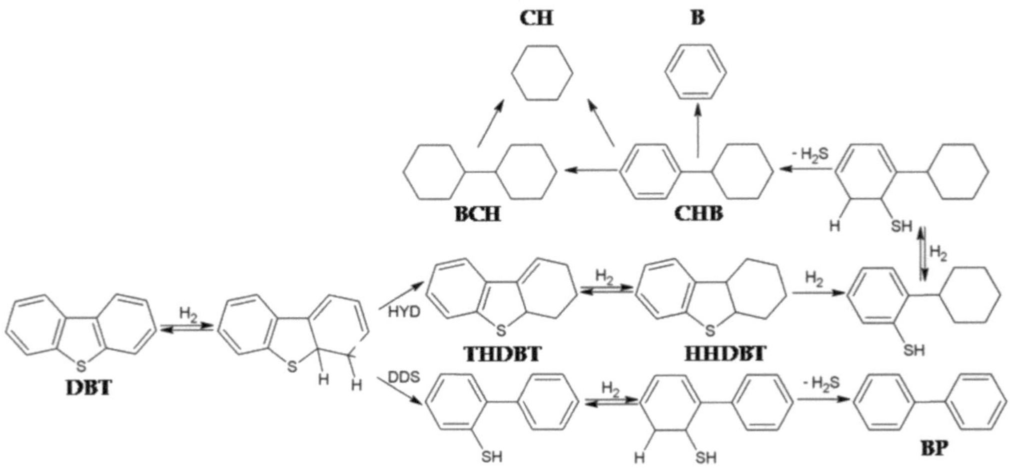


**Scheme 2.** Reaction scheme of HDS of DBT over Ni(Co)Mo(W) sulfided catalysts: HYD: hydrogenation route; DDS: direct desulfurization route; DBT: dibenzothiophene; BP: biphenyl; THDBT: tetrahydrodibenzothiophene; HHDBT: hexahydrodibenzothiophene; CHB: cyclohexylbenzene; BCH: Bicyclohexyl; CH: Cyclohexane; B: Benzene.

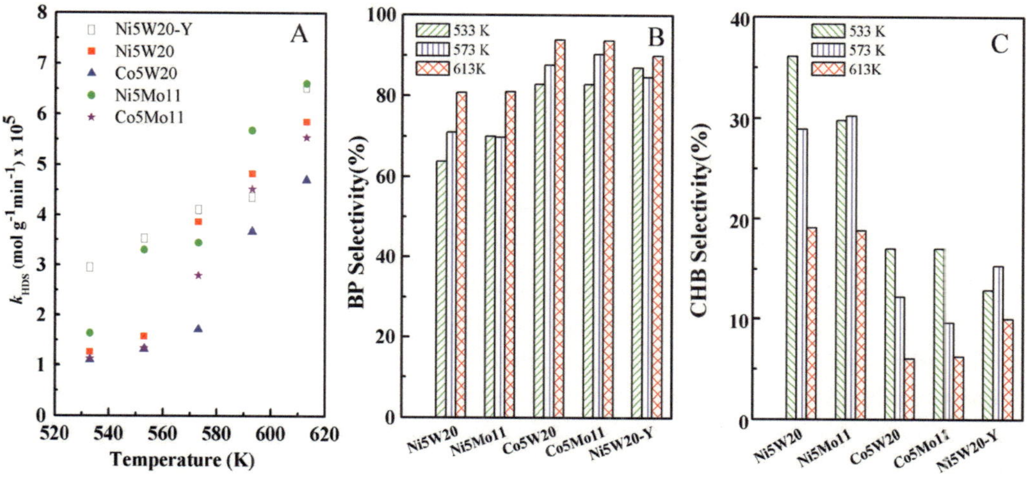


**Figure 1.** Influence of reaction temperature on HDS activity. Reaction conditions T= 533-613 K; P=30 bar, WHSV=32 h$^{-1}$, $H_2$= 100 cm$^3$ min$^{-1}$

The observed activity in the DBT HDS reaction with this set of catalysts is better than those reported for other catalysts prepared by using MCM-41 as support, such as CoMo (Song & Reddy, 1999), NiMo (Grzechowiak et al., 2006) and NiMo and CoMo supported on non proton-exchanged MCM-41 (Li et al., 2003). Moreover, the comparison with a catalyst supported on a commercial support such as $Al_2O_3$ (BET surface area: 302 m$^2$ g$^{-1}$ and pore volume: 0.33 cm$^3$ g$^{-1}$), denoted as Ni5W20-Y, shows that in spite of the higher activity at lower temperatures, at higher temperatures Ni5Mo11 and Ni5W20 catalysts reach and overcome the activity reported by the Ni5W20-Y catalyst.

With regards to the selectivity shown by these catalysts, Figure 1 also plots the selectivity to the direct desulfurization product (DDS), biphenyl (BP) (Figure 1.B); and to the hydrogenation (HYD) product, cyclohexylbencene (CHB) (Figure 1.C), of the sulfided catalysts as a function of reaction temperature. It is noticeable that the product distribution markedly changes with the promoter, since the formation of the hydrogenation product, CHB, is higher for both nickel promoted catalysts. The formation of CHB increases at the expense of BP. So, whatever the active phase may be (Mo, W), the presence of Ni as a promoter leads to a considerable increase in the HYD reaction. Further, and in agreement with the thermodynamic restrictions, given that the hydrogenation reaction is exothermic, a decrease in hydrogenation activity is observed with an increase in the reaction temperature for all catalysts (Farag et al., 2000). It is generally accepted that hydrogenation and desulfurization reactions take place on separate active sites (Li et al., 2002). It is assumed that the enhancement of the HYD pathway is attributed to the strong hydrogenation properties of the Ni species, and it is proposed that hydrogenation occurs on other sites (such as Ni atoms), while the removal of the sulfur atom from the dibenzothiophene ring takes place on the Ni-Mo(W) cluster. These results are in accordance with findings by Wang et al. (Wang et al., 2002), who compared the CHB/BP selectivity ratio for a series of Ni, Mo and Ni-Mo sulfided catalysts supported on MCM-41 and established that all the Ni-Mo/MCM-41 catalysts yield a higher ratio than single Mo or Ni/MCM-41, suggesting that there is a synergetic effect between Ni and Mo sulfides in the HYD pathway during HDS. In fact, according to Whiterhurst (Whiterhurst et al., 1998), the rate constant of HDS after hydrogenation of one aromatic ring of DBT is 33 times greater than that of DDS for this kind of catalyst. The enhanced hydrogenation capability of these catalysts has been previously observed, since Ni-W sulfides supported over Zr-MCM-41 exhibit up to 44% conversion in the hydrogenation of tetralin to decalins (Eliche-Quesada et al., 2003b).

The hydrogenation properties observed for these catalysts can also be ascribed to the presence of superficial zirconium species on the surface of Zr-MCM that along with its high surface area, seem to have an influence on the dispersion and specific electronic properties of the active species. In this sense, it is proposed that the presence of smaller slabs increases the number of rim sites which are responsible for hydrogenation reaction, according to the rim-edge model for HDS reaction (Whiterhurst et al., 1998). Our results are in agreement since the catalysts showing higher hydrogenating properties, Ni promoted catalysts, present better dispersion of Ni, Mo and W as extracted from XRD and XPS. The results are quite interesting since nickel promoted catalysts (Ni-Mo(W)) supported on Zr-MCM not only present high HDS activity but also a better hydrogenation capability leading to a gas-oil with improved quality such as higher cetane number.

By contrast, HDS over Co promoted catalysts supported on Zr-MCM, Co-Mo(W) sulfided catalysts, mainly follow the route of hydrogenolysis. The introduction of Co to Mo(W)-based catalysts enhances the direct extraction of sulfur atoms from the DBT molecules. It is assumed that Co sulfide acts by extracting sulfur atoms directly from the sulfur-containing molecules. This is essentially due to the low rate of HYD of Co-Mo(W) sulfides. In conclusion, with Ni promoted catalysts (Ni-Mo(W)), the HYD pathway is dominant over the hydrogenolysis pathway in the HDS reaction of DBT, whereas the direct DDS of DBT occurs with Co-Mo(W) catalysts. The comparison with the commercial catalyst reveals that the selectivities to hydrogenation product (CHB) are higher for the Zr-MCM-derivative (Figure 1), being the selectivity to Ni5W20-$Al_2O_3$ only a half of that found for its Zr-MCM counterpart, possibly due to the higher particle size formed over $Al_2O_3$.

## 3. Ruthenium sulfide catalysts

Supported ruthenium sulfide catalysts were studied in the hydrodesulfuration (HDS) of dibenzothiophene (DBT). The role of the support, sulfiding temperature, the presence of Cs as $RuS_2$-stabilyizing agent, as well as the ruthenium precursor salt employed was studied. Ruthenium was incorporated by the incipient wetness impregnation procedure adding an aqueous solution of ruthenium(III) chloride ($RuCl_3{\cdot}nH_2O$) to the pelletized support (0.85-1.00 mm) to obtain catalysts with 7 wt% of ruthenium. After air-dried, the samples were directly sulfided in situ at atmospheric pressure with a $N_2/H_2S$ (90/10%) flow of 60 $cm^{-3}$ $min^{-1}$ by heating from rt. to the sulfidation temperature (Ts) (2 h) at a heating rate of 10 K $min^{-1}$.

### 3.1. Influence of the material support

The supports employed were MCM-41 mesoporous silica (Rodríguez-Castellón et al., 2003); that doped with zirconium (Zr-MCM) (Rodríguez-Castellón et al., 2003); SBA-15 mesoporous silica (Gómez-Cazalilla et al., 2007); SBA doped with zirconium (Zr-SBA) and aluminium (Al-SBA) (Gómez-Cazalilla et al., 2007). Finally a commercial $\gamma$-$Al_2O_3$ was also employed as reference. The supports, precursor, sulfided and spent catalysts were characterized by a variety of experimental techniques in order to establish a clear catalyst performance-structure correlation.

The prepared catalysts are sulfided at 673 K and their catalytic activity evaluated in the HDS of DBT between 533 K and 613 K. The $k_{HDS}$ values (Eq. 1) versus reaction temperature are plotted in Figure 2. In all cases there is a direct improvement in the activity as a function of the reaction temperature. At low temperature, RuSiSBA and RuZrMCM are the most active catalysts, although all of them present $k_{HDS}$ values lower than $0.5.10^{-5}$ mol $g^{-1}$ $min^{-1}$. The influence of the support becomes more evident at higher temperature, where SBA-15 type supports provides more active catalysts than MCM-41 ones, with pure silica supported catalyst, RuSiSBA, the most active at 613 K. The catalyst prepared with a commercial support is the least active catalyst. The results presented here highlight the importance of the usage of mesoporous materials as supports, which could be probably ascribed to the better dispersion achieved on these materials.

By considering the selectivity data (Scheme 2), all the catalysts preferentially follow the DDS route, i.e., the formation of biphenyl is favoured in all cases (Figure 2). The influence of the reaction temperature reveals that its formation slightly increases with the temperature, due to thermodynamics considerations. The formation of the product coming from the HYD route, cyclohexylbenzene (CHB) decreases with the increase of the temperature (Figure 2), being the catalysts supported on Al-SBA-15 type materials, the least selective to this compound, and that supported on the commercial support, Ru-$Al_2O_3$, the catalyst with the best hydrogenating properties.

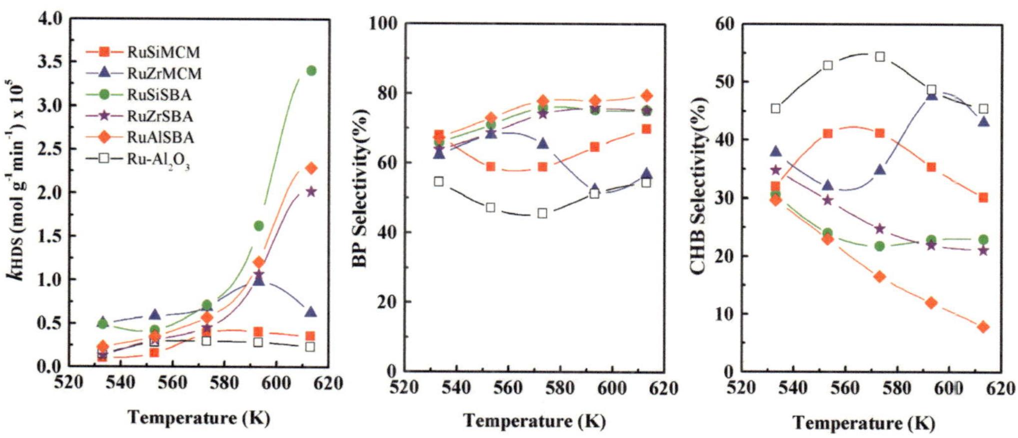


**Figure 2.** Influence of the support on the $RuS_2$ DBT HDS activity and selectivity as a function of reaction temperature. Experimental conditions: P=30 bar, WHSV=32 $h^{-1}$ and $H_2$ flow=100 $cm^3$ $min^{-1}$.

It can be clearly seen that the presence of heteroatoms into the silica mesoporous structure affects the catalytic response of the catalyst. Thus, the presence of zirconium in MCM-41, improves the HDS activity with regards to the catalyst supported on pure MCM-41 (RuZrMCM versus RuSiMCM), and also alters the selectivity, with a greater selectivity to CHB at higher temperatures. Nonetheless, an opposite trend is observed on SBA-based catalysts, since RuSiSBA presents the highest activity values, while the incorporation of aluminium (post-synthesis) and zirconium (direct synthesis) do not provoke any amelioration, only a slight increase in the formation of hydrogenation product is observed at lower temperatures for RuZrSBA. All these data points to a different Si and Zr environment in both mesoporous supports that has a strong influence on the $RuS_2$ phase formed after sulfidation. In this regard, the incorporation of zirconium into MCM-41 has a positive effect, since it could act as $RuS_2$ stabiliser avoiding its reduction under the experimental conditions employed, as well as incorporating acidic functions to the catalyst that could enhance the DBT HDS reaction. Nonetheless, when using SBA-15, in spite of increasing the acidity of the material, a depletion of catalytic activity is observed and therefore suggesting that it is not the acidity but other factors those governing the catalytic behaviour of these systems.

From catalytic data previously exposed, it is observed that $RuS_2$ phase on SBA-15 type support is more active for S-removal and that is why the influence of the sulfiding temperature has been evaluated on these systems, by comparing the HDS activity after sulfidation at 673 K and 773 K. The corresponding results are compiled in Figure 3. From this figure, the catalysts sulfided at 673 K are less active at all studied temperatures. Regarding catalysts sulfided at 773 K, those supported on Si-SBA and Zr-SBA present a similar behaviour with RuAlSBA catalyst the least active. This catalyst only shows a slight improvement at moderate temperatures after sulfiding at 773 K. With regards to the selectivity trend (data not shown here) it is only observed a slight increase of the HYD route by increasing the sulfidation temperature for RuSiSBA and RuZrSBA, what should be related to the formation of more active phase, while RuAlSBA is the most selective to BP at all studied reaction and sulfidation temperatures.

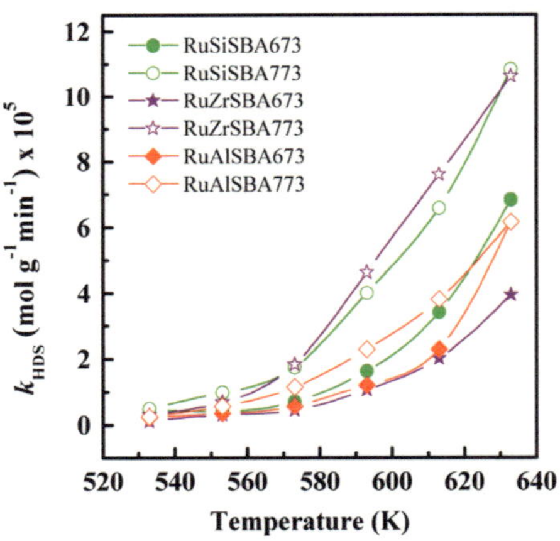


**Figure 3.** Influence of the sulfidation temperature on the catalytic behaviour of $RuS_2$ supported on SBA-15 type materials.

In order to find a possible explanation of catalytic results, catalysts characterization was performed. In this regard, XRD patterns of sulfided catalysts at 673 K supported on MCM-41 and γ-$Al_2O_3$ did not show the characteristic diffraction peaks of $RuS_2$, while XRD patterns of catalysts supported on SBA-15 presented such peaks at 2θ=31.8º, 45.7º and 54.2º (PDF Card Nº. 00-012-0737) at both sulfidation temperatures as observed in Figure 4.

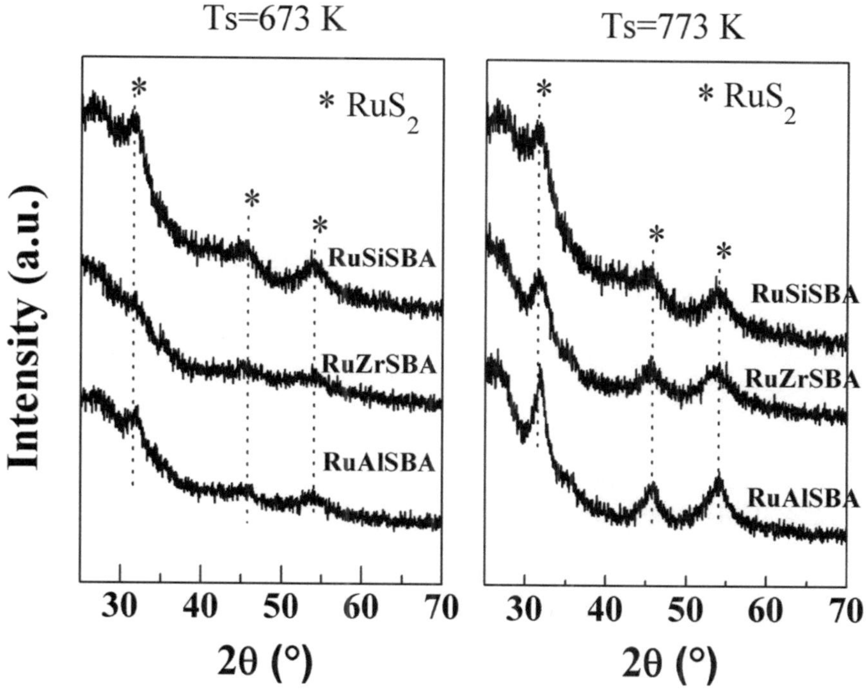


**Figure 4.** XRD patterns of SBA supported catalysts sulfided at 673 K and 773 K.

From this figure it is noticed that by sulfiding at 673 K, a poorly crystalline ruthenium sulfide phase is formed that undergoes reduction under hydrogen hydrotreating atmosphere, as seen from XRD of spent catalyst (data not shown). Therefore a sulfidation temperature of 673 K seems to be insufficient to form stable $RuS_2$ particles. Possibly particles with a stoichiometry of $RuS_{2-x}$ have been formed. These data are in agreement with the catalytic results, where the less active catalysts, RuSiMCM, RuZrMCM, Ru-$Al_2O_3$, do not show these peaks; while RuSiSBA presenting these lines better defined is the most active one. On the contrary, the catalysts sulfided at 773 K do not present the diffraction signals of metallic ruthenium after the catalytic run. Therefore, at 773 K, a greater proportion of the pyrite phase is formed, which is highly stable under reaction conditions. The presence of heteroatoms into the mesoporous structure provokes a greater interaction with the precursor and higher sulfidation temperatures are required as can be clearly seen from XRD patterns of RuZrSBA, presenting so a lower catalytic activity with regards to pure SBA.

The textural and acidic properties of the catalysts are compiled in Table 5.

| Sample | $S_{BET}$ ($m^2 g^{-1}$)[a] | $V_P$ ($cm^3 g^{-1}$)[a] | $d_P$ (nm)[a] | μmol $NH_3 \cdot g^{-1}$ [b] |
|---|---|---|---|---|
| SiMCM | 784 | 0.54 | 2.3 | 127 |
| RuSiMCM | 731 | 0.48 | 2.2 | 139 |
| ZrMCM | 608 | 0.49 | 3.0 | 474 |
| RuZrMCM | 501 | 0.40 | 2.6 | 504 |
| $Al_2O_3$ | 313 | 0.44 | 4.8 | 731 |
| Ru- $Al_2O_3$ | 280 | 0.35 | 3.9 | 780 |
| Si-SBA | 476 | 0.35 | 3.9 | 144 |
| RuSiSBA673 | 331 | 0.25 | 3.5 | 272 |
| Zr-SBA | 495 | 0.40 | 3.7 | 1081 |
| RuZrSBA673 | 327 | 0.26 | 3.7 | 798 |
| Al-SBA | 360 | 0.29 | 3.7 | 192 |
| RuAlSBA673 | 222 | 0.18 | 3.2 | 469 |

**Table 5.** Textural[a] and acidic properties[b] of supports and catalysts sulfided at 673 K

These data indicate that catalysts supported on MCM-41 present a much higher surface area and pore volume than those supported on SBA-15, although a much lower pore diameter; while the acidity values do not follow any trend. In all cases, an increase in the acidity is observed after the incorporation of $RuS_2$. It has been reported that the acidity of ruthenium sulfided catalysts is due to the presence of different species on the surface: coordinatively unsaturated sites (CUS) that provide Lewis acidity as well as $SH^-$ groups providing Brønsted acidity (Berhault et al., 1998). It cannot be established a clear correlation between textural/acidity properties and catalytic activity, although the higher pore diameter of SBA type support could favor a better distribution of the active phase inside the channels.

Information regarding the different sulfur species present on the catalyst, the degree of sulfidation attained as well as the stability of the active phase, can be obtained from $H_2$-TPRS profiles. Thus, the $H_2$-TPRS patterns of sulfided ruthenium catalysts are shown in Figure 5, where $H_2S$ removal signals are depicted as a function of temperature. De los Reyes et al. (De los Reyes et al., 1991) have pointed that the reduction of ruthenium sulfide based catalysts takes place in several steps. The first $H_2S$ removal at low temperatures ($T < 450$ K) is due to the surface sulfur excess which is formed during the sulfidation due to the lack of hydrogen, and/or could also be due to sulfur coordinated to surface Ru. In the second $H_2S$ removal, between 450 K and 570 K, the elimination of surface sulfur anions occurs. Some authors have found that this band tends to disappear by sulfiding at higher temperatures, suggesting that this band arises from the reduction of an amorphous or poorly crystallize $RuS_2$ phase at low sulfiding temperatures (Castillo-Villalón et al., 2008) and also ascribed to the release of labile sulfur (CUS). Finally, at $T > 573$ K, the elimination of bulk sulfur of the $RuS_2$-pyrite takes place leading to metallic ruthenium (Castillo-Villalón et al., 2008; De los Reyes et al., 1991).

Catalysts supported on MCM-41 type support possess a similar $H_2S$ desorption profile, with the exception of the temperature peak at lower temperatures, that is hardly observed for RuSiMCM and a higher intensity of the second desorption peak for RuZrMCM assigned to the presence of CUS sites. The catalytic activity of these two samples is different. These results point to the presence of zirconium as responsible for such a fact. The presence of zirconium has a manifold role; it could lead to a better dispersion of ruthenium sulfide and therefore providing a higher amount of labile sulfur (CUS) according to $H_2$-TPSR patterns. Secondly, it strengths the Ru-S bond avoiding its reduction under the experimental condition as well as the sinterization of the active phase, preserving so the active sites.

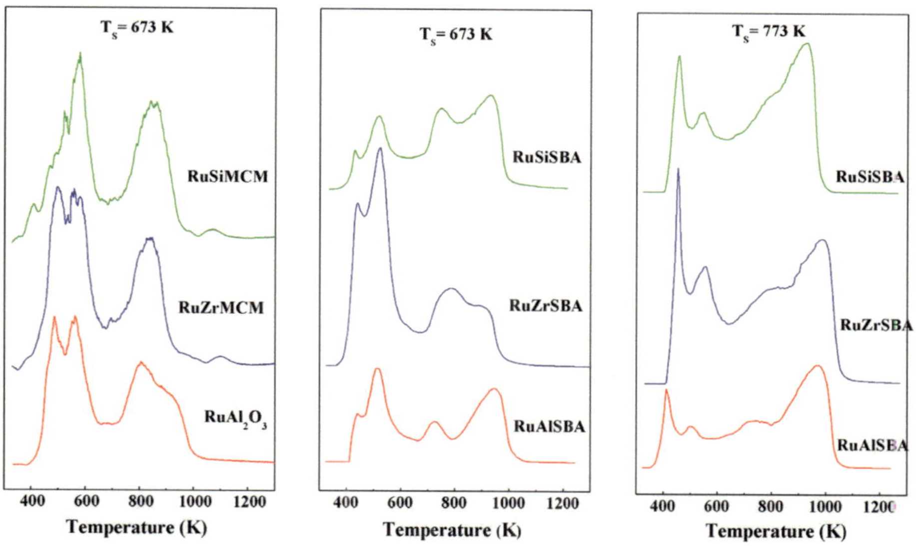


**Figure 5.** $H_2S$ desorption profiles for samples sulfided at 673 K (A and B) and samples sulfided at 773 K (C).

SBA-15 supported catalysts have also been sulfided at 773 K. By comparing the catalysts sulfided at 673 K and 773 K, in spite of presenting similar $H_2S$ bands, both the relative intensities and the peak maxima positions are different depending on the sulfidation temperature employed. In this sense, when sulfiding at 773 K the signals at $T < 573$ K decreases in intensity, while those at $T > 573$ K are more defined, what suggests that a higher amount of pyrite-type structure is formed, in accordance with previous works (De los Reyes et al., 1991) and also confirmed by XRD data. A similar profile at high temperatures has been reported by other authors (Castillo-Villalón et al., 2008), i.e., the asymmetry of this band, which can be decomposed into two contributions, is due to a surface reduction followed by bulk reduction. Castillo-Villalón et al. (Castillo-Villalón et al., 2008) have recently reported that in so far as the sulfidation temperature increases, the $H_2S$ evolving from species reduced at high temperature does, in detriment to some species that are reduced at low temperatures. Thus, by increasing the sulfidation temperature, some ruthenium sulfided species found when sulfiding at 673 K, are transformed into more stable

$RuS_2$ species when sulfiding at 773 K and that is why catalysts sulfided at 773 K present these bands more intense. Moreover, the peak maxima shift at higher temperatures, being this fact much more important in the bands at T > 573 K which are related to the reduction of $RuS_2$-pyrite phase. The formation of highly dispersed particles, as seen from TEM images, probably strongly interacting with the support, makes necessary a higher temperature to reduce them into the ruthenium metallic form (*vide infra*).

Regardless the material support employed, $H_2$-TPRS profiles of catalysts sulfided at 673 K do not present the bands at high temperatures with the characteristic reduction pattern of bulk $RuS_2$ in a pyrite-like structure. From these data, it can be concluded that at 673 K there is a great proportion of amorphous or poorly crystallize $RuS_2$ phase that is also less stable in HDS reactions. This fact should be much more important for catalysts supported on MCM type materials. XRD of spent catalysts reveals the formation of metallic ruthenium, while the diffraction bands of $RuS_2$ phase are not detectable. At 773 K, not only a greater formation of ruthenium sulfide in the form of pyrite but also the presence of amorphous ruthenium sulfide (intermediate temperatures) and sulfur excess on the surface (low temperature), are higher. The sample prepared on Al-SBA presents an increase in the $H_2S$ eliminated at higher temperatures, although this increment is less pronounced than for the other two samples.

Focusing on the catalysts sulfided at 773 K, which are much more active, TEM micrographs show the distribution of the active phase. Depending on the support employed, the location of $RuS_2$ is different. In general, no large ruthenium sulfur particles have been formed. The sulfided catalysts supported on Si-SBA and Zr-SBA show spherical particles with very small size, homogenously dispersed (Figure 6.A and 6.B) and mainly located inside the structure. The location of the ruthenium sulfide particles inside the pores of the mesoporous structure is clearly observed in Figure 6.A, where an alignment of the $RuS_2$ phase is observed. The distance between two rows corresponds to the $d_{100}$ parameter of the mesoporous structure. By measuring the distance between two rows, we obtain a value of 9.5 nm. If we compared this data with that calculated from the XRD pattern of the support at low angles, the $d_{100}$ reflection at $2\theta = 1.12^{\circ}$, we obtained an $a_0$ value of 9.1 nm. Therefore we can assess that ruthenium sulfide particles are homogeneously located inside the pores. On the other hand, the sulfided catalysts prepared from Al-SBA material support (Figures 6.C), present a different distribution of the active phase. We find small ruthenium sulfide particles inside the channels and larger particles located at the external surface of the support and scarcely interacting with it. A larger particle size is also found for these samples from XRD analysis.

From TEM analysis it can be seen that the small-sized particles are inside the pores of the carriers. Nonetheless the pores are not blocked by the incorporation of ruthenium, as long as, $N_2$ adsorption-desorption isotherms of sulfided catalysts are similar to the material support (isotherms of type IV), and only a slight decrease in the surface area is observed as a consequence of the location of $RuS_2$ particles inside the channels.

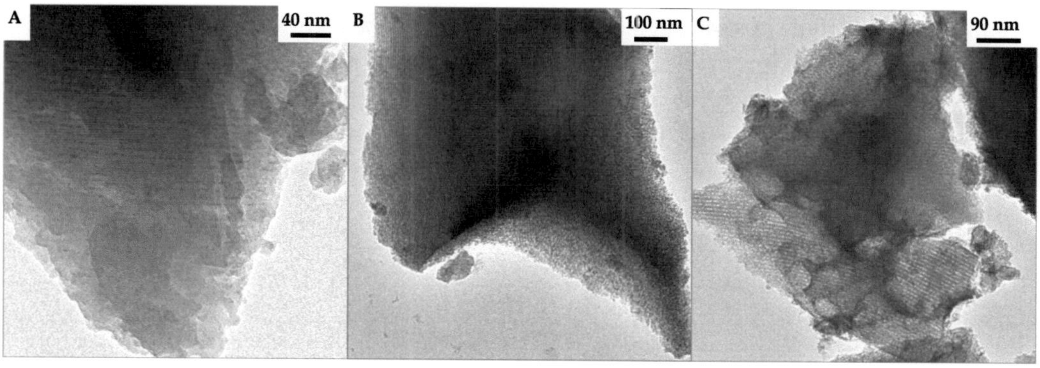


**Figure 6.** TEM micrographs for: A) RuSiSBA773, B) RuZrSBA773 and C) RuAlSBA773

The study of spent catalysts by TEM, reveals the great stability of $RuS_2$ particles during the test. While RuSiSBA and RuZrSBA present a homogenous distribution of small particles inside the pores, RuSiAlSBA shows a different distribution with zones presenting small and highly dispersed particles and mainly zones where agglomerates are present. Therefore, the better dispersion of the active phase is achieved when Si-SBA and Zr-SBA are used as supports.

As long as the catalytic performance of these systems is attributable to the formation of ruthenium sulfide, XPS spectra were recorded for sulfided and spent catalysts in order to elucidate the chemical state of the elements present. The Al 2*p*, O 1*s*, Si 2*p*, and Zr 3*d* core-level spectra were similar for sulfided and spent catalysts as well as their binding energies values maintained practically constant. Moreover, Table 6 lumps together the Ru $3p_{3/2}$ and S $2p_{3/2}$ binding energy values for all the samples, along with the S/Ru and Ru/X (X=Si+ (Si+Zr) or (Si+Al)) surface atomic ratios. Ruthenium species were analyzed by recording the Ru $3p_{3/2}$ spectrum of the samples and studied by an appropriate curve fitting. The signal is slightly asymmetric and can be decomposed into two contributions: the main one with its maximum between 460.8 and 461.4 eV; and a second one, much less intense whose maximum is between 463.2 and 463.8 eV. The main signal can be assigned to the $RuS_2$ phase. Mitchell et al. (Mitchell et al., 1987) reported, for ruthenium sulfide supported on alumina, Ru $3p_{3/2}$ binding energy values of 461.1-461-2 eV. On the other hand, the weak peak (463.2-463.8 eV) has been assigned to $RuCl_3$ species (Moulder et al., 1992). Catalysts supported on MCM-41 present Cl after the sulfidation procedure, therefore this band is due to $RuCl_3$. However, catalysts supported on SBA do not present the Cl 2*p* signal, and we suggest the presence of $Ru^{n+}$ species as responsible for such a band. The S $2p_{3/2}$ peak for sulfided catalysts is localized at BE ranged between 162.0 and 162.4 eV. This binding energy value is akin to that reported for sulfur forming disulfide polyanions $(S\text{-}S)^{2-}$ [8], i.e. sulfur forming $RuS_2$ with pyrite structure.

As far as spent catalysts are concerned, the Ru $3p_{3/2}$ signal hardly changes (data not shown), while the S 2*p* signal suffers a slight shift to lower binding energies and in some cases it is not detected. As reported for many authors (Berhault et al., 1997), during the catalytic run, due to the reducing atmosphere, sulfur can be eliminated from the catalyst surface and that is why a modification of the S 2*p* signal is observed. In fact, from XRD analysis of spent catalysts, those sulfided at 673 K showed the presence of metallic ruthenium, confirming this fact. Moreover, Navarro et al. (Navarro et al., 1996) reported that a large amount of sulfur vacancies are generated when recording photoelectron spectra under a highly energetic X-ray beam impinging on the sample. This subject should also be kept in mind.

| Catalyst | Binding Energy (eV) | | | Surface atomic ratios | |
|---|---|---|---|---|---|
| | Ru 3*p* | | S $2p_{3/2}$ | | |
| | $RuS_2$ | $RuCl_3/Ru^{n+}$ | $S_2^{2-}$ | Ru/X* | S/Ru |
| **Ts=673K** | | | | | |
| **RuSiMCM** | 461.3 | 463.7 | 163.0 | 0.009 | 2.5 |
| **RuZrMCM** | 461.3 | 463.6 | 163.7 | 0.016 | 2.9 |
| **Ru-$Al_2O_3$** | 460.6 | 463.4 | 163.2 | 0.095 | 3.1 |
| **RuSiSBA** | 461.4 | 463.2 | 162.4 | 0.006 | 6.2 |
| **RuZrSBA** | 461.4 | 463.2 | 162.3 | 0.012 | 2.4 |
| **RuAlSBA** | 461.0 | 463.2 | 162.0 | 0.093 | 3.5 |
| **Ts=773K** | | | | | |
| **RuSiSBA** | 461.3 | 463.5 | 162.4 | 0.004 | 7.0 |
| **RuZrSBA** | 461.1 | 463.8 | 162.3 | 0.010 | 2.3 |
| **RuAlSBA** | 460.8 | 463.2 | 162.0 | 0.076 | 3.4 |

*X =Si, Si+Zr, Si+Al or Al, as accordingly

**Table 6.** Spectral parameters of $RuS_2$ supported catalysts.

Quantitative XPS data (Table 6) show that Ru/X atomic ratio for samples supported on Al-SBA and $Al_2O_3$ are what indicates a higher concentration of active phase on the surface, while the Ru/(Si+Zr) and Ru/Si surface atomic ratios have the lowest values. From TEM analysis, catalysts supported on Si-SBA and Zr-SBA present the majority of the active phase highly dispersed and mainly located inside the pores. Since SBA porous structure possesses a wall thickness of ca. 50 Å, these particles are not detected by XPS due to the surface sensitivity nature of this technique. The lower values for RuSiSBA and RuZrSBA corroborate what observed from TEM, i.e., the preferential location of the active phase inside the pores forming small particles. The S/Ru atomic ratios showed values higher than the stochiometric one due to the presence of an excess of sulfur on the surface, $H_2S$ or $SH^-$ groups formed during sulfiding process, which is not forming the pyrite phase. The analysis of spent catalysts showed a decrease in the S/Ru atomic ratio, in accordance with all the experimental exposed here and the literature reports, i.e., surface sulfur elimination occur during the catalytic test.

To sum up, the results reported here highlights the important role that the material support plays on the stability of the active phase, i.e., SBA type mesoporous materials provide ruthenium sulfide catalysts which are more active and stable in the HDS reaction of DBT than MCM-41 mesoporous one. Characterization results reveal that the bigger pore diameter of the former could lead to a better filling of them with the ruthenium sulfide. Moreover, after the sulfidation processes the formation of $RuS_2$ with pyrite-type structure occurs in a greater extend on SBA-15 supported catalysts that is also more stable (XRD and $H_2$-TPSR), indicating a higher interaction of the precursor with SBA type material providing more stable $RuS_2$ particles, specially when the catalysts are sulfided at 773 K. The presence of heteroatoms depends on the type of mesoporous material employed, an increasing acidity is always observed, however no correlation is found between acidity and catalytic activity. While the incorporation of zirconium on MCM-41 seems to stabilise the $RuS_2$ phase and improving so the HDS behaviour; on SBA-15, no improvement is observed after heteroatoms addition, as long as Al doped support provides the least active catalysts; while Zr doped one (RuZrSBA) achieves the same level of activity than RuSiSBA only with the catalyst sulfided at 773 K. The presented data indicate that is the dispersion and distribution of the active phase what govern the catalytic behaviour of these systems and also the selectivity patterns. In this regard, the formation of small particles induces some preferential exposed planes, favouring the hydrogenation pathway as reported for other supported ruthenium sulfide catalysts (De los Reyes et al., 1991) what reveal the sensitivity to the structure of the $RuS_2$. Our results are in agreement with this statement, being RuAlSBA with the biggest particles sizes, the catalyst that provides the lowest values of selectivity to CHB, product formed in the hydrogenation route. Ishihara et al. (Ishihara et al., 1999) suggested that ruthenium atoms located on the surface and with anion vacancy are active in HDS. Thus, the bigger ruthenium sulfide crystals lead to a lower active surface and, as a consequence, to a lower activity.

The catalytic results reported here are similar to those reported for alumina supported ruthenium sulfide–cesium catalysts with metal loadings between 4 and 12 wt% (Ishihara et al.,1999). The strong interaction of $RuS_2$ particles with the pores leads to an equivalent performance to that obtained when alumina is doped with $Cs^+$ ions (Ishihara et al., 2004).

## 3.2. Influence of Cs as stabilizer agent of RuSiSBA catalyst

The addition of Cs to $RuS_2$ catalyst supported on Si-SBA-15 was studied in order to evaluate the role of Cs on the stabilization of $RuS_2$ phase, as reported in the literature. The influence of cesium content and sulfiding temperature (773 K and 873 K), as well as cesium precursor salt employed (cesium hydroxide and cesium chloride), was also studied. The quantity of ruthenium was maintained constant (7 wt%) while the amount of cesium variable, with Cs/Ru molar ratios of 0.1:1, 0.5:1, and 1:1. The catalysts will be denoted to as $x$Cs$y$Ru, being $x$:$y$ the Cs/Ru molar ratio.

The catalytic results obtained for this family of catalysts in the HDS of DBT reaction is plotted in Figure 7, where $k_{HDS}$ values at a reaction temperature of 613 K are plotted for catalysts containing different amount of cesium and sulfided at 773 K and 873 K. For comparison, the catalytic data without cesium are also plotted (RuSiSBA). From this figure it can be clearly seen that the presence of cesium and the increment in the sulfidation temperature do not have a beneficial effect in terms of HDS activity, since the highest $k_{HDS}$ values is achieved for pure RuSiSBA sulfided at 773 K. It seems that the sulfiding process at higher temperature does not have enough influence on the catalytic activity.

With regards to selectivity values (Scheme 2), the results obtained here show that all catalysts follow the DBT HDS reaction through the DDS route, being even biphenyl the unique reaction product found for the catalysts with higher cesium loadings, being in line with the results reported by Ishihara (Ishihara et al., 1999). It follows the order 1Cs1Ru (100%)=0.5Cs1Ru (100%) > 0.1Cs1Ru (90.9%) > RuSiSBA (69.4%). Therefore, the hydrogenation capability decreases under the presence of Cs. Only the catalyst with a 0.1Cs:1Ru molar ratio presented a CHB selectivity of 10%.

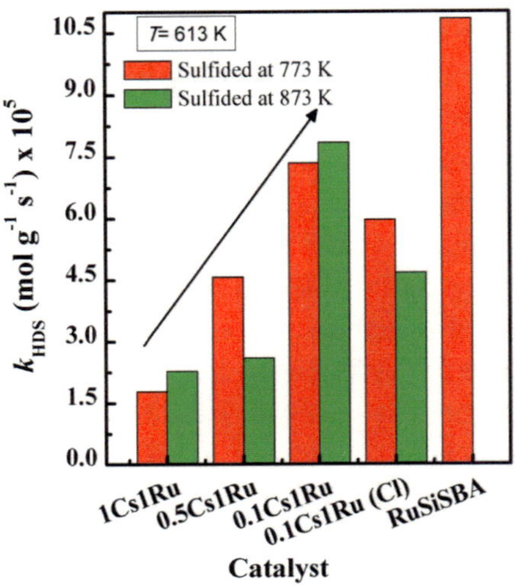


**Figure 7.** $k_{HDS}$ values as a function of cesium loading on the catalysts at a reaction temperature of 613 K

XRD results reveals that an agglomeration of the active phase occurs since the higher the cesium content the better defined the $RuS_2$ diffraction lines and therefore, the lower the dispersion. The same conclusion is extracted from textural properties, whose parameters decrease under the presence of cesium. It could be due to the formation $RuS_2$ agglomerates that provokes a blockage on the pores surface and hinders the access of nitrogen molecules. Moreover, the mean pore diameter also suffers a decrease but in a lesser extend. This is explained considering that agglomerates are blocking the entrance of some pores, however there are other ones where the metals are not deposited and are able to adsorb $N_2$ at 77 K, as can be clearly seen by TEM micrographs (*vide infra*).

Transmission electron microscopy elucidates the distribution of the active phase on the support. In general, it can be said that TEM analysis shows a heterogeneous distribution of the active phase, whose dispersion is totally dependent on cesium loading (Figure 8). The micrograph belonging to 1Cs1RuSTs catalyst is shown in Figure 8.A. At first glance, the micrographs show zones where there are big agglomerates on the external surface, assigned to the $RuS_2$ active phase according to EDAX analysis. Although there are some particles inside the pores, the dispersion of the active phase in the whole support is poor. Moreover, the analysis by EDAX in some zones of dispersed particles gave Cs/Cl atomic ratios close to 1 arising from the presence of CsCl compound. The sample with a Cs:Ru molar ratio of 0.5:1, presents a micrograph (Figure 8.B) where the dispersion of the $RuS_2$ phase is better, although agglomerates are still present but in a lesser extend than before. The catalyst possesses an alignment of the particles, indicating that they are mainly located inside the pores of the support. However, when diminishing the cesium loading until a Cs:Ru ratio of 0.1:1, the dispersion of the active phase increases conspicuously, as can be clearly seen from Figure 8.C, where particles lower than 10 nm are highly dispersed an located inside the channels. The data presented here reveal that the lower the cesium content, the better the dispersion of the active phase.

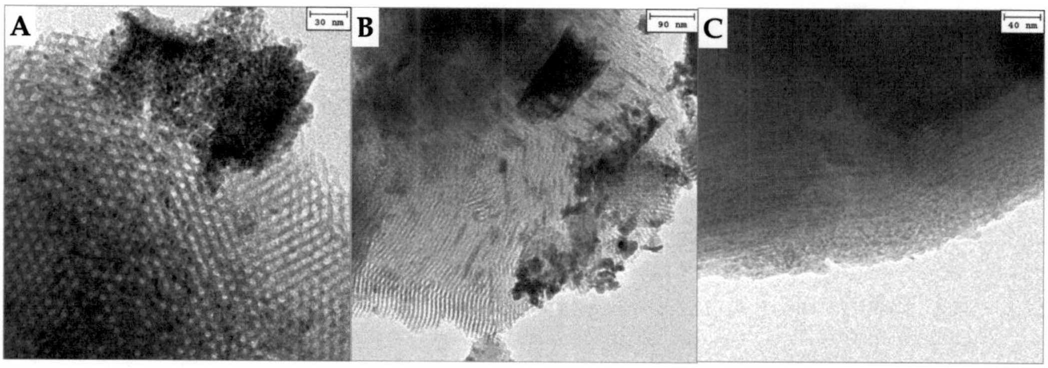


**Figure 8.** TEM micrograph for A) 1Cs1Ru catalyst, A) 0.5Cs1Ru catalyst and C) 0.1Cs1Ru catalysts sulfided at 773 K

In accordance with $H_2$-TPRS curves (Figure 9) the higher the cesium content the lesser the band intensities are. Moreover, the maxima of the curves are shifted to higher temperatures. In this regard, the 0.1Cs1Ru sample profile exhibits the most intense $H_2S$-release pattern that occurs at lower temperatures than that for 1Cs1Ru catalyst. It implies that the cesium content is the main reason of a low $H_2S$ elimination, i.e., a minor amount of labile sulfur is present on the catalysts.

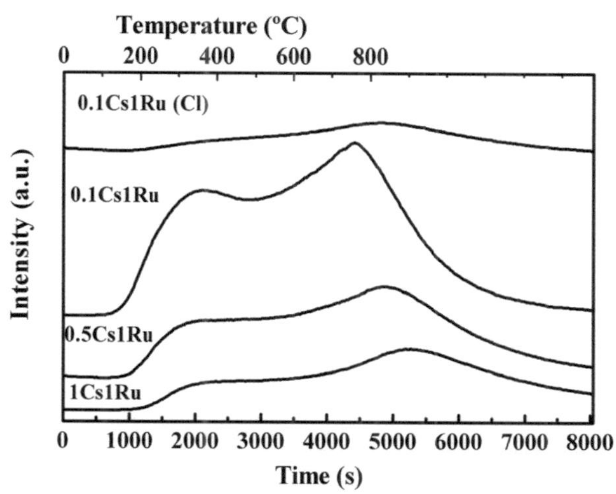


**Figure 9.** $H_2$-TPRS patterns of the catalysts sulfided at 773 K

The characterization and catalytic results indicate that the addition of cesium to a mesoporous material does not have a beneficial effect, mainly depending on the dispersion of the active phase attained. In this sense the higher the cesium content the lower the dispersion of the active phase and therefore the lower the catalytic activity in the DBT HDS reaction. Our results are contrary to those previously published in the literature. Ishihara, in a first work (Ishihara et al., 1996) studied the addition of alkali metal hydroxides to alumina-supported ruthenium catalysts and found that the cesium promoted catalyst was the most active. They reported that the location of cesium is close to ruthenium species, the dispersion of ruthenium species increases with an increase in the Cs/Ru ratio and furthermore the presence of cesium in close proximity to ruthenium atoms strengthens the bond of ruthenium and sulfur stabilizing ruthenium sulfide. In later works (Ishihara et al., 1998, 2003), they elucidated the behaviour of sulfur on the ruthenium catalysts and the role of cesium in HDS by radioisotope tracer methods, concluding that the mobility of sulfur on the catalysts decreased by the addition of cesium. On the contrary, the amount of labile sulfur on the catalyst increased with the amount of cesium added and reached the maximum at Ru:Cs=1:2 suggesting that Ru species in the catalyst was successfully dispersed on alumina. Further, it was reported that cesium promoted the C-S bond scission of DBT. With these premises and considering the characterization and activity results exposed here it can be pointed that the role of Cs to $RuS_2$ hydrotreating catalysts strongly depends on the support used. While the promoter effect of cesium on $\gamma$-$Al_2O_3$ is positive, on a mesoporous material such as SBA-15, the effect is negative. In this sense, the presence of cesium does not favour a good dispersion of the $RuS_2$ active phase, i.e., less cesium atoms are close to ruthenium atoms to stabilise the Ru-S bond and therefore the amount of labile sulfur also decreases by decreasing the dispersion. This is in agreement with our $H_2$-TPRS experiments that show an increase of the amount of $H_2S$ released with a decrease of cesium content in the catalysts, indicating that sulfur lability is inhibited in the presence of a large amount of Cs on the catalyst surface. The low dispersion of the active phase and the decreasing in the sulfur lability might explain the observed decrease in the catalytic activity with an increase Cs content in the catalysts (*vide supra*). Moreover, the presence of cesium altered the reaction mechanism in a way that only the product coming from the DDS route was obtained. By increasing the sulfiding temperature the catalytic activity did not improve considerably.

### 3.3. Influence of Cs precursor salt: CsOH versus CsCl

The catalyst with Cs:Ru molar ratio of 0.1:1 was also prepared from different cesium precursor salts, cesium hydroxide (0.1Cs1Ru) and cesium chloride (0.1Cs1Ru(Cl)), and sulfided at 773 K and 873 K. The catalytic results (Figure 7) showed that the catalyst prepared from cesium hydroxide exhibits a much higher HDS activity at both sulfided temperatures. The textural and structural properties of both catalysts are similar according to XRD, $S_{BET}$ and TEM analysis, i.e., they both present the same dispersion of the $RuS_2$ active phase. Notwithstanding the results extracted from TPRS results point to the different sulfur lability on both catalysts. The influence of the cesium precursor salt is more evident when the catalysts are sulfided at 773 K (Figure 7) where the differences found are very important, being the catalyst prepared from cesium hydroxide much more active.

The higher lability of sulfur in 0.1Cs1Ru sample was confirmed by DRIFT spectroscopy of adsorbed NO for both spent catalysts (Figures 10(a) and 10(b)). Thus, in Figure 10(a) the DRIFT spectra of NO adsorbed at room temperature onto both spent catalysts (after HDS at 613 K) are compared with that of NO adsorbed on pure SBA-15 support. As seen in this figure, both catalysts show two bands at 1905 and 1842 $cm^{-1}$ whereas the pure support shows one band centred at about 1860 $cm^{-1}$ and a shoulder at 1905 $cm^{-1}$. Additionally, all spectra show one band centred at 1875 $cm^{-1}$ due to NO adsorbed in its monomeric form in the gas phase (Dinerman et al., 1970). After subtraction of NO adsorbed on the pure support, the spectra of both spent catalysts show two bands at about 1900 and 1840 $cm^{-1}$ (Figure 10(b)) that could be tentatively ascribed to $(NO)_2$ dimmer species adsorbed on the Ru(Cs) sulfide phases. Interestingly, the spectrum of the most active catalyst in the HDS reaction at 613 K (0.1Cs1Ru) shows a band at about 1900 $cm^{-1}$ with a larger intensity than that of its counterpart prepared from cesium chloride, suggesting the presence of a larger amount of CUS sites. Moreover, TPO experiments revealed the higher amount of coke formed on 0.1Cs1Ru (Cl) catalyst, due to the presence of residual $Cl^-$ ions on the catalyst surface leads to an increase of the catalyst acidity which favours deactivation by coke formation.

The influence of the cesium precursor salt revealed that in spite of the same dispersion of the active phase, the usage of cesium hydroxide improved the amount of labile sulfur/the number of CUS sites and decreased the deactivation by coke.

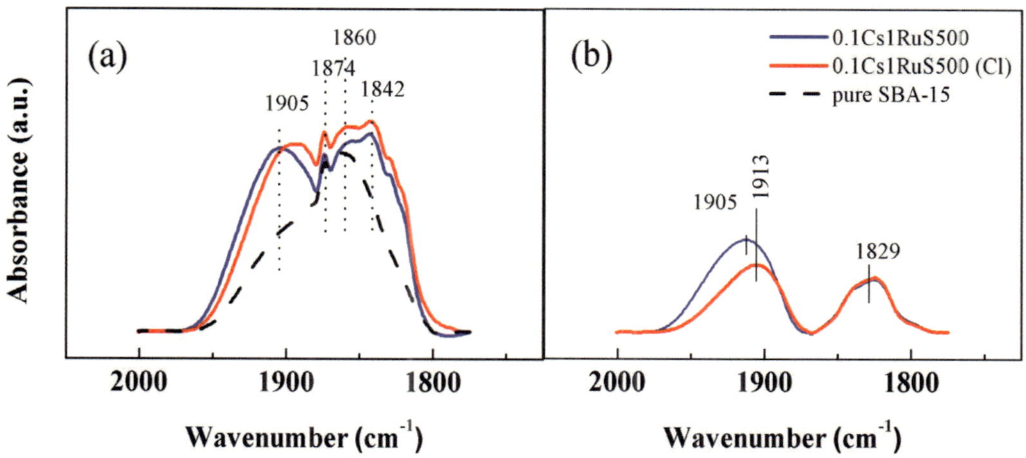


**Figure 10.** Influence of Cs precursor salt on the DRIFT spectra of NO adsorbed at room temperature for 10 min onto 0.1Cs1Ru (from CsOH) and 0.1Cs1Ru(Cl) (from CsCl) catalysts: (a) the spectra of spent catalysts (after HDS at 613 K) and pure SBA-15 support, (b) the difference spectra obtained after subtraction of NO adsorbed on pure support.

## Abbrevations

DRIFT Diffuse Reflectance Infrared Fourier Transform
EDAX Energy Dispersive X-ray analysis
TEM Transmission Electron Microscopy
TOS Time On Stream
TPD Temperature-Programmed Desorption
TPRS Temperature-Programmed Reduction of Sulfur
WHSV Weight Hourly Space Velocity
XPS X-ray Photoelectron Spectroscopy
XRD X-Ray Diffraction

## Author details

A. Infantes-Molina
*Instituto de Catálisis y Petroleoquímica, CSIC, Cantoblanco, Madrid, Spain*

A. Romero-Pérez, J. Mérida-Robles, A. Jiménez-López and E. Rodríguez- Castellón*
*Dpto. de Química Inorgánica, Cristalografía y Mineralogía. Facultad de Ciencias, Universidad de Málaga, Campus de Teatinos, Málaga, Spain*

D. Eliche-Quesada
*Departamento de Ingeniería Química, Ambiental y de los Materiales, EPS de Linares, Universidad de Jaén, Jaén,Spain*

## Acknowledgement

We gratefully acknowledge the support from the Ministry of Science and Innovation, Spain (MICINN, España) through the project MAT2009-10481, the regional government (JA) through the Excellence projects (P07-FQM-5070) and FEDER funds. A.R.P thanks the CONACyT (México) for its financial support (Scholarship No. 189933). A.I.M. also thanks the MICINN, Spain, for a Juan de la Cierva contract.

## 4. References

Alstrup, I., Chorkemdorff, I., Candia, R., Clausen, B.S., & Topsøe, H. (1982). A combined XPS and Mössbauer emission spectroscopy study of the state of cobalt in sulfided, supported, and unsupported CoMo catalysts. *J. of Catalysis*, Vol. 77, No. 2, (October 1982), pp. (397-409). ISSN: 0021-9517.

Arteaga, A., Fierro, J.L.G., Delannay, F., & Delmon, B. (1986). Simulated deactivation and regeneration of an industrial CoMo/$\gamma$-$Al_2O_3$ HDS catalyst. *Applied Catalysis*, Vol. 26, (1986), pp. (227-249). ISSN: 0166-9834.

* Corresponding Author

Bataille, B., Lemberton, J.-L., Michaud, P., Perot, G., Vrinat, M., Lemaire, M., Schulz, E., Breysse, M., & Kasztelan, S. (2000). Alkyldibenzothiophenes HDS Promoter Effect, Reactivity, and Reaction Mechanism. *J. of. Catalysis*, Vol. 191, No. 2, (April 2000), pp. (409-422). ISSN: 0021-9517.

Benitez, A., Ramírez, J., Fierro, J.L.G., & López-Agudo, A. (1996). Effect of fluoride on the structure and activity of $NiW/Al_2O_3$ catalysts for HDS of thiophene and HDN of pyridine. *Applied Catalysis A: General*, Vol. 144, No. 1–2, (September 1996), pp. (343-364). ISSN: 0926-860X.

Berhault, G., Lacroix, M., Breysse, M., Maugé, F., Lavalley, J.C., & Qu, L. (1997). Characterization of Acid–Base Paired Sites on Silica-Supported $RuS_2$ by Infrared Spectroscopy and Methyl Mercaptan Condensation Reaction. *J. of Catalysis*, Vol. 170, No. 1, (August 1997), pp. (37-45). ISSN: 0021-9517.

Berhault, G., Lacroix, M., Breysse, M., Maugé, F., Lavalley, J.C., Nie, H., & Qu, L. (1998). Characterization of Acidic Sites of Silica-Supported Transition Metal Sulfides by Pyridine and 2,6 Dimethylpyridine Adsorption: Relation to Activity in $CH_3SH$ Condensation. *J. of Catalysis*, Vol. 178, No. 2, (September 1998), pp. (555-565). ISSN: 0021-9517.

Castillo-Villalón, P., Ramírez, J., & Maugé, F. (2008). Structure, stability and activity of $RuS_2$ supported on alumina. *J. of Catalysis*, Vol. 260, No. 1, (November 2008), pp. (65-74). ISSN: 0021-9517.

Corma, A., Martinez, A., Martinezsoria, V., & Monton, J.B. (1995). Hydrocracking of Vacuum Gasoil on the Novel Mesoporous MCM-41 Aluminosilicate Catalyst. *J. of Catalysis*, Vol. 153, No. 1, (April 1995), pp. (25-31). ISSN: 0021-9517.

De los Reyes, J.A., Göbölös, S., Vrinat, M., & Breysse, M. (1990). Preparation and characterization of highly active ruthenium sulphide supported catalysts. *Catalysis Letters*, Vol. 5, No. 1, (January 1990), pp (17–24). ISSN: 1572-879X.

De Los Reyes, J.A., Vrinat, M., Geantet, C., & Breysse, M. (1991). Ruthenium sulphide catalysts supported on alumina: physicochemical characterization and catalytic properties in hydrotreating reactions. *Catalysis Today*, Vol. 10, No. 4, (November 1991), pp. (645-664). ISSN: 0920-5861.

De Los Reyes, J.A. (2007). Ruthenium sulfide supported on alumina as hydrotreating catalyst. *Applied Catalysis A: General*, Vol. 322, (April 2007), pp. (106-112). ISSN: 0926-860X.

Dinerman, C. E., Ewing, G.E. (1970). IR Spectrum, Structure, and Heat of Formation of Gaseous $(NO)_2$ .*J. of Chemical Physics,* 1970, Vol. 53, No. 2, pp. (626–632). ISSN: 0021-9606.

Eijsbouts, S., van Gestel, J.N.M., van Oers, E.M., Prins, R., van Veen, J.A.R., & de Beer, V.H.J. (1994). In situ poisoning of the thiophene HDS activity of carbon-supported transition metal sulfide catalysts by phosphorus. *Applied Catalysis A: General*, Vol. 119, No. 2, (November 1994), pp. (293-303). ISSN: 0926-860X.

Eliche-Quesada, D., Mérida-Robles, J., Maireles-Torres, P., Rodríguez-Castellón, E., & Jiménez-López, A. (2003a). Hydrogenation and Ring Opening of Tetralin on Supported Ni Zr-Doped Mesoporous Silica Catalysts. Influence of the Ni Precursor. *Langmuir,* Vol. 19, No. 12, (May 2003), pp (4985–4991). ISSN: 0743-7463.

Eliche-Quesada, D., Mérida-Robles, J., Maireles-Torres, P., Rodriguez-Castellón, E., Busca, G., Finocchio, E., & Jiménez-López, A. (2003b). Effects of preparation method and sulfur poisoning on the hydrogenation and ring opening of tetralin on NiW/Zr-doped mesoporous silica catalysts. *J. of Catalysis,* Vol. 220, No. 2, (December 2003), pp. (457-467). ISSN: 0021-9517.

Eliche-Quesada, D., Mérida-Robles, J., Maireles-Torres, P., Rodriguez-Castellón, E., & Jiménez-López, A. (2004). Superficial characterization and hydroconversion of tetralin over NiW sulfide catalysts supported on Zr-doped mesoporous silica. *Applied Catalysis A: General,* Vol. 262, No. 1, (May 2004), pp. (111-120). ISSN: 0926-860X.

Eliche-Quesada, D., Mérida-Robles, J., Rodriguez-Castellón, E., & Jiménez-López, A. (2005). Ru, Os and Ru–Os supported on mesoporous silica doped with Zr as mild thio-tolerant catalysts in the hydrogenation and hydrogenolysis/hydrocracking of tetralin. *Applied Catalysis A: General,* Vol. 279, No. 1–2, (January 2005), pp (209-221). ISSN: 0926-860X.

Farag H., Mochida I., & Sakanishi K. (2000). Fundamental comparison studies on HDS of DBT over CoMo-based carbon and alumina catalysts. *Applied Catalysis A: General,* Vol. 194–195, (March 2000), pp. (147-157). ISSN:0926-860X.

Gil, A., Díaz, A., Gandía, L.M., & Montes, M. (1994). Influence of the preparation method and the nature of the support on the stability of nickel catalysts. *Applied Catalysis A: General,* Vol. 109, No. 2, (March 1994), pp. (167-179). ISSN: 0926-860X.

Gómez-Cazalilla, M., Mérida-Robles, J.M., Gurbani, A., Rodríguez-Castellón, E., & Jiménez-López, A. (2007). Characterization and acidic properties of Al-SBA-15 materials prepared by post-synthesis alumination of a low-cost ordered mesoporous silica. *J. of Solid State Chemistry,* Vol. 180, No. 3, (March 2007), pp (1130-1140). ISSN: 0022-4596.

Gómez-Cazalilla, M., Infantes-Molina, A., Mérida-Robles, J., Rodríguez-Castellón, E., & Jiménez-López, A. (2009a). Cr Species as Captors of S Molecules on Ni-Based Hydrotreating Catalysts. *Energy Fuels,* Vol. 23, No. 1, (February, 2009), pp. (101–110). ISSN: 0887-0624.

Gómez-Cazalilla, M., Infantes-Molina, A., Moreno-Tost, R., Maireles-Torres, P.J., Mérida-Robles, J., Rodríguez-Castellón, E., & Jiménez-López, A. (2009b). Al-SBA-15 as a support of catalysts based on chromium sulfide for sulfur removal. *Catalysis Today,* Vol. 143, No. 1–2, (May 2009), pp. (137-144). ISSN: 0920-5861.

Grzechowiak J.R., Mrozinska K., Masalka A., Góralski J., & Tylus W. (2006). Effect of MCM-41 on the physicochemical properties of Mo and NiMo catalysts and their performance in DBT conversion. *Catalysis Today,* Vol. 114, No. 2–3, (May 2006), pp. (272-280). ISSN: 0920-5861.

Hsu, C.S., & Robinson, P.R., (2006). *Practical Advances in Petroleum Processing*, Vol. 1, Springer. ISBN: 0-387-25811-6. New York.

Infantes-Molina, A., Mérida-Robles, J., Rodríguez-Castellón, E., Pawelec, B., Fierro, J.L.G., & Jiménez-López, A. (2005). Catalysts Based on Co/Zr-doped Mesoporous Silica MSU for the Hydrogenation and Hydrogenolysis/Hydrocracking of Tetralin. *Applied Catalysis A: General*, Vol. 286, No. 2, (June 2005), pp. (239-248). ISSN: 0926-860X.

Ishihara, A., Nomura, M., & Kabe, T. (1992). HDS of DBT Catalyzed by Alumina-Supported Ruthenium Carbonyl-Alkali Metal Hydroxide Systems. *Chemistry Letters*. Vol. 21, No. 12, (August 1992), pp. (2285–2288). ISSN: 1348-0715.

Ishihara, A., Nomura, M., Takahama, N., Hamaguchi, K., & Kabe, T. (1996). HDS of DBT Catalyzed by Supported Metal Carbonyl Complexes (Part5), Catalysts for HDS Prepared from Alumina-supported Ru Carbonyl-Alkali Metal Hydroxide Systems. *Sekiyu Gakkaishi*, Vol. 39, No. 3, (October 1996), pp. (211-221). ISSN: 0582-4664.

Ishihara, A., Yamaguchi, M., Godo, H., Qian, W., Godo, M., & Kabe, T. (1997). HDS of DBT Catalyzed by Supported Metal Carbonyl Complexes (Part 8), HDS of $^{35}S$-Labeled DBT on Alumina-Supported Ru Sulfide-Cesium Catalysts. *Sekiyu Gakkaishi*, Vol. 1, No. 1, (June 1997), pp. (51-58). ISSN: 0582-4664.

Ishihara, A., Godo, H., Kanamori, R., Qian, W., & Kabe, T. (1999). HDS of $^{35}S$-labeled DBT on Alumina–Supported Ruthenium Sulfide–Cesium Catalysts. *Applied Catalysis A: General*, Vol. 182, No. 2, (June 1999), pp. (345-355). ISSN: 0926-860X.

Ishihara, A., Lee, J., Dumeignil, F., Wang, A., Qian, E.W., & Kabe, T. (2003). Elucidation of Sulfidation State and HDS Mechanism on Ruthenium–Cesium Sulfide Catalysts using $^{35}S$ Radioisotope Tracer Methods. *J. of Catalysis*, Vol. 217, No. 1, (July 2003), pp. (59-68). ISSN: 0021-9517.

Ishihara, A., Lee, J., Dumeignil, F., Yamaguchi, M., Hirao, S., Qian, E.W., & Kabe, T. (2004). Inhibiting Effect of $H_2S$ on the DBT HDS Activity of Ru-based Catalysts—Effect of the Cs Addition. *J. of Catalysis*, Vol. 224, No. 2, (June 2004), pp. (243-251). ISSN: 0021-9517.

Kloprogge, J.T., Welters, W.J.J., Booy, E., de Beer, V.H.J., van Santen, R.A., Geus, J.W., & Jansen J.B.H. (1993). Catalytic Activity of Nickel Sulfide Catalysts Supported on Al-pillared Montmorillonite for Thiophene HDS. *Applied Catalysis A: General*, Vol. 97, No. 1, (April 1993), pp. 77-85. ISSN: 0926-860X.

Kuo Y.J., Cocco, R.A., & Tatarchuk B.J. (1988). HYD and HDS over sulfided ruthenium catalysts: II. Impact of Surface Phase Behavior on Activity and Selectivity. *J. of Catalysis*, Vol. 112, No. 1, (July 1988), pp. (250-266). ISSN: 0021-9517.

Li, X., Wang, A., Wang, Y., Chen, Y., & Hu, Y. (2002). HDS of DBT over Ni–Mo Sulfides Supported by Protonexchanged Siliceous MCM-41. *Catalysis Letters*, Vol. 84, No. 1-2 (November 2002), pp. (107–113). ISSN: 1572-879X.

Li, X., Wang, A., Sun, Z., Li, C., Ren, J., Zhao, B., Wang, Y., Chen, Y., & Hu, Y. (2003). Effect of Surface Proton Exchange on HDS Performance of MCM-41-supported Catalysts.

*Applied Catalysis A: General*, Vol. 254, No. 2, (November 2003), pp. (319-326). ISSN: 0926-860X.

Liaw, S.J., Lin, R., Raje, A., & Davis, B.H. (1997). Hydrotreatment of coal-derived naphtha. Properties of zeolite-supported Ru sulfide catalysts. *Applied Catalysis A: General*, Vol. 151, No. 2, (April 1997), pp. (423-435). ISSN: 0926-860X.

Louwers, S.P.A., & Prins, R. (1992). Ni EXAFS Studies of the Ni-Mo-S Structure in Carbon-Supported and Alumina-Supported Ni-Mo Catalysts. *J. of Catalysis*, Vol. 133, No. 1, (January 1992), pp. (94-111). ISSN: 0021-9517.

Mangnus, P.J., Bos, A., & Moulijn, J. (1994). TPR of Oxidic and Sulfidic Alumina-Supported NiO, $WO_3$, and NiO-$WO_3$ Catalysts. *J. of Catalysis*, Vol. 146, No. 2, (April 1994), pp (437-448). ISSN: 0021-9517.

Mérida-Robles, J., Olivera-Pastor, P., Jiménez-López, A., & Rodríguez-Castellón E. (1996). Preparation and Properties of Fluorinated Alumina-Pillared $\alpha$-Zirconium Phosphate Materials. *The Journal of Physical Chemistry*. Vol. 100, No. 35, (August 1996), pp. (14726-14735). ISSN: 1932-7447.

Mitchell, P.C.H., Scott, C.E., Bonnelle, J.P., & Grimblot, J.G., (1987). Ru/Alumina and RuMo/Alumina Catalysts: An XPS study. *J. of Catalysis*, Vol. 107, No. 2, (October 1987), pp. ( 482-489). ISSN: 0021-9517.

Moulder, J.F., Stickle, W.F., Sobol, P.E., & Bomben, K.D. (October 1992). *Handbook of XPS*, Perkin-Elmer Corporation, ISBN:0-9627026-2-5. Minnesota, 1992.

Nava, R., Infantes-Molina, A., Castaño, P., Guil-López R., & Pawelec B. (2011). Inhibition of CoMo/HMS Catalyst Deactivation in the HDS of 4,6-DMDBT by Support Modification with Phosphate. *Fuel*, Vol. 90, No. 8, (August 2011), pp. (2726-2737). ISSN: 0016-2361.

Navarro, R., Pawelec, B., Fierro, J.L.G., & Vasudevan P.T. (1996). DBT HDS on Silica-Alumina-Supported Transition Metal Sulfide Catalysts. *Applied Catalysis A: General*, Vol. 148, No. 1, (December 1996), pp. (23-40). ISSN: 0926-860X.

Occelli, M.L., & Rennard, R.J. (1988). Hydrotreating Catalysts Containing Pillared Clays. *Catalysis Today*, Vol. 2, No. 2–3, (February 1988), pp.(309-319). ISSN: 0920-5861.

Okamoto, Y., Nakano, H., & Shimakawa, T. (1977). Stabilization effect of Co for Mo phase in Co-Mo/$Al_2O_3$ HDS catalysts studied with XPS. *J. of Catalysis*, Vol. 50, No. 3, (August 1977), pp. (447-454). ISNN: 0021-9517

Pawelec, B., Navarro, R.M., Campos-Martin, J.M., López-Agudo, A., Vasudevan, P.T., & Fierro, J.L.G. (2003). Silica–Alumina-Supported Transition Metal Sulfide Catalysts for Deep HDS. *Catalysis Today*, Vol. 86, No. 1–4, (November 2003), pp. (73-85). ISSN: 0920-5861.

Pecoraro, T.A., & Chianelli, R.R. (1981). HDS Catalysis by Transition Metal Sulfides. *J. of Catalysis*, Vol. 67, No. 2, (February 1981), pp. (430-445). ISSN: 0021-9517.

Quartararo, J., Mignard, S., & Kasztelan, S. (2000). HDS and Hydrogenation Activities of Alumina-Supported Transition Metal Sulfides. *J. of Catalysis*, Vol. 192, No. 2, (June 2000), pp (307-315). ISSN: 0021-9517.

Rodríguez-Castellón, E., Jiménez-López, A., Maireles-Torres, P., Jones, D.J., Rozière, J., Trombetta, M., Busca, G., Lenarda, M., & Storaro, L. (2003). Textural and Structural Properties and Surface Acidity Characterization of Mesoporous Silica-Zirconia Molecular Sieves. J. of Solid State Chemistry, Vol. 175, No. 2, (November 2003),pp. (159-169). ISSN: 0022-4596.

Ruette, F., & Ludeña, E.V. (1981). Molecular Orbital Calculations of the HDS of thiophene over a MoCo catalyst. J. of Catalysis, Vol. 67, No. 2, (February 1981), pp. (266-281). ISSN: 0021-9517.

Silva-Rodrigo, R., Calderón-Salas, C., Melo-Banda, J.A., Domínguez, J.M., & Vázquez-Rodríguez, A. (2004). Synthesis, characterization and comparison of catalytic properties of NiMo- and NiW/Ti-MCM-41 catalysts for HDS of thiophene and HVGO. *Catalysis Today*, Vol. 98, No. 1-2, (September 2004), pp. (123–129). ISSN: 0920-5861.

Song, C., & Reddy K.M. (1999). MCM-41 Supported Co–Mo Catalyst for HDS of DBT in Distillate Fuels. *Applied Catalysis A: General*, Vol. 176, No. 1, (January 1999), pp. (1-10). ISSN: 0926-860X.

Song, C., & Ma, X. (2003). New Design Approaches to Ultra-Clean Diesel Fuels by Deep Desulfurization and Deep Dearomatization. *Applied Catalysis B: Environmental*, Vol. 41, No. 1–2, (March 2003), pp. (207-238), ISSN: 0926-3373.

Topsoe, H., Clausen, B.S., Topsoe, N.-Y., Pedersen, E., Niemann, W., Müller, A., Bögge, H., & Lengeler, B. (1897). Faraday Transactions 1: Physical Chemistry in Condensed Phases. *J. of the Chemical Society*, Vol. 1, No. 83, (July 1987), pp. (2157-2167). ISSN:0300-9599.

Toulhoat, H., Raybaud, P., Kasztelan, S., Kresse, G., Hafner, J. (1999). Transition Metals to Sulfur Binding Energies Relationship to Catalytic Activities in HDS: Back to Sabatier with First Principle Calculations. *Catalysis Today*, Vol. 50, No. 3–4, (May 1999), pp. (629-636). ISSN: 0920-5861.

Vradman, L., Landau, M.V., Herskowitz, M., Ezersky, V., Talianker, M., Nikitenko, S., Koltypin, Y., & Gedanken, A. (2003). High Loading of Short $WS_2$ Slabs Inside SBA-15: Promotion with Ni and Performance in HDS and hydrogenation. *J. of Catalysis*, Vol. 213, No. 2, (January 2003), pp. (163-175). ISSN: 0021-9517.

Wang, A., Wang, Y., Kabe, T., Chen, Y., Ishihara, A., & Qian, W. (2002). HDS of DBT Over MCM-41-Supported Catalysts – II Sulfided Ni–Mo Catalysts. *J. of Catalysis*, Vol. 210, No. 2, (September 2002), pp. (319-327). ISSN: 0021-9517.

Welters, W.J.J., Vorbeck, G., Zandbergen, H.W., DeHaan, J.W., de Beer, V.H.J., & Van Sante, R.A. (1994). HDS Activity and Characterization of Zeolite-Supported Nickel Sulfide Catalysts. *J. of Catalysis*, Vol. 150, No. 1, (November 1994), pp. (155-169). ISSN: 0021-9517.

Whitehurst, D.D., Isoda, T., & Mochida, I. (1998). Present State of the Art and Future Challenges in the HDS of Polyaromatic Sulfur Compounds. *Advances in Catalysis*, Vol. 42, (1998), pp. (345-471). ISSN: 0-12-007844-9.

Yang, S. H., & Satterfield, C.N. (1983). Some Effects of Sulfiding of a $NiMoAl_2O_3$ Catalyst on its Activity for HDN of quinoline. *J. of Catalysis*, Vol. 81, No. 1, (May 1983), pp. (168-178). ISSN: 0021-9517.

Zepeda, T.A., Fierro, J.L.G., Pawelec, B., Nava, R., Klimova, T., Fuentes, G.A., & Halachev, T. (2005). Synthesis and Characterization of Ti-HMS and CoMo/Ti-HMS Oxide Materials with Varying Ti Content. *Chemistry of Materials*. Vol. 17, No. 16, (May 2005), pp. (4062–4073). ISSN: 0897-4756.

# Chapter 10

# Recent Advances in Transition Metal-Catalysed Homogeneous Hydrogenation of Carbon Dioxide in Aqueous Media

Wan-Hui Wang and Yuichiro Himeda

Additional information is available at the end of the chapter

http://dx.doi.org/10.5772/48658

## 1. Introduction

The excessive combustion of fossil fuels leads to enormous emissions of carbon dioxide which is the major greenhouse gases and significantly contributes to global warming. Since the middle of 20th century, the atmospheric concentration of $CO_2$ has risen remarkably. With the development of human society and increase in energy demand, emissions of $CO_2$ are increasing dramatically. To reduce the server environmental impact, scientists have paid considerable effort to prohibiting the increase of atmospheric $CO_2$ concentration. $CO_2$ is an attractive C1 resource because it is non-toxic, nonflammable, and abundant. Transforming of carbon dioxide to useful chemicals, fuels, and materials have attracted increasing attention because it could reduce the dependence on diminishing fossil oil as well as mitigate $CO_2$ increase. However, utilizing carbon dioxide is still a challenge research field due to its high stability ($\Delta G^{\circ}_{298}$ = −394.36 kJ mol$^{-1}$). In the last decades, the homogeneous catalytic hydrogenation of $CO_2$ has been widely studied. There are some reviews related to this subject.(Leitner et al., 1998; Jessop et al., 2004; Himeda, 2007; Jessop, 2007; Federsel et al., 2010b; Wang et al., 2011) Besides formic acid, theoretically, $CO_2$ can be hydrogenated to multiple compounds such as, formamides, formaldehyde, methanol, and methane (Eq 1). However, generation of these compounds typically require harsher conditions which make most homogeneous catalysts deactivating, increase the energy cost and make these reductions economically unfavourable. In this chapter, we will focus on the hydrogenation of $CO_2$ to formic acid or formate which is relatively easy to achieve. Especially, formic acid has recently been recognized as a feasible hydrogen vector. Hydrogenation of $CO_2$ to formic acid, combined with the reverse reaction (ie. decomposition of formic acid) is considered as one promising method of hydrogen storage (Eq 1, step 1).

$$CO_2 \underset{-H_2}{\overset{+H_2}{\rightleftharpoons}} HCO_2H \xrightarrow[-H_2O]{+H_2} HCHO \xrightarrow{+H_2} CH_3OH \xrightarrow[-H_2O]{+H_2} CH_4 \quad (1)$$

Complexes based on most of group VIII transition metals such as Pd, Ni, Rh, Ru, Ir et al. can be used to catalysis $CO_2$ hydrogenation. Among these catalysts, Rh, Ru, and most recently Ir complexes were found to be most effective. Besides transition-metal catalyst, solvent is also important for optimizing the reaction rate. The homogeneously catalytic hydrogenation of $CO_2$ to formic acid was firstly reported in 1976 by Inoue et al.(Inoue et al., 1976) They found the reaction was accelerated by adding small amounts of $H_2O$. However, in the early years, water-insoluble phosphine ligands are generally employed. Due to the insolubility of the phosphorous complexes in water, the homogeneous hydrogenation of $CO_2$ generally proceeded in organic solvents, such as DMSO, with water less than 20%. Until 1993, Leitner et al. reported the first water soluble rhodium catalysts which achieve the high turnover number (TON) of 3440 under relatively mild conditions.(Gassner & Leitner, 1993) Noyori and Jessop et al. have demonstrated supercritical $CO_2$ is an effective solvent due to the enormous concentration of $CO_2$ and $H_2$ and obtained highest catalytic performance at that time.(Jessop et al., 1994; Jessop et al., 1996; Munshi et al., 2002) Compared to reduction of $CO_2$ in organic solvent and supercritical $CO_2$, the homogeneous hydrogenation of $CO_2$ to formic acid in the green solvent—water has recently achieved great success and attracted much more attention. Despite $H_2$ is less soluble in water, it is still considered to be a preferred solvent because water is abundant, inexpensive, and eco-friendly. More importantly, hydrogenation of $CO_2$ in water is considerably favoured ($\Delta G° = -4$ kJ mol$^{-1}$) compared to the reaction in gas phase ($\Delta G° = +32.9$ kJ mol$^{-1}$). In addition, excellent activity usually requires basic additives, such as NaOH, $NaHCO_3$, $Na_2CO_3$ and amines, which can absorb the generated proton and make the reaction thermodynamically favourable (Scheme 1).

| Reaction | $\Delta G^0$ (kJ mol$^{-1}$) | $\Delta H^0$ (kJ mol$^{-1}$) | $\Delta S^0$ (J mol$^{-1}$ K$^{-1}$) |
|---|---|---|---|
| $CO_2(g) + H_2(g) \longrightarrow HCO_2H(l)$ | 32.9 | -31.2 | -215 |
| $CO_2(aq) + H_2(aq) \longrightarrow HCO_2H(aq)$ | -4 | - | - |
| $CO_2(aq) + H_2(aq) + NH_3(aq) \longrightarrow HCO_2^-(aq) + NH_4^+(aq)$ | -9.5 | -84.3 | -250 |

**Scheme 1.** The thermodynamics of hydrogenation of carbon dioxide to formic acid/formate.

In this chapter, we review the state-of-the-art in homogeneous $CO_2$ hydrogenation to formic acid or formate in water; discuss the design and synthesis of highly effective water soluble complexes, as well as the catalytic mechanism. We also present the latest strategy for recycle and reuse of homogenous catalyst.

## 2. Hydrogenation with Ru and Rh complexes

In the pioneering work of Inoue et al., the famous Wilkinson catalyst ($RhCl(PPh_3)_3$) and the Ru analogue ($RuCl(PPh_3)_3$) were used and showed much better results than other catalysts of Pd, Ni, and Ir.(Inoue et al., 1976) Following this work, a variety of Rh and Ru catalyst based on various phosphorus ligands were developed and applied in the hydrogenation of $CO_2$. Recently, N-based ligands have been also investigated for this purpose and achieved great success. Most of the highly efficient Ru and Rh catalysts as well as some Ir complexes and their performance are listed in Table 1.

### 2.1. Phosphorous ligands

In 1993, Leitner et al. reported the first homogenous hydrogenation of $CO_2$ with water soluble rhodium–phosphane complexes in aqueous solutions.(Gassner & Leitner, 1993) The reaction was carried out in an aqueous solution of amine at room temperature under 40 atm of $H_2/CO_2$ (1/1). Using dimethylamine as an additive, the Rh complex [RhCl(tppts)$_3$]/tppts (tppts = tris(3-sulfonatophenyl)phosphine) can provide 1.76 M of formic acid with TON of 3440 which is the highest at that time. In 1999, Joó et al. have reported the hydrogenation using inorganic base such as $NaHCO_3$ and $CaCO_3$ instead of the organic amine as additive.(Joó et al., 1999) Among the different series of Ru and Rh catalysts, [RhCl(tppms)$_3$] (tppms = 3-sulfonatophenyldiphenylphosphine) exhibited the better activity than others, and gave a turnover frequency (TOF) of 262 $h^{-1}$. Based on the equilibrium of $CO_2 + H_2O \rightarrow HCO_3^- + H^+$, they proposed that $HCO_3^-$ may be the real substrate in the catalytic cycle. In 2000, Laurenczy et al. reported hydrogenation with moderate activity using [$RuCl_2$(PTA)$_4$] (PTA = 1,3,5-triaza-7-phosphaadamantane) in an aqueous solution at 25-80 °C under 20 bar $CO_2$ and 60 bar $H_2$. In contrast to the other works, they found that slightly acidic and neutral conditions are preferable for the reaction rate. In case of 10%$HCO_3^-$/90%$CO_2$, they obtained the maximum TOF of 807 $h^{-1}$ and they suggested that the real substrate of hydrogenation in this system is the bicarbonate anion.(Laurenczy et al., 2000)

In 2003, Joó et al. reported the hydrogenation of bicarbonate with [$RuCl_2$(tppms)$_2$]$_2$ (tppms = sodium diphenylphosphinobenzene-3-sulfonate) at 50 °C and 10 bar $H_2$, a TOF of 54 $h^{-1}$ was obtained.(Elek et al., 2003) Their results suggest that bicarbonate is more reactive than hydrated $CO_2$. Interestingly, in presence of 5 bar $CO_2$ the reaction was about 10% slower than that without $CO_2$. This result is in contrast to the reaction with [$RuCl_2$(PTA)$_4$], which increased significantly with increasing of $CO_2$ pressure. Using this complex [$RuCl_2$(tppms)$_2$]$_2$, they have achieve a TOF of 9600 $h^{-1}$, the highest rate in pure aqueous solutions at that time, at 80 °C under $H_2/CO_2$ (60/35 bar) in a 0.3 M $NaHCO_3$ solution.

Most recently, Beller et al. reported hydrogenation of bicarbonate in $H_2O$/THF with in situ catalyst of [$RuCl_2(C_6H_6)$]$_2$/dppm (dppm = 1,2-bis(diphenylphosphino)methane).(Boddien et al., 2011) In the presence of $CO_2$, the reaction gave higher TON than that in the absence of $CO_2$. In addition, the catalyst can also catalyse the dehydrogenation of formate. Consequently, they pronounced the first hydrogen storage based on interconversion of formate and bicarbonate. Soon after that, Joó et al. used [$RuCl_2$(tppms)$_2$]$_2$/tppms to catalyse the hydrogenation of bicarbonate as well as the dehydrogenation of formate, and constructed a simple, rechargeable hydrogen storage device.(Papp et al., 2011) Similar with the system reported by Beller et al., but Joó's system is in pure aqueous solutions and require no organic solvent.

## 2.2. Nitrogenous ligands

In 2003, Himeda et al. announced the homogenous hydrogenation of $CO_2$ in water with a series of 2,2′-bipyridine- and 1,10-phenathroline-based Ru and Rh catalysts including [Cp*Rh(bpy)Cl]Cl (Cp* = pentamethylcyclopentadinyl; bpy = 2,2′-bipyridine), [Cp*Rh(4,4′-Me-bpy)Cl]Cl (4,4′-Me-bpy = 4,4′-dimethyl-2,2′-bipyridine), $[(\eta^6\text{-}C_6Me_6)Ru(phen)Cl]Cl$ (phen = 1,10-phenathroline) etc.(Himeda et al., 2007a) Among these catalysts, complex **4** based on 4,7-dihydroxyl-1,10-phenanthroline (DHPT) exhibited high activity and reached a TON of 2400 in a 1 M $KHCO_3$ solution under 4 MPa $H_2/CO_2$ (1/1) at 80 °C after 21 h. Soon after that, Ogo et al. reported a mechanistic study of the hydrogenation with similar complexes under acidic conditions. They synthesized a water-soluble ruthenium hydride complex $[(\eta^6\text{-}C_6Me_6)Ru^{II}(bpy)H](SO_4)$ from the reaction of an aqua complex $[(\eta^6\text{-}C_6Me_6)Ru^{II}(bpy)(OH_2)](SO_4)$ with $NaBH_4$ in water.(Hayashi et al., 2003) The hydride complex was found to be active in reaction with $CO_2$, but the reaction rate obtained by UV spectroscopy was demonstrated to be very slow. One year later, they achieved the hydrogenation of $CO_2$ to HCOOH in acidic solutions (pH 2.5-5.0) under $H_2$ (5.5 MPa) and $CO_2$ (2.5 MPa) at 40 °C with ruthenium complexes $[(\eta^6\text{-}C_6Me_6)Ru^{II}(bpy)(OH_2)](SO_4)$ and $[(\eta^6\text{-}C_6Me_6)Ru^{II}(4,4'\text{-}OMe\text{-}bpy)(OH_2)](SO_4)$ (4,4′-MeO-bpy = 4,4′-dimethoxyl-2,2′-bipyridine).(Hayashi et al., 2004) The TON was over 50 after 70 h. The reaction rate reached a maximum value at 40 °C and decreased with further increasing of temperature due to the decomposition of HCOOH at higher temperature. In contrast to the inactivity of this kind of complexes, Himeda et al. achieved significantly higher activity with **1-4** (Figure 1 and Table 1) by introducing two strong electron-donating groups into the bipyridine ligands. (Himeda et al., 2004, 2006, 2011) More interestingly, much higher activity was obtained with the iridium analogue (vide infra).

In 2010, Peris et al. used strong electron-donating bis-NHCs ligand (**5** and **6**) to mimic the bipyridine ligand and achieved a high TON of 23,000 with complex **6** at 40 atm $H_2/CO_2$ (1/1) and 200 °C in a 1 M KOH solution for 75 h.(Sanz et al., 2010a) It is worth note that they also achieved transfer hydrogenation of $CO_2$ with *i*PrOH using the Ru complex **6**, and obtained the highest TON of 874 so far reported for this type of reaction.

# 3. Hydrogenation with Ir complexes

Although research into iridium catalysts dates back to 1976(Inoue et al., 1976), the promising catalytic activity of iridium complexes has only recently been discovered. In the pathbreaking work of Inoue et al., the iridium complex $H_3Ir(PPh_3)_3$ exhibited a lower activity than Rh and Ru analogues. About 20 years later, an iridium catalyst was again applied to $CO_2$ hydrogenation and similar result was observed.(Joó et al., 1999) Rhodium(I) and

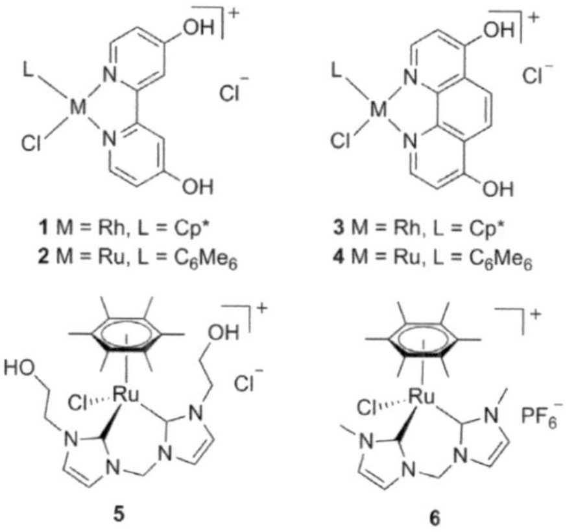


**Figure 1.** Ru and Rh catalysts with nitrogenous ligands.

| Catalyst | Solvent | Additive | $P(H_2/CO_2)$ /MPa | T/°C | t/h | TOF[a] /$h^{-1}$ | TON |
|---|---|---|---|---|---|---|---|
| **Ruthenium** | | | | | | | |
| $RuH_2(PPh_3)_4$ | $C_6H_6$ | $NEt_3/H_2O$ | 2.5/2.5 | rt | 20 | 4 | 87 |
| $RuH_2(PMe_3)_4$ | $scCO_2$ | $NEt_3$ | 8.5/12 | 50 | 1 | 1400 | 1400 |
| $RuCl(OAc)(PMe_3)_4$ | $scCO_2$ | $NEt_3/C_6F_5OH$ | 7/12 | 50 | 0.3 | 95,000 | 32,000 |
| $RuCl_2(PTA)$ | $H_2O$ | $NaHCO_3$ | 6/0 | 80 | - | (807) | - |
| | $H_2O$ | $NaHCO_3$ | 6/3.5 | 80 | 0.03 | 9600 | 320 |
| $[RuCl_2(tppms)_2]_2$ | $H_2O$ | $NaHCO_3$ | 0.2/0.8 | 50 | 1 | 50 | |
| $[RuCl_2(C_6Me_6)]$/ dppm | $H_2O$ | $NaHCO_3$ | 5/0 | 130 | 2 | 800 | 1600 |
| K[RuCl(EDTA-H)] | $H_2O$ | - | 0.3/1.7 | 40 | 0.5 | 250 | - |
| **2** | $H_2O$ | KOH | 3/3 | 120 | 8 | (4400) | 13,620 |
| **4** | $H_2O$ | KOH | 3/3 | 120 | 24 | (3600) | 15,400 |
| **Rhodium** | | | | | | | |
| $RhCl(tppts)_3$ | $H_2O$ | $NHMe_2$ | 2/2 | 81 | 0.5 | 7300 | - |
| | $H_2O$ | $NHMe_2$ | 2/2 | rt | 12 | 287 | 3400 |
| **1** | $H_2O$ | KOH | 2/2 | 80 | 12 | (790) | 1800 |
| **3** | $H_2O$ | KOH | 2/2 | 80 | 32 | (270) | 2400 |
| **Iridium** | | | | | | | |
| **8** | $H_2O$ | KOH | 3/3 | 120 | 57 | (42,000) | 190,000 |
| **9** | $H_2O$ | KOH | 3/3 | 120 | 48 | (33,000) | 222,000 |
| **9** | $H_2O$ | KOH | 0.5/0.5 | 30 | 30 | (3.5) | 81 |
| **10** | $H_2O$/THF | KOH | 4/4 | 200 | 2 | 150,000 | 300,000 |
| | $H_2O$/THF | KOH | 4/4 | 120 | 48 | 73,000 | 3,500,000 |
| **15** | $H_2O$ | KOH | 3/3 | 200 | 75 | 2500 | 190,000 |

**Table 1.** Hydrogenation of $CO_2$ to formic acid/formate. *a*. The data in parenthesis are initial TOF.

ruthenium(II) complexes with a water-soluble phosphine ligand, tppms, showed a TOF up to 262 $h^{-1}$ in an aqueous solution under mild conditions. However, the iridium complex, [IrCl(CO)(tppms)$_2$], gave no formate product under the same conditions. In 2008, Gonsalvi and Laurenczy et al. reported a half-sandwich iridium complex, **7**, bearing a water-soluble phosphine ligand PTA.(Erlandsson et al., 2008) Compared to the ruthenium and rhodium analogous,(Horváth et al., 2004) it gave a much lower TOF of 22.6 $h^{-1}$ at 100 °C. These preliminary studies implied that iridium complexes only provide low catalytic activity for the hydrogenation of $CO_2$. However, recent research made a breakthrough and the high catalytic ability of iridium complexes was demonstrated resulting in renewed attention on iridium complexes. The representative iridium catalysts, **7–15**, are presented in Figure 2 and the catalytic results are listed in Table 2.

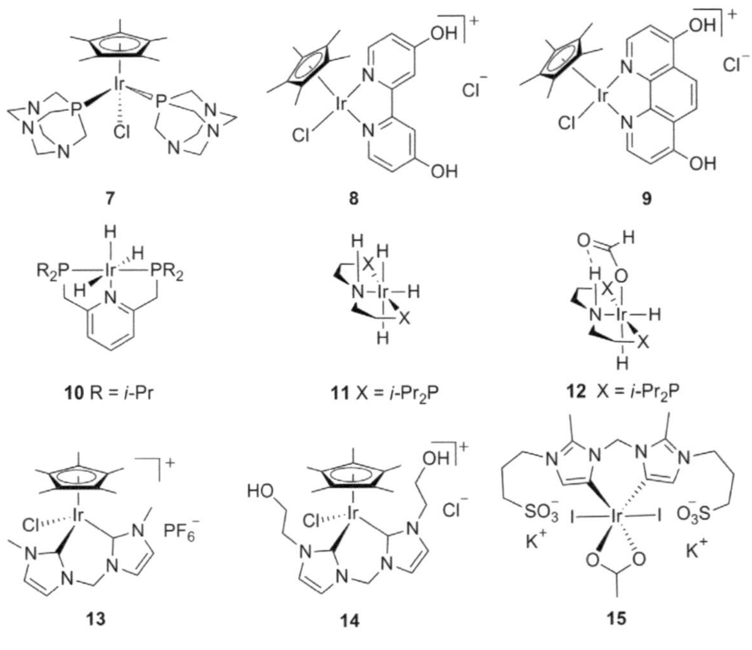


**Figure 2.** Representative iridium catalysts for the hydrogenation of $CO_2$.

Himeda and co-workers achieved a highly efficient iridium catalyst for the hydrogenation of $CO_2$ in $H_2O$ through sophisticated ligand design.(Himeda et al., 2004, 2005, 2006, 2007b; Himeda, 2007) At first, they focused on a half-sandwich bipyridine (bpy) rhodium complex, [Cp*Rh(bpy)X]$^+$, as a prototype catalyst.(Himeda et al., 2003) Preliminary studies showed that this catalyst successfully hydrogenated $CO_2$ in water but in low rate. Based on the rationale that electron-donating ligands should improve the catalytic activity of the complex, a tunable dihydroxylbipyridine (DHBP) ligand was introduced. The acid-base equilibrium between the hydroxyl and oxyanion forms enabled switching of the polarity and electron-donating ability of the ligand thus affecting the catalytic activity and water-solubility of the complex (Figure 3).

| Catalyst | Solvent | Additive | T/°C | Pressure /MPa | t/h | Initial $TOF^a/h^{-1}$ | TON |
|---|---|---|---|---|---|---|---|
| **7** | $H_2O$ | - | 100 | 10 | - | 23 | - |
| **8** | $H_2O$ | KOH | 120 | 6 | 57 | 42,000 | 190,000 |
| **8** | $H_2O$ | KOH | 60 | 0.1 | 50 | 33 | 376 |
| **9** | $H_2O$ | KOH | 60 | 0.1 | 50 | 32 | 444 |
| **9** | $H_2O$ | KOH | 30 | 0.1 | 30 | 3.5 | 81 |
| **9** | $H_2O$ | KOH | 120 | 6 | 48 | 33,000 | 222,000 |
| **10** | $H_2O$/THF | KOH | 200 | 5 | 2 | (150,000) | 300,000 |
| **10** | $H_2O$/THF | KOH | 120 | 6 | 48 | (73,000) | 3,500,000 |
| **12** | $H_2O$ | KOH | 185 | 5.5 | 24 | (14,500) | 348,000 |
| **12** | $H_2O$ | KOH | 185 | 5.5 | 1 | (18,780) | 18,780 |
| **13** | $H_2O$ | KOH | 80 | 6 | 18 | - | 1600 |
| **14** | $H_2O$ | KOH | 200 | 4 | 20 | - | 9500 |
| **15** | $H_2O$ | KOH | 200 | 6 | 75 | (2500) | 190,000 |

**Table 2.** Hydrogenation of $CO_2$ to formic acid/formate using iridium catalysts. *a*. The data in parenthesis are average TOF.

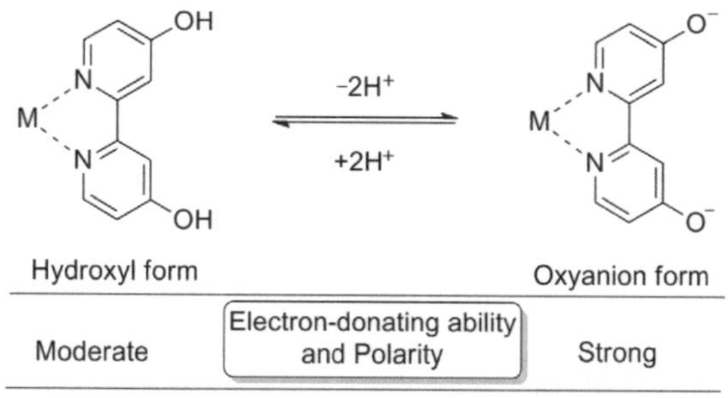


**Figure 3.** Acid-base equilibrium between hydroxyl and oxyanion forms.

Under basic conditions, the hydroxyl group can be deprotonated to generate an oxyanion, which is a much stronger electron donor. Therefore, high catalytic activity was achieved by introducing two electron-donating hydroxyl groups onto the bipyridine ligand. Table 3 shows the effect of the hydroxyl group in the half-sandwich bipyridine catalyst, $[(C_nMe_n)M(L)Cl]^+$ (M = Rh, Ir, n = 5; M = Ru, n = 6), on the hydrogenation of $CO_2$. Significant activation of the catalysts was observed. The TONs of the iridium catalysts with hydroxyl groups were 52–103 times greater than those of the unsubstituted catalysts. The electronic substituent effect was investigated using $[(C_nMe_n)M(4,4'\text{-}R_2\text{-}2,2'\text{-bpy})Cl]^+$ (M = Ir, Rh, Ru; R = OH, OMe, Me, H). Note that under basic conditions the hydroxyl group (Hammett constant, $\sigma_p^+ = -0.92$) was deprotonated to generate an oxyanion, which is a much stronger donor and has a $\sigma_p^+$ of −2.30. The Hammett plots show a good correlation between the initial TOFs and the $\sigma_p^+$ values which indicate their electron donating ability (Figure 4). This result suggests that strong donating ability of the substituents lead to high activity of the complexes. On the other hand, the substituent effects on the rhodium and ruthenium complexes, **1** and **2**, were moderate compared to the effect on iridium complex **8** (Figure 4).(Himeda et al., 2011) It is apparent that the remarkable activation of the iridium DHBP catalyst can be attributed to the strong electron-donating ability of the oxyanion. The maximal catalytic activity (TOF = 42,000 $h^{-1}$, TON = 190,000) of the iridium DHBP catalyst was obtained at 6 MPa and 120 °C. Moreover, the catalyst **8** allowed the reaction proceeding at atmospheric pressure. These results indicate that the corresponding hydride complex can easily be generated as an active species at atmospheric pressure.

| Catalyst[a] | | TON | | | |
|---|---|---|---|---|---|
| | L: | bpy | DHBP | phen | DHPT |
| Rh-L | | 216[b] | 1800 | 220 | 2300 |
| Ru-L | | 68 | 4400 | 78[c] | 5100 |
| Ir-L | | 105[c] | 5500 | 59 | 6100 |

**Table 3.** Substituent effect of the ligand on the TON for hydrogenation of $CO_2$. The reaction was carried out with the catalyst (0.1 mM) in a 1 M KOH solution under 4 MPa ($CO_2/H_2$ = 1:1) at 80 °C for 20 h. *a.* Rh-L = [Cp*Rh(L)Cl]Cl, Ru-L = [($C_6Me_6$)Ru(L)Cl]Cl, Ir-L = [Cp*Ir(L)Cl]Cl. *b.* [Catalyst] = 0.2 mM.

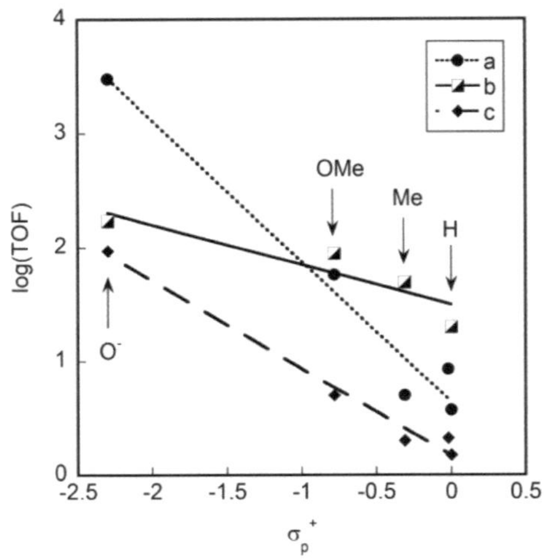


**Figure 4.** Correlation between initial TOFs and $\sigma_p^+$ values of substituents (R) for the hydrogenation of $CO_2$ catalyzed by [($C_nMe_n$)M(4,4′-$R_2$-2,2′-bpy)Cl]Cl. a) M = Ir, n = 5; b) M = Rh, n = 5; c) M = Ru, n = 6; R = OH, OMe, Me, H. The reactions were carried out in an aqueous 1 M KOH solution under 4 MPa ($CO_2$:$H_2$ = 1:1) at 80 °C for 20 h.

In 2009, Nozaki and co-workers designed Ir(III) complexes **10** in which alkylphosphine-based pincer ligands were employed as efficient electron donors. These complexes were used for the hydrogenation of $CO_2$ in $H_2O$/THF. The PNP-Ir trihydride complex, **10**, showed the highest TON (3,500,000) and TOF (150,000 $h^{-1}$) to date.(Tanaka et al., 2009) In 2011, Hazari and co-workers investigated $CO_2$ insertion into PNP-Ir hydrides using a computational method.(Schmeier et al., 2011) They evaluated the nucleophilicity of the hydride through its calculated NBO charge and found a strong correlation between the NBO charge of the hydride and the thermodynamics of $CO_2$ insertion. Using this simple model, they predicted that complex **11** is favorable for $CO_2$ insertion. Furthermore, they experimentally isolated air- and moisture-stable complex **12**. When **12** was used for the hydrogenation of $CO_2$, maximum TON of 348,000 and TOF of 18,780 $h^{-1}$ were obtained, which is comparable to the best system reported by Nozaki.

N-Heterocyclic carbenes (NHCs), which have a high electron-donating ability, have also been introduced as ligands in iridium complexes for the hydrogenation of $CO_2$. Most recently, Peris and co-workers reported a series of IrCp*(NHC) complexes. A bis-NHC Ir complex, **13**, showed modest activity (TON of 1600) for the hydrogenation of $CO_2$ to HCOOK.(Sanz et al., 2010b) To improve the water solubility of the complex, hydroxyl groups were introduced to the side carbon chains. Consequently, complex **14** gave a higher TON of 9500 under optimized conditions.(Sanz et al., 2010a) Furthermore, blocking the C2-position of imidazole with a methyl group and coordinating to the C5 position led to a higher electron-donating ability of the ligand. In addition, the introduction of sulfonate groups into the bis-NHC ligand increased the water solubility of the complex. As a result, a TON of 190,000 was achieved with complex **15**.(Azua et al., 2011) Interestingly, these complexes also succeeded in the transfer hydrogenation of $CO_2$ to formate using *i*PrOH as a hydrogen donor.

## 4. Hydrogenation with other metal complexes

As mention above, the homogenous catalysts for hydrogenation of $CO_2$ into formic acid are typically restricted to complexes of the precious or platinum-group metals Rh, Ru and Ir. Other metals are less investigated due to the low efficiency. Hence the development of nonprecious metal based homogeneous catalyst is limited. Most catalysis using this kind of complexes were carried out in organic solvent and only few examples were in aqueous media, but not pure water. In the original work of Inoue, a non-platinum-group metal catalyst, $Ni(dppe)_2$ (dppe = 1,2-bis(diphenylphosphino)ethane), have been studied. It was proved to be inefficient with only a low TON of 7. Two year later, Evans and Newell studied the homogeneous catalytic reduction of $CO_2$ to formate esters in alcohols with $[HFe_3(CO)_{11}]^-$, but only obtained a low TOF (0.06 $h^{-1}$) and TON (< 6). (Evans & Newell, 1978) In 1994, Yamamoto et al. have studied the Pd based complex. They have synthesized and characterized the first carbon dioxide coordinated palladium(0) complex, $Pd(\eta^2\text{-}CO_2)(PMePh_2)_2$. In addition, using the Pd complexes, $PdCl_2L_2$ (L = $PMe_3$; $PMePh_2$; $PPh_3$), they obtained formic acid in 12% yield in benzene/$H_2O$ under 100 atm $H_2/CO_2$ (1/1) at room temperature.(Sakamoto et al., 1994)

Nonprecious metal was almost not concerned in the following years until 2003. Jessop and co-workers investigated a number of inexpensive metals such as Cu, Fe, Mn, Mo, Ni, and Zn with a high-throughput screening method in the hydrogenation of $CO_2$ in DMSO.(Tai et al., 2003) They found the combination of $FeCl_3$ and $NiCl_2$ with dcpe ligand (dcpe = $Cy_2PCH_2CH_2PCy_2$) gave better results (TON up to 117, TOF up to 15.6 $h^{-1}$) than other metals. In 2010, Beller and Laurenczy et al. have reported different iron precursors and various nitrogen- and phosphine-ligands for the homogeneous hydrogenation of $CO_2$ and bicarbonate to formate in MeOH. The best iron catalyst, $Fe(BF_4)_2/PP_3$ ($PP_3$ =

$P(CH_2CH_2PPh_2)_3$), could reduce bicarbonate to formate in a TON of 610 which was comparable with that using $[RuCl_2(C_6H_6)_2]/PP_3$. It also could transform $CO_2$ and $H_2$ to formate esters and formamide in the presence of the corresponding alcohols and amines with TON up to 292 and 727, respectively.(Federsel et al., 2010a) Very recently, Beller et al. reported hydrogenation of sodium bicarbonate and $CO_2$ with in situ generated cobalt catalyst. They obtained a high TON of 3877 using the $Co(BF_4)_2{\cdot}6H_2O$ and $PP_3$ in a sodium formate at 120 °C for 20 h. This catalytic productivity is six times as high as the TON for the iron catalyst. They also found other phosphine ligands, such as triphenylphosphine, xantphos, 1,2-bisdiphenylphosphinoethane, and 1,1,1-tris(diphenylphosphinomethyl) ethane, showed no activity.(Federsel et al., 2012)

Inspired by the iridium pincer complexes, in 2011, Milstein et al. reported the most active iron(II) pincer complex trans-[(*t*Bu-PNP)Fe(H)$_2$(CO)] which showed similar activity to noble metal catalysts.(Langer et al., 2011) Hydrogenation of sodium bicarbonate to formate in $H_2O$/THF (10/1) have achieved a TON of 320 at 80 °C under 8.3 bar $H_2$. Hydrogenation of $CO_2$ in a 2 M NaOH solution gave a TON of 788 and TOF of 156 $h^{-1}$ at total pressure of 10 bar ($H_2/CO_2$ = 6.7/3.3) for 5 h.

## 5. Mechanism of $CO_2$ hydrogenation

To understand the reaction process of $CO_2$ hydrogenation and design better catalyst, the study of the mechanism has always been the focus for chemists.(Hutschka et al., 1997; Getty et al., 2009) In the $CO_2$ hydrogenation, there are several aspects need to pay attention to, such as activation of dihydrogen and $CO_2$ involving ligand and metal, as well as the important effect of solvent and additive.

Along with the development of reaction in water, the exploration of the water effect has been on-going. The accelerating effect of small amounts of added water in organic solvents has been observed in active Pd(Inoue et al., 1976), Rh(Tsai & Nicholas, 1992)] and Ru(Jessop et al., 1996) etc. systems. Note that in some case adding small amount of water to organic solvent could not improve the performance of the reaction system, even showed prohibit effect due to the deactivation of the hydrophobic catalyst.(Leitner et al., 1994) Therefore understanding the mechanism and then developing appropriate catalyst that can be applied in water is essential to high effective catalytic system. Nicholas et al. (Tsai & Nicholas, 1992) have proposed that water acts as an ancillary ligand and form hydrogen bond with oxygen of the $CO_2$ to facilitate $CO_2$ insertion (Figure 5A). Lau et al. found the reaction rate is enhanced by adding 20% water in THF by using $TpRu(PPh_3)(CH_3CN)H$ [Tp = hydrotris(pyrazolyl)borate] as a catalyst.(Ng et al., 2004) They also have studied the promoting effect of water with the same Ru catalyst and proposed a mechanism to illustrate the water effect (Figure 5B). As suggested by their calculation, the incorporation of water could activate the $CO_2$ molecule and significantly reduce the reaction barrier.(Yin et al., 2001) In the most recent work of Nozaki et al. they investigated the reaction mechanism by density functional theory (DFT) calculation. They found that adding one or two water molecules the reaction barrier is markedly lowered compared to that in gas phase.

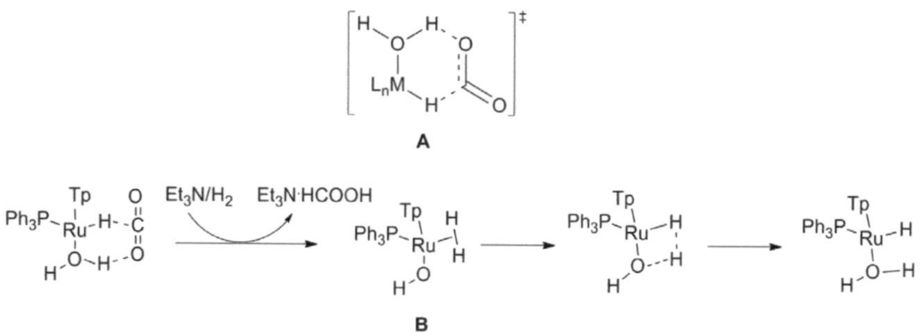


**Figure 5.** Water effect in the hydrogenation of $CO_2$.

Lau et al. have demonstrated the intramolecular N-H---H-Ru hydrogen bond in the Ru complexes catalysed hydrogenation.(Chu et al., 1998) Although the reaction rate was very slow, their research gave insight into the mechanism of hydrogen activation: intramolecular heterolytic cleavage of the dihydrogen was aided by the pendant amino group. The design principle has been employed by Hazari et al. in the designing of PNP Ir(III) catalyst and demonstrated the feasibility of such activation method.(Schmeier et al., 2011) With DFT calculations, they demonstrated that the intramolecular hydrogen bond in complex **12** facilitates $CO_2$ insertion (Figure 6).

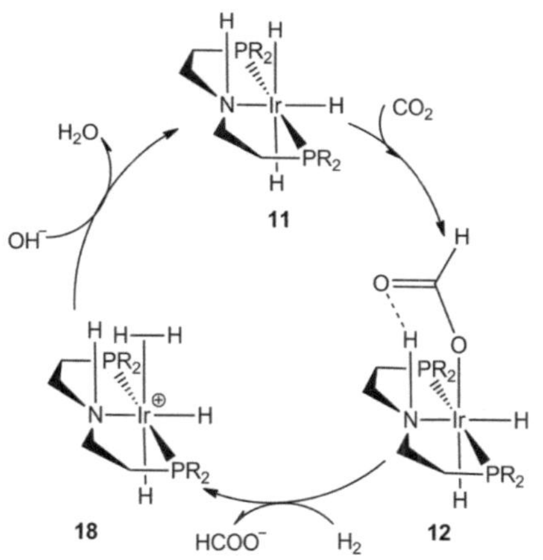


**Figure 6.** Reaction mechanism of $CO_2$ hydrogenation proposed by Hazari et al.

Sasaki's theoretical calculation(Ohnishi et al., 2005) and Jessop's experimental results(Tai et al., 2002) have demonstrated the strong electron-donating power of the ligand resulting in high activity of the complexes. The following design of complexes generally applied this principle. As abovementioned, Himeda et al. developed DHBP catalyst by introducing the hydroxyl group to bpy; Nozaki et al. designed the complex with PNP-based pincer ligand as a strong donor; Peris et al. used the NHC as a strong electron-donating ligand for new catalyst design. All the examples have verified the importance of the donor powder of the ligand in activating the complex.

In 2009, Nozaki et al. reported the PNP pincer ligated Ir(III) complexes, **10**, and achieved the highest TON (3,500,000) and TOF (150,000 $h^{-1}$) to date.(Tanaka et al., 2009) They also proposed a mechanism for the catalytic reaction (Figure 7): the insertion of $CO_2$ into **10** gives formato complex **16**, which undergoes dissociation of the formato ligand under basic conditions. Simultaneously, deprotonative dearomatization of the PNP ligand by $OH^-$ leads to intermediate **17**, which is hydrogenated to regenerate the trihydride complex **10**.

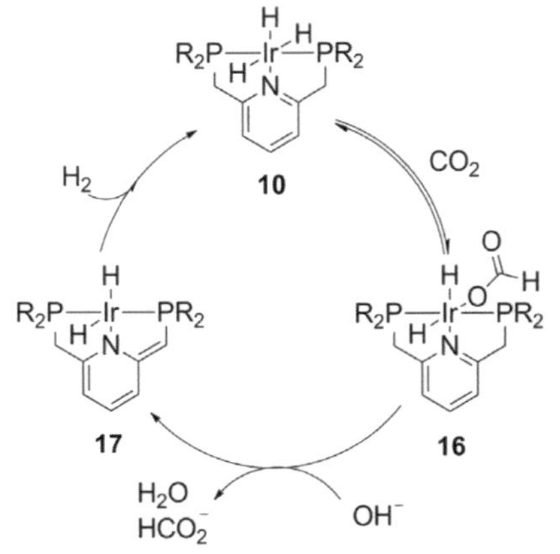


**Figure 7.** Catalytic mechanism for the hydrogenation of $CO_2$ proposed by Nozaki.

After Nozaki and co-workers reported the excellent PNP Ir(III) complexes, several groups have investigated the mechanism of $CO_2$ hydrogenation with these catalysts. Ahlquist et al. used a simple (PNP)$IrH_3$ structure to study the mechanism with DFT calculation.(Ahlquist, 2010) Their research suggested that the deprotonation by the hydroxide is the rate-limiting step (Figure 7). This calculation agreed with the experimental observation that higher basicity leads to a higher rate. Most recently, Yang reinvestigated this mechanism using the DFT method and proposed a different reaction pathway (Figure 8).(Yang, 2011) He suggested that direct $H_2$ cleavage by $OH^-$ is more favourable than the Nozaki-postulated $H_2$ cleavage model. Using this new reaction pathway, the calculation gave a low overall enthalpy barrier of 77.8 kJ $mol^{-1}$ for the formation of HCOOH from $H_2$ and $CO_2$.

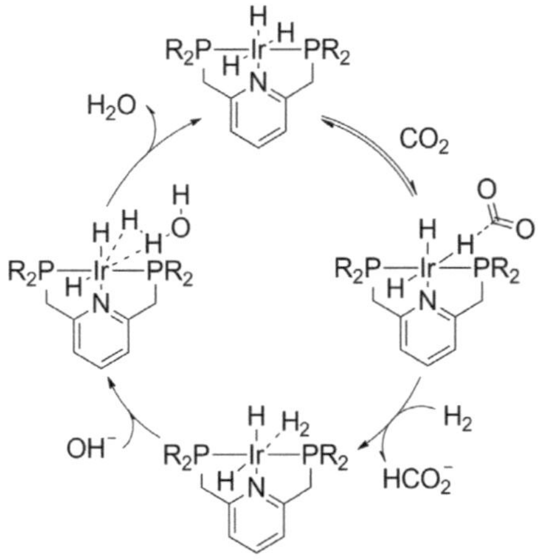


**Figure 8.** Catalytic mechanism of $CO_2$ hydrogenation proposed by Yang.

The catalytic hydrogenation mechanism of nitrogen-based complexes has been less investigated. In 2006, Ogo et al. determined that the different rate-determining step for bpy-based Ru and Ir complexes by the observation of the saturation behaviour of the TON with increasing $P_{H_2}$ and $P_{CO_2}$ respectively.(Ogo et al., 2006) The rate-determining step of [(η⁶-$C_6Me_6$)Ru(bpy)($OH_2$)]($SO_4$) and [(η⁶-$C_6Me_6$)Ru(4,4′-OMe-bpy)($OH_2$)]($SO_4$) was suggested to be the reaction of aqua complexes with $H_2$. In contrast, the Ir analogous was supposed to be the $CO_2$ insertion into the iridium hydride complexes which were isolated and characterized by NMR, ESI-MS, and IR. The different mechanism of Ru and Ir complexes may help to understand the excellent performance of other iridium complexes.

## 6. Catalyst immobilization and recycle

Homogeneous catalyst has exhibited high catalytic activity in the hydrogenation of $CO_2$. For further practical application, the recycle and reuse of the catalyst is an important issue that needs to be resolved because most of the catalysts contain noble metal.

In 2011, Baffert et al. reported a series of silica supported ruthenium-N-heteroclyclic carben species for hydrogenation of $CO_2$.(Baffert et al., 2011) Using pyrrolidine as an additive, $Ru_{cym}$ and M-$Ru_{cym}$ (Figure 9) showed low catalytic activity. By introducing basic phosphorous ligand $PMe_3$, the activity of M-RuP (Figure 9) is improved and showed comparable TON with the parent catalyst [$RuCl_2(PMe_3)_4$]. However, the supported catalyst are unstable due to the weak Ru-NHC linkage, 50% of Ru was found leached into reaction solution.

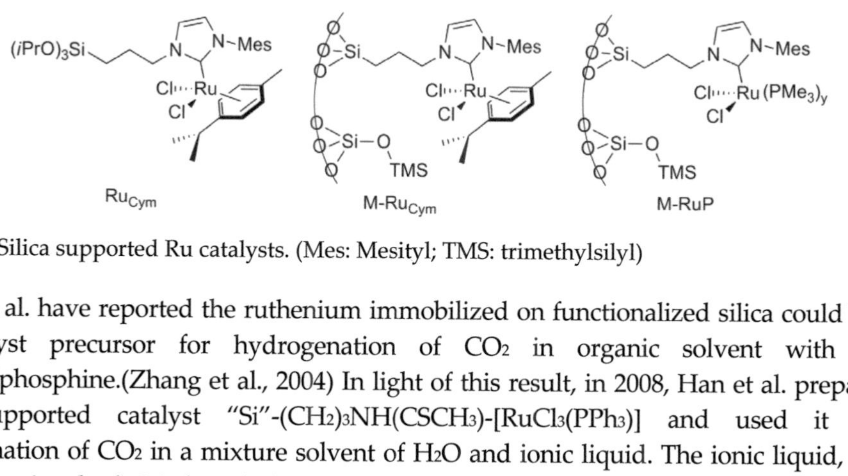


**Figure 9.** Silica supported Ru catalysts. (Mes: Mesityl; TMS: trimethylsilyl)

Zheng et al. have reported the ruthenium immobilized on functionalized silica could be used as catalyst precursor for hydrogenation of $CO_2$ in organic solvent with adding triphenylphosphine.(Zhang et al., 2004) In light of this result, in 2008, Han et al. prepared the silica supported catalyst "Si"-($CH_2$)$_3$NH($CSCH_3$)-[$RuCl_3(PPh_3)$] and used it to the hydrogenation of $CO_2$ in a mixture solvent of $H_2O$ and ionic liquid. The ionic liquid, 1-(N,N-dimethylaminoehtyl)-2,3-dimethylimidazolium trifluoromethanesulfonate ([mammin][TfO]), has a tertiary amino group which makes it can acts as a basic additive as well as a solvent. After the reaction, the immobilized catalyst could be simply recycled by filtration. The filtrate was first warmed to 110 °C to remove the water and then heated to 130 °C to separate the formic acid. Since the ionic liquid is non-volatile and stable below 220 °C, it can be separated and reused after the distillation. The catalyst and ionic liquid could be reused four times without decrease of TOF (~44 $h^{-1}$). With ICP-AES analysis, they found no significant loss of Ru during the recycling process. (Zhang et al., 2008)

Soon after that, they report another type of ionic liquid, 1,3-di(N,N-dimethylaminoethyl)-2-methylimidazolium trifluoromethanesulfonate ([DAMI][TfO]), which has two tertiary amino groups on the side chain of the cation. Using the silica supported Ru catalyst, "Si"-$(CH_2)_3NH(CSCH_3)$-$[RuCl_3(PPh_3)]$, $CO_2$ was hydrogenated to formic acid in the presence of water and [DAMI][TfO]. They found TOF increased with increasing the amount of water added, and a weight ratio of water to ionic liquid is suitable between 1 and 2.5. Under the optimal conditions, a TON of 1840 and a TOF of 920 $h^{-1}$ was obtained at 80 °C under 18 MPa of $H_2/CO_2$ (1/1) for 2h. The ionic liquid and catalyst can be reused at least over four cycles without significant decrease of TOF.(Zhang et al., 2009)

The catalyst recycle usually require a solid supporter, and suffer from loss of catalytic activity due to the insolubility of the catalyst in the reaction solution. Himeda et al. reported an interesting method for catalyst recycling without a supporter. It was realized by utilising the tunable solubility of the complex along with changing pH of the reaction solution. The catalyst **9** based on DHPT showed a similar TOF and a slightly improved TON(Himeda et al., 2005) than DHBP catalyst **8** (Table 2). The abovementioned acid-base equilibrium not only changes the electronic properties of the complex but also affects its polarity and thus its water solubility. As shown in Figure 10, DHPT catalyst **9** exhibited negligible solubility (ca. 100 ppb) in a weakly acidic formate solution. Therefore, recycling of **9** was investigated in batch-wise cycles based on the concept shown in Figure 11. When the added KOH was completely consumed, the catalyst precursor spontaneously precipitated due to its decreased water solubility as a result of the lower pH value. Thus, the reaction was terminated and formed a heterogeneous system that could be filtered to recover the precipitated catalyst. The iridium complex remaining in the filtrate was found to be less than 2% of the catalyst loading (0.11 ppm). Since the catalytic action was "turned off", the reverse reaction (i.e., the decomposition of formic acid) was prevented in the separation step. Additionally, the pure product (i.e., the formate salt) could be isolated simply by evaporating the filtrate.

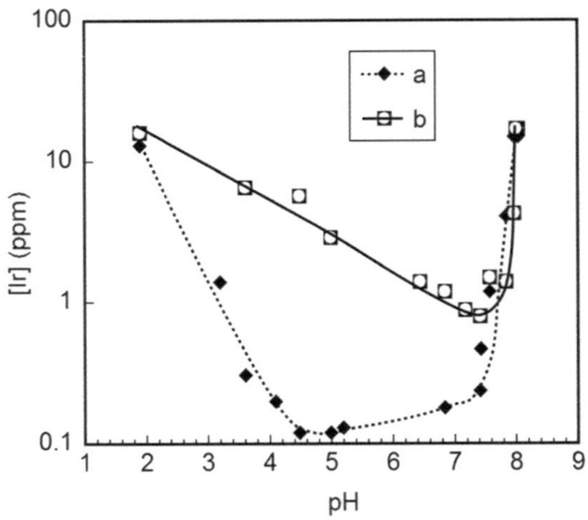


**Figure 10.** Solubility of a) DHBP catalyst **8** and b) DHPT catalyst **9** in a 1 M aqueous formate solution.

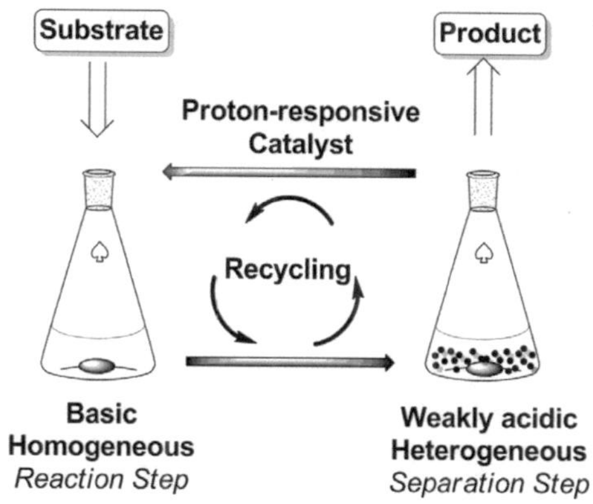


**Figure 11.** Recycling of proton-responsive catalyst with tunable solubility.

The recovered catalyst retained a high catalytic activity across four cycles, as shown in Table 4. It is clear that the three components (i.e., catalyst, product, and solvent) can be easily separated without significant waste. The sophisticated design of the catalyst provided a proton-responsive catalyst with pH-tunable catalytic activity and water solubility. These results suggest that by carefully considering reaction profiles, the design and use of innovative homogeneous catalytic systems such as tailor-made catalysts can overcome the disadvantages of homogeneous catalysis.

| Cycle | Loaded/recovered cat./ppm | Recovery efficiency/% | Leaching iridium[a]/ppm | Final conc. of formate/M |
|---|---|---|---|---|
| 1 | 9.0 | - | 0.11 | 0.105 |
| 2 | 8.4 | 93 | 0.22 | 0.104 |
| 3 | 7.7 | 92 | 0.42 | 0.103 |
| 4 | 7.0 | 91 | 0.61 | 0.103 |

**Table 4.** Catalyst recycling studies for the conversion of $CO_2$ into formate using iridium-DHPT catalyst 9. Reaction conditions: DHPT catalyst (2.5 μmol), 6 MPa of $H_2/CO_2$ (1:1), 0.1 M KOH solution (50 mL), 60 °C for 2 h. *a*. Determined by ICP-MS analysis.

## 7. Conclusion

After decades of research, chemists have achieved great success in homogeneous hydrogenation of carbon dioxide. With appropriate catalysts and optimum conditions, some of the results are close to commercialization. However, to hydrogenate carbon dioxide efficiently, economically, and eco-friendly, several critical problems remain to be solved. The first is the high activity of the catalyst, which is generally required in order to lower the overall energy barrier for the conversion of thermodynamically stable $CO_2$. Present catalysts usually need high temperature and pressure to achieve high activity. Consequently, the energy cost is increased. The highly efficient catalyst that can work under mild conditions is highly requisite. Therefore, further research and understanding the mechanism and delicate design of the catalyst with multi-functional ligand are necessary. The second is the prevention of waste generation (e.g., organic solvents and additives) during the reaction. The third is the recycle and reuse of the catalyst, which is important to increase the cost efficiency. The research of the catalyst recycling is still in the preliminary stage and suffers from a lot of problems. Better performance can be expected with the development of new immobilizing method, such as using ionic liquid.

Since much more effort has been paid to homogeneous hydrogenation of carbon dioxide over the last decade, we can expect more exciting results in the near future. We also believe transformation and utilization of carbon dioxide, especially for fuels production, will decrease its emission and reduce the reliance on fossil sources.

## Author details

Wan-Hui Wang and Yuichiro Himeda
*National Institute of Advanced Industrial Science and Technology, Tsukuba, Ibaraki, Japan*

## Acknowledgement

We thank the Japanese Ministry of Economy, Trade, and Industry for financial support.

## 8. References

Ahlquist, M. S. G. (2010). Iridium catalyzed hydrogenation of $CO_2$ under basic conditions—Mechanistic insight from theory. *J. Mol. Catal. A: Chem.*, Vol.324, No.1-2, pp. 3-8, ISSN 13811169

Azua, A.; Sanz, S. & Peris, E. (2011). Water-soluble IrIII N-heterocyclic carbene based catalysts for the reduction of $CO_2$ to formate by transfer hydrogenation and the deuteration of aryl amines in water. *Chem.-Eur. J.*, Vol.17, No.14, pp. 3963-3967, ISSN 1521-3765 (Electronic)

Baffert, M.; Maishal, T. K.; Mathey, L.; Coperet, C. & Thieuleux, C. (2011). Tailored ruthenium-N-heterocyclic carbene hybrid catalytic materials for the hydrogenation of carbon dioxide in the presence of amine. *ChemSusChem*, Vol.4, No.12, pp. 1762-1765, ISSN 1864-564X (Electronic)

Boddien, A.; Gärtner, F.; Federsel, C.; Sponholz, P.; Mellmann, D.; Jackstell, R.; Junge, H. & Beller, M. (2011). $CO_2$-"Neutral" Hydrogen Storage Based on Bicarbonates and Formates. *Angew. Chem. Int. Ed.*, Vol.50, No.28, pp. 6411-6414, ISSN 1521-3773

Chu, H. S.; Lau, C. P.; Wong, K. Y. & Wong, W. T. (1998). Intramolecular N–H···H–Ru Proton–Hydride Interaction in Ruthenium Complexes with (2-(Dimethylamino)ethyl)cyclopentadienyl and (3-(Dimethylamino)propyl)cyclopentadienyl Ligands. Hydrogenation of $CO_2$ to Formic Acid via the N–H···H–Ru Hydrogen-Bonded Complexes. *Organometallics*, Vol.17, No.13, pp. 2768-2777, ISSN 0276-7333

Elek, J.; Nadasdi, L.; Papp, G.; Laurenczy, G. & Joó, F. (2003). Homogeneous hydrogenation of carbon dioxide and bicarbonate in aqueous solution catalyzed by water-soluble ruthenium(II) phosphine complexes. *Appl. Catal. A-Gen.*, Vol.255, No.1, pp. 59-67, ISSN 0926-860X

Erlandsson, M.; Landaeta, V. R.; Gonsalvi, L.; Peruzzini, M.; Phillips, A. D.; Dyson, P. J. & Laurenczy, G. (2008). (Pentamethylcyclopentadienyl)iridium-PTA (PTA = 1,3,5-Triaza-7-phosphaadamantane) Complexes and Their Application in Catalytic Water Phase Carbon Dioxide Hydrogenation. *Eur. J. Inorg. Chem.*, Vol.2008, No.4, pp. 620-627, ISSN 14341948

Evans, G. O. & Newell, C. J. (1978). Conversion of $CO_2$, $H_2$, and alcohols into formate esters using anionic iron carbonyl hydrides. *Inorg. Chim. Acta*, Vol.31, pp. L387-L389, ISSN 0020-1693

Federsel, C.; Boddien, A.; Jackstell, R.; Jennerjahn, R.; Dyson, P. J.; Scopelliti, R.; Laurenczy, G. & Beller, M. (2010a). A well-defined iron catalyst for the reduction of bicarbonates and carbon dioxide to formates, alkyl formates, and formamides. *Angew. Chem. Int. Ed.*, Vol.49, No.50, pp. 9777-9780, ISSN 1521-3773 (Electronic)

Federsel, C.; Jackstell, R. & Beller, M. (2010b). State-of-the-art catalysts for hydrogenation of carbon dioxide. *Angew. Chem. Int. Ed.*, Vol.49, No.36, pp. 6254-6257, ISSN 1521-3773 (Electronic)

Federsel, C.; Ziebart, C.; Jackstell, R.; Baumann, W. & Beller, M. (2012). Catalytic hydrogenation of carbon dioxide and bicarbonates with a well-defined cobalt dihydrogen complex. *Chem.-Eur. J.*, Vol.18, No.1, pp. 72-75, ISSN 1521-3765 (Electronic)

Gassner, F. & Leitner, W. (1993). $CO_2$ Activation 3. Hydrogenation of Carbon Dioxide to Formic Acid Using Water-Soluble Rhodium Catalysts. *J. Chem. Soc.-Chem. Commun.*, No.19, pp. 1465-1466, ISSN 0022-4936

Getty, A. D.; Tai, C.-C.; Linehan, J. C.; Jessop, P. G.; Olmstead, M. M. & Rheingold, A. L. (2009). Hydrogenation of Carbon Dioxide Catalyzed by Ruthenium Trimethylphosphine Complexes: A Mechanistic Investigation Using High-Pressure NMR Spectroscopy. *Organometallics*, Vol.28, No.18, pp. 5466-5477, ISSN 0276-7333

Hayashi, H.; Ogo, S.; Abura, T. & Fukuzumi, S. (2003). Accelerating effect of a proton on the reduction of $CO_2$ dissolved in water under acidic conditions. Isolation, crystal structure, and reducing ability of a water-soluble ruthenium hydride complex. *J. Am. Chem. Soc.*, Vol.125, No.47, pp. 14266-14267, ISSN 0002-7863

Hayashi, H.; Ogo, S. & Fukuzumi, S. (2004). Aqueous hydrogenation of carbon dioxide catalysed by water-soluble ruthenium aqua complexes under acidic conditions. *Chem. Commun.*, No.23, pp. 2714-2715, ISSN 1359-7345

Himeda, Y. (2007). Conversion of $CO_2$ into formate by homogeneously catalyzed hydrogenation in water: Tuning catalytic activity and water solubility through the acid-base equilibrium of the ligand. *Eur. J. Inorg. Chem.*, No.25, pp. 3927–3941, ISSN 1434-1948

Himeda, Y.; Miyazawa, S. & Hirose, T. (2011). Interconversion between Formic Acid and $H_2/CO_2$ using Rhodium and Ruthenium Catalysts for $CO_2$ Fixation and $H_2$ Storage. *ChemSusChem*, Vol.4, No.4, pp. 487-493, ISSN 1864-5631

Himeda, Y.; Onozawa-Komatsuzaki, N.; Sugihara, H.; Arakawa, H. & Kasuga, K. (2003). Transfer hydrogenation of a variety of ketones catalyzed by rhodium complexes in aqueous solution and their application to asymmetric reduction using chiral Schiff base ligands. *J. Mol. Catal. A-Chem.*, Vol.195, No.1-2, pp. 95–100, ISSN 1381-1169

Himeda, Y.; Onozawa-Komatsuzaki, N.; Sugihara, H.; Arakawa, H. & Kasuga, K. (2004). Half-sandwich complexes with 4,7-dihydroxy-1,10-phenanthroline: Water-soluble, highly efficient catalysts for hydrogenation of bicarbonate attributable to the generation of an oxyanion on the catalyst ligand. *Organometallics*, Vol.23, No.7, pp. 1480–1483, ISSN 0276-7333

Himeda, Y.; Onozawa-Komatsuzaki, N.; Sugihara, H.; Arakawa, H. & Kasuga, K. Japan Patent 3968431 (filed on Jan. 21, 2003), 2007a.

Himeda, Y.; Onozawa-Komatsuzaki, N.; Sugihara, H. & Kasuga, K. (2005). Recyclable catalyst for conversion of carbon dioxide into formate attributable to an oxyanion on the catalyst ligand. *J. Am. Chem. Soc.*, Vol.127, No.38, pp. 13118–13119, ISSN 0002-7863

Himeda, Y.; Onozawa-Komatsuzaki, N.; Sugihara, H. & Kasuga, K. (2006). Highly efficient conversion of carbon dioxide catalyzed by half-sandwich complexes with pyridinol ligand: The electronic effect of oxyanion. *J. Photochem. Photobiol. A-Chem.*, Vol.182, No.3, pp. 306–309, ISSN 1010-6030

Himeda, Y.; Onozawa-Komatsuzaki, N.; Sugihara, H. & Kasuga, K. (2007b). Simultaneous tuning of activity and water solubility of complex catalysts by acid-base equilibrium of ligands for conversion of carbon dioxide. *Organometallics*, Vol.26, No.3, pp. 702–712, ISSN 0276-7333

Horváth, H.; Laurenczy, G. & Kathó, Á. (2004). Water-soluble ($\eta^6$-arene)ruthenium(II)-phosphine complexes and their catalytic activity in the hydrogenation of bicarbonate in aqueous solution. *J. Organomet. Chem.*, Vol.689, No.6, pp. 1036-1045, ISSN 0022328X

Hutschka, F.; Dedieu, A.; Eichberger, M.; Fornika, R. & Leitner, W. (1997). Mechanistic aspects of the rhodium-catalyzed hydrogenation of $CO_2$ to formic acid - A theoretical and kinetic study. *J. Am. Chem. Soc.*, Vol.119, No.19, pp. 4432-4443, ISSN 0002-7863

Inoue, Y.; Izumida, H.; Sasaki, Y. & Hashimoto, H. (1976). Catalytic fixation of carbon dioxideto formic acid by transition-metal complexes under mild conditions. *Chem. Lett.*, pp. 863-864, ISSN

Jessop, P. G., Homogeneous hydrogenation of carbon dioxide. In *Handbook of Homogeneous Hydrogenation*, De Vries, J. G.; Elsevier, C. J., Eds. Wiley-VCH: Weinheim, 2007; Vol. 1, pp 489-511.

Jessop, P. G.; Hsiao, Y.; Ikariya, T. & Noyori, R. (1996). Homogeneous catalysis in supercritical fluids: Hydrogenation of supercritical carbon dioxide to formic acid, alkyl formates, and formamides. *J. Am. Chem. Soc.*, Vol.118, No.2, pp. 344-355, ISSN 0002-7863

Jessop, P. G.; Ikariya, T. & Noyori, R. (1994). Homogeneous Catalytic-Hydrogenation of Supercritical Carbon-Dioxide. *Nature*, Vol.368, No.6468, pp. 231-233, ISSN 0028-0836

Jessop, P. G.; Joó, F. & Tai, C.-C. (2004). Recent advances in the homogeneous hydrogenation of carbon dioxide. *Coord. Chem. Rev.*, Vol.248, No.21-24, pp. 2425-2442, ISSN 00108545

Joó, F.; Laurenczy, G.; Nadasdi, L. & Elek, J. (1999). Homogeneous hydrogenation of aqueous hydrogen carbonate to formate under exceedingly mild conditions - a novel possibility of carbon dioxide activation. *Chem. Commun.*, No.11, pp. 971-972, ISSN 1359-7345

Langer, R.; Diskin-Posner, Y.; Leitus, G.; Shimon, L. J.; Ben-David, Y. & Milstein, D. (2011). Low-Pressure Hydrogenation of Carbon Dioxide Catalyzed by an Iron Pincer Complex Exhibiting Noble Metal Activity. *Angew. Chem. Int. Ed.*, pp. 9948-9952, ISSN 1521-3773 (Electronic)

Laurenczy, G.; Joó, F. & Nadasdi, L. (2000). Formation and characterization of water-soluble hydrido-ruthenium(II) complexes of 1,3,5-triaza-7-phosphaadamantane and their catalytic activity in hydrogenation of $CO_2$ and $HCO_3^-$ in aqueous solution. *Inorg. Chem.*, Vol.39, No.22, pp. 5083-5088, ISSN 0020-1669

Leitner, W.; Dinjus, E. & Gassner, F. (1994). Activation of Carbon Dioxide 4. Rhodium-Catalyzes Hydrogenation of Carbon Dioxide to Formic Acid. *J. Organomet. Chem.*, Vol.475, No.1-2, pp. 257-266, ISSN 0022-328X

Leitner, W.; Dinjus, E. & Gassner, F., In *Aqueous-Phase Organometallic Catalysis, Concepts and Applications*, Cornils, B.; Herrmann, W. A., Eds. Wiley-VCH: Weinheim, 1998; pp 486-498.

Munshi, P.; Main, A. D.; Linehan, J. C.; Tai, C. C. & Jessop, P. G. (2002). Hydrogenation of carbon dioxide catalyzed by ruthenium trimethylphosphine complexes: The accelerating effect of certain alcohols and amines. *J. Am. Chem. Soc.*, Vol.124, No.27, pp. 7963-7971, ISSN 0002-7863

Ng, S. M.; Yin, C. Q.; Yeung, C. H.; Chan, T. C. & Lau, C. P. (2004). Ruthenium-catalyzed hydrogenation of carbon dioxide to formic acid in alcohols. *Eur. J. Inorg. Chem.*, No.9, pp. 1788-1793, ISSN 1434-1948

Ogo, S.; Kabe, R.; Hayashi, H.; Harada, R. & Fukuzumi, S. (2006). Mechanistic investigation of $CO_2$ hydrogenation by Ru(II) and Ir(III) aqua complexes under acidic conditions: two catalytic systems differing in the nature of the rate determining step. *Dalton Trans.*, No.39, pp. 4657-4663, ISSN 1477-9226 (Print)

Ohnishi, Y. Y.; Matsunaga, T.; Nakao, Y.; Sato, H. & Sakaki, S. (2005). Ruthenium(II)-catalyzed hydrogenation of carbon dioxide to formic acid. theoretical study of real catalyst, ligand effects, and solvation effects. *J. Am. Chem. Soc.*, Vol.127, No.11, pp. 4021-4032, ISSN 0002-7863

Papp, G.; Csorba, J.; Laurenczy, G. & Joó, F. (2011). A Charge/Discharge Device for Chemical Hydrogen Storage and Generation. *Angew. Chem. Int. Ed.*, Vol.50, No.44, pp. 10433-10435, ISSN 1521-3773

Sakamoto, M.; Shimizu, I. & Yamamoto, A. (1994). Synthesis of the first carbon dioxide coordinated palladium(0) complex, $Pd(\eta^2\text{-}CO_2)(PMePh_2)_2$. *Organometallics*, Vol.13, No.2, pp. 407-409, ISSN 0276-7333

Sanz, S.; Azua, A. & Peris, E. (2010a). '($\eta^6$)-arene)Ru(bis-NHC)' complexes for the reduction of $CO_2$ to formate with hydrogen and by transfer hydrogenation with *i*PrOH. *Dalton Trans.*, Vol.39, No.27, pp. 6339-6343, ISSN 1477-9234 (Electronic)

Sanz, S.; Benítez, M. & Peris, E. (2010b). A New Approach to the Reduction of Carbon Dioxide: $CO_2$ Reduction to Formate by Transfer Hydrogenation in *i*PrOH. *Organometallics*, Vol.29, No.1, pp. 275-277, ISSN 0276-7333

Schmeier, T. J.; Dobereiner, G. E.; Crabtree, R. H. & Hazari, N. (2011). Secondary coordination sphere interactions facilitate the insertion step in an iridium(III) $CO_2$

reduction catalyst. *J. Am. Chem. Soc.,* Vol.133, No.24, pp. 9274-9277, ISSN 1520-5126 (Electronic)

Tai, C. C.; Chang, T.; Roller, B. & Jessop, P. G. (2003). High-pressure combinatorial screening of homogeneous catalysts: Hydrogenation of carbon dioxide. *Inorg. Chem.,* Vol.42, No.23, pp. 7340-7341, ISSN 0020-1669

Tai, C. C.; Pitts, J.; Linehan, J. C.; Main, A. D.; Munshi, P. & Jessop, P. G. (2002). In situ formation of ruthenium catalysts for the homogeneous hydrogenation of carbon dioxide. *Inorg. Chem.,* Vol.41, No.6, pp. 1606-1614, ISSN 0020-1669

Tanaka, R.; Yamashita, M. & Nozaki, K. (2009). Catalytic hydrogenation of carbon dioxide using Ir(III)-pincer complexes. *J. Am. Chem. Soc.,* Vol.131, No.40, pp. 14168-14169, ISSN 1520-5126 (Electronic)

Tsai, J. C. & Nicholas, K. M. (1992). Rhodium-Catalyzed Hydrogenation of Carbon-Dioxide to Formic-Acid. *J. Am. Chem. Soc.,* Vol.114, No.13, pp. 5117-5124, ISSN 0002-7863

Wang, W.; Wang, S.; Ma, X. & Gong, J. (2011). Recent advances in catalytic hydrogenation of carbon dioxide. *Chem. Soc. Rev.,* Vol.40, No.7, pp. 3703-3727, ISSN 1460-4744 (Electronic)

Yang, X. (2011). Hydrogenation of Carbon Dioxide Catalyzed by PNP Pincer Iridium, Iron, and Cobalt Complexes: A Computational Design of Base Metal Catalysts. *ACS Catal.,* Vol.1, No.8, pp. 849-854, ISSN 2155-5435

Yin, C.; Xu, Z.; Yang, S.-Y.; Ng, S. M.; Wong, K. Y.; Lin, Z. & Lau, C. P. (2001). Promoting Effect of Water in Ruthenium-Catalyzed Hydrogenation of Carbon Dioxide to Formic Acid. *Organometallics,* Vol.20, No.6, pp. 1216-1222, ISSN 0276-7333

Zhang, Y.; Fei, J.; Yu, Y. & Zheng, X. (2004). Silica immobilized ruthenium catalyst used for carbon dioxide hydrogenation to formic acid (I): the effect of functionalizing group and additive on the catalyst performance. *Catal. Commun.,* Vol.5, No.10, pp. 643-646, ISSN 15667367

Zhang, Z.; Hu, S.; Song, J.; Li, W.; Yang, G. & Han, B. (2009). Hydrogenation of $CO_2$ to formic acid promoted by a diamine-functionalized ionic liquid. *ChemSusChem,* Vol.2, No.3, pp. 234-238, ISSN 1864-564X (Electronic)

Zhang, Z.; Xie, Y.; Li, W.; Hu, S.; Song, J.; Jiang, T. & Han, B. (2008). Hydrogenation of carbon dioxide is promoted by a task-specific ionic liquid. *Angew. Chem. Int. Ed.,* Vol.47, No.6, pp. 1127-1129, ISSN 1521-3773 (Electronic)

Chapter 11

# Hydrogenation of Carbon Oxides on Catalysts Bearing Fe, Co, Ni, and Mn Nanoparticles

T.F. Sheshko and Yu. M. Serov

Additional information is available at the end of the chapter

http://dx.doi.org/10.5772/47743

## 1. Introduction

One way of obtaining synthetic liquid fuels and valuable chemical compounds on the basis of non-oil raw materials (coal, natural gas, biomass) is the synthesis of hydrocarbons from CO and $H_2$, which takes place with the participation of catalysts containing transition metals of Group VIII, known as the Fischer-Tropsch synthesis [1-3]. Although there are other methods for hydrocarbon mixtures of non-oil raw materials (for example, hydrogenation of coal and biomass pyrolysis and semi-coking coal), the priority development of the Fischer-Tropsch process clearly demonstrates its vitality and promise, as determined by an enormous source of raw materials - coal in the energy equivalent an order of magnitude higher than the oil.

In addition to carbon monoxide for the hydrogenation reaction is possible, and repeatedly described the synthesis of hydrocarbons from mixtures containing carbon dioxide [3-6] and hydrogen according to the equation of general form:

$$nCO_2 + (2n + m)\ H_2 \rightarrow CnH_{2m} + 2n\ H_2O$$

The use of carbon dioxide is one of the most promising directions of development of efficient catalytic systems that allow atmospheric pressure to convert the process emissions containing both CO and $CO_2$ in the olefins.

Global trend, most pronounced in industrialized countries, it became tougher environmental legislation. It is aimed primarily at reducing harmful emissions, which led to a sharp increase in the number of works connected with the search for technologies that could be returned to the commercialization of gas by-products. Process emissions include both mono-and carbon dioxide, and hydrogen. Develop and implement new and improved catalysts, allowing atmospheric pressure to convert these emissions into olefins is one of the most promising directions of development of nanotechnology and can make a significant contribution to solving the problem of reduction of pressure on the environment.

That is why the aim of this study was to investigate the possibility of obtaining olefins from a mixture of carbon oxides with hydrogen at atmospheric pressure, as well as determine the effect of the composition and characteristics of the catalysts containing transition metal nanoparticles on their activity and selectivity to olefins.

## 2. Experimental

### 2.1. The method of catalytic experiments

Catalytic activity in hydrogenation reactions of carbon dioxide, and mixtures of carbon oxides was determined by applying a mixture of genius with oxides of carbon and hydrogen in the ratio of the components [$CO_2$: $H_2$], [($CO_2$ + CO):$H_2$] = 1:2 and 1:4. Experiments were conducted in a flow catalytic apparatus at atmospheric pressure and flow rate 1.5 - 5.0 l/h, in the temperature range 423-723 K. The catalysts (weight 0.3-0.4 g) were placed in a quartz reactor with a quartz filter to prevent particle entrainment. Analyses of the products were performed chromatographically in a column of stainless steel filled with Porapack Q at 393K using a thermal conductivity detector and flame ionization. The rate of formation of reaction products W (mol/h*$g_{-af}$) was measured after reaching steady state and calculated per unit mass of metal (active) phase of the catalyst.

The catalysts were subjected to reduction treatment stream of hydrogen at a temperature of 623 K and flow rate of hydrogen 1.5-2.0 l/h before the catalytic.

### 2.2. The method of adsorption experiments.

The chemisorption of carbon oxides was investigated by thermodesorption. The adsorption of carbon oxides and hydrogen occurred when the reaction vessel was filled to atmospheric pressure with gases at temperatures of 293, 473, and 573 K. Thermodesorption was performed in a regime for a linearly programmed 293 to 823 K temperature increase in a helium flow. When needed, the products of thermal desorption from the reactor fed into nitrogen trap to remove water and carbon dioxide. The composition and amount of the desorbed gases were controlled using a Crystall 2000M chromatograph. All spectra of the thermoprogrammed desorption (TPD) were treated in terms of the coordinates of the Polyani–Wigner equation [7-10]:

$$-\frac{d\Theta_S}{dt} = \nu_n \Theta_S^n exp\left(-\frac{E_{a,des}}{RT}\right) \tag{1}$$

where $\theta_s$ is the surface coverage in terms of single layer fractions or the number of molecules adsorbed onto 1 $cm^2$ of the sample; $E_{des}$ is the activation energy of desorption; T is the temperature of the surface; n is the kinetic order of desorption; and $\nu_n$ is the frequency factor, $s^{-1}$.

To determine the activation energy of desorption according to the change in the surface concentration of an adsorbate during the thermodesorption process, the following integral equations for the first and the second order desorption were developed by Ehrlich [7,8]:

$$\frac{\ln\frac{\Theta_0}{\Theta_s}}{T^2} = \ln\frac{\nu_1 R}{\beta E_{des}} - \frac{E_{des}}{RT} \qquad \text{at } n = 1, \tag{2}$$

$$\ln\frac{\frac{1}{\Theta_s} - \frac{1}{\Theta_0}}{T^2} = \ln\frac{\nu_2 R}{\beta E_{des}} - \frac{E_{des}}{RT} \qquad \text{at } n = 2, \tag{3}$$

where $\beta$ is the linear temperature rate of increase and $\Theta_0$ is the maximum surface coverage that is proportional to the total area of the desorption peak. If the linear dependence were observed in presenting the experimental data in terms of Eq. (2) coordinates, the order of the desorption process would come first; if presented in terms of (3), the order would come second. It was according to this that the activation energy was found.

## 2.3. Methods of catalysts preparation.

Two series of catalysts were investigated. The first series was prepared matrixing in silicon oxide $SiO_2$ powders of iron and cobalt, which are obtained by electrochemical method[1]. We used high-purity silica particles which were the ultra-thin filament length of 1 mm and a diameter of 20-30 nm. The process of obtaining ultrathin crystals of metals occurs in a two-layer bath on the cathode surface, periodically or continuously wetted with a solution of the surfactant in an organic liquid. The lower layer is spun metal salt solution. The working surface of the cathode is located below the layer interface. Deposition of metal powder on the cathode is the formation of crystals, loosely connected with each other, between which there are gaps filled with an electrolyte solution, the metal oxides and hydroxides. By moving the cathode surface, or other mechanical action whiskers separated from the cathode and pass into the upper layer. Together with some of the top layer of liquid, powder extract from the electrolyzer and squeeze. This technology allows to obtain, depending on the electrolysis conditions whiskers of thickness 15-55 nm and lengths from 1 to several tens of micrometers.

The catalysts of the second series were obtained by plasma decomposition of a mixture of metals carbonyls vapors with hydrogen[2]. Hydrogen was the carrier and plasma gas. Couples metals carbonyls fed into the plasma-chemical reactor, the flow of hydrogen was bubbled through the liquid. Metal carbonyls decomposed in the plasma of a pulsed high-voltage capacitor, the frequency of 10 Hz, voltage pulses of 1 to 5 kW up to 100 microseconds. Speed of hardening produced ultrafine powders of metals reached $10^7$K/s. According to the data from transmission electron microscopy on an EMB100L instrument, the size of particles in the metallic phase was 2–25 nm on average. To synthesize bimetallic systems, nickel and

[1] The powders obtained by the Novocherkassk State University.

[2] The powders were obtained at the Moscow Academy of Steel and Alloys.

manganese were preliminarily deposited on aluminum oxide from aqueous solutions of their nitrates and exposed to a mixture of argon and oxygen at 623K. The aluminum oxide with the deposited metal was then placed in the plasma–chemical reactor, in which the iron and cobalt carbonyls were decomposed in hydrogen plasma. Templating the metallic particles into a supporter instantaneously during their formation prevents the separation of nanoparticles of the active phase from the supporter during the operation of the catalyst.

### 2.4. Methods of studying of physical and chemical properties of catalysts.

The catalysts surface layer was determined by X-ray photoelectron spectroscopy (XPS) at spectrometer XSAM-800 Kratos with a magnesium or aluminum anodes. The calibration of the spectra was carried out by C1s - carbon line (Eb = 285.0 eV). The pressure in the analyzer chamber was $10^{-9}$Pa. Identification of the elements was carried out by survey spectra. For quantitative analysis of surface composition using standard software processing and synthesis of the spectra.

The shapes and dimensions of particles in the powders were investigated by transmission electron microscopy (EMV-100L).

Specific surfaces of the powders were determined by low-temperature adsorption of nitrogen on the instrument GH-1 using the method of thermal desorption of nitrogen from the surface of solids. Constant composition gas mixture of nitrogen and helium was passed through the adsorbent at liquid nitrogen temperature prior to the establishment of adsorption equilibrium. Then, raising the temperature, nitrogen desorbed into a stream of gas mixture. Changing the concentration of the mixture at adsorption and desorption was recorded using thermal conductivity detector with programmable integrator ANALITIC-8C. The specific surface $S_s$ was calculated by the equation:

$$S_s = \frac{S_o V_m}{g} \qquad (4)$$

where $V_m$ - gas volume that covers the surface of the adsorbent is a dense monolayer, cm$^3$; $S_0$ - surface covered with 1 cm$^3$ of adsorbate in a monomolecular layer, m$^2$/m$^3$; g-mass of adsorbent, grams.

Surface covered with a monomolecular layer of adsorbate is determined by the equation:

$$S_0 = \frac{N_A \sigma}{\tilde{V}} \qquad (5)$$

Where $N_A$ - Avogadro's number; $\tilde{V}$ - volume of 1 mole of gas (cm3/mol); σ- the surface area of adsorbed gas molecules in a dense monolayer, for nitrogen $\sigma = 1,62 * 10^{-18}$m$^2$.

$$S_0(N_2) = 4{,}35\ m^2/m^3$$

The value of Vm is determined by the equation of multilayer adsorption isotherm BET and counted in a special program integrator ANALITIC-8C.

The data on the composition and specific surface areas of the catalysts are given in Table1.

| Series | № | Catalysts | $S_{s.}$, $m^2/g$ |
|---|---|---|---|
| I | 1.1 | Fe(7.5%)-Co(2.5%)/$SiO_2$ | 4.0 |
| | 1.2 | Fe(6.5%)-Co(3.5%)/$SiO_2$ | 2.3 |
| | 1.3 | Fe(5%)-Co(5%)/$SiO_2$ | 1.3 |
| | 1.4 | Fe(2.5%)-Co(7.5%)/SiO2 | 1.0 |
| II | 2.1 | Ni(5%)/$Al_2O_3$ | 45 |
| | 2.2 | Fe(5%)/$Al_2O_3$ | 92 |
| | 2.3 | Co(5%)/$Al_2O_3$ | 25 |
| | 2.4 | Mn(5%)/$Al_2O_3$ | 20 |
| | 2.5 | Fe(5%)-Ni(5%)/$Al_2O_3$ | 65 |
| | 2.6 | Fe(5%)-Co(5%)/$Al_2O_3$ | 20 |
| | 2.7 | Fe(5%)-Mn(5%)/$Al_2O_3$ | 17 |
| | 2.8 | Fe(5%)-Co(2,5%)-Mn(2,5%)/$Al_2O_3$ | 18 |
| | 2.9 | Fe(5%)-Ni(2,5%)-Mn(2,5%)/$Al_2O_3$ | 63 |
| | 2.10 | Fe(5%)-Co(2,5%)-Ni(2,5%)/$Al_2O_3$ | 65 |

**Table 1.** The composition and specific surface areas of the catalysts.

## 3. Results and discussion

### 3.1. Physico-chemical properties of the catalysts

According to electron microscopy, the first series of samples (Table 1) of different particle shape. Each was a single crystal alloy with different percentages of iron and cobalt. Samples of catalysts № 1.1 and № 1.3 were a whiskers length from 0.1 to several micrometers and a diameter of 20 nm, the sample № 1.2 had a dendritic structure of intergrown crystals and particles of the sample № 1.4 were characterized by a hexagonal structure. Figure 1 shows a micrograph of crystals of iron, obtained by electrochemical method.

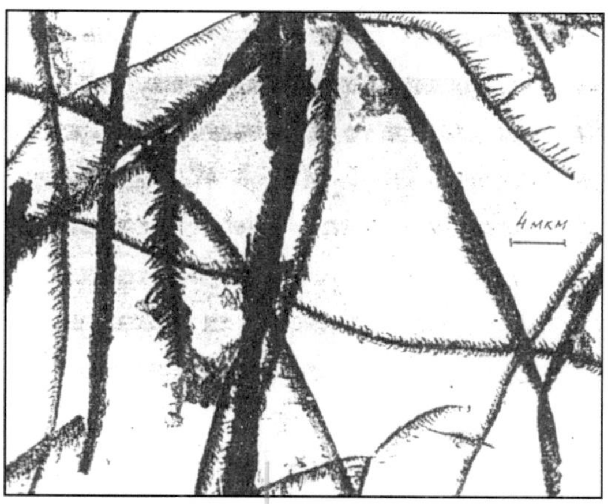


**Figure 1.** Micrograph of a iron micro-thin crystals, obtained by electrolysis (increase of 25000).

For the second series of samples it was found that powders contain both individual particles and their agglomerates. The particle size of the metallic phase averaged - 20 nm. Figure 2 shows a micrograph of a sample of iron powder. The smallest particle size of 2-5 nm is not facet and form large agglomerates that are difficult to beat at ultrasonic dispersion. The particles with sizes of 5-15 nm were weak faceting, while larger particles with sizes of 15-25 nm faceted clear.

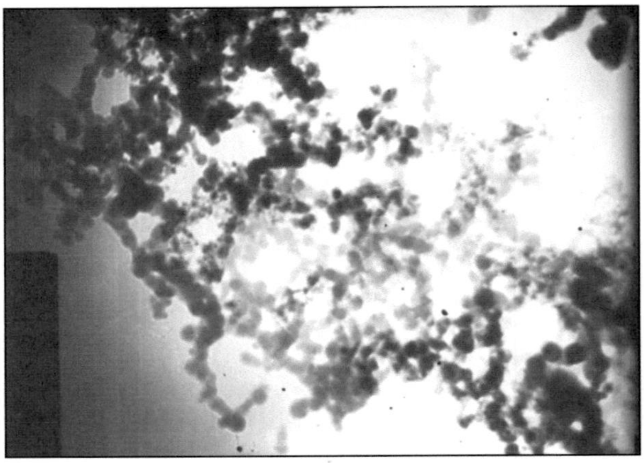

**Figure 2.** Micrograph of a iron sample obtained by plasma method (increase of 120000).

XRD ("Drone-7") and XPS were found in samples of the second series of catalysts (Table 1), $\alpha$-and $\gamma$-phases of iron, $\alpha$-and $\gamma$-phases of cobalt, $\alpha$-and $\beta$-phase nickel, small amounts of oxides, FeO, $Fe_3O_4$, $Fe_2O_3$, CoO, MnO, $Mn_2O_3$, $Mn_3O_4$, NiO, $Ni_2O_3$, solid solutions of $FeMn_3$, $FeMn_4$, as well as traces of $CoFe_2O_4$. Chemical analysis of these powders showed that they contained up to 3% of free and up to 2% fixed carbon, and about 80% of the metal.

## 3.2. The results of thermodesorption experiments and analysis.

The method of thermodesorption established the presence of molecular and dissociative adsorption of both carbon oxides. When the joint adsorption of carbon oxides competition between CO and $CO_2$ did not occur. Installed desorption forms, orders and activation energies of desorption are shown in Table 2.

Analysis of the experimental and literature data [10-13] allowed to interpret the desorption form as follows:

$\alpha$-CO - low-temperature linear ;
$\beta$-CO - high-bridging;
$\gamma$-CO - dissociative adsorption $CO_2$ : $CO_2$ $\rightarrow$ $CO_{ads} + O_{ads}$;
$\alpha$- $CO_2$ - low-temperature molecular;
$\beta$-$CO_2$ - high-molecular or at $CO_{ads} + O_{Me}$ $\rightarrow$ $CO_2$;
$\gamma$-$CO_2$ - or molecular, or CO $\rightarrow$ $C_{ads} + O_{ads}$ ; $CO_{ads} + O_{ads}$ $\rightarrow$ $CO_2$;
$\kappa$-$CO_2$ - carbonate-carboxylate complexes $n(CO+CO_2)/mMe$.

| $T_{des}$, K | | 293-353 | 353-423 | 423-473 | 473-503 | 573-623 | 623-673 | 673-723 | 773-823 |
|---|---|---|---|---|---|---|---|---|---|
| CO | | α- $CO_2$ | α-CO | β-$CO_2$, β-CO | - | - | - | - | - |
| n | | 1 | 1 | 2 | | | | | |
| $E_{a,des}$, kJ / mol | Fe/$Al_2O_3$ | 27 | 32 | 49 | | | | | |
| | Ni/$Al_2O_3$ | 36 | 35 | - | | | | | |
| | Fe-Co/ $Al_2O_3$ | - | 61 | 82 | | | | | |
| | Fe-Mn/ $Al_2O_3$ | - | - | 94,50 | | | | | |
| $CO_2$ | | α- $CO_2$ | | β-CO | γ-$CO_2$ | κ-$CO_2$ | - | - | γ-CO |
| n | | | | 2 | 1 | $1<n<2$ | | | 2 |
| $E_{des}$, kJ / mol | Fe/$Al_2O_3$ | - | | - | 41 | 52 | | | 121 |
| | Ni/$Al_2O_3$ | 36 | | 114 | - | 211 | | | - |
| | Fe-Co/ $Al_2O_3$ | | | 110 | 45 | 52 | | | |
| CO+$CO_2$ | | | α-CO | β-$CO_2$, β-CO | γ-$CO_2$ , β-CO | κ-$CO_2$ | κ-$CO_2$ | β-CO, κ-$CO_2$ | γ-CO |
| n | | | 1 | 2, 1 | 1 | 2 | 2 | 1,$1<n<2$ | 2 |
| $E_{des}$, kJ / mol | Fe/$Al_2O_3$ | | | 30, 25 | 46, 118 | 51 | 57 | 126 | 122 |
| | Ni/$Al_2O_3$ | | 43 | 82,148 | 118 | | 94 | | |
| | Fe-Co/ $Al_2O_3$ | | 60 | 85,112 | 45, 120 | | | | |

**Table 2.** Forms, orders (n), and desorption activation energies ($E_{a,des}$) for the Fe/$Al_2O_3$, Ni/$Al_2O_3$, Fe-Co/$Al_2O_3$ and Fe-Mn/ $Al_2O_3$ II Series catalysts after the adsorption of CO, $CO_2$, and a mixture of CO and $CO_2$ at 293 K

It should be noted that the degree of dissociation of CO increased in a series of Ni, Co, Fe, Mn, and pre-adsorption of hydrogen on the surface hardening is caused due Me-C and contributed to the dissociative chemisorption of carbon monoxide.

As shown previously [6,8], and we have reiterated, on the surface of metals, capable of dissolving hydrogen chemisorbed possible existence of two forms: one of which is associated with only one metal atom ($H_I$), and another - strongly-adsorbed ($H_{II}$) - with a few. Chromatographic analysis of the primary products of desorption revealed that the surface of the catalysts occurred removing the weakly bound hydrogen, probably in the form of $H_I$ and hydrocarbons ($CH_4$, $C_2H_4$, $C_3H_6$).

According to the works reported by Popova et al and Sokolova et al [13,14] on transition metals desorption of CO in the mode of programmed temperature increase leads to dissociation of the adsorbed gas on the carbon and oxygen and subsequent formation of a new form strongly-adsorbed CO desorption which occurs with higher activation energies.

When the temperature of desorption to the catalytic domain strongly adsorbed active carbon particles formed during the dissociative adsorption of CO on the catalyst surface begin to interact with hydrogen to form $H_{II}$ light hydrocarbons, which are allocated to the gas phase.

Pre-saturation of the iron-containing catalysts with hydrogen led to a change in the electron density of the catalyst surface, due to the introduction of hydrogen atoms in it. As a result, the TPD spectrum obtained after exposure of the catalyst with a mixture of CO and $CO_2$ (Figure 3), no peaks $T_{des}$ = 433K (CO formed by dissociative adsorption of $CO_2$) and $T_{des}$ = 523K ($\gamma$-$CO_2$), and the intensity the other peaks was lower than that of the similar peaks in the case of the initial sample. Similar results were obtained in experiments with a nickel catalyst (Figure 1).

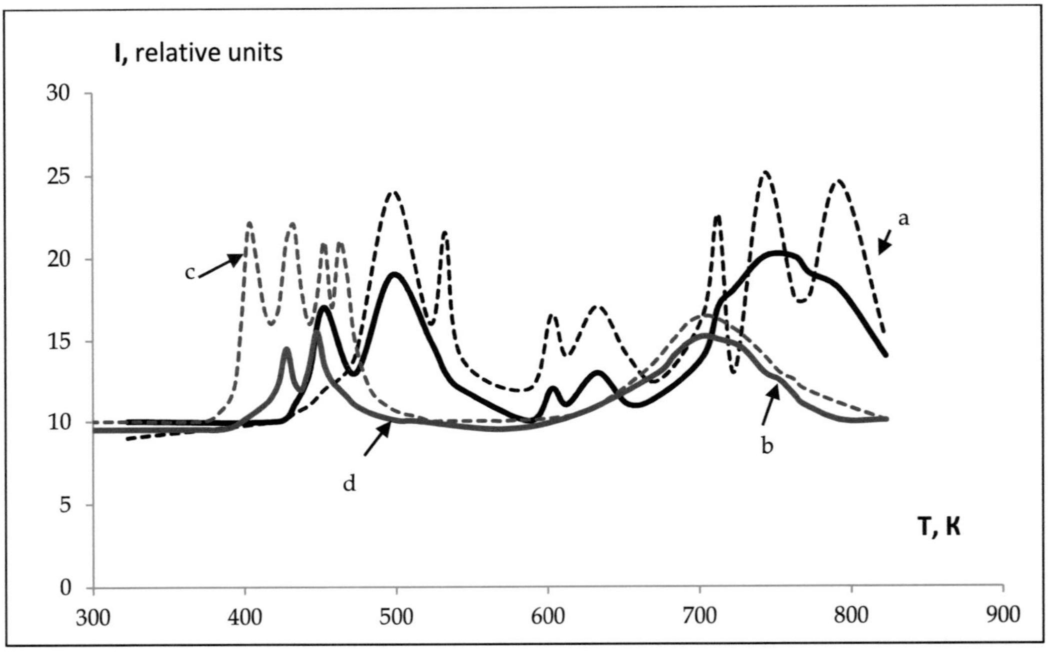


**Figure 3.** TPD spectra from the $Fe/Al_2O_3$ (a) and $Fe\text{-}H/Al_2O_3$ (b) , $Fe\text{-}Ni/Al_2O_3$(c) and $Fe\text{-}Ni\text{-}H/Al_2O_3$ (d) surfaces after coadsorption of CO and $CO_2$ at 293 K.

According to the works reported by Tananaev et al [15], in the presence of adsorbed hydrogen on the surface of ($H_I$) decreases the number of d-electrons of the metal involved in the formation of a bond M-C, with the result that the relationship is weakened and, therefore, communication is strengthened CO. Using a model two-dimensional electron gas is shown in [18] that modify the metal adsorption on the surface due to the redistribution of electron density ($n_e$) surface, and the partial diffusion of the adsorbate in the metal volume slightly reduces this effect. It can be assumed that the hydrogen $H_I$ is negatively charged and partially pulls the electrons in a metal. Its adsorption reduces the electron density, which leads to an increase in the number of weakly bound CO. It is known that dissolved in the metal hydrogen ($H_{II}$), on the contrary, very strongly protonated, and its electron becomes itinerant electrons of the metal, it is likely that hydrogen increases the electron density of the metal surface and leads to an increase in temperature desorption peaks decrease in intensity of thermodesorption and increase in the desorption activation energy, as well as the disappearance of the products of desorption from the surface of weakly bound low-temperature form $\alpha$-$CO_2$, that is to modifying of the surface state of catalysts, and as a consequence, a qualitative change in the desorption of products). Chemisorbed hydrogen $H_I$ is probably a modifier that reduces the $n_e$ part of the free surface, and reduces both the total and induced CO adsorption. Since in our experiments, the temperature range lying below the Curie point, it is natural to assume that the hydrogen in the form of $H_I$ was negatively charged, which confirms the above, the effect of this form of hydrogen on the electron density $n_e$.

### 3.3. Catalytic properties of systems containing Fe, Co, Ni, and Mn nanoparticles.

Performing the reaction on the first series bimetallic catalysts (Table 1) at a ratio of the mixture [$CO_2$ : $H_2$] from 1:1 to 1:4 showed that the major reaction products were methane, ethane, ethylene, propane, propylene, carbon monoxide and water. Hydrogenation of carbon dioxide to methane (Fig. 4) becomes noticeable at 523-553 K and in all the samples of catalysts increases with increasing temperature. However, for samples № 1.2 and № 1.4 of the first series, which differ from the samples № 1.1 and № 1.3 the structure of metal particles at 613 K was observed at most. Further increase in temperature caused a drop in the rate of methane formation and the appearance of the products of carbon monoxide. Rate of formation of $C_2$-$C_4$ hydrocarbons for the first series of catalysts passes through a maximum, which is slightly shifted depending on the content of cobalt in the catalyst and the ratio of $CO_2$: $H_2$

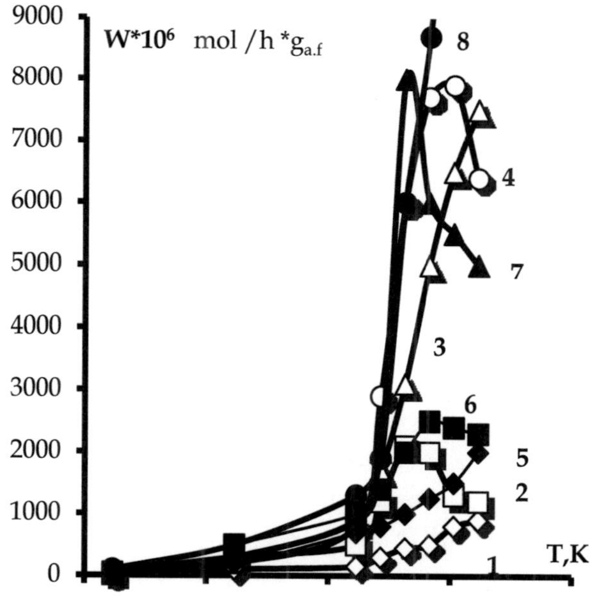


**Figure 4.** The temperature dependence of the rate of methane formation (W) on I group catalysts: with ratios of $CO_2$:$H_2$ = 1:1 in samples №1.1 (1), №1.2 (2), № 1.3 (3), № 1.4 (4); with ratios of $CO_2$: $H_2$ = 1: 4 to № 1.1 (5), №1.2 (6), № 1.3 (7), № 1.4 (8).

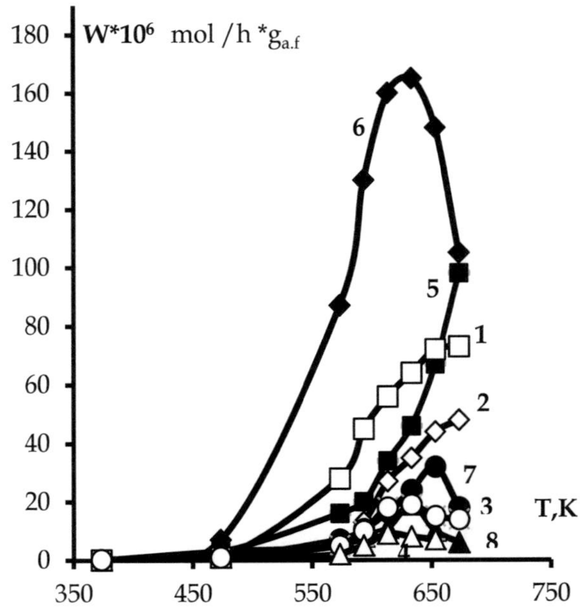


**Figure 5.** Dependence of the yields (W) of $C_2$-$C_3$ hydrocarbons on temperature on the sample №1.3 of the first series at a ratio of $CO_2$:$H_2$ = 1:1: $C_2H_4$ (1), $C_2H_6$ (2), $C_3H_6$ (3), $C_3H_8$ (4) at a ratio of $CO_2$: $H_2$ = 1:4 $C_2H_4$ (5), $C_2H_6$ (6), $C_3H_6$ (7), $C_3H_8$ (8).

The reaction to the lack of hydrogen ($CO_2$: $H_2$ = 1:1) increased the number of formed olefins and paraffins - declined. These patterns are shown in Fig. 5 for example, sample №3, containing equivalent amounts of iron and cobalt.

The total selectivity to olefins was highest for the № 2 catalyst and it was 30.2% at 573 K in the lack of hydrogen (Table 3). Increasing the hydrogen content in the mixture to 1:4 led to an increase in the number of produced hydrocarbons, but the drop in selectivity to olefins.

The increase in cobalt content in the catalyst led to an increase in the rate of formation of all products: as paraffins and olefins (Fig. 4, Table. 3). Probably, with increasing cobalt content in the single-crystal structure of iron-cobalt alloy is an increase in the number of active centers (growth $\ln k_o$). Changing the crystal structure of the catalysts (samples number 1.2 and number 1.4) leads to the creation of the crystals on the surface of larger number of high-active centers, which occurs at lower temperatures, dissociative adsorption of carbon dioxide with the formation of active carbon reacts with hydrogen [6, 15]. However, for sample №1.3 with the same content of both metals and having a thread-like crystal structure, a reduction in the rate of formation of olefins, probably due to the energy factor (Table 3). Since the specific surface area of the first group catalysts are similar, we can assume that their properties are determined only by the number of active centers and the crystal structure.

| series | № | Catalyst / $CO_2$:$H_2$= | $W_{CnH2n}$*$10^6$ mol /h *$g_{a.f}$ | | $S_{CnH2n}$, % | | Ea($_{CnH2n}$), kJ/mol | | $\ln k_0$ | |
|---|---|---|---|---|---|---|---|---|---|---|
| | | | 1:1 | 1:4 | 1:1 | 1:4 | 1:1 | 1:4 | 1:1 | 1:4 |
| I | 1.1 | Fe(7.5%)-Co(2.5%)/$SiO_2$ | 2,50 | 5,90 | 10.3 | 4.6 | 27,6 | 32,3 | -5,9 | -5,1 |
| | 1.2 | Fe(6.5%)-Co(3.5%)/$SiO_2$ | 15,50 | 39,90 | 30.2 | 7.95 | 27,3 | 38,7 | -5,4 | -3,2 |
| | 1.3 | Fe(5%)-Co(5%)/$SiO_2$ | 84,00 | 29,90 | 20.9 | 2.8 | 60,0 | 44,2 | 1,9 | 1,1 |
| | 1.4 | Fe(2.5%)-Co(7.5%)/$SiO_2$ | 57,20 | 100,50 | 18.8 | 4.3 | 33,2 | 29,2 | -3,4 | -4,2 |
| II | 2.5 | Fe(5%)-Ni(5%)/$Al_2O_3$ | - | - | - | - | - | - | - | - |
| | 2.6 | Fe(5%)-Co(5%)/$Al_2O_3$ | - | - | - | - | - | - | - | - |
| | 2.7 | Fe(5%)-Mn(5%)/$Al_2O_3$ | 3,7 | 2,3 | 6,8 | 7,0 | 55,8 | 36,5 | -2,4 | 1,0 |

**Table 3.** The rate of formation, the total selectivity to olefins at 573K, the activation energy of formation of olefins and the logarithm of the pre-exponential factor.

It is known that the formation of olefins in the hydrogenation reaction of carbon dioxide at atmospheric pressure is not a characteristic of the deposited cobalt catalysts [4, 16-18] and on iron catalysts, the formation of olefins takes place at high pressures [4, 10]. However, the nanoparticles of the first group of samples have a large number of defects as a result of electrochemical synthesis, and for them there is practically no interaction between the metal carrier. This is likely to lead to the formation of active forms of carbon. Even more significantly, at atmospheric pressure and at any ratio of $CO_2$:$H_2$ highest selectivity to unsaturated hydrocarbons have been the catalysts with high iron content (Table 3). The increase in cobalt content causes an increase in the rates of formation of all products. The combination in a single crystal of the active phase of iron catalyst, which is responsible for the formation of olefins and cobalt, which has high catalytic activity, allowing to obtain a significant rate of formation of reaction products and high selectivity.

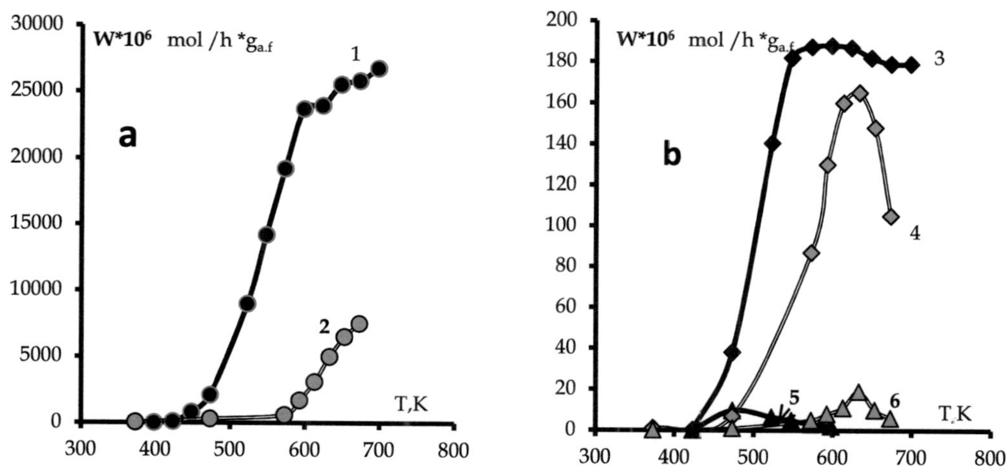


**Figure 6.** Dependence of the yields (W) of hydrocarbons on temperature when conducting the hydrogenation reaction with a $CO_2$ : $H_2$ ratio of 1 : 4: a) $CH_4$ on the catalyst № 2.6 Fe-Co/$Al_2O_3$ (1), on the catalyst №1.3 Fe-Co/$SiO_2$ (2); b) $C_2H_6$ (3, 4) and $C_3H_8$ (5,6) on the catalysts №2.6 Fe-Co/$Al_2O_3$(3,5) and № 1.3 Fe-Co/$SiO_2$(4,6).

Saturated hydrocarbons (methane, ethane and propane) were the main products during the hydrogenation of $CO_2$ to the sample №2.6 Fe-Co/$Al_2O_3$ of the II series catalysts under the same conditions and at the same ratios of reagents as for the samples of I group. The maximum amount of methane on the Fe-Co-sample was observed at 648K and was in 1489 * $10^{-6}$ mol /h *$g_{a.f}$, and a further increase in temperature resulted in the deactivation of the catalyst (Fig. 6).

Replacement of Co by Mn in the II Series catalysts resulted in the appearance of the reaction products of unsaturated compounds. Increasing the ratio of $CO_2$:$H_2$ to stoichiometric lead to an increase in both the rate of formation of olefins and selectivity for olefins (Table 4). It should be noted that the second series of samples with spherical particles of metal, templated in $Al_2O_3$, the number of generated hydrocarbons was higher than in the first series of samples with $SiO_2$ as carrier.

From the above data (Fig. 6, Table 3) that the catalytic activity of powders having a filamentary crystal structure (first series) is much lower than the activity of crystalline samples with an average particle diameter of 20 nm (second series). This may be due to the presence of point defects, dislocations, shear planes, edges and vertices of a crystalline metal catalysts of the second series, which leads to an increase in the number of active centers of the surface.

Introduction of CO in the reaction mixture $CO_2$:$H_2$ has increased the yield of olefins as ferromanganese, and the Fe-catalysts of II Series (Table 4). Quantitative characteristics of hydrogenation products were determined by the composition of the catalyst (Table 4). The highest yield of methane was observed when using Fe-Co/$Al_2O_3$. Substitution of cobalt for nickel or manganese resulted in a decrease in activity for methane formation in a series of Fe-Co/$Al_2O_3$ - Fe-Ni/ $Al_2O_3$ - Fe-Mn/ $Al_2O_3$.

Despite the low catalytic activity of ferromanganese catalyst for the formation of hydrocarbons, olefins selectivity on average 4-4.5 times higher than the figure obtained in the iron-cobalt and iron-nickel samples.

Comparison of the apparent activation energies of methane and ethylene, as well as the pre-exponential factor logarithm lnK0, indirectly characterizing the number of active centers of the surface, with data on activity and selectivity of Fe-Co, Fe-Ni and Fe-Mn binary systems nanopowder showed that differences in catalytic properties probably due to a different number and nature of the surface active sites, as well as the influence of the nature of the second component (Table 4). Thus, according to [3, 19-24], manganese primarily responsible for the formation of olefins and nickel and cobalt activate methane. This, apparently, explains the marked increase in selectivity to unsaturated products when replacing Co or Ni on Mn. At the same time Mn/ $Al_2O_3$ showed no catalytic activity for carbon oxides hydrogenation. However, the addition of Mn to Fe/ $Al_2O_3$ caused a significant increase in selectivity to olefins, in particular up to 50% at 573K.

Since Mn do not give olefins, this reaction is probably due to the fact that during the synthesis of the catalyst $Fe_xMn_y$ particles are formed. Moreover, Mn has a $\delta^+$ charge, since the ionization potentials of iron and manganese are different ($J_{Fe}$ = 7,893 eV, $J_{Mn}$ = 7,435 eV).

| № | 2.1 | 2.2 | 2.3 | 2.4 | 2.5 | 2.6 | 2.7 |
|---|---|---|---|---|---|---|---|
| | Ni/Al2 O3 | Fe/Al2O3 | Co/Al2O3 | Mn/Al2O 3 | Fe- Ni/Al2O3 | Fe- Co/Al2O3 | Fe- Mn/Al2O3 |
| $S_{s.}$, m²/g | 45 | 92 | 25 | 20 | 65 | 20 | 17 |
| $W_{CH_4}$*10⁶ mol /h *$g_{a.f}$ | 241 | 80 | 9,2 | 0 | 243 | 808 | 20,8 |
| $W_{CnH2n}$*10⁶ mol /h *$g_{a.f}$ | 0 | 12,2 | 0,5 | 0 | 14,3 | 138,8 | 24,7 |
| $S_{CnH2n}$, % | 0 | 11,8 | 5,1 | 0 | 5,6 | 14,0 | 54,3 |
| Ea($_{CH4}$), kJ/mol | 54,0 | 57,0 | | - | 86,8 | 65,2 | 58.3 |
| lnKo | 5,69 | 2,9 | | - | 9,38 | 6,42 | 3,46 |
| Ea($_{CnH2n}$), kJ/mol | - | 63 | | - | 64,6 | 48,9 | 31,7 |
| lnKo | - | 2,5 | | - | 5,2 | 1,32 | -4,0 |

**Table 4.** The yields of methane ($W_{CH4}$) and olefins ($W_{CnH2n}$), and the selectivity towards olefins ($S_{CnH2n}$, %) upon the hydrogenation of a mixture with the ratio [($CO_2$ + CO) : $H_2$] = 1 : 2 on II Series catalysts at 573 K.

The same effect was observed, but to a lesser extent Fe-Ni/$Al_2O_3$ and Fe-Co/$Al_2O_3$ (Table 4), which may also be associated with the occurrence of a positive charge on Ni and Co, but smaller than Mn, since the ionization potentials are close ($J_{Ni}$ = 7,635 eV, $J_{Co}$ = 7,87 eV).

Content analysis of the reactants in the gas phase at the joint hydrogenation of carbon oxides, showed that at room temperature, intensive adsorption as CO, $CO_2$ as the catalyst surface. The nature of the curves is identical for all samples studied and presented as an example Fe-Co/$Al_2O_3$ (Fig. 7) composition of the gas phase (CO+$CO_2$):$H_2$ has not changed

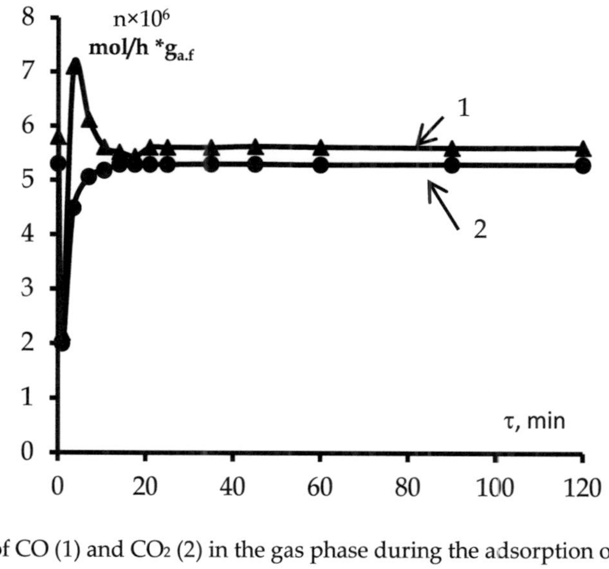


**Figure 7.** Amounts of CO (1) and $CO_2$ (2) in the gas phase during the adsorption on the catalyst № 2.6 Fe-Co/$Al_2O_3$

since the establishment of adsorption equilibrium and up to a temperature of 523 K. The transition temperature in the catalytic domain was accompanied by a decrease in the amount of CO and $CO_2$ formation (Fig. 8).

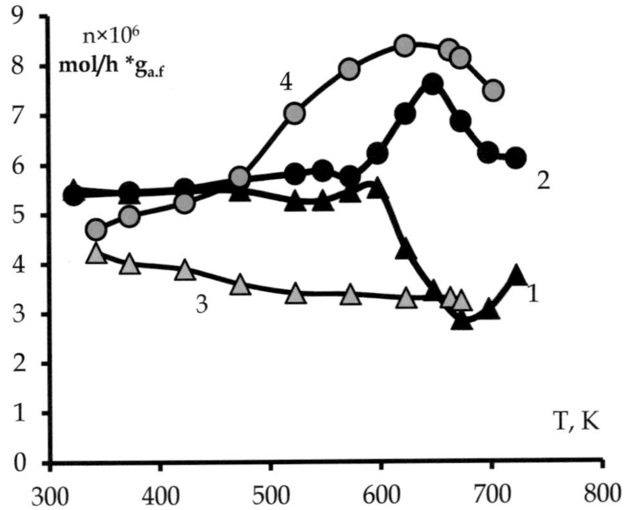


**Figure 8.** Amounts of carbon monoxide (1, 3) and dioxide (2, 4) in the gas phase during the hydrogenation reaction with a($CO+CO_2$): $H_2$ ratio of 1 : 4 on the catalysts № 2.6 Fe-Co /$Al_2O_3$ (1,2) and № 2.7 Fe-Mn/$Al_2O_3$ (3,4).

The formation of carbon dioxide is possible when an adsorbed oxygen atom released upon the dissociative adsorption of a CO molecule reacts with carbon monoxide adsorbed in the molecular form,

$$CO(ads.) + O(ads.) \rightarrow CO_2.$$

The probability of the occurrence of the disproportionation reaction of carbon monoxide:

$$2CO \rightarrow CO_2 + C$$

and of the reaction of steam conversion of CO:

$$CO + H_2O = CO_2 + 3H_2$$

is also high

These processes occur most intensively if the reaction is performed on an iron–manganese catalyst (Fig. 8). Replacing manganese with nickel results in a considerable reduction in the amount of formed $CO_2$ (Fig. 6), due possibly to the different quantities of the linear and bridged forms of adsorbed CO [25-27].

The trend of curves describing the temperature dependence of carbon monoxide concentration in a reaction mixture does not differ appreciably for reactions with a deficient or stoichiometric amount of hydrogen on the investigated catalysts, since it is possible that only those carbon particles formed as a result of the dissociative adsorption of carbon oxides are involved in the reaction.

Thus, it is logical to assume that if the introduction of CO in the reaction mixture reduces the yields of saturated hydrocarbons and the resulting increase in the amounts of olefins, it is likely, during the flow process of adsorption of carbon oxides formed on the surface of carbon atoms with different catalytic activities: one - there are mostly in dissociative adsorption of CO and their interaction with surface hydrogen, leading to the formation of unsaturated hydrocarbons, and points arising from the chemisorption of $CO_2$ are responsible for the formation of both paraffins and olefins.

The ratio of saturated and unsaturated hydrocarbons in the hydrogenation products is determined basically by the amount of atomic hydrogen that is capable of migrating toward the surface's active centers, and by the structure of these centers [3, 6, 23]. As already mentioned, the adsorption of hydrogen on metals that are capable of dissolving it, the formation on the surface of the two forms of $H_2$ is possible [6]. One of them is weakly associated with only one metal atom ($H_I$), and another - strongly-adsorbed $H_{II}$ - with a few. CHx-radicals are formed from the active carbon, which is the product of dissociative adsorption of carbon oxides. However, the selectivity for olefins is probably determined by the ratio of forms $H_I$: $H_{II}$ on the catalyst surface and the increase in $H_I$ concentration of hydrogen increases the yield of olefins.

The chemical nature of the metal is also responsible for the appearance of some form of adsorbed substances. Among the iron-group metals, it is a linear complex of Co-CO, according to the works reported by Nieuwenhuys [28], the electron density at the carbon atom and the smallest such set is easier to interact with other chemisorbed particles. Therefore, of all the studied catalysts, regardless of how they receive the greatest activity was shown the iron-containing samples with the addition of cobalt.

In addition, the transition from bulk samples to nanoparticles and an increase in dispersion of the catalyst in its distribution on the supporter increases the coordinative unsaturation of metal atoms [29]. As a consequence, the binding energy of the metal varies not only with carbon but also a redistribution of hydrogen and the ratio $H_I$: $H_{II}$ regions in favor of the weakly bound hydrogen. This gives rise to additional forms of adsorption of carbon dioxide and an increase in selectivity to olefins.

Differences in the catalytic activity of the bimetallic samples could be associated with different rates of the diffusion of weakly bound hydrogen ($H_I$) on the catalyst surface through the contact boundaries between the Fe, Co, and Mn particles (a spillover effect), or with the existence of a jumpover effect [2, 3, 6, 11, 16]when CHx radicals formed in one centers carry centers from one gas phase to another, where their further hydrogenation by atomic hydrogen ($H_I$) to methane takes place. And the co-interaction of these $CH_x$ radicals leads to the formation as paraffins and olefins. The high selectivity to olefins ferromanganese catalyst may be due to the almost complete absence on the manganese surface of adsorbed hydrogen [9,12, 16]. And it contributes to the interaction between a $CH_x$-radicals.

To confirm this assumption, the idea [30,31] of the layered filling of the reactor has been used. The reaction mixture initially came to the catalyst layer of the iron nanoparticles templating in $ZrO_2$, and then held an insulating layer of $SiO_2$, and came on a layer of manganese in the $SiO_2$. The composition of products at the reactor outlet cardinally changed over the composition of products obtained by using only Fe / $ZrO_2$: significantly reduced the amount of methane, propane and butanes, and many times has increased the amount of olefins (Table 5).

| Catalyst | T,K | $W_{CH_4}$ | $W_{C_2H_4}$ | $W_{C_3H_8}$ | $W_{C_3H_6}$ | $W_{C_4H_8}$ | $S_{C_nH_{2n}}$ |
|---|---|---|---|---|---|---|---|
| Fe/$ZrO_2$ | 573 | 130,1 | - | 25,1 | 25,6 | 7,0 | 18,0 |
| | 623 | 869,1 | - | 118,0 | 169,0 | 29,3 | 16,7 |
| | 673 | 2084 | - | 348,0 | 511,0 | 69,0 | 19,2 |
| Mn/$SiO_2$ | 573 | 10,3 | - | 0,10 | 0,60 | 0,03 | 10,5 |
| | 623 | 13,5 | - | 0,24 | 0,90 | 0,04 | 6,4 |
| | 673 | 16,7 | - | 3,80 | 0,40 | 0,05 | 2,1 |
| Fe/$ZrO_2$+ Mn/$SiO_2$ | 573 | 33,0 | 1,0 | 9,0 | 70,0 | 52,0 | 73,0 |
| | 623 | 110,0 | 4,0 | 13,0 | 108,0 | 68,0 | 50,4 |
| | 673 | 230,0 | 2,0 | 18,0 | 134,0 | 82,0 | 46,7 |

**Table 5.** The yields (W*$10^6$mol /h *$g_{a.f}$) .) and the selectivity (S,%) towards olefins upon the hydrogenation of CO with the ratio CO : $H_2$ = 1 : 2 at atmospheric pressure.

The results confirm the suggestion that the iron generates CHx-radicals, which then transferred through the gas phase to the surface of manganese. It is their further recombination, mainly to olefins.

In carrying out the hydrogenation in the № 2.8 Fe-Co-Mn/$Al_2O_3$ and № 2.9 Fe-Ni-Mn/$Al_2O_3$ catalysts of the reaction products other than methane were also detected ethylene, ethane and propylene. At a stoichiometric ratio of carbon oxides and hydrogen, methane yields on these samples did not differ, but in an hydrogen's excess, a sharp increase in the amount of methane produced by Fe-Ni-Mn/$Al_2O_3$ than Fe-Co-Mn/$Al_2O_3$ (Table 6). Analysis of the rates of formation of other products showed that the catalyst containing nickel dominated saturated hydrocarbons, whereas on the Fe-Co-Mn sample was higher than the selectivity to unsaturated (Table 6).

The study of reactions on catalysts № 2.8 Fe-Co-Mn/$Al_2O_3$ and № 2.9 Fe-Ni-Mn/$Al_2O_3$ also identified a number of deviations from the above described processes. First, there was an increase in the apparent activation energy of formation of reaction products during the hydrogenation of an excess of hydrogen, in contrast to the bi-metal and Fe-Co-Ni system №2.10, where the trend is discernible only with respect to olefins, and for the saturated hydrocarbons were characterized, on the contrary, it decline. Secondly, the increase of hydrogen content in the reaction mixture, or virtually no effect on the selectivity, or even lead to its growth in olefins and some reduction in methane (Table 6), whereas for the catalysts described above, the nature of these relationships were reversed.

| № | Catalyst | $S_{CH4}$, % | $S_{CnH2n}$, % | Ea($_{CH4}$), kJ/mol | Ea($_{CnH2n}$), kJ/mol |
|---|---|---|---|---|---|
| $[(CO_2 + CO) : H_2] = 1 : 2$ | | | | | |
| 2.8 | Fe-Co-Mn/$Al_2O_3$ | 62,9 | 26,2 | 90,6 | 56,2 |
| 2.9 | Fe-Ni-Mn/$Al_2O_3$ | 83,6 | 9,6 | 96,1 | 55,2 |
| 2.10 | Fe-Co-Ni/$Al_2O_3$ | 81,8 | 0,0 | 95,1 | 52,1 |
| $[(CO_2 + CO) : H_2] = 1 : 2$ | | | | | |
| 2.8 | Fe-Co-Mn/$Al_2O_3$ | 62,1 | 31,1 | 95,3 | 59,1 |
| 2.9 | Fe-Ni-Mn/$Al_2O_3$ | 83,3 | 11,0 | 106,2 | 81,9 |
| 2.10 | Fe-Co-Ni/$Al_2O_3$ | 95,3 | 0,7 | 90,9 | 30,4 |

**Table 6.** The selectivity towards methane ($S_{CH4}$) and olefins ($S_{CnH2n}$, % ) upon the carbon oxides hydrogenation on II Series catalysts at 573 K.

These data may indicate the formation under the influence of reaction medium on the surface of these polymetallic manganese catalyst of new centers responsible for the formation of olefins and requires more energy expenditure.

## 4. Conclusion

In the synthesis of hydrocarbons involved the active carbon formed by dissociative adsorption of carbon oxides on the catalyst surface. The differences in catalytic activity and selectivity of bimetallic samples are due, probably, different rates of the weakly bound hydrogen spillover ($H_I$), as well as speed $CH_x$ - radicals jump-over effect from one center to another, where they undergo further hydrogenation. The increase in dispersion of the catalyst can lead to a redistribution of the ratio $H_I$: $H_{II}$ in favor of the weakly bound hydrogen, which causes an increase in selectivity to olefins

## Author details

T. F. Sheshko and Yu. M. Serov
*Peoples' Friendship University of Russia, Russia*

## 5. References

[1] G. P. Van der Laan and A. A. C. M. Beenackers, 1999, "Kinetics and selectivity of the Fischer-Tropsch synthesis: a literature review," Catalysis Reviews, vol. 41, no. 3-4, pp. 255–318.

[2] Lapidus A.L., Krylova A.J., 2000, "The mechanism of formation of liquid hydrocarbons from CO and $H_2$ on cobalt catalysts ", Russian chemical journal, vol. XLIV, no. 1, pp. 4–18. [in Russian].

[3] Qinghong Zhang, Jincan Kang, Ye Wang. 2010, "Development of Novel Catalysts for Fischer–Tropsch Synthesis: Tuning the Product Selectivity. Review". vol. 2, no. 9, pp. 1030–1058.

[4] Tomoyuki T., Inui T. 1991 "Effective conversion of carbon dioxid." Catal. Today.. vol.10. no 1. pp.95-106

[5] Suzuki T., Mayama Y., Nayashi S. 1991. "Hydrogenation of carbon dioxide over iron catalyst". Kinet and Katal Lett. vol. 44. no 2. pp. 78-93

[6] Serov Yu. M., 1999. "Composite membranes for hydrogen recovery from gas mixtures, catalytic systems for vapor and carbon dioxide conversion of methane, detoxification of exhaust gases and the hydrogenation of carbon oxides ". Extended Abstract of Doctoral Dissertation in Chemistry (RUDN, Moscow, 1999). *[in Russian].*

[7] G. Erlich, Catalysis. Physical Chemistry of Heterogeneous Catalysis (Mir, Moscow, 1967) [in Russian].

[8] Wedler, G., Colb, K. G., Heinrich, W., McElhiney, G., "The interaction of hydrogen and carbon monoxide on polycrystalline iron films", Applications of Surf. Sci., 2(1), 1978, pp. 85-101.

[9] Sheshko T.F., Serov Yu. M., 2011 «Interaction of Carbon Oxides with the Surface of Catalysts Containing Iron and Nickel Nanoparticles» J. Russ.Phys. Chem. A 85, pp778-783.

[10] Popova N. M., Babenkova L. V. and Savel'eva G. A., 1985 "Application of Thermodesorption. Method to Adsorption and Catalysis (Nauka, Alma_Ata,) [in Russian].

[11] Mikhalenko I. I. and Yagodovskii V. D., Zh. Fiz. Khim. 79, 1540 (2005) [Russ. J. Phys. Chem. A 79, 1363(2005)].

[12] Edussuriya M. 1993 "Adsorption and hydrogenation of CO on powders of iron and ruthenium" (RUDN, Moscow, 1993). P. 153 [in Russian].

[13] Popova N.M., Babenkova L.V., Saveliev G.A. 1979. "Adsorption and interaction of simple gases with metals of Group VIII" . Alma-Ata: Nauka, 1979. P.278.

[14] Sokolova N.P. 1993 Russ. J. Phys. Chem. A 67, 10(1993).

[15] Tananaev I.V., Fedorov I.B., Kalashnikov E.G. 1987 "Progress of physical chemistry energy-environments". Successes of chemistry. 1987. T.56, N2. P.193.

[16] Mochoki A. 1991. "Fotmation of carbonaceous deposit and its effect of carbon monoxide hydrogenation on iron-based catalyst". Appl. Catal.. vol.70. no.2. pp. 253-267

[17] Rabo J.A., Risck A.P., Poutsma M.L 1978. "Reactions of carbon monoxide and hydrogen on Co,Ni,Ru and Pd metals". J. Catal. vol.53. pp.295-311.

[18] Catalysis in C1-chemistry. 1983. Edited by KEIM W. Institut fur Technische Chemie und Petrolchemie., D. Reidel Publishing Co., Dordrecht, Holland, 1983. 296p

[19] Gordon D., Watherbey C., Calvin H.B. 1987. "Hydrogenation of $CO_2$ on Group VIII Metals". J. Catal. vol.77. pp.82-91.

[20] Babenkova L V, Popova N M, Blagoveshchenskaya I N, 1985 "The Mechanism of the Reaction of Hydrogen with the Metals of the Iron Sub-group", Russ. Chem. Rev., vol.54. no2, pp. 105–116.

[21] Wenter J., Kaminsky M., Geoffroy G.L., Vannice M.A. 1987. "Carbon-supported Fe-Mn and K-Fe-Mn clusters for the synthesis of C2-C4 olefins from CO and $H_2$. Activity and selectivity maintenance and regenerability". J. Catal. vol. 105. pp.155-160.

[22] Andrew Campos, Nattaporn Lohitharn, Amitava Roy, Edgar Lotero, James G. Goodwin Jr., James J. Spivey 2010, "An activity and XANES study of Mn-promoted, Fe-based Fischer–Tropsch catalysts". Appl. Catal. A: General, Vol.375, pp. 12-16

[23] F. Morales, E. de Smit, F. M. F. de Groot, T. Visser, B. M. Weckhuysen. 2007. "Effects of manganese oxide promoter on the CO and $H_2$ adsorption properties of titania-supported cobalt Fischer–Tropsch catalysts". J.Catal. № 246, pp.91 –99.

[24] Slivinskii EV, Kliger, EA, L. Kuzmin, Abramov AV, Kulikov EA 2003, "Strategy for sustainable use of natural gas and other carbon compounds in the production of synthetic liquid fuels and petrochemical intermediates." Journal of Russ. Chem. Society nam. D.I. Mendeleev. . T. XLVII, № 6, pp.14-20. *[in Russian].*

[25] Sheshko T.F., Serov Yu.M., 2011, "Interaction of Carbon Oxides with the Surface of Catalysts Containing Iron and Nickel Nanoparticles". Russian Journal of Physical Chemistry, Vol. 85, no. 5, pp. 867–873.

[26] Ismailov, M.A., Akhverdiev R.B., Haji-Kasumov V.S., Matyshak V.A. 1992. "The structure and reactivity of surface complexes formed in the oxidation of CO with oxygen and NO on catalysts based on activated $Al_2O_3$". Kinetics and Catalysis. T.33, № 3, pp.611-617. *[in Russian].*

[27] Firsov A.A., Khomenko T.I., Silchenkova O.N., Korchak V.N. 2010. "The oxidation of CO in the presence of hydrogen in the oxides CuO, CoO and $Fe_2O_3$, deposited on $ZrO_2$". Kinetics and Catalysis. T.51. Number 2. pp.317-329.

[28] Nieuwenhuys B.E. 1983 « Adsorption and Reaction of CO, NO, $H_2$ and $O_2$ on Group YIII Metal Surface ». Surf. Sc. V.126, N1-3. pp.307-337.

[29] Levshin N.L. 2000. "The influence of adsorption-desorption processes on phase transitions in solids," Doctoral Dissertation in. phys. and Math., Moscow: Moscow State University.

[30] Matyshak V.A., Korchak V.N., Burdeinaya T.N., Tret'yakov V.F., Zakirova A.G., Lermontov A.S., Lunin V.V. 2008 "The mechanism of selective $NO_X$ reduction by hydrocarbons in excess oxygen on oxide catalysts: VII. The nature of synergism on a mechanical mixture of catalysts". Kinetics and Catalysis. 2008. V. 49. № 3. - pp. 413-420.

[31] Matyshak V.A., Ismailov I.T., Tretyakov V.F., Silchenkova O.N. 2011. " On the mechanism of conversion of ethanol to copper oxide catalysts for According to IR-spectroscopy *in sit* ". Russian Congress on Catalysis "RussCatalysis" 3 - 7 October, Moscow. Abstracts. V.II, p. 61

Section 3

# Special Topics in Hydrogenation

Chapter 12

# Photocatalytic Hydrogenation on Semiconductor Particles

Shigeru Kohtani, Eito Yoshioka and Hideto Miyabe

Additional information is available at the end of the chapter

http://dx.doi.org/10.5772/45732

## 1. Introduction

Photocatalytic hydrogenation on semiconductor particles is a quite unique methodology compared to the conventional hydrogenations such as catalytic hydrogenation on metals or homogeneous catalysis using metal complexes. The catalysis can be defined as a light-driven redox reaction at a solid/liquid or a solid/gas interface. The semiconductor photocatalysis have been mainly developed by researchers in the areas of photochemistry, electrochemistry, and heterogeneous catalysis. Since Fujishima and Honda have discovered the photoelectrochemical water splitting upon titanium dioxide ($TiO_2$) photoelectrode in the early 1970's [1], the heterogeneous photocatalysis, especially using $TiO_2$, has been applied to various fields such as storage of solar energy [2,3], environmental purification [3,4], and specific surface properties of self-cleaning, self-sterilizing, and anti-fogging induced by incident light [5]. The photocatalysis has also attracted much attention with respect to inducing characteristic organic transformations since 1970's [6-11]. Among those, the photocatalytic hydrogenation of ethene and ethyne on $TiO_2$ was already reported in 1975 [12]. Since then, the reductive photocatalysis has been applied to the hydrogenation for several organic compounds having various double or triple bonds.

Herein, we summarize the progress in photocatalytic hydrogenations covering the literatures available up to 2011. At first, we introduce the fundamentals of semiconductor photocatalysis in order to understand the mechanistic principles. Next, we review the reports on applications to the photocatalytic hydrogenation. Here, we refer to scope and limitation of the photocatalytic hydrogenation on semiconductor particles. Advantages and disadvantages using this method are also compared to those of the conventional hydrogenations. Finally, prospects of the photocatalytic hydrogenation are discussed.

## 2. Fundamentals

Some metal oxides (e.g. titanium dioxide ($TiO_2$) etc.) and sulfides (e.g. cadmium sulfide (CdS), zinc sulfide (ZnS) etc.) are regarded as semiconductors, in which electrons ($e^-$) photogenerated in conduction band (CB) and holes ($h^+$) simultaneously generated in valence band (VB) play important roles in electroconductivity as well as chemical reactivities on the surface, i.e. $e^-$ and $h^+$ can induce redox reactions as shown in Fig. 1. This reaction system is so called "a micro-photoelectrochemical cell", consisting of an anode and a cathode within a small particle. The hydrogenation generally proceeds as the reduction of an electron acceptor (**A**) followed by the protonation as depicted in Fig. 1. The photocatalytic reduction of **A** can be carried out in the presence of a large excess amount of electron donor (**D**) such as alcohols or amines, and in the absence of molecular oxygen ($O_2$). The aim using the electron donor (**D**) is to scavenge hole generated in VB, thereby diminishing the degree of recombination between $e^-$ and $h^+$ within the semiconductor particles. Thus, it is important for the reductive hydrogenation of organic substrates to choice an appropriate electron donor (**D**) as hole scavenger. Moreover, it is necessary to purge $O_2$ gas from the reaction system in order to improve the reduction efficiency, since $O_2$ acts as a competitive electron acceptor.

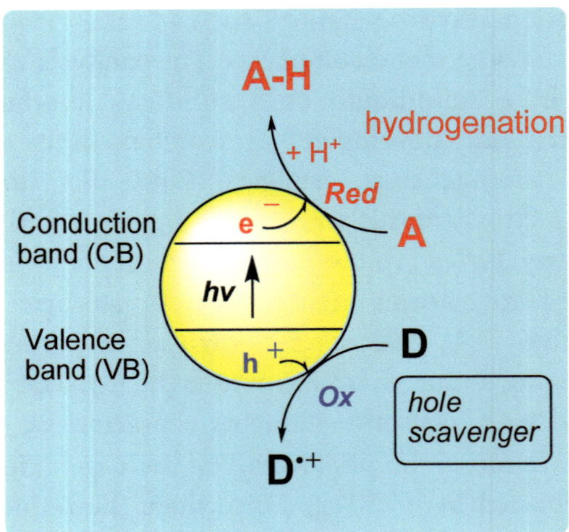


**Figure 1.** Mechanistic principle of photocatalytic hydrogenation on a semiconductor particle

Important points in the semiconductor photocatalyst materials are band gap energy between CB and VB and potential energy levels of CB and VB. The band gap energy defines the minimum photon energy absorbed by the semiconductor materials: Band gap (eV) = 1240/wavelength (nm). Therefore, if a photocatalyst possesses less than 3.1 eV band gap, it can absorb photons with visible light region (> 400 nm). Fig. 2 indicates the band gaps and the band levels (V vs. standard hydrogen electrode (SHE) at pH 7) of $TiO_2$ (anatase), $TiO_2$ (rutile), ZnS, and CdS semiconductor materials as measured photoelectrochemically [8]. $TiO_2$ has four polymorphs: rutile (tetragonal, the most stable phase), anatase (tetragonal), brookite (orthorhombic), and $TiO_2$ (B) (monoclinic), in which rutile and anatase have been mainly used as photocatalysts. The band gap of anatase (3.2 eV) is slightly larger than that of rutile (3.0 eV), because the CB level of anatase is located at 0.2 eV more negative than that of rutile as depicted in Fig. 2. In order to proceed the photocatalytic reaction effectively, the bottom level of CB has to be more negative than a reduction potential of **A** (**A/A**$^{-\cdot}$), while the top level of VB has to be more positive than an oxidation potential of **D** (**D**$^{+\cdot}$**/D**) as shown in Fig. 2.

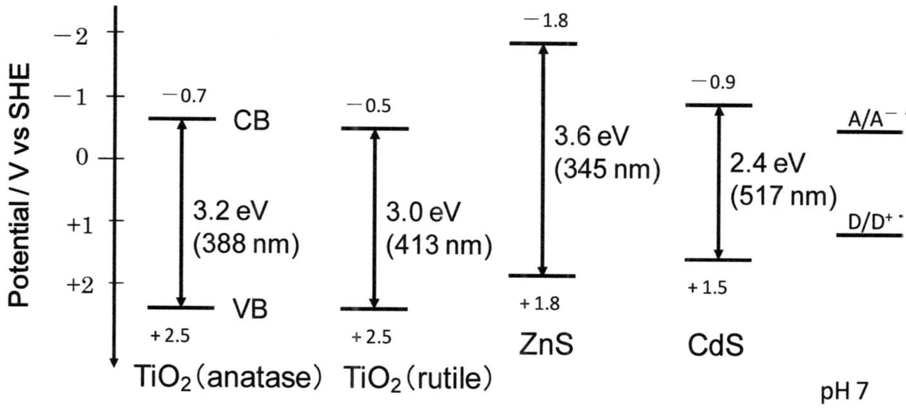


**Figure 2.** Band gaps and band edge positions of some semiconductor photocatalysts

It should be also noted that the band levels usually shift with a change in pH (-0.059 V/pH) for oxide materials. In addition to this pH dependence, surface impurities, adsorbed compounds, and the change to organic solvents would induce band shifts. For an example, the negative band shift is observed when water is replaced by organic solvents such as acetonitrile and alcohols [13]. The negative CB level is essential for the photoreductive hydrogenation reaction. From this point of view, ZnS and CdS having the negative CB level are effective candidates for the photocatalytic hydrogenation. However, these sulfides can be oxidized by $h^+$ in the absence of an appropriate hole scavenger: CdS or ZnS + $2h^+ \rightarrow Cd^{2+}$ or $Zn^{2+}$ + S, which is so called "photocorrosion" [8]. Thus, the choice of a suitable hole scavenger (solvent) is particularly important for hydrogenation using ZnS or CdS.

It is important to estimate a quantum yield (QY) in photocatalytic reactions. QY is defined as the number of events which occurs per photon absorbed by a photocatalyst as follows:

$$QY = \frac{(number\ of\ reacted\ electrons\ or\ holes)}{(number\ of\ absorbed\ photons)}$$

However, it is difficult to determine the real number of photons absorbed by a photocatalyst in a dispersed system because of light scattering by the photocatalyst powder. Therefore, in general, an apparent QY as described below is applied to the actual photocatalytic reactions.

$$Apparent\ \ QY = \frac{(number\ of\ reacted\ electrons\ or\ holes)}{(number\ of\ incident\ photons)}$$

The apparent QY is estimated to be smaller than the real QY because the number of absorbed photons is usually smaller than that of incident photons.

Loading of fine particles, usually nanometer size of noble metals (e.g. platinum (Pt) or silver (Ag) etc.) or metal oxide (nickel oxide (NiO) etc.), on the photocatalyst surface often improves the reaction efficiency as a co-catalyst due to the following reasons: (1) the particles enhance the charge separation between $e^-$ and $h^+$ and prevent the charge recombination, and (2) the particles act as effective reactive sites in the photocatalytic transformation processes. For an example, Pt fine particles loaded on photocatalyst greatly improve the chemical efficiency of hydrogen evolution from water or alcohol under irradiation.

## 3. Applications

### 3.1. Hydrogenation of alkenes and alkynes

Photocatalytic hydrogenation of ethene and ethyne firstly reported by Boonstra and Mutsaers in 1975 [12]. Some hydrogenated products have been found upon the UV illumination (320 – 390 nm) of $TiO_2$ in an atmosphere of ethene or ethyne. Main products from ethene were methane, ethane, propane, and *n*-butane, whereas, in the case of ethyne, the products were methane, ethene, ethane, and propane. In these reactions, the hydrogen source has been thought to be the surface Ti-OH group on $TiO_2$.

Anpo and co-workers investigated the photocatalytic hydrogenation of various alkenes and alkynes in the presence of water vapor upon $TiO_2$ [14-17]. Major products on the UV illuminated $TiO_2$ powders were the compounds formed by hydrogenation accompanied by C=C or C≡C bond fission [14, 15]. In contrast, a significant enhancement of hydrogenation products without the bond fission has been observed on the UV illuminated Pt-loaded $TiO_2$ (Pt/$TiO_2$) powders [16]. Especially, the Pt-loaded rutile $TiO_2$ predominantly catalyzed the hydrogenation of propyne ($CH_3C{\equiv}CH$) to afford $C_3H_6$ and $C_3H_8$ ($C_3/C_2$ ratio = 5.60) as listed in Table 1. Anpo et al. further investigated the size effects on Pt/$TiO_2$ particles in the photocatalytic hydrogenation of $CH_3C{\equiv}CH$ with $H_2O$ [17]. Quantum yields of the whole photoreactions became smaller with increasing the particle size of Pt/$TiO_2$ (rutile), whereas selectivity of the hydrogenation products without vs. with the bond fission ($C_3H_8/C_2H_6$) reversely became larger and reached at $C_3H_8/C_2H_6$ = 99 upon the particle size of 200 nm.

| Catalysts Type | Products ($10^{-9}$ mol m$^{-2}$ h$^{-1}$) | | | | | $C_3/C_2$ ratio |
|---|---|---|---|---|---|---|
| | $CH_4$ | $C_2H_6$ | $C_2H_4$ | $C_3H_6$ | $C_3H_8$ | |
| $TiO_2$ (P25[a]) | 1.90 | 12.3 | 0.40 | 1.10 | 0.50 | 0.12 |
| Pt[b]/$TiO_2$ (P25[a]) | 2.40 | 15.7 | 0.50 | 15.7 | 0.40 | 0.86 |
| $TiO_2$ (rutile) | 0.55 | 3.27 | 0.10 | 0.48 | 0.19 | 0.17 |
| Pt[b]/$TiO_2$ (rutile) | 0.22 | 1.60 | 0.23 | 10.3 | 1.10 | 5.60 |

[a] A composition ratio of rutile/anatase is ca. 1/4. [b] 4wt%

**Table 1.** The amount of products formed in the photocatalytic hydrogenation of $CH_3C{\equiv}CH$ on the four type of $TiO_2$ in the presence of water vapor at 300 K [16].

The hydrogenation of C-C multiple bonds of alkenes and alkynes was also examined in alcohols using Pt/$TiO_2$ powders by Yamataka et al. [18]. In this case, some saturated alkanes were produced in good yields (> 50% after 24 h), while the alcoholic solvents were concurrently oxidized to the corresponding carbonyl compounds. Baba et al. have synthesized bimetal-loaded $TiO_2$ such as Pd/$TiO_2$/Ni or Pt/$TiO_2$/Cu and applied those to the hydrogenation of ethene [19]. The hydrogenation to ethane was efficiently occurred upon the bimetal/$TiO_2$, on which the role of the latter metals (Ni or Cu) was to suppress the hydrogen production as side reaction. Titanium-silicon (Ti/Si) binary oxides were prepared and utilized for the photocatalytic hydrogenation of alkenes and alkynes with $H_2O$ by Anpo's group [20-22]. The reactivity and selectivity for the hydrogenation of $CH_3C{\equiv}CH$ were investigated as a function of the Ti content [21, 22]. It was found that the hydrogenation with bond fission producing $C_2H_6$ and $CH_4$ was predominant in regions of low Ti content, whereas the hydrogenation yielding $C_3H_6$ proceeded in regions of high Ti content. They have revealed that tetrahedrally coordinated titanium oxide species played a significant role in the efficient photoreaction with a high selectivity for the hydrogenation with the bond fission, while the catalysts involving the aggregated and octahedrally coordinated titanium dioxide species showed a high selectivity for the hydrogenation producing $C_3H_6$, being similar to reaction using the powdered $TiO_2$ catalysts. Molybdenum oxide or sulfide complexes supported on $TiO_2$ were developed by Kunts and applied to the photocatalytic hydrogenation of ethyne [23, 24]. The sulfur systems were somewhat more efficient in the hydrogenation, and also favored the 2 electron-transferred product ethene rather than the 4 electron-transferred product ethane.

Yanagida et al. reported that visible light response CdS nanocrystallites (2 – 5 nm) catalyzed the efficient hydrogenation of electron-deficient alkenes with triethylamine (TEA) (Table 2) [25]. This reaction is the first example of visible light induced hydrogenation of alkenes on semiconductor particles. The hydrogenation (two electron reduction) was always accompanied with *cis-trans* isomerization (one electron reduction /oxidation process). The yield of the hydrogenation becomes favorable with increasing light intensity [26]. TEA was used for a sacrificial electron donor, in which the oxidation reaction proceeded as follows: $Et_3N + H_2O + 2h^+ \rightarrow Et_2NH + CH_3CHO + 2H^+$.

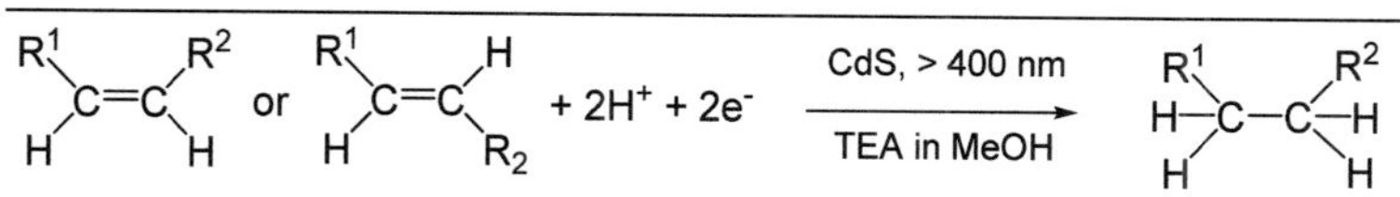


| R¹ | R² | Time / h | Conversion / % | Yield / % | $-E_{red}$[a] / V |
|---|---|---|---|---|---|
| $CO_2CH_3$ | $CO_2CH_3$ (*cis*) | 2 | 100 | 70 | 1.56 |
| $CO_2CH_3$ | $CO_2CH_3$ (*trans*) | 2 | 100 | 60 | 1.60 |
| *p*-$CNC_6H_4$ | CN (*cis*) | 3 | 70 | 47 | 1.75 |
| *p*-$CNC_6H_4$ | CN (*trans*) | 3 | 68 | 41 | 1.73 |
| *p*-$CNC_6H_4$ | $CO_2CH_3$ (*cis*) | 3 | 100 | 76 | 1.75 |
| *p*-$CNC_6H_4$ | $CO_2CH_3$ (*trans*) | 3 | 100 | 92 | 1.75 |
| $C_6H_5$ | $CO_2CH_3$ (*trans*) | 3 | 10 | trace | 1.98 |

[a] Polarographic half-wave reduction potential vs. SCE in MeOH.

**Table 2.** CdS-catalyzed photohydrogenation of alkenes with triethylamine in MeOH [25].

## 3.2. Hydrogenation of imine, azo, and azide compounds

Hydrogenation of the C=N bond of imine intermediates upon the $Pt/TiO_2$ and CdS photocatalysts was observed during some inter- or intramolecular deaminocondensations in one-pot reaction systems as reported by Ohtani and co-workers [27, 28]. Fig. 3 illustrates the photocatalytic transformation of primary amine to symmetrical or asymmetrical secondary amines via the photocatalytic deaminocondensation in water or alcohols. In water [27], a primary amine was firstly oxidized by two holes to form an imine ($R^1CH{=}NH$) followed by hydrolysis to give a corresponding aldehyde ($R^1CHO$). Next, the aldehyde was condensed with another amine to yield the imine intermediate ($R^1CH{=}NCH_2R^1$). Finally, the imine was photocatalytically hydrogenated to produce the symmetrical secondary amine ($R^1CH_2NHCH_2R^1$). On the other hand, in alcoholic solvent [28], the alcohol molecule (sacrificial hole scavenger) was primarily oxidized by two holes to afford the corresponding carbonyl compound ($R^2R^3C{=}O$) which was also condensed with amine to yield the imine intermediate ($R^2R^3C{=}NCH_2R^1$). The imine was further hydrogenated to form the asymmetrical secondary amine ($R^2R^3CHNHCH_2R^1$). This mechanistic principle has been used for several inter- or intramolecular reaction systems as depicted in Scheme 1. Furthermore, photocatalytic deaminocyclization of L-lysine to pipecolinic acid was examined using $TiO_2$ and CdS, and concluded that these two photocatalysts exhibited the different stereochemistry [29, 30].

Kisch el al. found that photocatalytic hydrogenation of azobenzene to hydrazobenzen on ZnS or CdS proceeded as a side reaction of a photocatalyzed addition of 3,4-dihydropyran to azobenzene giving 1-(3,4-Dihydro-2H-pyran-4-yl)-1,2-diphenylhydrazine (DPDH) (Scheme 2) [8, 31]. The formation of the hydrazobenzen was strongly favored when Pt-loaded ZnS or Pt-loaded CdS was used as the photocatalyst. This would be caused by the two electron-transfer process preferentially occurred on the platinum fine particles.

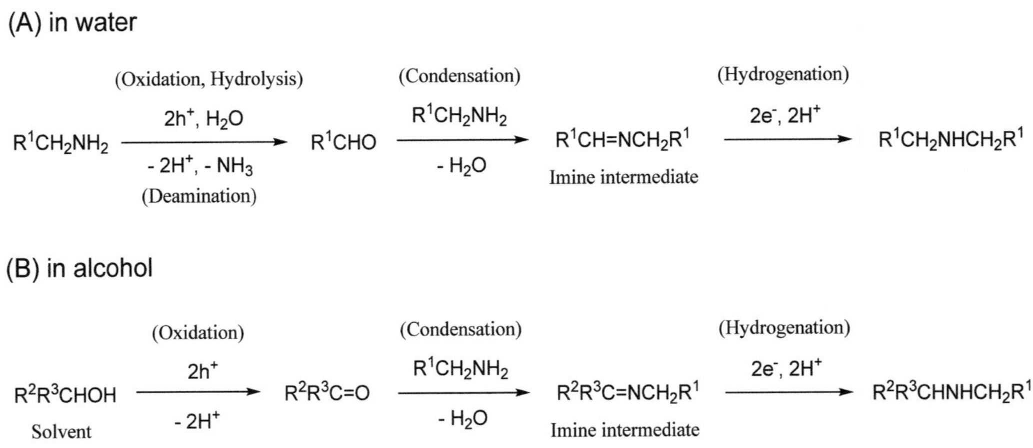


**Figure 3.** Deaminocondensation reactions in one pot system containing hydrogenation process.

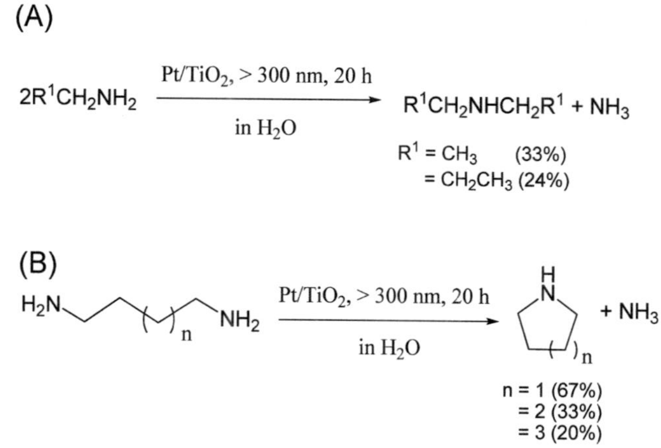


**Scheme 1.** Inter- (A) or intramolecular (B) deaminocondensation reactions [27].

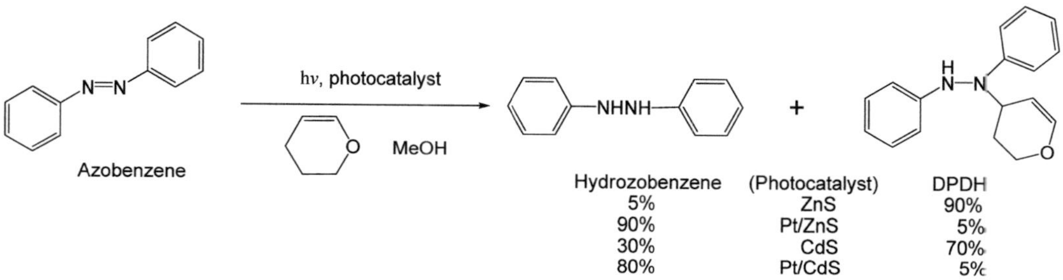


**Scheme 2.** Pt-loading effect on photocatalytic hydrogenation and addition reactions upon ZnS or CdS photocatalysts [8, 31] .

Photocatalytic hydrogenation of aromatic azides (Ar-$N_3$) to amines upon CdS and cadmium selenide (CdSe) was examined by Warrier et al. [32]. The wide scope of the reaction was confirmed with compounds containing electron-withdrawing (-$NO_2$, COOR, COR) and electron-donating groups (-OMe, -R, -Cl) at the para-, meta-, and ortho-positions. These reactions took place with high quantum yields (near 0.5). Sodium formate was used as electron donor which was oxidized to $CO_2$ during the reaction.

## 3.3. Hydrogenation of carbonyl compounds

Aldehydes and ketones are two classes of compounds that possess a reactive carbonyl group. The first report on photocatalytic hydrogenation of carbonyl compound is, to our knowledge, the hydrogenation of pyruvate to lactate under irradiation of aqueous suspension of $TiO_2$ reported by Cuendet and Grätzel in 1987 [33]. Later, Li and co-workers reported P25 $TiO_2$-catalyzed hydrogenation of benzaldehyde forming benzylalcohol in ca. 80% yield [34]. This reaction was recently applied to a micro-reaction system by Matsushita et al. [35, 36]. Kohtani et al. recently demonstrated that the P25 $TiO_2$ powder exhibited the excellent photocatalytic activity to hydrogenate several aromatic ketones into corresponding secondary alcohols under the combination of UV light irradiation and deaerated conditions in ethanol as summarized in Table 3 [37]. Acetaldehyde was simultaneously produced in the oxidation of ethanol by $h^+$ generated in the VB or those trapped at the surface sites on $TiO_2$. The desired secondary alcohols were obtained in almost quantitative yields for more than ten examples: e.g. acetophenone (Ar = $C_6H_5$, R = Me) to 1-phenylethanol in 97% yield (highlighted in gray color in Table 3). They also found that most of the reaction rates depend on the reduction potential ($E_{red}$) of substrates, except for 2,2,2-trifluoroacetophenone [37]. In general, aldehydes are more reactive than ketones because of electronic and steric factors. Electrochemically, $E_{red}$ of aldehydes are more positive than those of ketones. However, the hydrogenations of aldehydes were accompanied with competitive formation of by-products such as diethyl acetal etc., leading to erosion of chemical efficiency.

Aliphatic aldehydes and ketones were hydrogenated by using the ZnS photocatalyst, because the CB level of ZnS is sufficiently negative as indicated in Fig. 2 [38, 39]. In contrast, the photocatalytic hydrogenation did not occur upon $TiO_2$, since the CB level of $TiO_2$ is too positive to reduce the aliphatic ketones [37]. Yanagida and co-workers reported that acetaldehyde was reduced and oxidized upon the UV irradiated ZnS nano-crystallites (2 – 5 nm) in water as depicted in Scheme 3 [38]. The reduction products were ethanol and $H_2$ generated by the photocatalytic reduction of $H_2O$, while acetaldehyde was oxidized by the photogenerated holes to produce acetic acid, diacetyl, and acetoin. The apparent quantum yield of ethanol (main product) formation was 0.25 at 313 nm. The similar result was observed in the photocatalytic reaction of propionaldehyde on the ZnS nano-crystallite [38]. The ZnS nano-crystallites were further applied to the photocatalytic hydrogenation of aliphatic ketones such as acetone, 2-butanone, 3-pentanone, 2-hexanone, cyclopentanone, and cyclohexanone in the presence of sacrificial hole scavengers of $S^{2-}$ and $SO_3^{2-}$ (Oxidation reaction: $S^{2-} + SO_3^{2-} + 2h^+ \rightarrow S_2O_3^{2-}$) [39]. Most of examined ketones were photocatalytically hydrogenated at comparable rates to give the alcohols in almost quantitative yields except for acetone. The apparent quantum yield for the formation of 2-butanol was 0.27 at 313 nm.

Yanagida et al. applied the visible light response CdS nanocrystallite to the photoreduction of aromatic ketones [25, 26, 40]. Two types of the reduction products, secondary alcohols from two electron-transfer process and pinacols from one electron-transfer process, were observed as listed in Table 4, while the oxidation of TEA as a sacrificial electron donor by the VB holes afforded diethylamine and acetaldehyde. The yields of the hydrogenation involving two electron-transfer process were preferable for substrates possessing the electron- withdrawing group (-CN or –Cl) as shown in Table 4, and further became favorable with increasing light intensity [26] and with decreasing the particle size [40].

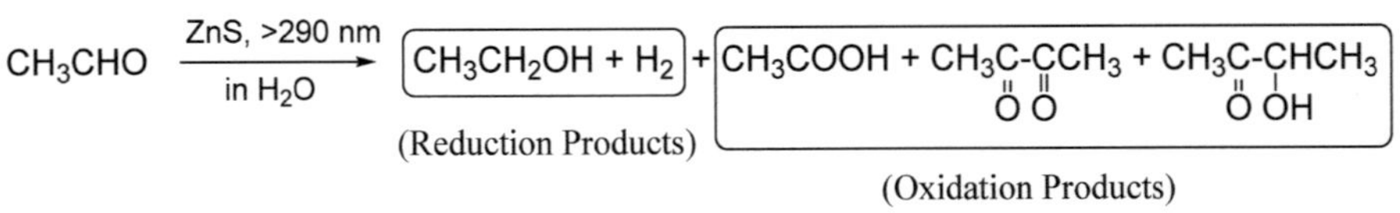


**Scheme 3.** Photocatalytic reduction and oxidation reactions of acetaldehyde in water on ZnS.